Environmental Organic Chemistry for Engineers

Environmental Organic Chemistry for Engineers

Dr. James G. Speight

ELSEVIER

AMSTERDAM • BOSTON • HEIDELBERG • LONDON
NEW YORK • OXFORD • PARIS • SAN DIEGO
SAN FRANCISCO • SINGAPORE • SYDNEY • TOKYO
Butterworth-Heinemann is an imprint of Elsevier

Butterworth-Heinemann is an imprint of Elsevier
The Boulevard, Langford Lane, Kidlington, Oxford OX5 1GB, United Kingdom
50 Hampshire Street, 5th Floor, Cambridge, MA 02139, United States

Notices
Knowledge and best practice in this field are constantly changing. As new research and experience
broaden our understanding, changes in research methods, professional practices, or medical
treatment may become necessary.

Practitioners and researchers must always rely on their own experience and knowledge in
evaluating and using any information, methods, compounds, or experiments described herein.
In using such information or methods they should be mindful of their own safety and the safety
of others, including parties for whom they have a professional responsibility.

To the fullest extent of the law, neither the Publisher nor the authors, contributors, or editors,
assume any liability for any injury and/or damage to persons or property as a matter of products
liability, negligence or otherwise, or from any use or operation of any methods, products,
instructions, or ideas contained in the material herein.

Library of Congress Cataloging-in-Publication Data
A catalog record for this book is available from the Library of Congress

British Library Cataloguing-in-Publication Data
A catalogue record for this book is available from the British Library

ISBN: 978-0-12-804492-6

Working together
to grow libraries in
developing countries

www.elsevier.com • www.bookaid.org

Publisher: Joe Hayton
Acquisition Editor: Ken McCombs
Editorial Project Manager: Peter Jardim
Production Project Manager: Mohana Natarajan
Cover Designer: Victoria Pearson

Typeset by SPi Global, India

Contents

Author Biography

Dr. James G. Speight CChem., FRSC, FCIC, FACS earned his BSc and PhD degrees from the University of Manchester, England. He also holds a DSC in The Geological Sciences (VINIGRI, St. Petersburg, Russia) and a PhD in Petroleum Engineering, Dubna International University, Moscow, Russia). Dr. Speight is the author of more than 50 books in petroleum science, petroleum engineering, and environmental sciences. Formerly the CEO of the Western Research Institute (now an independent consultant), he has served as Adjunct Professor in the Department of Chemical and Fuels Engineering at the University of Utah and in the Departments of Chemistry and Chemical and Petroleum Engineering at the University of Wyoming. In addition, he has also been a Visiting Professor in Chemical Engineering at the following universities: the University of Missouri-Columbia, the Technical University of Denmark, and the University of Trinidad and Tobago.

Dr. Speight was elected to the Russian Academy of Sciences in 1996 and awarded the Gold Medal of Honor that same year for outstanding contributions to the field of petroleum sciences. He has also received the Scientists without Borders Medal of Honor of the Russian Academy of Sciences. In 2001, the Academy also awarded Dr. Speight the Einstein Medal for outstanding contribution and service in the field of geological sciences.

Preface

The latter part of the 20th century saw the realization arise that all chemicals can acts as environmental pollutants and, in addition, there came the realization that emissions of carbon dioxide (CO_2), methane (CH_4), nitrous oxide (N_2O), and chlorofluorocarbons (CFCs) to the atmosphere had either a direct or indirect impacts on the climate as well as on depletion of the ozone layer. As a result, unprecedented efforts were then made to reduce all global emissions in order to maintain a *green perspective*. Furthermore, operations have been designed to reduce the direct emissions of organic chemical products and organic chemical by-products into the air (*the atmosphere*), water (*the aquasphere*), and soil (*the terrestrial biosphere*), and to recycle and reuse these chemicals and chemical wastes as much as possible.

Advanced technologies for the rapid, economical, and effective elimination of industrial and domestic chemical wastes have been developed and employed on a large scale and, in fact, advanced technologies for the control and monitoring of chemical pollutants at regional and global levels continue to be developed and implemented. Satellite-based instruments are able to detect, to quantify, and to monitor a wide range of chemical pollutants. In addition, an understanding of the fate and consequences of chemicals in the environment (Chapters 6 and 7) has increased dramatically and there are now available the means of predicting many of the environmental, ecological, and biochemical consequences of the inadvertent introduction of organic chemicals into the environment with much greater precision.

Organic chemistry is the chemistry of carbon, an element that forms strong chemical bonds to other carbon atoms as well as to many other elements such as hydrogen, nitrogen, oxygen, sulfur, the halogens (fluorine, chlorine, bromine, and iodine), as well as a variety of different metals, such as nickel, vanadium iron, and copper that occur as organo-metallic derivatives in many crude oils.

Thus from this definition, environmental organic chemistry is not just the organic chemistry of the environment that is represented by simple equations but it also includes, in the context of this book, the influence of organic chemicals on the environment which in turn includes: (1) the study of the structure of organic compounds, (2) the physical properties of organic compounds, (3) the chemical properties of organic compounds, and (4) the reactivity or organic compounds with the goal of understanding the behavior of organic compounds not only in the pure form (when possible), but also in aqueous and nonaqueous solutions as well as the chemistry of complex mixtures.

Organic chemistry is a subject that becomes easier as the researcher works with it. The topics covered in this book are the basis topics that serve to introduce the reader not only to organic chemistry but to the effect of organic chemicals on various ecosystems. Basic rules of nomenclature are presented. Understanding the mechanism of how a reaction takes place is particularly crucial in this and of necessity, the book brings a logic and simplicity to the reactions of the different functional groups. This in turn transforms a list of apparently unrelated facts into a sensible theme. Thus, this chapter will serve as an introduction to the physico-chemical properties of organic chemicals and their effect on the floral and faunal environments.

Thus, the intent of this book is to focus on the various organic chemical issues that are the focus of any environmental chemistry program. Thus, this book will serve as an information source to the engineers in presenting details of the various aspects of organic chemicals as they pertain to pollution of the environment. To accomplish this goal, the book focuses on the various aspects of environmental science and engineering. The initial section (Chapters 1–5) presents an introduction to, and a description of the nomenclature of organic compounds and the properties of these materials. The remaining part of the book (Chapters 6–9) presents information relevant to the behavior of organic chemicals and cleanup of the environment.

Dr. James G. Speight,
Laramie, Wyoming
July, 2016

Chapter 1

Chemicals and the Environment

1 INTRODUCTION

Contamination by organic chemicals is a global issue, and such toxic chemicals are found practically in all ecosystems because at the end of the various organic chemical lifer cycles the chemicals have either been recycled for further use or sent for disposal as waste. However, it is the inappropriate management of such waste (e.g., through haphazard and unregulated burning) poses negative impacts on human health and the environment.

This text relates to an introduction to the planned and unplanned effects of organic chemicals on the environment. Chemicals are an essential component of life, but some chemicals can severely damage the floral (plant life) and faunal (animal life) environment. There is an increase in health problems that can be partially explained by the use of chemicals, and many man-made chemicals are found in the most remote places in the environment. Specific groups of chemicals, such as biocides, pesticides, pharmaceuticals, and cosmetics, are covered by various pieces of legislation. In addition, the challenges posed by endocrine disruptors (i.e., chemicals that interfere with the hormone system causing adverse health effects) are also being addressed. However, in order to successfully manage the environment and protect the flora and fauna from such chemicals, a knowledge of chemical, specifically organic chemicals (in the context of this book), is a decided advantage.

The chief reason for studying this subject is the effects of organic chemicals not only on the environment but also on human health, which may be caused by unforeseen side effects of a chemical substance during its production, transport, use, and disposal. These effects provide the motivation for the build of scientific knowledge on the effects of organic chemicals on the floral and faunal environments. Ideally, scientists should be able to predict the possible (if not, likely) likely effects of an organic chemical directly on the environment before the chemical substance is released, enabling a more realistic appraisal to be made of any effects. A first approximation to predicting a potentially harmful organic chemical may involve the following criteria: (1) biologically nonessential, (2) toxic in larger amounts, (3) unlikely to form highly stable inert compounds in nature, (4) persistent in the environment, (5) biochemically active, and (6) environmentally mobile in any of the biogeochemical cycles.

Environmental Organic Chemistry for Engineers. http://dx.doi.org/10.1016/B978-0-12-804492-6.00001-0

Thus, like any technical discipline, the nonchemist is faced with understanding many new terms that are related to the chemical discipline and which need to be understood to place them in context, and this is especially true of organic chemistry. Some terms may seem familiar, but in chemistry, especially organic chemistry, the terms may have meanings that are not quite the same as when used in popular commentaries. The best and common examples are found in the area of pharmaceutical chemistry (a subdiscipline of organic chemistry) where the trade name and the alternate trade name of the pharmaceutical product bear little resemblance to any formalized system of nomenclature. In organic chemistry and, indeed, in all subdisciplines of chemistry, the terms need to have definite and specific meanings to make the subject matter understandable, and this is the *raison d'être* for the acceptable systems of nomenclature. One of the purposes of any system of chemical nomenclature is to provide definitions for many of these terms in a form and at a level that will make the meaning clear to those with limited backgrounds in chemistry as well as to those technical persons in other fields who need to deal with chemistry.

The International Union of Pure and Applied Chemistry (IUPAC—an international organization of chemists and national chemistry societies and the world authority on chemical nomenclature and terminology with a secretariat in Research Triangle Park, North Carolina; http://iupac.org/who-we-are/) makes the final determination of terminology and nomenclature in chemistry. Among other things, this organization authorizes and establishes systematic rules for naming compounds so that any chemical structure can be defined uniquely. Compounds are frequently called by common names or trade names, often because the IUPAC names may be long and complex and difficult to understand for the nonchemist, but the IUPAC name permits a chemist to know the structure of any chemical compound based on the rules of the terminology, while the common name or trade name requires remembering what structure goes with what name, which may be outside of the realms of chemical reality.

Engineering students, on the other hand, undergo a much different training system when compared to the students of chemistry. While some (but not all) engineering disciplines (especially the chemical engineer) may require some background knowledge of chemistry, the practicing engineer is more concerned with practical applications (such as reactor construction and operation), and there are differences in, for example, reactor novelty and reactor scale—areas that are not always pertinent to the chemist outside of laboratory chemistry. Thus, a chemist is more likely to be engaged in developing new compounds and materials using novel or new synthetic routes, while a chemical engineer is more likely to be working with existing substances. A chemist may be involved in the synthesis of a few grams of a new compound, while a chemical engineer will focus on scale up of the synthetic process in order to produce the chemical (say, on a tonnage scale) at a profit. Thus, the chemical engineer will be more concerned with heating and cooling large reaction vessels, pumps and piping to transfer chemicals, plant design, plant operation, and process optimization, while a chemist will be more concerned with establishing the parameters of the reaction from which the engineer will design the plant. However, these differences are generalizations and there is often much overlap.

To summarize: chemistry is the study of matter in its many forms as well as the manner in which these forms react with each other. Chemistry can be used to study the molecular size and the structure of the smallest of ions that exist in the human body to the much-larger scale inner workings of the core of the Earth and even with the faraway study of the chemical composition of the rocks on Mars. To understand basic chemistry is to have a healthy understanding of the complexities of the modern world. This understanding also brings forward the realization that combining chemicals released to the atmosphere by human activities can have serious health effects. And this is where an understanding of organic chemistry—which came into being as a scientific discipline in the 19th century—can play an important role in dealing with the various issues of the environment.

Thus, throughout the pages of this book, the reader will be presented with the definitions and explanations of terms related to organic chemistry and the means by which organic chemistry can be understood and used.

2   THE ENVIRONMENT

When an organic chemical is introduced into the environment, it becomes distributed among the four major environmental compartments: (1) air, (2) water, (3) soil, and (4) biota (living organisms). Each of the first three categories can be further subdivided in floral (plant) environments and faunal (animal, including human) environments.

The fraction of the chemical that will move into each compartment is governed by the physicochemical properties of that chemical. In addition, the distribution of organic chemicals in the environment is governed by physical processes such as sedimentation, adsorption, and volatilization, and the chemicals can then be degraded by chemical and/or biological processes. Chemical processes generally occur in water or the atmosphere and follow one of four reactions: oxidation, reduction, hydrolysis, and photolysis. Biological mechanisms in soil and living organisms utilize oxidation, reduction, hydrolysis, and conjugation to degrade chemicals. The process of degradation will largely be governed by the compartment (water, soil, atmosphere, biota) in which the organic chemical is distributed, and this distribution is governed by the physical processes already mentioned (i.e., sedimentation, adsorption, and volatilization).

The impact on the environment of the changes in the chemical state of organic chemicals is only partially elucidated but will be significant in many cases. Changes in the atmospheric abundance of radiatively active gases could lead to substantial drift in the Earth's climate, including changes in temperature and precipitation, and in the frequency of occurrence of extreme events (such as hurricanes). Reduction in the ozone column abundance leads to enhanced levels of UV-B radiation at the surface with potentially harmful effects on living organisms, including phytoplankton in the ocean, and increased frequency of skin cancers affecting humans. Ecosystem damage and health problems also result from regional and global pollution. Acidic precipitation is believed to have suppressed life in several lakes of North America and Europe and, together with enhanced ozone levels, to have damaged forests in those same parts of the world.

In fact, when assessing the impact of organic chemicals on the environment, the most critical characteristics are: (1) the types of chemicals discharged, which depends on the type of industries and processes used and (2) the amount and concentration of the organic chemicals. Solid wastes (containing organic chemicals) and/or gaseous emission generated from industrial sources also contribute to the amount and concentration of organic chemicals in the environment.

This has led to the introduction of various *emissions factors* that are representative value that attempts to relate the quantity of a pollutant released to the atmosphere with an activity associated with the release of that pollutant. These factors are usually expressed as the weight of pollutant divided by a unit weight, volume, distance, or duration of the activity emitting the pollutant (such as the kilogram of pollutant per kilogram of produced product). Such factors facilitate estimation of emissions from various sources of air pollution. In most cases, these factors are simply averages of all available data of acceptable quality and are generally assumed to be representative of long-term averages for all facilities in the source category (i.e., a population average). The general equation for emissions estimation is:

$$E = A \times \mathrm{EF} \times (1 - \mathrm{ER}/100)$$

In this equation, E = emissions, A = activity rate, EF = emission factor, and ER = overall emission reduction efficiency, %.

A detailed understanding of the observed degradation in the "health of the planet" requires that atmospheric chemistry be studied in the broader context of "global change," and that the Earth system be viewed as a nonlinear interactive system consisting of the atmosphere, the ocean, and the continental biosphere. That the chemical composition of the atmosphere has been maintained far away from the thermodynamic equilibrium conditions encountered and that the two major gases surrounding the planet are nitrogen (78%) and oxygen (21%), as opposed to carbon dioxide (CO_2) are a direct consequence of oxidation and reduction processes associated with the energy metabolism of various forms of life.

Changes encountered in the Earth system occur at different scales in time and space. For example, the formation of a tornado requires only a few minutes, while the response of the ocean to greenhouse warming is characterized by time scales of several decades. A variety of different time scales need also to be considered in the case of chemical processes in the atmosphere. For example, the chemical lifetime of a radical such as OH (which plays a key role in the chemistry of the atmosphere) is typically a few seconds, while that of most chlorofluorocarbon molecules lie in the range of several decades to a century. In the past, changes in the chemical composition of the atmosphere associated with glacial-interglacial transitions have occurred over hundreds to several thousands of years although sometimes associated with rapid fluctuations indicative of the Earth system's nonlinear nature.

2.1   Structure of the Atmosphere

The atmosphere of the Earth is largely transparent and allows incoming sunlight to reach the surface of the Earth, provided the sunlight is not reflected or absorbed by clouds. A fraction of the light that does reach the planet is absorbed, based on the degree of reflectivity of the surface. Most of the absorbed light is converted to energy, and heat is radiated back out at infrared wavelengths. Although the atmosphere allows most of the visible light through, many of the gases, such as water vapor, carbon dioxide, and methane, absorb infrared radiation, converting it to rotational and vibrational energy. This raises the energy content of the atmosphere and thus the average temperature. The more greenhouse gases present, the greater the chances of the infrared light being absorbed before it escapes into space. Thus, if all other influences are kept constant, increased levels of greenhouse gases will necessarily produce increased atmospheric temperatures. However, the impact of greenhouse gases differs based on the chemistry of the individual gases. For example, methane is much more potent than carbon dioxide because it absorbs more infrared radiation. In addition, the impact of a gas is also influenced by the lifetime of the gas in the atmosphere—water vapor falls back out quickly as precipitation, and methane is typically oxidized to carbon dioxide within decades of its appearance in the atmosphere.

Physically, the atmosphere is the thin and fragile envelope of air surrounding the Earth that is held in place around the Earth by gravitational attraction and which has a substantial effect on the environment. The total dry mass of the atmosphere (annual mean), three quarters of which is within approximately 36,000 ft of the surface, is estimated to be in excess of 5×10^{21} tons (Trenberth and Guillemot, 1994).

The atmosphere also contains oxygen used by most organisms for respiration and carbon dioxide used by plants, algae, and cyanobacteria for photosynthesis. The atmosphere helps protect living organisms from genetic damage by solar ultraviolet (UV) radiation, solar wind, and cosmic rays. Its current composition is the product of billions of years of biochemical modification of the paleoatmosphere by living organisms.

In general, air pressure and density decrease with altitude in the atmosphere. However, temperature has a more complicated profile with altitude and may remain relatively constant or even increase with altitude in some regions. Because the general pattern of the temperature/altitude profile is constant and measurable by means of instrumented balloon soundings, the temperature behavior provides a useful metric to distinguish atmospheric layers. In this way, the atmosphere can be divided (called atmospheric stratification) into five main layers. Excluding the exosphere, the Earth has four primary layers, which are the troposphere, stratosphere, mesosphere, and thermosphere.

Generally, the atmosphere is defined by the homosphere and the heterosphere which, in turn, are defend by whether the atmospheric gases are well mixed. The surface-based homosphere includes the troposphere, stratosphere, mesosphere, and the lowest part of the thermosphere, where the chemical composition of the atmosphere does not depend on molecular weight because the gases are mixed by turbulence. This relatively homogeneous layer ends at the *turbopause* found at approximately 62 miles (330,000 ft), which places it approximately 12 miles (66,000 ft) above the mesopause. Above this altitude lies the heterosphere, which includes the exosphere and most of the thermosphere. Here, the chemical composition varies with altitude. This is because the distance that particles can move without colliding with one another is large compared with the size of motions that cause mixing. This allows the gases to stratify by molecular weight, with the heavier ones, such as oxygen and nitrogen, present only near the bottom of the heterosphere. The upper part of the heterosphere is composed almost completely of hydrogen, the lightest element. The planetary boundary layer is the part of the troposphere that is closest to Earth's surface and is directly affected by it, mainly through turbulent diffusion. During the day the planetary boundary layer usually is mixed well, whereas at night it becomes stably stratified with weak or intermittent mixing. The depth of the planetary boundary layer ranges from as little as approximately 300 ft on clear, calm nights to 10,000 ft or more during the afternoon in dry regions.

However, structurally, the atmosphere consists of a number of layers that differ in properties such as composition, temperature, and pressure. From highest to lowest, the five main layers are: (1) the troposphere, (2) the stratosphere, (3) the mesosphere, (4) the thermosphere, and (5) the exosphere. Approximately three-quarters (75%,v/v) of the atmosphere's mass resides within the troposphere and is the layer within which the weather systems develop. The depth of this layer varies between 548,000 ft at the equator and 23,000 ft over the polar regions. The stratosphere extends from the top of the troposphere to the bottom of the mesosphere, contains the ozone layer which ranges in altitude between 49,000 and 115,000 ft, and is where most of the UV radiation from the Sun is absorbed. The top of the mesosphere ranges from 164,000 to 279,000 ft and is the layer, wherein most meteors burn up. The thermosphere extends from 279,000 ft to the base of the exosphere at approximately 2,300,000 ft altitude and contains the ionosphere, a region where the atmosphere is ionized by incoming solar radiation.

2.1.1 The Troposphere

The troposphere is the lowest layer of atmosphere of the Earth and the layers to which changes can greatly influence the floral and faunal environments. Atmosphere of the Earth: it extends from Earth's surface to an average height of approximately 12 km although this altitude actually varies from approximately 30,000 ft at the polar regions to 56,000 ft) at the equator, with some variation due to weather. The troposphere is bounded above by the tropopause, a boundary marked in most places by a temperature inversion (i.e., a layer of relatively warm air above a colder one), and in others by a zone which is isothermal with height.

Although variations do occur, the temperature usually declines with increasing altitude in the troposphere because the troposphere is mostly heated through energy transfer from the surface. Thus, the lowest part of the troposphere (i.e., Earth's surface) is typically the warmest section of the troposphere, which promotes vertical mixing. The troposphere contains approximately 80% of the mass of the atmosphere of the Earth. The troposphere is denser than all its overlying atmospheric layers because a larger atmospheric weight sits on top of the troposphere and causes it to be most severely compressed. Fifty percent of the total mass of the atmosphere is located in the lower 18,000 ft of the troposphere.

Nearly all atmospheric water vapor or moisture is found in the troposphere, so it is the layer where most of Earth's weather takes place. It has basically all the weather-associated cloud genus types generated by active wind circulation although very tall cumulonimbus thunder clouds can penetrate the tropopause from below and rise into the lower part of the stratosphere. Most conventional aviation activity takes place in the troposphere, and it is the only layer that can be accessed by propeller-driven aircraft.

In addition, the atmosphere is generally described in terms of layers characterized by specific vertical temperature gradients. The troposphere is characterized by a decrease of the mean temperature with increasing altitude. This layer, which contains approximately 85–90% (v/v) of the atmospheric mass, is often dynamically unstable with rapid vertical exchanges of energy and mass being associated with convective activity. Globally, the time constant for vertical exchanges is of the order of several weeks. Much of the variability observed in the atmosphere occurs within this layer, including the weather patterns associated, for example, with the passage of fronts or the formation of thunderstorms. The planetary boundary layer is the region of the troposphere where surface effects are important, and the depth is on the order of 3300 ft but varies significantly with the time of day and with meteorological conditions. The exchange of chemical compounds between the surface and the free troposphere is directly dependent on the stability of the boundary layer.

2.1.2 The Stratosphere

Above the troposphere, the atmosphere becomes very stable, as the vertical temperature gradient reverses in a second atmospheric region—the *stratosphere*—which extends from the top of the troposphere at approximately 39,000 ft above the surface of the Earth to the stratopause at an altitude of approximately 164,000–180,000 ft. The atmospheric pressure at the top of the stratosphere is approximately 1/1000th the pressure at sea level, which contains 90% of the atmospheric ozone. A typical residence time for material injected in the lower stratosphere is 1–3 years. The stratosphere is the second-lowest layer of atmosphere of the Earth and lies above the troposphere and is separated from it by the tropopause.

The stratosphere contains the ozone layer, which is the part of atmosphere that contains relatively high concentrations of that gas. In this layer ozone concentrations are approximately 2–8 ppm, which is much higher than in the lower atmosphere but still very small compared to the main components of the atmosphere. It is mainly located in the lower portion of the stratosphere from approximately 49,000 to 115,000 ft, though the thickness varies seasonally and geographically. Approximately 90% (v/v) of the ozone in Earth's atmosphere is contained in the stratosphere.

The stratosphere defines a layer in which temperatures rise with increasing altitude. This rise in temperature is caused by the absorption of UV radiation from the Sun by the ozone layer, which restricts turbulence and mixing. Although the temperature is typically on the order of −60°C (−76°F) at the tropopause, the top of the stratosphere is much warmer and may be near 0°C (32°F). The stratospheric temperature profile creates very stable atmospheric conditions, so the stratosphere lacks the weather-producing air turbulence that is prevalent in the troposphere. Consequently, the stratosphere is almost completely free of clouds and other forms of weather. However, polar stratospheric or nacreous clouds are occasionally seen in the lower part of this layer of the atmosphere where the air is coldest, and this is the highest layer that can be accessed by conventional jet-powered aircraft.

2.1.3 The Mesosphere

The mesosphere is the third highest layer of atmosphere and occupies the region above the stratosphere and below the thermosphere. This layer extends from the stratopause at an altitude of approximately 160,000 ft to the mesopause at approximately 260,000–80,000 ft above sea level. Temperatures drop with increasing altitude to the mesopause that marks the top of this middle layer of the atmosphere. It is the coldest place on Earth and has a temperature on the order of −85°C (−120°F).

Just below the mesopause, the air is so cold that even the very scarce water vapor at this altitude can be sublimated into polar-mesospheric noctilucent clouds. These are the highest clouds in the atmosphere and may be visible to the naked eye if sunlight reflects off them approximately an hour or two after sunset or a similar length of time before sunrise. They are most readily visible when the Sun is around 4–16 degrees below the horizon. The mesosphere is also the layer where most meteors burn up upon atmospheric entrance.

2.1.4 The Thermosphere

The thermosphere is the second highest layer of Earth's atmosphere and extends from the mesopause (which separates it from the mesosphere) at an altitude of approximately 260,000 ft up to the thermopause at an altitude that ranges from 1,600,000 to 3,300,000 ft. In the thermosphere, the temperature increases to reach maximum values that are strongly dependent on the level of solar activity. Vertical exchanges associated with dynamical mixing become insignificant, but molecular diffusion becomes an important process that produces gravitational separation of species according to their molecular or atomic weight.

The uneven distribution of radiative heating in the Earth system produces a meridional circulation of air (circulation in the north-south direction), with rising motion at low latitudes and sinking motion at mid- and high latitudes. This meridional overturning of air masses is modified substantially by the Earth's rotation, especially outside the tropics, where the mean circulation is nearly circumpolar (along latitude circles). A small meridional component, however, transfers heat and transports trace constituents from equatorial to polar regions. The zonal flow (circulation in the west-east direction) is perturbed by orographic features at the Earth's surface (e.g., mountains) and latent heat release associated with the formation of clouds, as well as synoptic weather systems in the troposphere. Chemical constituents (including organic chemicals) are redistributed in the atmosphere by transport processes caused by such dynamical disturbances.

The height of the thermopause varies considerably due to changes in solar activity. Because the thermopause lies at the lower boundary of the exosphere, it is also referred to as the exobase. The lower part of the thermosphere, from 260,00 to 1,800,000 ft above the surface of the Earth surface, contains the *ionosphere*.

The ionosphere is a region of the atmosphere that is ionized by solar radiation and is responsible for auroras (the *aurora borealis* in the northern hemisphere and the *aurora australis* in the southern hemisphere). The ionosphere increases in thickness and moves closer to the Earth during daylight and rises at night allowing certain frequencies of radio communication a greater range. During daytime hours, it stretches from approximately 160,000 to 3,280,000 ft and includes the mesosphere, thermosphere, and parts of the exosphere. However, ionization in the mesosphere largely ceases during the night, so auroras are normally seen only in the thermosphere and lower exosphere. The ionosphere forms the inner edge of the magnetosphere. It has practical importance because it influences, for example, radio propagation on Earth.

The temperature of the thermosphere gradually increases with height. Unlike the stratosphere beneath it, wherein a temperature inversion is due to the absorption of radiation by ozone, the inversion in the thermosphere occurs due to the extremely low density of its molecules. The temperature of this layer can rise as high as 1500°C (2700°F), though the gas molecules are so far apart that its temperature in the usual sense is not very meaningful. This layer is completely cloudless and free of water vapor. However nonhydrometeorological phenomena such as the aurora borealis and *aurora australis* are occasionally seen in the thermosphere.

The atmosphere becomes thinner and thinner with increasing altitude, with no definite boundary between the atmosphere and outer space. The Kármán line, at 328,000, or 1.57% of the radius of the Earth's radius is often used as the border between the atmosphere and outer space. The International Space Station orbits in this layer, between 1,200,000 and 1,373,000 ft.

2.1.5 The Exosphere

The exosphere is the outermost layer of Earth's atmosphere (i.e., the upper limit of the atmosphere) and extends from the exobase, which is located at the top of the thermosphere. The exosphere begins variously from approximately 2,300,000 to 3,280,000 ft above the surface, where it interacts with the magnetosphere, to space. Each of the layers has a different lapse rate, defining the rate of change in temperature with height. Initial atmospheric composition is generally related to the chemistry and temperature of the local solar nebula during planetary formation and the subsequent escape of interior gases.

The escape to space of hydrogen atoms (estimated to be on the order of 25,000 tons/year) from the *exosphere*, the atmospheric region in contact with the interplanetary medium, has most certainly contributed to substantial changes in the chemical composition of the atmosphere over geological history. This escape flux increases due to the increase in methane abundance (Ehhalt, 1986).

The exosphere layer is mainly composed of extremely low densities of hydrogen, helium, and several heavier molecules including nitrogen, oxygen, and carbon dioxide closer to the exobase. The atoms and molecules are so far apart that they can travel hundreds of kilometers without colliding with one another. Thus, the exosphere no longer behaves like a gas, and the particles constantly escape into space. The exosphere contains most of the satellites orbiting Earth.

2.1.6 Composition

The three major constituents of air, and therefore of Earth's atmosphere, are nitrogen, oxygen, and argon. Thus, the atmosphere of the Earth is a mixture of chemical constituents—the most abundant of them are nitrogen (N_2, (78%, v/v)) and oxygen (O_2, (21%, v/v)). These gases, as well as the noble gases (argon, neon, helium, krypton, xenon), possess very long lifetimes against chemical destruction and, hence, are relatively mixed well throughout the entire homosphere (below approximately 295,000 ft altitude). Minor constituents, such as water vapor, carbon dioxide, ozone, and many others, also play an important role despite their lower concentration. These constituents influence the transmission of solar and terrestrial radiation in the atmosphere and are therefore linked to the physical climate system; they are key components of biogeochemical cycles; in addition, they determine the *oxidizing capacity* of the atmosphere and, hence, the atmospheric lifetime of biogenic and anthropogenic traces gases.

Water vapor accounts for approximately 0.25% (w/w) of the atmosphere by mass. The concentration of water vapor (a greenhouse gas) varies significantly from around 10 ppm by volume in the coldest portions of the atmosphere to as much as 5% by volume in hot, humid air masses, and concentrations of other atmospheric gases are typically quoted in terms of dry air (without water vapor). The remaining gases are often referred to as trace gases, among which are the greenhouse gases, principally carbon dioxide, methane, nitrous oxide, and ozone. The spatial and temporal distribution of chemical species in the atmosphere is determined by several processes, including surface emissions and deposition, chemical and photochemical reactions, and transport. Surface emissions are associated with volcanic eruptions, floral and faunal activity on the continents as well as in the ocean, as well as anthropological activity such as biomass burning, agricultural practices, and industrial activity. Chemical conversions are achieved by a multitude of reactions whose rate constants are measured in the laboratory. Transport is usually represented by large-scale advective motion (displacements of air masses in the quasi-horizontal direction), and by smaller scale processes, including convective motions (vertical motions produced by thermal instability and often associated with the presence of large cloud systems), boundary layer exchanges, and mixing associated with turbulence. Wet deposition results from precipitation of soluble species, while the rate of dry deposition is affected by the nature of the surface (e.g., type of soils, vegetation, ocean, etc.).

2.1.7   Chemical Activity

Chemical compounds released at the surface by natural and anthropogenic processes are oxidized in the atmosphere before being removed by wet or dry deposition. Key chemical species of the troposphere include organic compounds such as methane and nonmethane hydrocarbons as well as oxygenated organic species and carbon monoxide, nitrogen oxides (which are also produced by lightning discharges in thunderstorms), as well as nitric acid and peroxyacetyl nitrate (PAN, an unstable secondary pollutant present in photochemical smog and which decomposes into peroxyethanoyl radicals and nitrogen dioxide):

Peroxyacetyl nitrate

Other chemical species include: hydrogen compounds (specifically the hydroxyl radical, OH, and the hydroperoxy radical, HO_2, as well as hydrogen peroxide, H_2O_2, and ozone, O_3) and sulfur compounds [dimethyl sulfide (DMS), sulfur dioxide, SO_2, and sulfuric acid, H_2SO_4].

The hydroxyl radical (OH) deserves additional consideration since it has the capability of reacting with and efficiently destroying a large number of organic chemical compounds, and hence of contributing directly to the oxidation capacity (reactivity) of the atmosphere. Ozone also plays an important role in the troposphere—together with water vapor ozone is the source of the hydroxy radical and, in addition, it contributes to climate forcing—climate forcing is any influence on climate that originates from outside the climate system itself and is a major cause of climate change, which includes the temperature rise of the earth during an interglacial period which exists at the present. The presence of this gas in the troposphere results not only from the intrusion of ozone-rich stratospheric air masses through the tropopause; it is also produced by photochemical reactions involving hydrocarbon derivatives, nitrogen oxides (NO_x), and carbon monoxide (CO). One major question is to what extent the oxidizing capacity of the atmosphere has changed as a result of human activities. Finally, the release of sulfur compounds at the surface of the Earth surface and the subsequent oxidation of the sulfur compounds in the atmosphere leads to the formation of small liquid or solid particles that remain in suspension in the atmosphere. These aerosol particles affect the radiative balance of the atmosphere directly, by reflecting and absorbing solar radiation, and indirectly, by influencing cloud microphysics. The release to the atmosphere of sulfur compounds has increased dramatically, particularly in regions of Asia, Europe, and North America as a result of human activities, specifically coal combustion (Speight, 2013).

Gases that are not rapidly destroyed (such as by interaction with the hydroxyl radical) or removed by clouds and rain in the troposphere are transported upward into the stratosphere, where they are dissociated by short-wave UV radiation to produce fast-reacting radicals. Chlorofluorocarbons or nitrous oxide are examples of such long-lived gases that, when subject to photolysis in the stratosphere, provide a source of chlorine or nitrogen oxides, respectively. Such fast-reacting radicals initiate catalytic cycles that lead to the destruction of ozone, before being converted into chemical reservoirs that are gradually removed from the stratosphere. The abundance of ozone results from a delicate balance between these destruction mechanisms and the natural production of O_3 through the photolysis of molecular oxygen. There is strong evidence that suggests that the depletion of stratospheric ozone observed over the past decade is the direct consequence of the release in the atmosphere of industrially manufactured chlorofluorocarbons.

Filtered air includes trace amounts of many other chemical compounds. Many substances of natural origin may be present in locally and seasonally variable small amounts as aerosols in an unfiltered air sample, including dust of

TABLE 1.1 Composition of the Atmosphere

Gas	Formula	Volume (ppm)	Volume (%)
Nitrogen	N_2	80,840	78.084
Oxygen	O_2	209,460	20.946
Argon	Ar	9340	0.9340
Carbon dioxide	CO_2	397	0.0397
Neon	Ne	18.18	0.001818
Helium	He	5.24	0.000524
Methane	CH_4	1.79	0.000179
Water vapor[a]	H_2O	10–50,000[D]	0.001–5[D]

[a]Water vapor is not included in above dry atmosphere and is approximately 0.25% (v/v) over the full atmosphere but does vary considerably.

mineral and organic composition, pollen, and spores, sea spray, and volcanic ash. Various industrial pollutants also may be present as gases or aerosols, such as chlorine, (elemental or in compounds), fluorine compounds, and elemental mercury vapor. Sulfur compounds such as hydrogen sulfide and sulfur dioxide (SO_2) may be derived from natural sources or from industrial air pollution. By volume, dry air contains (subject to minor rounding of the data) 78.09% nitrogen, 20.95% oxygen, 0.93% argon, 0.039% carbon dioxide, and small amounts of other gases (Table 1.1). Air also contains a variable amount of water vapor, approximately 1% (v/v) at sea level and 0.4% (v/v) throughout the entire atmosphere. Air content and atmospheric pressure vary at different layers, and air suitable for use in photosynthesis by terrestrial plants and breathing of terrestrial animals is found only in the troposphere.

2.2   The Aquasphere

Water (the Aquasphere, the aquatic biome) makes up the largest part of the biosphere, covering nearly 75% of the Earth's surface. Clean freshwater resources are essential for drinking, bathing, cooking, irrigation, industry, and for plant and animal survival. Due to overuse, pollution, and ecosystem degradation the sources of most freshwater supplies—groundwater (water located below the soil surface), reservoirs, and rivers—are under severe and increasing environmental stress. The majority of the urban sewage in developing countries is discharged untreated into surface waters such as rivers and harbors. Approximately 65% (v/v) of the global freshwater supply is used in agriculture and 25% (v/v) is used in industry. Freshwater conservation therefore requires a reduction in wasteful practices like inefficient irrigation, reforms in agriculture and industry, and strict pollution controls worldwide.

Aquatic regions house numerous species of plants and animals, both large and small. In fact, this is where life began billions of years ago when amino acids first started to come together. Without water, most life forms would be unable to sustain themselves and the Earth would be a barren, desert-like place. Although water temperatures can vary widely, aquatic areas tend to be more humid and the air temperature on the cooler side. The aquasphere can be broken down into two basic regions, (1) freshwater—ponds and rivers and (2) marine regions.

2.2.1 Freshwater Regions

A freshwater region is an area where the water has a low salinity (a low salt concentration, usually on the order pf <1%, w/w). Plants and animals in freshwater regions are adjusted to the low salt content and would not be able to survive in areas of high salt concentration (such as the ocean). There are different types of freshwater regions: ponds and lakes, streams and rivers, and wetlands. The following sections describe the characteristics of these three freshwater zones.

Ponds and Lakes

Ponds and lakes vary on aerial extent (without placing hard and fast boundaries on such waters) from just a few square years to many square miles. Scattered throughout the surface of the Earth, several are remnants from the last Ice Age after which the ice melted approximately 10,000–12,000 years ago—the final part of the Quaternary glaciation lasted from approximately 110,000 to 12,000 years ago and occurred during the last 100,000 years of the Pleistocene epoch.

Many ponds are seasonal, lasting just a couple of months (such as sessile pools), while lakes may exist for hundreds of years or more. Ponds and lakes may have limited diversity of floral and faunal species since they are often isolated from one another and from other water sources, such as rivers and oceans which have a larger floral and faunal diversity. Lakes and ponds are divided into three different zones, which are usually determined by depth and distance from the shoreline.

The topmost zone near the shore of a lake or pond is the *littoral zone*—this zone is the warmest since it is shallow and can absorb more of the Sun's heat. It sustains a fairly diverse community, which can include several species of algae (like diatoms), rooted and floating aquatic plants, grazing snails, clams, insects, crustaceans, fishes, and amphibians. In the case of the insects, such as dragonflies and midges, only the egg and larvae stages are found in this zone. The vegetation and animals living in the littoral zone are food for other creatures such as turtles, snakes, and ducks. The near-surface open water surrounded by the littoral zone is the *limnetic zone*. The limnetic zone is well lighted (like the littoral zone) and is dominated by plankton, both phytoplankton and zooplankton. Plankton are small organisms that play a crucial role in the food chain. Without aquatic plankton, there would be few living organisms in the world, and certainly no humans. A variety of freshwater fish also occupy this zone. Plankton have short life spans, and when plankton die, the remains fall into the deepwater part of the lake/pond, the *profundal zone*. This zone is much colder and denser than the other two and little light penetrates all the way through the limnetic zone into the profundal zone. The fauna are heterotrophs—organisms which eat other dead organisms and use oxygen for cellular respiration.

Temperature varies in ponds and lakes seasonally. For example, during the summer, the temperature can range from 4°C (39°F) near the bottom of the pond or lake to 22°C (72°F) at the top. During the winter, the temperature at the bottom of the pond or lake can be 4°C (39°F) while the top is 0°C (39°F). In between the two layers, there is a narrow zone (the *thermocline*) where the temperature of the water changes rapidly. Outside of the summer and winter season (during the spring and autumn/fall seasons), there is a tendency for the top layers and the bottom layers to mix, usually due to winds, which results in a uniform water temperature of around 4°C (39°F), but this is very dependent upon the climatology of the region. This mixing also circulates oxygen throughout the lake but if the lake or pond did not freeze during the winter, thus the top layer can be expected to be warmer than the bottom of the pond or lake.

Streams and Rivers

Streams and rivers are bodies of flowing water that are ubiquitous throughout the surface of the Earth and which move predominantly move naturally in one direction (unless there are upsets such as earthquakes). These water courses start at the headwaters which arise from springs, from snowmelt, or even from lakes after which the water flow to the mouth (or delta) of the river into another water channel, typically the ocean.

The characteristics of a river or stream change during the journey from the source to the mouth. For example, the temperature of the water is cooler at the source than it is at the mouth. The water is also clearer and has higher oxygen levels, and freshwater fish such as trout and heterotrophs can be found there. Toward the middle part of the stream/river, the width increases, as does species diversity—numerous aquatic green plants and algae can be found. Toward the mouth of the river/stream, the water becomes murky from all the sediments that it has picked up upstream, decreasing the amount of light that can penetrate through the water. Since there is less light, there is less diversity of flora, and because of the lower oxygen levels, fish that require less oxygen, such as catfish and carp, can be found.

Wetlands

Wetlands are areas where water covers the soil, or is present either at or near the surface of the soil all year or for varying periods of time during the year, including during the growing season. Water saturation (hydrology) largely determines how the soil develops and the types of plant and animal communities living in and on the soil. Wetlands may support both aquatic and terrestrial species. The prolonged presence of water creates conditions that favor the growth of specially adapted plants (hydrophytes) and promote the development of characteristic wetland (hydric) soils. Wetlands vary widely because of regional and local differences in soils, topography, climate, hydrology, water chemistry, vegetation, and other factors, including human disturbance. Indeed, wetlands are found from the tundra to the tropics and on every continent except Antarctica. Two general categories of wetlands are recognized: coastal or tidal wetlands and inland or nontidal wetlands.

Tidal wetlands are, as the name suggests, located found along coast lines and are often linked to river estuaries where sea water mixes with fresh water to form an environment of varying salinity. The salt water and the fluctuating water level (due to tidal action) combine to create a rather difficult environment for most plants. Consequently, many shallow coastal areas are nonvegetated mud flats or sand flats. Some plants, however, have successfully adapted to this environment—certain types of grasses and grass-like plants that are able to adapt to the saline conditions form the tidal salt marshes that are found along a coast line. Moreover, mangrove swamps, with salt-loving shrubs or trees, are common in tropical climate areas. Some tidal freshwater wetlands form beyond the upper edges of tidal salt marshes where the influence of salt water ends.

Nontidal wetlands are most common on floodplains along rivers and streams (riparian wetlands), in isolated depressions surrounded by dry land, along the margins of lakes and ponds, and in other low-lying areas where the groundwater intercepts the soil surface or where precipitation sufficiently saturates the soil (vernal pools and bogs). Inland wetlands include marshes and wet meadows dominated by herbaceous plants, swamps dominated by shrubs, and wooded swamps dominated by trees.

Many of wetlands are seasonal (they are dry one or more seasons every year), and, particularly in the arid and semiarid areas, may be wet only periodically (i.e., less than seasonal). The quantity of water present and the timing of its presence in part determine the functions of a wetland and its role in the environment. Even wetlands that appear dry at times for significant parts of the year—such as vernal pools (temporary pools of water that provide habitat for distinctive plants and animals)—often provide critical habitat for wildlife adapted to breeding exclusively in these areas.

Thus, wetlands are areas of standing water that support aquatic plants. Marshes, swamps, and bogs are all considered wetlands. Plant species adapted to the very moist and humid conditions (hydrophytes) include pond lilies, cattails, sedges, tamarack, and black spruce as well as marsh flora which include species such as cypress and gum. Wetlands have the highest species diversity of all ecosystems and are attractive to many species of amphibians, reptiles, birds (such as ducks and waders), and furbearers can be found in the wetlands. Wetlands are not considered freshwater ecosystems as there are some, such as salt marshes, that have high salt concentrations and which support different species of animals, such as shrimp, shellfish, and various grasses.

Wetlands—in which high rainfall, appropriate topography, or low evapo-transpiration lead to permanent or seasonal flooding—are an exception to the relationship between emissions and climatic conditions. Wetlands are disproportionately found in cooler latitudes, and northern wetlands are thought to share in importance with methane emissions over tropical wetlands simply because of the greater area (albeit with lower fluxes) of northern wetlands. These areas become anoxic (deficient in oxygen) because microbial activity consumes all the oxygen—anoxic conditions lead to reduced productivity, reduced organic matter decomposition, and greatly enhanced methane production. Wetlands may also contribute significant amounts of reduced sulfur-containing gases to the atmosphere.

The Oceans

The oceans (sometime referred to as *marine regions*) cover approximately 75% of the surface of the Earth and also (under the general term *marine regions*) includes coral reefs, and estuaries—estuaries are areas where freshwater streams or rivers merge with the ocean. This mixing of waters with such different salt concentrations creates a very interesting and unique ecosystem. Microflora (such as algae) and macroflora (such as seaweeds, marsh grasses, and mangrove trees) can be found here. Estuaries support a diverse fauna, including a variety of worms, oysters, crabs, and waterfowl. Marine algae supply much of the oxygen supply of the Earth and, at the same time, remove a substantial amount of carbon dioxide from the atmosphere. In addition, the evaporation of the seawater provides rainwater for the land.

The largest of all of the marine region ecosystems sand are very large bodies of water that dominate the surface of the Earth. Like ponds and lakes, the ocean regions are separated into separate zones: intertidal, pelagic, abyssal, and benthic. All four zones have a great diversity of species—often claimed (with some justification) to be richest diversity of species even though the oceans may contain fewer species than exist on land-based ecosystems.

The ocean, which covers approximately 75% of the surface of the Earth, is coupled to the atmosphere from both a physical and a biogeochemical perspective. The basic structure of the ocean is set by the geographic patterns of surface heating and freshwater input (precipitation—rainfall, snowfall—minus water lost due to evaporation), which influences the salinity distribution in the ocean.

In general, there is net warming of the ocean surface in the tropics and subtropics and net cooling at of the oceans in the temperate and polar latitudes.

In addition, the ocean can be divided into two general regions: (1) a warm, surface pool—typically 18°C (64°F—i.e., approximately 3280 ft thick) and (2) the deep water—typically: 3°C (37°F)—that outcrops to the surface at high latitudes and forms the bulk of the ocean volume. Unlike the atmosphere, heating of the ocean surface stabilizes the water and also prevents rapid exchange (or mixing) between the surface and deep water. In fact, contact between the surface and deep waters is limited to localized polar regions where losses of heat and freshwater lead to sinking and deep water formation. The resulting thermohaline circulation (part of the large-scale ocean circulation that is driven by global density gradients that are created by surface heat and fresh-water flow) is especially important over long timescales (such as glacial cycles).

The other component of the ocean circulation is produced by the drag of the surface winds on the ocean. The zonal wind patterns over the ocean result in rotating circulation patterns in the subtropical and subpolar regions and are also responsible for the circumpolar current in the southern oceans (i.e., oceans to the south of the equator). Typically, the wind-driven surface currents move heat and trace species (such as organic chemical compounds) from the tropics to the poles, and approximately an equal amount of solar energy received in the tropics is transported toward the pole by oceanic and atmospheric circulations. Wind forcing also causes divergence of the surface water and upwelling along both the equator and coastal regions on the eastern margin of ocean basins. The upwelling of cooler, nutrient-rich waters in these areas greatly enhance ocean productivity. The ocean circulation, which exhibits variations on different timescales (including perturbations such as the El Niño events that occur in the equatorial Pacific on an average of four years and produce massive warming of the coastal waters off Peru and Ecuador with torrential rainfall in the region), greatly affects biogeochemical cycles as well as the global climate.

From a chemical point of view, the ocean influences the atmosphere through the exchanges of trace gases across the air-sea interface. The transfer of carbon dioxide from the atmosphere to the ocean is controlled by the two competing factors of temperature: warming of surface waters, which releases carbon dioxide to the atmosphere, and biological productivity. Photosynthesis by marine phytoplankton converts dissolved carbon dioxide into organic carbon, leading to a reduction in surface carbon dioxide values and a carbon dioxide flow into the ocean. The amount of carbon dioxide dissolved in seawater is quite large due to its high solubility and its reactivity with water to form carbonic acid and its dissociation products. The ocean, therefore, serves as a major reservoir for carbon dioxide, approximately 65 times larger than the atmosphere, and oceans have played an important role in the evolution of atmospheric carbon dioxide over the geological history of the Earth and is a primary sink for anthropogenic carbon dioxide.

Other chemical species are released by the ocean, such as reduced sulfur, certain hydrocarbon derivatives, and carbon monoxide. The largest oceanic source of sulfur is provided by dimethyl sulfide (CH_3SCH_3), which is produced by various, but specific, types of phytoplankton. These emissions appear to be most intense in regions where the net primary productivity of the ocean is highest, modified somewhat by poorly understood large-scale patterns in the distribution of phytoplankton species.

The *intertidal zone* is where the ocean meets the land—sometimes this zone is a submerged zone and at other times exposed, as the ocean ebbs (outgoing tide) and flows (incoming tide). Because of this, the floral and faunal communities are constantly changing. For example, on rocky coasts, the zone is stratified vertically and, where only the highest tides reach, there are only a few species of algae and mollusks. In those areas usually submerged during high tide, there is a more diverse array of algae and small animals, such as herbivorous snails, crabs, sea stars, and small fishes. At the bottom of the intertidal zone, which is only exposed during the lowest tides, many invertebrates, fishes, and seaweed can be found. The intertidal zone on sandier shores is not as stratified as in the rocky areas—waves action maintain the mud and sand in a state of constant motion and, thus, very few algae and plants can establish themselves—the fauna includes worms, clams, predatory crustaceans, crabs, and shorebirds.

The *pelagic zone* includes those waters further from the land and the open ocean. The pelagic zone is generally cold though it is hard to give a general temperature range since, just like ponds and lakes, there is thermal stratification with a constant mixing of warm and cold ocean currents. The flora in the pelagic zone include surface seaweeds. The faunal animals include many species of fish and some mammals, such as whales and dolphins. Many feed on the abundant plankton.

The *benthic zone* is the area below the pelagic zone but does not include the very deepest parts of the ocean. The bottom of the zone consists of sand, slit, and/or dead organisms. Here temperature decreases as depth increases toward the abyssal zone, since light cannot penetrate through the deeper water. Flora are represented primarily by seaweed while the fauna, since it is very nutrient-rich, include all sorts of bacteria, fungi, sponges, sea anemones, worms, sea stars, and fishes.

The deep ocean is the *abyssal zone* and in this region is very cold (typically: approximately $3°C$, $347°F$), highly pressured, high in oxygen content, but low in nutritional value. The abyssal zone supports many species of invertebrate species and fishes. The mid-ocean ridges (spreading zones between tectonic plates), often with hydrothermal vents, are found in the abyssal zone along the ocean floor. Chemosynthetic bacteria thrive near these vents because of the large amounts of hydrogen sulfide and other minerals produced from the vents. These bacteria are thus the start of the food web as they are eaten by invertebrates and fishes.

Coral reefs are widely distributed in warm shallow waters and can occur as barriers along continents, such as the Great Barrier Reef located in the Coral Sea off the coast of Queensland, Australia. The Great Barrier Reef is the largest coral reef system in the world and is composed of over 2900 individual reefs and 900 islands stretching for more than 1400 miles and over an area of approximately 133,000 square miles. The dominant organisms in coral reefs are corals. Corals are interesting since they consist of both algae and tissues of animal polyp. Since reef waters tend to be nutritionally poor, corals obtain nutrients through the algae via photosynthesis and also by extending tentacles to obtain plankton from the water. Besides corals, the faunal animals include several species of microorganisms, invertebrates, fishes, sea urchins, octopuses, and sea stars.

2.3 The Terrestrial Biosphere

The terrestrial biosphere (*land*) is important to atmospheric chemistry as a source and sink for many compounds—a major activity within atmospheric chemistry has been (and remains) the determination of such flows. The structure of the biosphere is controlled by the interaction of climate with the patterns of soils and topography resulting from geological processes on a range of time scales, and further modified by the biogeographic distribution of organisms. Climate patterns are reflected in productivity (annual carbon fixation through photosynthesis), with warmer and wetter regions having higher productivity— the rate of nitrogen cycling follows similar trends—and as a result, trace gas emissions are usually higher (in some cases higher by one or more orders of magnitude) in the tropics than in the mid-to-high-latitude regions. This is clearly true for all soil trace gas fluxes and may be true for plant-mediated fluxes.

For example, large quantities of hydrocarbons such as isoprene (C_5H_8) are produced by the foliage of the abundant vegetation in productive ecosystems. Biomass burning fluxes are highest in tropical savanna ecosystems, which are warm and have sufficient rainfall during the wet seasons to accumulate significant biomass, which burns readily during the dry seasons. Large quantities of atmospheric carbon dioxide, carbon monoxide, hydrocarbon derivatives, and nitrogen oxides (NO_x) are produced as a result of the combustion of biomass. Soil and nitrogen oxide relationships are enhanced when the soil is rapidly wetted, dried, and wetted again in succession, and so may be higher in regions of sporadic rainfall, despite higher overall rates of nitrogen cycling, and nitrogen oxide emissions in moist areas.

The structure of the biosphere is important for the understanding issues related to the source of trace gas sources and sinks for trace gases. However, the spatial structure of the biosphere cannot be conveniently described as the outcome of a series of physical calculations (unlike the atmosphere and oceans) but rather requires the use of large databases describing fine-scale structures within ecosystems. Not only does the terrestrial biosphere play an important role in the functioning of the global climate system, but pollution can have major impacts on the biosphere itself. Climate change is projected to impact agricultural production, forestry, natural ecosystems, and biodiversity through changes to the soil (Gouin et al., 2013).

Soil, a mixture of mineral, plant, and animal materials, is essential for most plant growth and is the basic resource for agricultural production. In the process of developing the land and clearing away the vegetation that holds water and soil in place, erosion has caused devastation on a worldwide scale. The rapid deforestation taking place in the tropics is especially damaging because the thin layer of soil that remains is extremely fragile and quickly washes away when exposed to the heavy tropical rain storms.

Technically, soil is a mixture of mineral constituents—the inorganic components of soil are principally produced by the weathering of rocks and minerals—plant materials, and animal materials, that forms during a long process that may take thousands of years and it is an unconsolidated, or loose, combination of inorganic and organic materials. Soil is necessary for most plant growth and is essential for all agricultural production. The organic materials are composed of debris from plants and from the decomposition of animals as well as the many tiny (microscopic) life forms that inhabit the soil. The chemical composition and physical structure of soils is determined by a number of factors such as: the kinds of rocks, minerals, and other geologic materials from which the soil is originally formed. The vegetation that grow in the soil are also important.

Food sources grown on soils are predominately composed of carbon, hydrogen, oxygen, phosphorous, nitrogen, potassium, sodium, and calcium. Plants take up these elements from the soil and configure them into the plants that are recognized as food-plants. Each plant has unique nutritional requirements that are obtained through the roots from the soil. Nutrients are stored in soil on "exchange sites" of the organic and clay components. Calcium, magnesium, ammonium, potassium, and the vast majority of the micronutrients are present as cations in soils of varying acidity and alkalinity (varying under most soil pH).

2.3.1 Composition of Soil

Soil comprises a mixture of inorganic and organic components: minerals, air, water, and plant and animal material. Mineral and organic particles generally compose approximately 50% of the volume of soil. The other 50% consists of open areas (pores) that are of various shapes and sizes. Networks of pores hold water within the soil and also provide a means of water transport. Oxygen and other gases move through pore spaces in soil, and the pores also serve as passageways for small animals and provide room for the growth of plant roots.

The mineral component of soil consists of an arrangement of particles that are less than 2.0 mm in diameter. Technically, soil is composed of particles that fall into three main mineral groups (1) sand, (2) silt, and (3) clay each of which is determined by particle size: sand, 0.05–2.00 mm; silt, 0.002–0.05 mm; and clay, <0.002 mm. Depending upon the parent rock materials from which these mineral were derived, the assorted mineral particles ultimately release the chemicals on which plants depend for survival, such as potassium, calcium, magnesium, phosphorus, sulfur, iron, and manganese.

Organic materials constitute another essential component of soils. Some of the organic material arises from the residue of plants, such as the remains of the roots of plants deep within the soil, or materials that fall on the ground, such as leaves on a forest floor or even a dead animal. These materials become part of a cycle of decomposition and decay, a cycle that provides important nutrients to the soil. In general, soil fertility depends on a high content of organic materials.

Soils are also characterized according to how effectively they retain and transport water. Once water enters the soil from rain or irrigation, gravity comes into play, causing water to trickle downward. Soil differs in the capacity to retain moisture against the pull exerted by gravity and plant roots. Coarse soil, such as soil consisting of mostly of sand, tend to hold less water than do soils with finer textures, such as those with a greater proportion of clays.

Water also moves through soil pores by capillary action, which is the type of movement in which the water molecules move because they are more attracted to the pore walls (adhesion) than to one another (cohesion). Such movement tends to occur from wetter to drier areas of the soil. The attraction of water molecules to each other is an example of cohesion.

2.3.2 Soil Pollution

Unhealthy soil management methods have seriously degraded soil quality, caused soil pollution, and enhanced erosion. In addition to other human practices, the use of chemical fertilizers, pesticides, and fungicides has disrupted the natural processes occurring within the soil resulting in soil pollution. Soil pollution is a buildup of toxic chemical compounds, salts, pathogens, or radioactive materials that can affect plant and animal life. The concern over soil contamination stems primarily from health risks, both of direct contact and from secondary contamination of water supplies. All kinds of soil pollutants originate from a source. The source is particularly important because it is generally the logical place to eliminate pollution. After a pollutant is released from a source, it may act upon a receptor. The receptor is anything that is affected by the pollutant. The following subunit describes some of the most common sources of soil pollution.

Some of the most common toxic soil pollutants include organic chemicals, oils, tars, pesticides, biologically active materials, combustible materials, asbestos, and other hazardous materials. These substances commonly arise from the rupture of underground storage tanks; application of chemical fertilizers, pesticides, and fungicides; percolation of contaminated surface water to subsurface strata; leaching of wastes from landfills or direct discharge of industrial wastes to the soil. Pesticides that are used in agricultural practices pollute the soil directly by affecting the organisms that reside in it. Pesticides include many types of chemicals that are spread around in the environment to kill some specific sort of pest, usually insects (insecticides), weeds (herbicides), or fungi (fungicides).

Organic pollutants enter the soil via atmospheric deposition, direct spreading onto land, contamination by wastewater, and waste disposal. Organic contaminants include pesticides and many other components, such as oils, tars, chlorinated hydrocarbons, PCBs, and dioxins. The use of pesticides may lead to: (1) destruction of the soil's micro-flora and fauna, leading to both physical and chemical deterioration; (2) severe yield reduction in crops; and (3) leaching of toxic chemicals into groundwater and potentially threatening drinking water resources.

Existence of the ecosystems requires existence of plants. Humans and animals cannot survive without plants. Soil is not only a source of nutrition but also a place for plants to stand. Pollution of agricultural soils is known to reduce agricultural yield and increase levels of these toxic heavy metals in agricultural products, and thus to their introduction into the food chain. Vegetables and crop plants grown in such soils take up these toxic elements and pose health risk to humans and animals feeding on these plants. The major concern approximately soil pollution is that there are many sensitive land uses where people are in direct contact with soils such as residences, parks, schools, and playgrounds. Other contact mechanisms include contamination of drinking water or inhalation of soil contaminants which have vaporized. There is a very large set of health consequences from exposure to soil contamination depending on pollutant type, pathway of attack, and vulnerability of the exposed population.

Organic pollutants which are directly applied into soils or deposited from the atmosphere may be taken up by plants or leached into water bodies. Ultimately they affect human and animal health when taken up through the food they eat and the water they drink. More recently research has revealed that many chemical pollutants, such as DDT and polychlorobiphenyls (PCBs), mimic sex hormones and interfere with the reproductive and developmental functions of the human body—the substances are known as *endocrine disrupters*. Although, soil might be affected less by pollution compared to water or air but cleaning polluted soil is more difficult, complex, and expensive than cramming water and air.

As part of the biosphere, forests are very important for maintaining ecological balance and provide many environmental benefits. In addition to timber and paper products, forests provide wildlife habitat, prevent flooding and soil erosion, help provide clean air and water, and contain tremendous biodiversity. Forests are also an important defense against global climate change. Forests produce life-giving oxygen and consume carbon dioxide, the compound that is claimed to be the most responsible for global warming through photosynthesis, thereby reducing the effects of global warming.

3 ORGANIC CHEMISTRY AND THE ENVIRONMENT

The 20th century came into being in much the same manner as the 19th century ended insofar as there was a continuation of the less-than-desirable disposal methods for chemical waste, which included gaseous waste, liquid waste, and solid waste. As the 20th century evolved, the use and disposal of chemicals expanded by several orders of magnitude and this expansion seemed to be unstoppable. In fact, it was not only industrial waste that was disposed of in a manner that was dangerous-to-the-environment but also the disposal of household chemicals (in considerable quantities when measured on a city-wide basis) that were used to paint, clean, and maintain homes and gardens. At the time, there was not the realization that many of these products were toxic to the flora and fauna (including humans) of the environment, whether or not they are used or disposed of improperly. However, during the latter quarter of the 20th century and by beginning of the 21st century, there came the realization that chemicals (some in large concentrations, other in small concentrations) were toxic and the unabated disposal of chemicals had to change. This awakening of an (almost global) environmental consciousness led to the legislation in many countries that chemical disposal must be organized and carried out by legislatively sanctioned methods, and the unabated and dangerous disposal of chemicals must cease.

The chemicals industry (which, within the context of this book, includes the fossil fuels industry) and its products provide many real and potential benefits, particularly related to improving and sustaining human health and nutrition as well as, on the economic side, financial capital through new opportunities for employment. At the same time that benefits accrue, the production and use of chemicals creates risks to the environment at all stages of the production cycle. The generation and intentional and unintentional release of the produced chemicals (and the process by-products) has contributed to environmental contamination and degradation at multiple levels—local, regional, and global—and in many instances the impact will, more than likely, continue to be felt for generations.

As a result, it is now (some observers would use the word *finally* instead of the word *now*) recognized and legislated that any process waste (including hazardous and nonhazardous wastes) should never be discarded without proper guidance and authority. The effects of these errant, irresponsible, irregular (and often illegal) methods of disposal were being observed in the atmosphere (the occurrence of smog in cities such as London and Manchester in the late 1950s is often cited as examples), the waterways (dead fish floating in rivers, streams, lakes, and oceans), and landfills (or anywhere that the waste was dumped on to solid ground (leading to objectionable odors and poisonous run-off material). Many forms of disposal (which are considered to be illegal by modern disposal standards or protocols) continued unchecked and unmonitored during the early part of the 20th century. Then in the 1950s, Rachel Carson (a marine biologist) wrote pamphlets on conservation and natural resources and edited scientific articles, but in her free time turned her government research into lyric prose, first as an article *Undersea* (1937, for the Atlantic Monthly), and then in a book, *Under the Sea-Wind* (1941) followed by *The Sea Around Us* in 1952, *The Edge of the Sea* in 1955, and Silent Spring in 1962. In her book *Silent Spring*, Carson warned of the dangers to all natural floral and faunal ecosystems from the misuse of pesticides such as DDT (dichlorodiphenyltrichloroethane).

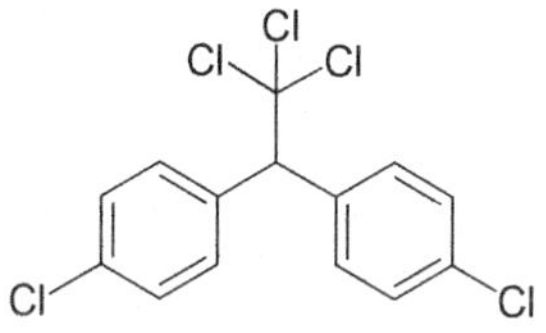

DDT (dichlorodiphenyltrichloroethane)

In *Silent Spring*, Carson also questioned the scope and direction of modern science, which many observers consider her (justifiably) to be the initiator of the modern environmental movement (Carson, 1962; Lear, 2015). The book dealt with many environmental problems associated with the unmonitored use chlorinated pesticides and initiated an extensive examination approximately safety of many different types of chemicals and the disposal methods of unwanted chemicals, waste chemicals, and chemical byproducts that can cause environmental pollution. As a result, the use of DDT (and similar pesticidal chemicals) was not only initially discouraged but was banned in 1972.

Following from this, and with the establishment of environmental protection agencies or departments by various levels, for example, local, state, and federal governments (in the United States and many other countries), serious consideration was given to the need for investigation of the methods by which chemicals were affecting the environment. As a result, methods were devised for handling chemical wastes with minimal effect on the environment (Carson and Mumford, 1988, 1995). In fact, during the last five decades it has become increasingly clear that the chemical and allied industries (including the pharmaceutical industries and the fossil fuel industries) can cause serious environmental problems if methods of disposal for unwanted chemicals and chemical wastes remain unchecked. As a result, many of these industries generate large amounts of chemical waste and have been subjected to strict legislation that requires minimization or, preferably, elimination of the various waste streams (Sheldon, 2010).

Furthermore, because of the importance of chemical contamination of the environment which involves a study of the effects of chemicals on the environment, the recent subdiscipline of *environmental chemistry* with the subcategories of *environmental organic chemistry* and *environmental inorganic chemistry* has arisen and evolved. Both environmental organic chemistry and environmental inorganic chemistry are now a component (optional or otherwise) of many chemistry degree courses in universities and are included in environmental science courses and environmental engineering courses as elements of increasing substance. These relatively new disciplines focus on the various environmental factors which govern the processes that determine the fate of organic chemicals and inorganic chemicals in ecosystems. The information discovered is then combined with the properties of the compound and applied to a quantitative assessment of the environmental behavior of a wide variety of chemicals (Mackay et al., 2006). Furthermore, research opportunities in environmental chemistry and environmental engineering have continued to be an educational growth area as new programs evolve to respond to local, national, regional, or global problems of various environmental issues at both fundamental and applied levels. In concert with these educational advances, the chemical industry is faced ever-increasing challenges from regulations pertaining to the environmental safety and environmental acceptability of its products.

Prior to the environmental revolution and during its infancy, *organic chemistry* related to the study of compounds from living organisms and as the subject matter evolved, the meaning was expanded to include the chemistry of carbon compounds. Relating to the current theme of this book, organic chemistry (a subdiscipline of *chemistry*) is the chemistry of carbon compounds—generally excluding the chemistry of carbon monoxide (CO) and carbon dioxide (CO_2)—and involves the scientific study of the structure, properties, and reactions of organic compounds, that is, matter in its various forms that are based on carbon atoms.

Thus, organic chemistry is the chemistry of carbon, an element that forms strong chemical bonds to other carbon atoms as well as to many other elements such as hydrogen, nitrogen, oxygen, sulfur, the halogens (fluorine, chlorine, bromine, and iodine), as well as a variety of different metals, such as nickel, vanadium, iron, and copper that occur as organometallic derivatives in many crude oils (Reynolds, 1998; Speight, 2014). Carbon always forms four covalent bonds (four shared pairs of electrons) that may be present as four single bonds per atom, or two single bonds and one double bond, or one single bond and one triple bond. With the ability of carbon to bond in different ways, an important part of organic chemistry concerns the structure of compounds. Organic chemistry is important because the vital biological molecules in living systems are largely organic compounds and, because of its versatility in forming covalent bonds, the number of known carbon compounds can only be conservatively estimated to be on the order of one million, possibly more. Furthermore, in the context of this book, environmental organic chemistry addresses the influence of organic chemicals on the environment which includes: (1) the study of the structure of organic compounds, (2) the physical properties of organic compounds, (3) the chemical properties of organic compounds, and (4) the reactivity or organic compounds with the goal of understanding the behavior of organic compounds not only in the in the pure form (when possible) but also in aqueous and nonaqueous solutions as well as the chemistry of complex mixtures to reflect the manner in which such chemicals exist and react in the environment.

For example, the presence of one or more functional groups (specific groups, moieties) consisting of atoms or bonds within organic molecules that are responsible for the characteristic chemical reactions and behavior of those organic molecules and which can dictate (1) the reactivity of the compound, (2) the manner by which the compound reacts to external influences, (3) the manner by which the chemical can dissipate into the atmosphere, (4) the solubility of the compound in aqueous systems, as well as (5) the manner by which the compound can adhere to, and remain in, the soil.

Simple, physical properties are properties that do not change the chemical nature of matter, while chemical properties are properties that do change the chemical nature of matter. Examples of physical properties are: color, smell, freezing point, boiling point, melting point, infra-red spectrum, attraction (paramagnetic) or repulsion (diamagnetic) to magnets, opacity, viscosity, and density. In addition, measuring each of these properties will not alter the basic nature of the substance. Examples of chemical properties are: heat of combustion, reactivity with water, acidity-alkalinity, and electromotive force. The more properties that can be used to identify a chemical, the better the nature of the chemical becomes known. These properties can to understand how this substance will behave under various conditions. In concert with the chemical properties physical properties of organic compounds, the physical properties that are typically of interest to the environmental scientist and environmental engineer include both quantitative and qualitative properties (Table 1.2).

Quantitative information includes the melting point (and the solidification point or the freezing point), the boiling point, as well as the tendency of the compound to evaporate under ambient conditions while qualitative properties include odor, solubility, and color. For example, organic compounds typically have recordable melting points and/or boiling points, but, in contrast, many mixtures will melt or boil over a range depending upon the composition of the mixture. In addition, inorganic chemicals generally melt (often at extremely high temperatures), and many do not exhibit recordable boiling points and tend to undergo thermal degradation. For organic chemicals, the melting point and the boiling point can provide valuable information not only on the purity and identity of an organic compound as well as a chemical assessment of the behavior of the compound in the environment. Furthermore, some organic compounds have a sublimation point, the temperature at which the organic compound undergoes sublimation (evaporation without melting thereby omitting the intermediate liquid phase).

As an example of the use of properties and the forewarning that properties can offer in terms of behavior, p-dichlorobenzene (*para*-dichlorobenzene, 1,4-dichlorobenzene, ClC_6H_4Cl), which is used as an insecticide for moth control,

TABLE 1.2 Important Common Physical Properties of Organic Compounds

State: Gas
Density
Critical temperature, critical pressure (for liquefaction)
Solubility in water, selected solvents
Odor threshold
Color
Diffusion coefficient

State: Liquid
Vapor pressure-temperature relationship
Density; specific gravity
Viscosity
Miscibility with water, selected solvents
Odor
Color
Coefficient of thermal expansion
Interfacial tension

State: Solid
Melting point
Density
Odor
Solubility in water, selected solvents
Coefficient of thermal expansion
Hardness/flexibility
Particle size distribution/physical form, such as fine powder, granules, pellets, lumps
Porosity

contains two chlorine atoms that are located in positions directly opposite to each other in the benzene ring and which sublimes (passes from the solid phase to the gas phase without the intervention of a liquid phase) readily at or near room temperature (Rossberg et al., 2006):

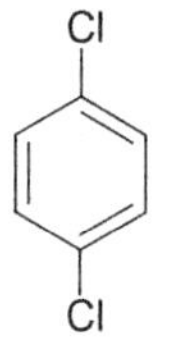

Thus, this compound can give the appearance of disappearing over a short period of time (depending upon the amount of the compound) which may present the false impression that the compound does not harm the environment but merely evaporates. Nevertheless, the compound after sublimation is most likely to condense on the nearest cool surface, thereby giving rise to further pollution problems.

In addition, solubility of organic compounds in water as well as solubility in organic solvents is also important properties that must be acknowledged. Organic compounds tend to dissolve in organic solvents such as ether (diethyl ether, $C_2H_5OC_2H_5$) and in paraffinic solvents such as the various types of petroleum-derived naphtha and kerosene as well as in a variety of aromatic solvents (aromatic naphtha, aromatic kerosene). Solubility in the different solvents depends not only upon the solvent type but also on the type and number of the functional groups present in the organic compound. Thus, nonacidic and nonbasic (neutral) organic compounds tend to be hydrophobic (water-hating, water repellant) insofar as they are less soluble in water than in organic solvents. Exceptions include organic compounds that contain polar (ionizable) functions (functional groups that can be converted into positively- or negatively charged ions) as well as low-molecular-weight alcohol (ROH) derivatives, amine (RNH_2) derivatives, and carboxylic acid (RCO_2H) derivatives.

Finally, modern organic chemistry is a dynamic discipline and it is evolving rapidly, and the concepts are applicable to all aspects of organic chemistry, especially the organic chemistry of the environment. Thus, organic compounds can be described in terms of simple carbon-based molecular structures in which atoms are held together by chemical bonds. This concept of organic chemistry has persisted for more almost 200 years and seems unlikely to be superseded, no matter how much the discipline is refined and modified (Smith, 2013).

4   USE AND MISUSE OF CHEMICALS

The use of chemicals for domestic and commercial purposes increased phenomenally during the 19th and 20th centuries but although brining benefits also had negative impacts on human health and safety as well as on the integrity of terrestrial and marine ecosystems and on air and water quality. The general lack of definitive plans to manage the use of chemicals threatened the sustainability of the environment. Whatever the chemical, there are risks to its use—known and unknown—and some chemicals, including heavy metals, persistent organic pollutants, and PCBs present risks that have been known for decades. On the other hand, there has been the release of chemicals into the environment, many of which are long lived and transform into by-products whose behavior, synergies, and impacts are not well known (Jones and De Voogt, 1999).

Nevertheless, organic chemicals are a significant contributor to the human lifestyle (Appendix: Table A1) and as long as there is sound chemical management across the lifecycle of a chemical—from extraction or production to disposal—it is possible (under current legislative guidelines) and essential to avoid risks to the floral and faunal environments. Nevertheless, there are always two sides to the statement: chemicals are a blessing but also can be curse. Just as there are benefits to the use of chemicals, they must be with respect so as to minimize any harmful impact from exposure of the environment to the organic chemicals.

Thus, organic chemicals while being considered to be the *chemicals of life* can also be the *chemicals of harm*. Understanding organic chemistry, perhaps not to the extent of the dyed-in-the-wool organic chemist, is a part of understanding the use and effects of organic chemicals. To many nonchemists, chemicals tend to be seen as frightening, and (often without justification) the general perception is that all chemicals are dangerous and use of chemicals should be avoided—as a by-the-way, water is a chemical. The important aspect of organic chemicals is that they are essential for life, but there is the necessity to treat organic chemicals with respect and caution. Some organic chemicals can be hazardous and should always be handled with care, as evidenced by the advisory (warning) statements on the packaging of the various chemicals which are presented as a matter of safety. The risk faced from exposure to an organic chemical is based on the intrinsic danger multiplied by the exposure to the chemical both in terms of the amount of the chemical and the time of exposure. A simple example is the chemical curare (an alkaloid—a nitrogen containing natural product), which is a common name for various plant extracts which are used as arrow-tip poisons (often fatal) originating in Central America and South America. On the other hand, it has also been used as a muscle relaxant (in extremely small dosages) but with some risk to the patient. Nevertheless, it has been possible for the medical community to adjust the dosage from a death-dealing quantity (on an arrowhead) to a medicinal quantity under strict supervision (EB, 2015).

Organic chemicals, and their various derivatives, are widely used in many sectors of the modern world including the chemicals industry, the fossil fuels industry, agriculture, mining, water purification, and public health. However, not only the dedicated use of organic chemicals but also the production, storage, transportation, and removal of these substances can pose risks to the environment if safe handling protocols are not followed. Developing an effective management system for organic chemicals requires addressing the specific challenges that arise because of the individual chemicals and chemical mixtures because the irregular management of obsolete organic chemicals and chemical mixtures, stockpiles, and waste presents serious threats to the environment. As the use of organic chemicals and production increases, chemical management, which already has limited resources and capacity, will be further constrained and overburdened and may fail if not regulated. Measures and systems need to be developed to reduce exposure to negative impacts and to reduce vulnerability of the environment.

The initial moves in the development of an efficient management system is to ensure that there are education programs that prepare professionals to enter the field of environmental technology as well education programs that prepare individuals to meet the challenges of environmental management in the forthcoming decades (Speight and Singh, 2014). There is no single discipline by which these challenges can be met—young professionals should be skilled in the sciences, the engineering technologies, and the relevant subdisciplines that enable them to cross-over from one discipline to another as the occasion demands.

Along with the increased use of chemicals, specifically the use of organic chemicals in the context of this book has come the realization that many widely used organic compounds are more toxic to the environment than was previously suspected. Some are carcinogenic and some may contribute to the destruction of the ozone layer in the upper atmosphere, which protects all life from the sun's strong UV radiation, while other organic chemicals are concentrated and persist in living tissue with an, as yet, unknown effect. Nonetheless, the modern world has adapted to the use of synthetic organic chemicals, and there are continuing debates that crude oil the largest source of organic chemicals—while in good supply at the present time—may be in short supply in the next 50–100 years, and there will be the need to rely on alternate sources of energy, which are not immune from causing damage to the environment (Speight, 2011, 2014; Lee et al., 2014; Speight and Islam, 2016). There have been several suggestions—not taken in any serious form so far—that coal once again become king in terms of chemicals production. Thus, in order to develop an effective management system that protects the environment from organic chemicals, there is the need to recognize that the modern world relies on both natural and synthetic chemicals which can be tailored to serve specific purposes. In fact, the gasification of coal (or, for that matter, other carbonaceous material such as biomass) to produce synthesis gas (a mixture of carbon monoxide and hydrogen) is an established process from which a variety of organic chemicals can be synthesized (Davis and Occelli, 2010; Chadeesingh, 2011; Speight, 2013).

In fact, the rise of industry has been centered on the development of organic chemistry that has resulted, for example, in the creation of synthetic polymers for the more effective and efficient production of goods. Occasionally, organic chemicals are actually designed to damage faunal life (in theory, excluding human life), and in so doing, they offer benefits to humanity. For example, pesticides and herbicides are designed to destroy various organisms so that crop yields may be higher while antibiotics are designed to destroy harmful bacteria.

However, although chemicals supply a variety benefits, they also present the potential for misuse—stepping aside from the issue of organic chemicals (pesticides, herbicides, and fungicides) and the environment to the more specific issues of chemicals and the human environment, drugs such as morphine (which affect the human fauna of the Earth) is an important multiuse organic chemical and is used as a pain-killer in medicine as well as in, as well as to create other pharmaceutical derivatives such as codeine, another pain-killer.

In addition, heroin is an opioid drug that is synthesized from morphine, a naturally occurring substance extracted from the seed pod of the Asian opium poppy plant. Heroin, the (the 3,6-diacetyl ester of morphine:

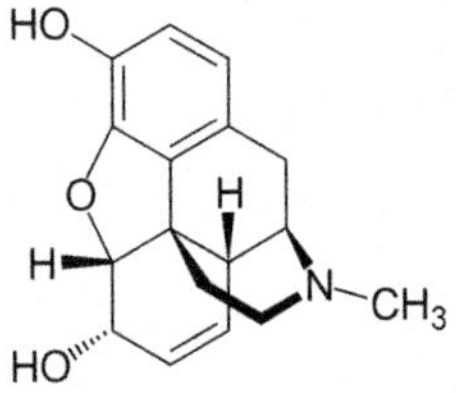

Morphine: conversion of the two hydroxy (—OH) functions to acetyl functions (—OChCH$_3$) produced heroin; the thicker lines indicate bond coming out of the plane of the paper while hash-mark lines indicate bond behind the plane of the paper.

Heroin usually appears as a white or brown powder or as a black sticky substance (*black tar heroin*). In fact, the illegal production of heroin remains an incredibly profitable industry, at the expense of the wellbeing of thousands of addicts. When absorbed into the brain, both codeine and heroin are converted back into morphine. However, heroin crosses the blood-brain barrier more rapidly than either morphine or codeine, thus producing a more immediate and potent response. While there are risks accompanying the use of morphine due to its addictive potential, the risks are even higher in the case of heroin. Nevertheless, a ban on opium production is not a plausible response to this problem because the opium poppy is used for a spectrum of purposes. Therefore, the regulation of opium and the production of morphine is an important and difficult task, one which cannot be ignored if it is hoped to lessen the burden of drug addiction in many nations. Another drug, the chemical known as methamphetamine:

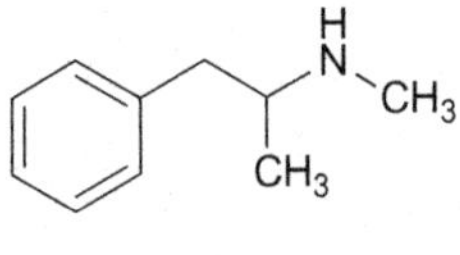

is a highly addictive synthetic drug, derived from a medicinal plant also causes severe addition and alteration of man behavior. Thus, the same knowledge that has led to cures and treatments for disease has also accelerated illegal drug use (typically misuse of organic chemicals). Similarly, important industrial chemicals can be used to create chemical agents of warfare agents, and, as a result, the response to the abuse of chemicals, known as the Chemical Weapons Convention (http://www.cwc.gov/), seeks to monitor and prevent the development of chemical weapons.

The preceding paragraphs are used as attention-getters to demonstrate how easily organic chemicals (especially, in the human context) in the form of illegal drugs can be misused, and the same reasoning can be applied to the misuse of other organic chemicals that bring harm to the environment. As the knowledge and understanding of organic chemicals increases—as well as the variety of routes of synthesis of these chemicals—so does the ease of synthesis and production of such compounds. This makes understating the different aspects of the use of these chemicals a greater task than ever, and the legislated regulation of multiuse chemicals become an essential part of the organic chemicals industry. However, as an understanding of organic chemistry increases, so must a sense of responsibility related to the use of organic chemicals as well as the effects of these chemicals on the floral and faunal environments, and the various protocols for the disposal of organic chemicals. Laws relating to the regulation and use of such chemicals must be vigorously policed and updated, especially since there is the continual search for new synthetic routes to the chemicals (accompanied by the efforts of the would-be multiuse chemical abusers) so as to avoid being outstripped by the development of science and suffer the resulting environmental consequences of this negligence.

When an event occurs that is detrimental to the floral and faunal environments, the allocation of chemical responsibility is often a difficult process. The issues relating to the responsibility for the development and dispersal of organic chemicals continue to be debated. Many observers would argue that the users are to blame as are the producers and the chemists who discovered the synthesis reactions by which the chemicals are produced. However, given that the use and disposal of chemicals is a global problem, the responsibility to deal with the problem must fall to policy makers in the various levels of government (local, state, and federal) who are involved in the creation of regulatory laws, it is important to create codes of conduct to guide behavior and actions with regard to this complex problem. In addition, governments who fail to create responsible regularity laws must also share some of the blame for the misuse of chemicals. The politicians cannot consider themselves immune from blame when the necessary laws are not passed or are not policed.

There are two types of codes related to the use/misuse of organic chemicals and their subsequent disposal: (1) enforceable codes of conduct and (2) aspirational codes of conduct. An *enforceable codes* of conduct deals with the necessary protocols for regulation and enforcement of the code, while an *aspirational code* of conduct presents the ideals of performance so that those bound to the code may be reminded of their obligations to perform ethically and responsibly. Nevertheless, there are many observers who are in serious doubt about the practical effectiveness of such codes, which may even prescribe ambiguous (and often unattainable) ideals which can be circumvented if the producers and/or the users of the chemicals wish to do so.

Typically, the value of a code of conduct is usually most clearly evident to the creators and writers of the code. Those who must consider every word and phrase included in the code must also explain the importance of expressing the meaning of the code in an unambiguous, straight-forward, understandable, and effective way. Furthermore, it is also essential to involve the various groups with different interests and perspectives at the time when the code is being formulated so as to inform the various groups of the issues addressed in the code as well as to remind all participants and the users of the responsible use of organic chemicals. In doing so, a code of conduct can be written to be highly effective which should assist the scientists, the engineers, and the public of the issues at hand. From this understanding should come the responsibilities and the guidelines for each party to act in a responsible and ethical manner.

Thus, effective management of organic chemicals to protect all types of flora and fauna from all chemicals should carry with it the reminder that to ensure the proper use of chemistry and chemicals there is the need to develop and hold to strict codes of conduct that establish guidelines for ethical scientific development and protection of the environment.

5 CHEMICALS IN THE ENVIRONMENT

Organic chemicals and organic chemical waste (such as organic hazardous waste) are pose substantial or potential threats to the floral and faunal environments. In the United States, the treatment, storage, and disposal of any type of waste (but for the purposes of this text, organic waste and organic hazardous waste are regulated under the Resource Conservation and Recovery Act (RCRA). In this Act, which hazardous wastes are defined 40 CFR 261 and are also divided into two major categories: (1) characteristic wastes and (2) listed wastes.

5.1 Indigenous Chemicals

While natural gas, crude oil (including the more recent *tight oil*) are naturally occurring compounds (i.e., natural products), and there are environmental issues associated with the development and use of these resources (Speight and Lee, 2000; Speight, 2005; Speight and Arjoon, 2012; Speight, 2013 2014, 2016). In addition, there are other chemicals in the ecosystems of the Earth that can also pose a threat to the environment and which are often ignored by the various environmentally conscious groups. Organic matter or organic material, natural organic matter, is matter composed of organic compounds that has come from the remains of organisms such as plants and animals and their waste products in the environment. Basic structures are created from naturally occurring chemicals such as cellulose, and lignin, as well as various proteins, lipids, and carbohydrates, and these chemicals are typically fall under the umbrella of naturally occurring organic matter.

Naturally occurring organic matter—usually referred to as natural products—are chemicals produced by living organisms that are found in nature and include (Bhat et al., 2005; Cseke et al., 2006). Natural products may also be referred to as biomolecular organic chemicals. Biomolecular organic chemistry (sometimes shortened to *biomolecular chemistry*) is a major category within organic chemistry, and many complex multifunctional group molecules are important in living organisms. Some are long-chain biopolymers, and these include peptides, deoxyribonucleic acid (DNA), ribonucleic acid (RNA), and polysaccharide derivatives, such as starch derivatives in animals and cellulose derivatives in plants. The other main classes are amino acids (monomer building blocks of peptides and proteins), carbohydrates (which includes the polysaccharides), nucleic acids (which include DNA, RNA), and the lipids. In addition, animal biochemistry contains many small molecule intermediates such as isoprene, the most common hydrocarbon in animals. Isoprene derivatives in animals form the important steroid structure (cholesterol) and steroid hormone compounds. In plants, isoprene form terpene derivatives, terpenoid derivatives, alkaloid derivatives, and biopolymers (such as poly-isoprenoid derivatives) that are present in the latex of various species of plants, which is the basis for making rubber.

5.2   Nonindigenous Chemicals

In relation to the environmental effects of organic chemicals, consideration must also be given to the effect of the so-called *harmless chemicals* (*indigenous chemicals, natural products chemicals*) these harmless but nonindigenous chemicals on the environment. Within the local environment, these chemicals will be present in a measurable concentration, but the flora and fauna present in that ecosystem may be fatally susceptible to the concentrations of such chemicals when they are present in a concentration that is above the indigenous concentration of the chemicals. For example, a sprinkling salt on a meal may add to the taste of the meal, but it is unadvisable for human to attempt to consume several ounces of salt with that same meal. Not only would the taste be ruined, but the high concentration of salt could have a serious health effects (even death) on the consumer.

Characteristic hazardous wastes are materials that are known or tested to exhibit one or more of the following four hazardous traits: (1) ignitability, (2) reactivity, (3) corrosivity, and (4) toxicity. Thus, chemicals in the environment can be designated as hazardous or nonhazardous (Carson and Mumford, 2002), generally as a category of wastes. Listed hazardous wastes are materials specifically listed by regulatory authorities as a hazardous waste which are from nonspecific sources, specific sources, or discarded chemical products. These wastes may be found in different physical states such as gaseous, liquids, or solids. A hazardous waste is a special type of waste because it cannot be disposed of by common means like other by-products of our everyday lives.

Depending on the physical state of the waste, treatment and solidification processes might be required.

Household hazardous waste (HHW) (also referred to as domestic hazardous waste or home-generated special materials) is waste that is generated from residential households. HHW only applies to wastes that are the result of the use of materials that are labeled for and sold for *home use*. Wastes generated by a company or at an industrial setting are not HHW. The following list includes categories often applied to HHW, and it is important to note that many of these organic chemical categories overlap and that many household wastes can fall into one or more categories: (1) paints and solvents, (2) automotive wastes such as used motor oil and glycol antifreeze, and (3) pesticides, which include insecticides, herbicides, and fungicides.

More specific to the present text are the chemicals designated as hazardous waste (Appendix) (US EPA, 2015). Proper management of chemicals and chemical waste is an essential part of maintaining a sustainable environment. The Resource Conservation and Recovery Act (RCRA), passed in 1976, created the framework for hazardous and nonhazardous chemical solid waste management programs, and only materials that meet the definition of solid waste under RCRA can be classified as hazardous wastes, which are subject to additional regulation. EPA developed detailed regulations that define what materials qualify as solid wastes and hazardous wastes.

If a chemical has a listing as a hazardous waste, there will be a narrative description of a specific type of waste that United States Environmental Protection Agency (US EPA) considers to be sufficiently dangerous to warrant regulation. Hazardous waste listings describe (1) wastes from specific processes, (2) wastes from very specific sectors of industry, or (3) wastes in the form of very specific chemical formulations. Before developing a hazardous waste listing, the US EPA thoroughly studies a particular waste stream and the threat it can pose to human health and the environment. If the waste poses enough of a threat, the US EPA includes a precise description of that waste on one of the hazardous waste lists in the regulations. Thereafter, any waste fitting that narrative listing description is considered hazardous, regardless of its chemical composition or any other potential variable.

For example, one of the current hazardous waste listings includes API separator sludge from the petroleum refining industry. An API separator is a device commonly used by the petroleum refining industry to separate contaminants from refinery wastewater (Speight, 2005, 2014). After studying the petroleum refining industry and typical sludge samples from API separators, the US EPA determined the sludge were dangerous enough to warrant regulation as hazardous waste under all circumstances. The listing therefore designates all petroleum refinery API separator sludge as hazardous. Chemical composition or other factors about a specific sample of API separator sludge are not relevant to its status as a listed hazardous waste under the RCRA program.

The US EPA has studied and listed as hazardous hundreds of specific industrial waste streams. These wastes are described or listed on four different lists, which are found in the regulations in Part 261, Subpart D. These first two lists are: (1) the F list, which designates as hazardous the particular wastes from many common business, government, industrial, or manufacturing processes, and because the processes producing these wastes can occur in different sectors, the F list wastes are known as waste from nonspecific sources, and (2) the K list, which designates as hazardous particular waste streams from specific sectors of industry and, hence the K list wastes are known as wastes from specific sources (Appendix: Table A2, Table A3).

Additional classification of waste now also includes two other lists: (1) the P-Code list and (2) the U-Code list (Appendix: Tables A4 and A5). These two codes cover an extensive list of wastes, and generators of such (hazardous) waste are required to comply with extensive and complex rules and regulations promulgated by federal, state, and local regulatory agencies. The Resource Conservation and Recovery Act (RCRA) requires a *cradle to grave* system of accounting for hazardous waste; the Department of Transportation requires compliance with Federal Motor Carrier Safety Regulations during transportation of hazardous waste. Furthermore, it is often required that state department of environmental quality license, inspect, and regulate generators, haulers, and disposal facilities handling hazardous waste. These modern safety guidelines, regulations, and procedures presented are intended to help generators comply with governmental rules and regulations designed to protect human health and the environment. Strict compliance with these regulations ensures the waste is managed, transported, and disposed of safely and properly while reducing potential liability to waste generator.

The P and U list waste description involves two key factors. First, a P or U listing applies only if one of the listed chemicals is discarded unused—the P list and U list do not apply to process wastes, as do the F list and the K list. The P list and the U list apply to unused chemicals that become wastes for a number of reasons. For example, some unused chemicals are spilled by accident, while other chemicals are intentionally discarded because they do not meet specification and cannot serve the purpose for which they were originally produced. Some chemicals are discarded because the facility no longer uses that product or process line. The second key factor governing the applicability of the P list or U list is that the listed chemical must be discarded in the form of a commercial chemical product. The phrase "commercial chemical product" applies to a chemical that has the generic name of that chemical. For example, heptachlor is the generic (or common) name for 4,7-methano-1H-indene, 1,4,5,6,7,8,8-heptachloro-3a,4,7,7a-tetrahydro-. It may include the chemical in pure form, in commercial grade form, or that is an active ingredient in a chemical formulation that has the generic name of the chemical. Manufacturing chemical intermediates that have the generic name of the chemical also are commercial chemical products.

Finally, a chemical is an active ingredient in a formulation if that chemical serves the function of the formulation. For instance, a pesticide made for killing insects may contain a poison such as heptachlor as well as various solvent ingredients which act as carriers or lend other desirable properties to the poison. Although all of these chemicals may be capable of killing insects, only the heptachlor serves the primary purpose of the insecticide product. The other chemicals involved are present for other reasons, not because they are poisonous. Therefore, heptachlor is the *active* ingredient in such a formulation even though it may be present in low concentrations—this formulation would carry the P059 waste code.

6   CHEMISTRY AND ENGINEERING

There is often the question: why teach engineers chemistry and why teach chemists engineering? Both questions can be answered by understanding the need to establish process knowledge (reactor construction, reactor parameters) for the chemist and to establish chemical knowledge (reaction parameters, feedstock, and product properties) for engineers. This will help to establish a link between the disciplines which can then (in the context of this book) be applied to the development of pathways for a sustainable environment. Furthermore, this cross-fertilization of chemistry with the various engineering disciplines is especially useful when many technical issues cannot be dealt with successfully by a chemist of by an engineer working individually. The need is for teamwork in which professionals from both the chemical and engineering disciplines (and the related subdisciplines) work together for a better environment, sometimes referred to as a *green environment* which in turn is brought about by the application of chemistry and engineering to solving environmental issues. Furthermore, when applied to the development of a sustainable environment (and to add some confusion to the terminology), chemistry and engineering not only referred to as *environmental chemistry* and *environmental engineering* but also as *green chemistry* and *green engineering*. As an historical aside, environmental engineering (formerly known as *sanitary engineering*) originally developed as a subdiscipline of civil engineering.

The term *green chemistry* is often used in the context of environmental science to which can be added, in the current context, *green engineering*. By way of explanation, *green chemistry* and *green engineering* focus on the environmental concerns related to the use of materials, the generation of energy, and the various production cycles and can also be used to demonstrate the means by which the fundamental chemical principles, engineering principles, as well as the various chemical and engineering methodologies can be applied to the protection of the environment (Anastas and Kirchhoff, 2002). The principles of both disciplines (chemistry and engineering) are central to chemical education and to engineering education because professionals entering the environmental field need to develop the tools and skills to support the concept of global sustainability. As a result, future chemists and engineers will acquire the technical knowledge to design products and chemical processes though an increased awareness of environmental impact and understand the importance of sustainable strategies to protect the environment.

Thus, both environmental chemistry and environmental engineering course (which should be taught as cross-over inter-related courses) have the potential to add considerable enhancement to chemistry learning and to engineering leading to an improved and understanding of the chemical and engineering concepts. Incorporation of the principles of both chemistry and engineering into course material can be coupled with specific inserts that will complement the chemistry curriculum and complement the engineering curriculum and which will serve as a reminder that the practice of chemistry and the practice of engineering can lead to important developments in environmental technology (Braun et al., 2006).

The implementation of the principles of environmental chemistry for engineers and environmental engineering for chemists in any university curriculum will not only contribute to the general aims of science and engineering education but also to important elements in the development of scientific and engineering literacy and knowledge (Van Eijck and Roth, 2007). Cross-over studies will help the fledgling chemist and the fledgling engineer make the necessary connections among the disciplines of chemistry and engineering which, in turn, will contribute to the education of chemical and engineering professionals and bring about practices related to protection of the environment (Karpudewan et al., 2012).

Furthermore, cross-over studies will provide the required knowledge and awareness that lead to development of technologies that are necessary to achieve the ultimate goal of environmental protection. Teaching environmental chemistry and environmental engineering at different levels of the chemistry and engineering degree programs education has received significant attention recently (Andraos and Dicks, 2012; Eilks and Rauch, 2012; Burmeister and Eilks, 2012; Burmeister et al., 2012; Mandler et al., 2012; Karpudewan et al., 2012). The importance of this type of education, beyond the basics of chemical and engineering learning, relates to the ability to participate in the development of sustainable environmental practices (Eilks and Rauch, 2012).

Thus, an understanding of the chemical types that contribute to pollution can lead to an understanding of the chemical and physical methods (and the related process parameters) for mitigating pollution. Mitigation of such effects is not only a matter of knowing the elemental composition of the pollutant but also a matter of understanding the bulk properties as they relate to the chemical or physical composition of the material relating to the behavior of (in the context of this book) the organic chemical in the environment.

REFERENCES

Anastas, P.T., Kirchhoff, M.M., 2002. Origins, current status, and future challenges of green chemistry. Acc. Chem. Res. 35 (9), 686–694.

Andraos, J., Dicks, A.P., 2012. Green chemistry teaching in higher education: a review of effective practices. Chem. Educ. Res. Pract. 13, 69–79.

Bhat, S.V., Nagasampagi, B.A., Sivakumar, M., 2005. Chemistry of Natural Products, second ed. Springer, Berlin.

Braun, B., Charney, R., Clarens, A., Farrugia, J., Kitchens, C., Lisowski, C., Naistat, D., O'Neil, A., 2006. Completing our education. Green chemistry in the curriculum. J. Chem. Educ. 83 (8), 1126–1128.

Burmeister, M., Eilks, I., 2012. An example of learning about plastics and their evaluation as a contribution to education for sustainable development in secondary school chemistry teaching. Chem. Educ. Res. Pract. 13 (2), 93–102.

Burmeister, M., Rauch, F., Eilks, I., 2012. Education for sustainable development (ESD) and chemistry education. Chem. Educ. Res. Pract. 13 (2), 59–68.

Carson, R., 1962. Silent Spring. Houghton Mifflin Company; Houghton Mifflin Harcourt International, Geneva, IL.

Carson, P., Mumford, C., 1988. The Safe Handling of Chemicals in Industry. vols. 1 and 2. John Wiley & Sons, New York.

Carson, P., Mumford, C., 1995. The Safe Handling of Chemicals in Industry, vol. 3. John Wiley & Sons, New York.

Carson, P., Mumford, R., 2002. Hazardous Chemicals Handbook, second ed. Butterworth-Heinemann, Oxford.

Chadeesingh, R., 2011. The Fischer-Tropsch process. In: Speight, J.G. (Ed.), The Biofuels Handbook. The Royal Society of Chemistry, London, pp. 476–517 (Part 3, Chapter 5).

Cseke, L.J., Kirakosyan, A., Kaufman, P.B., Warber, S., Duke, J.A., Brielmann, H.L., 2006. Natural Products From Plants. CRC Press, Taylor and Francis Group, Boca Raton, FL.

Davis, B.H., Occelli, M.L., 2010. Advances in Fischer-Tropsch Synthesis, Catalysts, and Catalysis. CRC Press, Taylor & Francis Group, Boca Raton, FL.

EB, 2015. Curare. Encyclopedia Britannica. Chicago, Illinois.

Ehhalt, D.H., 1986. On the consequence of a tropospheric CH_4 increase to the exospheric density. J. Geophys. Res. 91, 2843.

Eilks, I., Rauch, F., 2012. Sustainable development and green chemistry in chemistry education. Chem. Educ. Res. Pract. 13 (2), 57–58.

Gouin, T., James, Y., Armitage, M., Cousins, I.T., Muir, D.C.G., Ng, C.A., Reid, L., Tao, S., 2013. Influence of global climate change on chemical fate and bioaccumulation: the role of multimedia models. Environ. Toxicol. Chem. 32 (1), 20–31.

Jones, K.C., De Voogt, P., 1999. Persistent organic pollutants (POPs): state of the science. Environ. Pollut. 100, 209–221.

Karpudewan, M., Ismail, Z., Roth, W.M., 2012. Ensuring sustainability of tomorrow through green chemistry integrated with sustainable development concepts (SDCs). Chem. Educ. Res. Pract. 13 (2), 120–127.

Lear, L., 2015. Rachel Carson: Witness for Nature. Houghton Mifflin Company; Houghton Mifflin Harcourt International, Geneva, IL.

Lee, S., Speight, J.G., Loyalka, S., 2014. Handbook of Alternative Fuel Technologies, second ed. CRC Press, Taylor & Francis Group, Boca Raton, FL.

Mackay, D., Shiu, W., Ma, K., Lee, S., 2006. Handbook of Physical-Chemical Properties and Environmental Fate for Organic Chemicals, second ed. CRC Press, Taylor & Francis Group, Boca Raton, FL.

Mandler, D., Mamlok-Naaman, R., Blonder, R., Yayon, M., Hofstein, A., 2012. High school chemistry teaching through environmentally oriented curricula. Chem. Educ. Res. Pract. 13 (2), 80–92.

Reynolds, J.G., 1998. Metal and heteroatoms in heavy crude oils. In: Speight, J.G. (Ed.), Petroleum Chemistry and Refining. Taylor & Francis/CRC Press, Taylor and Francis Group, Philadelphia, PA/Boca Raton, FL, pp. 63–102 (Chapter 3).

Rossberg, M., Lendle, W., Pfleiderer, G., Tögel, A., Dreher, E.L., Langer, E., Rassaerts, H., Kleinschmidt, P., Strack, H., Cook, R., Beck, U., Lipper, K.A., Torkelson, T.R., Löser, E., Beutel, K.K., Mann, T., 2006. Chlorinated hydrocarbons. Ullmann's Encyclopedia of Industrial Chemistry. John Wiley & Sons, Hoboken, NJ.

Sheldon, R., 2010. Introduction to green chemistry, organic synthesis and pharmaceuticals. In: Dunn, P.J., Wells, A.S., Williams, M.T. (Eds.), Green Chemistry in the Pharmaceutical Industry. Wiley-VCH Verlag GmbH, Weinheim.

Smith, M.B., 2013. March's Advanced Organic Chemistry: Reactions, Mechanisms, and Structure, seventh ed. John Wiley & Sons, Hoboken, NJ.

Speight, J.G., 2005. Environmental Analysis and Technology for the Refining Industry. John Wiley & Sons, Hoboken, NJ.

Speight, J.G., 2011. An Introduction to Petroleum Technology, Economics, and Politics. Scrivener Publishing, Salem, MA.

Speight, J.G., 2013. The Chemistry and Technology of Coal, third ed. CRC Press, Taylor and Francis Group, Boca Raton, FL.

Speight, J.G., 2014. The Chemistry and Technology of Petroleum, fifth ed. CRC Press, Taylor and Francis Group, Boca Raton, FL.

Speight, J.G., 2016. Deep Shale Oil and Gas. Gulf Professional Publishing, Elsevier, Oxford.

Speight, J.G., Arjoon, K.K., 2012. Bioremediation of Petroleum and Petroleum Products. Scrivener Publishing, Salem, MA.

Speight, J.G., Islam, M.R., 2016. Peak Energy—Myth or Reality. Scrivener Publishing, Salem, MA.

Speight, J.G., Lee, S., 2000. Environmental Technology Handbook, second ed. Taylor & Francis/CRC Press, Taylor and Francis Group, New York/Boca Raton, FL.

Speight, J.G., Singh, K., 2014. Environmental Management of Energy From Biofuels and Biofeedstocks. Scrivener Publishing, Salem, MA.

Trenberth, K.E., Guillemot, C.J., 1994. The total mass of the atmosphere. J. Geophys. Res. 99, 23079–23088.

US EPA, 2015. List of Lists: Consolidated List of Chemicals Subject to the Emergency Planning and Community Right-to-Know Act (EPCRA), Comprehensive Environmental Response, Compensation and Liability Act (CERCLA) and Section 112(r) of the Clean Air Act. Report No. EPA 550-B-15-001, Office of Solid Waste and Emergency Response, United States Environmental Protection Agency, Washington, DC.

Van Eijck, M., Roth, W.M., 2007. Improving science education for sustainable development. PLoS Biol. 5, 2763–2769.

Chapter 2

Organic Chemistry

1 INTRODUCTION

Although a good deal is already known about the influence of molecular structure on the toxicity to human beings of drugs and certain other chemicals, much less is known about the influence of molecular structure on the environmental persistence of a chemical. For wildlife, persistence is probably the most important criterion for predicting potential harm because there is inevitably some wild species or other which is sensitive to any compound and any persistent chemical, apparently harmless to a limited number of toxicity-test organisms, will eventually be delivered by biogeochemical cycles to a sensitive target-species in nature. This means that highly toxic, readily biodegradable substances may pose much less of an environmental problem, than a relatively harmless persistent chemical which may well damage a critical wild species. The study of chemical effects in the environment resolves itself into a study of (1) the levels of a substance accumulating in air, water, soils including sediments and biota, including man, and (2) when the threshold action-level has been reached, effects produced in biota which constitute a significant adverse response (i.e., environmental dose-response curve). In order to predict trends in levels of a chemical, much more information is needed about rates of injection, flow, and partitioning between air, water, soils, and biota; and loss via degradation (environmental balance-sheets). These dynamic phenomena are governed by the physicochemical properties of the molecule.

Fluid mechanics and meteorology may in future provide the conceptual and technical tools for producing predictive models of such systems. Most of our knowledge of effects derives from acute toxicology and medical studies on man, but since environmental effects are usually associated with chronic exposure, studies are being increasingly made of long-term continuous exposure to minute amounts of a chemical. The well-known difficulty of recognizing such effects when they occur in the field is aggravated by the fact that many of the effects are nonspecific and are frequently swamped by similar effects deriving from exposure to such natural phenomena as famines, droughts, cold spells, etc. Even when a genuine effect is recognized, a candidate causal agent must be found and correlated with it. This process must be followed by experimental studies, unequivocally linking chemical cause and adverse biological effect. All three stages are difficult and costly, and it is not surprising that long delays

Environmental Organic Chemistry for Engineers. http://dx.doi.org/10.1016/B978-0-12-804492-6.00002-2

43

are often experienced between the recognition of a significant adverse effect and a generally agreed chemical cause. There is often ample uncertainty to allow under-reaction as well as over-reaction to potential hazards, both backed up by scientific evidence.

This can only come about by the collection and assimilation of the technical knowledge of organic chemistry as it relates to the properties and behavior or organic chemicals (Patrick, 2004). Note that it is expected that the engineer will accumulate as much chemical knowledge as the professional organic chemist—just as the chemist would shudder at having to be proficient in one or more of the engineering disciplines. But, the accumulation of sufficient knowledge to (at first) understand the behavior of organic chemicals in the environment followed by the ability to make a reasonable prediction (based on properties) of the behavior of organic chemicals in the environment. Failure to recognize the mutually interactive roles of chemist and engineer will hinder the development of a united environmental management policy that will apply to the comments of the biosphere (Chapter 1).

Finally, organic chemistry is a subject in that it becomes easier as the individual researcher works with it. The topics covered in this book are the basis topics that serve to introduce the reader not only to organic chemistry but the effect of organic chemicals on various ecosystems. Basic rules of nomenclature are presented. Understanding the mechanism of how a reaction takes place is particularly crucial in this and of necessity, the book brings a logic and simplicity to the reactions of the different functional groups. This in turn transforms a list of apparently unrelated facts into a sensible theme. Thus, this chapter will serve as an introduction to the physicochemical properties of organic chemicals and their effect on the floral and faunal environments.

2 CLASSIFICATION OF ORGANIC MOLECULES

Organic chemicals exist as gases, liquids, or solids but are not typically classified on the basis of the physical state (gas, liquid, or solid) of the compound but more on data related to the chemical and physical properties of the chemicals (Table 2.1). Typically, at ambient temperature and pressure—gasses and liquids take on the shape of the container in the bulk phase, while solids have definite shapes and volume and are held together by strong intermolecular and interatomic forces. For many solid organic chemicals, these forces are strong enough to maintain the atoms in definite ordered crystals, while solids with little or no crystal structure are termed amorphous.

Organic gases have weaker attractive forces between individual molecules and therefore diffuse rapidly and assume the shape of the container, but the volume of the gas is affected by temperature and pressure (often simply considered as container size). The molecules of organic liquids are separated by relatively small distances, and the attractive forces between molecules tend to hold firm within a definite volume (container size) at a fixed temperature. Thus, organic

TABLE 2.1 Important Common Physical Properties of Organic Compounds

State: Gas

Density

Critical temperature, critical pressure (for liquefaction)

Solubility in water, selected solvents

Odor threshold

Color

Diffusion coefficient

State: Liquid

Vapor pressure-temperature relationship

Density; specific gravity

Viscosity

Miscibility with water, selected solvents

Odor

Color

Coefficient of thermal expansion

Interfacial tension

State: Solid

Melting point

Density

Odor

Solubility in water, selected solvents

Coefficient of thermal expansion

Hardness/flexibility

Particle size distribution/physical form, such as fine powder, granules, pellets, lumps

Porosity

solvents are covalent compounds in which molecules are much closer together than in a gas, and the intermolecular forces are therefore relatively strong. A useful property of organic liquids is their ability to dissolve gases, dissolve or mix with other liquids, and dissolve organic solids. The solutions produced may be end products, e.g., paints and disinfectants or the process itself may serve a useful function, such as removal of pollutant gas from air or a gas stream by absorption, leaching of a constituent from bulk solid. The properties of a solution can differ significantly from the individual constituents of the solution and typically depends upon the properties of the solvent and the solute(s) as well as the concentration of the solute(s) in the solvent. When the molecules of a covalent (organic) solute are physically and chemically similar to those of a liquid (organic) solvent, the intermolecular forces of each are the same and the solute and solvent will usually mix readily to form the solution.

Organic compounds are classified according to the presence of functional groups in the molecule (Tables 2.2–2.5), and the functional group typically

TABLE 2.2 General Classes of Hydrocarbons

Chemical Class	Group	Formula	Structural Formulae
Alkane	Alkyl	$R(CH_2)_nH$	
Alkene	Alkenyl	$R_2C{=}CR_2$	
Alkyne	Alkynyl	$R_1C{\equiv}CR_2$	$R_1C{\equiv}CR_2$
Benzene derivative	Phenyl	RC_6H_5	

TABLE 2.3 General Classes of Oxygen Compounds

Chemical Class	Group	Formula	Structural Formula
Alcohol	Hydroxyl	ROH	
Ketone	Carbonyl	$RCOR'$	
Aldehyde	Aldehyde	$RCHO$	
Acyl halide	Haloformyl	$RCOX$	
Carbonate	Carbonate ester	$ROCOOR$	

TABLE 2.3 General Classes of Oxygen Compounds—cont'd

Chemical Class	Group	Formula	Structural Formula
Carboxylate	Carboxylate	RCOO⁻	
Carboxylic acid	Carboxyl	RCOOH	
Ester	Ester	RCOOR′	
Methoxy	Methoxy	ROCH₃	

TABLE 2.4 General Classes of Nitrogen Compounds

Class	Group	Formula	Structural Formula
Amide	Carboxamide	RCONR₂	
Amines	Primary amine	RNH₂	
	Secondary amine	R₂NH	
	Tertiary amine	R₃N	
	4° ammonium ion	R₄N⁺	
Imine	Primary ketimine	RC(=NH) R′	

Continued

TABLE 2.4 General Classes of Nitrogen Compounds—cont'd

Class	Group	Formula	Structural Formula
	Secondary ketimine	RC(=NR)R′	
	Primary aldimine	RC(=NH)H	
	Secondary aldimine	RC(=NR′)H	
Imide	Imide	$(RCO)_2NR′$	
Azide	Azide	RN_3	
Azo compound	Azo (Diimide)	$RN_2R′$	
Cyanates	Cyanate	ROCN	
	Isocyanate	RNCO	
Nitrate	Nitrate	$RONO_2$	
Nitrile	Nitrile	RCN	$RC\equiv N$
	Isonitrile	RNC	$RN^+\equiv C^-$
Nitrite	Nitroso-oxy	RONO	
Nitro compound	Nitro	RNO_2	

TABLE 2.5 General Classes of Sulfur Compounds

Chemical Class	Group	Formula	Structural Formula
Thiol	Sulfhydryl	RSH	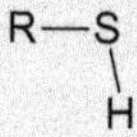 R—S, H
Sulfide (Thioether)	Sulfide	RSR'	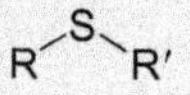 R—S—R'
Disulfide	Disulfide	RSSR'	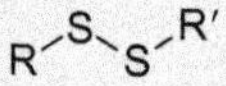 R—S—S—R'
Sulfoxide	Sulfinyl	RSOR'	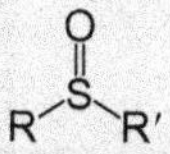 $\overset{O}{\underset{R\quad R'}{\parallel S}}$
Sulfone	Sulfonyl	RSO₂R'	R—S(=O)(=O)—R'
Sulfinic acid	Sulfino	RSO₂H	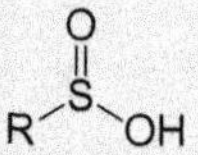 R—S(=O)—OH
Sulfonic acid	Sulfo	RSO₃H	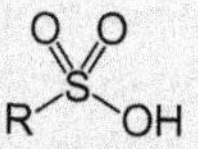 R—S(=O)(=O)—OH
Thiocyanate	Thiocyanate	RSCN	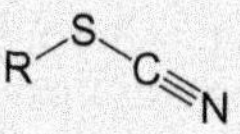 R—S—C≡N
	Isothiocyanate	RNCS	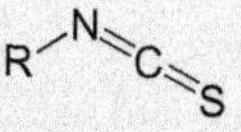 R—N=C=S
Thione	Carbonothioyl	RCSR'	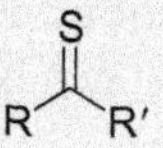 R—C(=S)—R'
Thial	Carbonothioyl	RCSH	 R—C(=S)—H

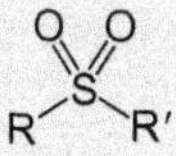

dictates the behavior of the organic compound in the environment. A functional group is a molecular moiety, and the reactivity of that functional group is assumed to be the same in a variety of molecules, within some limits and assuming that steric effects do not interfere. Thus, most functional groups feature heteroatoms (atoms other than C and H). The concept of functional groups is major concept in organic chemistry, both as a means to classify the structure of organic compounds and for predicting physical and chemical properties.

For example, when comparing the properties of ethane (CH_3CH_3) with the properties of propionic acid ($CH_3CH_2CO_2H$) which is simply due to the insertion of a carboxylic acid functional group (CO_2H) into the ethane molecule in place of a hydrogen atom, the change is spectacular. Alternatively, the replacement of a methyl group (CH_3) into the ethane molecule by the carboxylic acid function to produce acetic acid (CH_3CO_2H) (or the replacement of a hydrogen in the methane molecule by the carboxylic acid function) produces equally spectacular changes in the properties.

	Methane	Ethane	Acetic Acid	Propionic Acid
Chemical formula	CH_4	C_2H_2	$C_2H_4O_2$	$C_3H_6O_2$
Molecular formula	CH_4	C_2H_6	CH_3CO_2H	$CH_3CH_2CO_2H$
Molar mass	16.04	30.07	60.05	74.08
Physical state	Gas	Gas	Liquid	
Color	Colorless	Colorless	Colorless	Colorless
Odor	None	None	Pungent	Pungent
Density	0.656	0.5446	1.049	0.988
Melting point (°C):	−182.5	−182.8	16	−20.5
Boiling point (°C):	−161.5	−88.5	118	114
Solubility in water	22.7 mg/L	56.8 mg/L	Miscible	Miscible

While some observers might consider these comparisons to be too extreme (perhaps even unfair or biased comparisons), they do serve to illustrate the point that insertion of a functional group into a hydrocarbon molecule (in place of a hydrogen atom) can cause major changes in the properties that reflect the influence of the function group.

Thus, it is not surprising that, in organic chemistry, molecules are classified on the basis of their functional groups (Tables 2.2–2.5). Alcohols (Table 2.3), for example, tend to be hydrophilic (water-loving insofar as alcohols mix readily with water or are readily soluble in water), usually form esters by reaction with acids, and usually can be converted to the corresponding halides (an ester of hydrochloric acid) by reaction with the corresponding halogen or by reaction with the hydrogen halide:

$$ROH + Cl_2 \rightarrow RCl + HCl$$
$$ROH + HCl \rightarrow RCl + H_2O$$

2.1 Hydrocarbons

Hydrocarbons are often considered to be the simplest organic compounds (being composed of carbon and hydrogen *only*—the general application of the name hydrocarbon to hydrocarbonaceous mixtures which contain atoms other than carbon and hydrogen is incorrect), but the overwhelming complexity of the different hydrocarbon types (often referred to as *hydrocarbon series*) often make identification of the constituents of hydrocarbon mixtures a difficult task, as is often evidenced in the fossil fuel industry (Speight, 2013, 2014). In addition, the spills of hydrocarbon mixtures (such as petroleum and petroleum products) are complicated even further by the variation in the molecular weight of the constituents as well as by the presence of nonhydrocarbon compounds of sulfur, oxygen, nitrogen in the petroleum and in some of the petroleum products (Speight, 2014).

The complexity of organic chemicals is due to the tetravalent carbon atom which, with four valence electrons, is capable of bonding to four other atoms. Additionally, carbon is capable of forming not only a single bond ($\equiv$C—C$\equiv$), with one pair of shared valence electrons, but also a double bond (>C=C<, two pairs of shared electrons) or even a triple bond (—C$\equiv$C—, three pairs of shared electrons). Another special property of the carbon atom is its ability to form long chains of carbon atoms that constitute strings of carbon atoms and other noncarbon atoms. Furthermore, though sometimes carbon forms a typical molecule (e.g., carbon dioxide, CO_2, is one carbon atom with two oxygen atoms), it is also capable of forming molecules that are really not molecules in the way that the word is typically used in chemistry. Graphite, for instance, is a series of *sheets* of carbon atoms bonded tightly in a hexagonal (six-sided) pattern, while a diamond is simply a huge molecule composed of carbon atoms bonded together by covalent bonds.

Thus, carbon, together with other elements, forms so many millions of organic compounds, and fortunately, it is possible to classify organic compounds into specific groups. The largest and most significant group is that class of organic compounds known as hydrocarbons—chemical compounds whose molecules that contain carbon and hydrogen only. Every molecule in a hydrocarbon is built upon a *skeleton* composed of carbon atoms, either in closed rings or in long chains. The chains may be straight (linear) or branched (nonlinear), but in each case—rings or chains, straight-chains or branched chains—the carbon bonds not used in bonding to other carbon atoms are taken up by hydrogen atoms, e.g., *n*-pentane and *iso*-pentane:

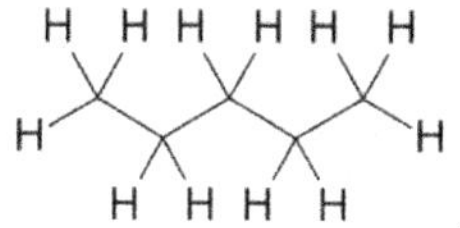

n-pentane

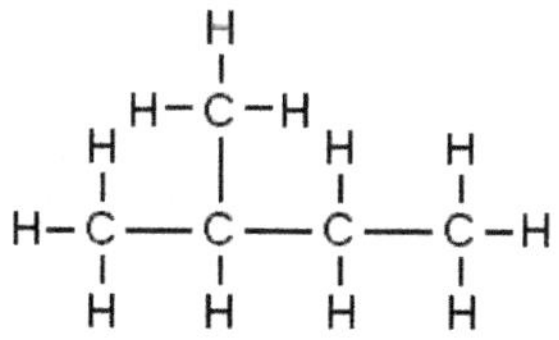

iso-pentane (2-methyl butane)

In theory, there is no limit to the number of possible hydrocarbons, and carbon chains can form into apparently limitless molecular shapes since hydrogen is a good partner atom. Hydrogen is a particularly small atom and can bond to one of carbon's valence electrons without getting in the way of the other three—hydrogen does not sterically hinder the other atoms.

As a result of the numerous potential compounds that can be formed from carbon and hydrogen, it has been well established that hydrocarbon compounds are

TABLE 2.6 General Classes of Hydrocarbons

Class	Compound Types
Saturated hydrocarbons	*n*-Paraffins
	iso-Paraffins and other branched paraffins
	Cycloparaffins (naphthenes)
	Condensed cycloparaffins (including steranes, hopanes)
	Alkyl side chains on ring systems
Unsaturated hydrocarbons	Olefins
Aromatic hydrocarbons	Benzene systems
	Condensed aromatic systems
	Condensed aromatic-cycloalkyl systems
	Alkyl side chains on ring systems
Saturated heteroatomic systems	Alkyl sulfides
	Cycloalkyl sulfides
	Alkyl side chains on ring systems
Aromatic heteroatomic systems	Furans (single-ring and multiring systems)
	Thiophenes (single-ring and multiring systems)
	Pyrroles (single-ring and multiring systems)
	Pyridines (single-ring and multiring systems)
	Mixed heteroatomic systems
	Amphoteric (acid-base) systems
	Alkyl side chains on ring systems

composed of (1) saturated hydrocarbon derivatives, often referred to as paraffin hydrocarbons and (2) aromatic hydrocarbon derivatives (Table 2.6). Furthermore, the saturated hydrocarbon derivatives can be sub-divided into the following three classes: (1) aliphatic hydrocarbons and (2) cycloaliphatic hydrocarbons.

2.1.1 Aliphatic Hydrocarbons

Aliphatic hydrocarbons are subdivided into three groups of homologous series, according to their state of saturation or unsaturation (i.e., the ratio of hydrogen to carbon: (1) paraffin hydrocarbons, which are hydrocarbons without a double bond (—C=C—) or without a triple bond, —C≡C—, (2) olefins or alkenes, which contain one or more double bonds, such as diolefins of which butadiene is

TABLE 2.7 The Different Hydrocarbon Series

Number of Carbon Atoms	Alkane (Single Bond)	Alkene (Double Bond)	Alkyne (Triple Bond)	Cycloalkane
1	Methane	–	–	–
2	Ethane	Ethene (ethylene)	Acetylene (ethyne)	–
3	Propane	Propene (propylene)	Propyne (methylacetylene)	Cyclopropane
4	Butane	Butene (butylene)	Butyne	Cyclobutane
5	Pentane	Pentene	Pentyne	Cyclopentane
6	Hexane	Hexene	Hexyne	Cyclohexane
7	Heptane	Heptene	Heptyne	Cycloheptane
8	Octane	Octene	Octyne	Cyclooctane
9	Nonane	Nonene	Nonyne	Cyclononane
10	Decane	Decene	Decyne	Cyclodecane

an example, $CH_2=CHCH=CH_2$, and (3) alkynes, which have one or more triple bonds $R_1C\equiv CR_2$ of which acetylene, $HC\equiv CH$, is the simplest example of the series (Tables 2.7 and 2.8). All series show a regular increase in properties, such as melting point and boiling point with molecular weight (Table 2.9).

The remainder of the hydrocarbon group is classed according to molecular structure—the hydrocarbons can be: (1) *straight-chain*, such as *n*-pentane, $CH_3CH_2CH_2CH_2CH_3$; (2) *branched chain*, such as *iso*-pentane; and (3) *cyclic* hydrocarbons, such as cyclopentane which is a ring system consisting of five methylene $(-CH_2-)$ groups:

Iso-pentane

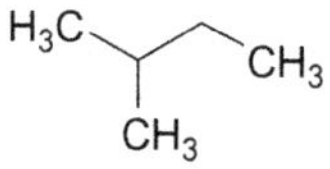

Neopentane

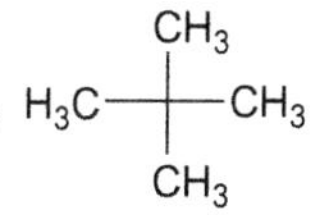

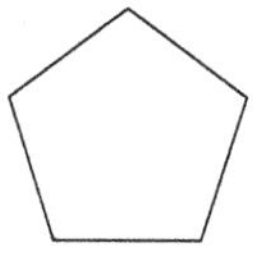

Cyclopentane

TABLE 2.8 Examples of Hydrocarbons from the Different Series

Formula	Name	Structural Formula	Classification
CH_4	Methane	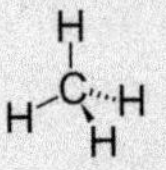	Alkane
C_2H_2	Acetylene (Ethyne)	$H-C≡C-H$	Alkyne
C_2H_4	Ethylene (Ethene)	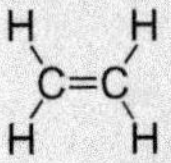	Alkene
C_2H_6	Ethane	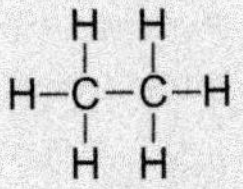	Alkane
C_3H_4	Propyne	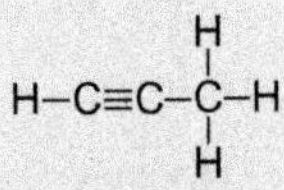	Alkyne
C_3H_6	Propene	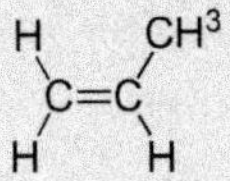	Alkene
C_3H_8	Propane	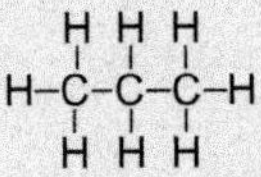	Alkane
C_4H_6	1-Butyne	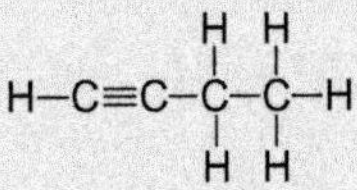	Alkyne
C_4H_8	Butene	e.g., 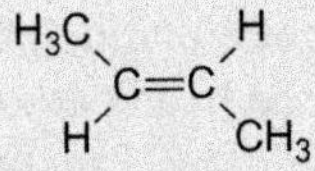	Alkene
C_4H_{10}	Butane	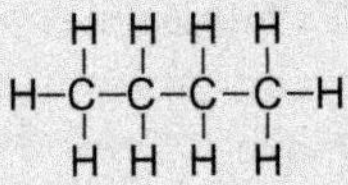	Alkane
C_6H_{10}	Cyclohexene	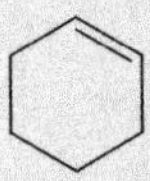	Cycloalkene
C_5H_{12}	n-pentane		Alkane
C_7H_{14}	Cycloheptane	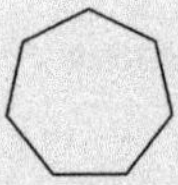	Cycloalkane
C_7H_{14}	Methylcyclohexane	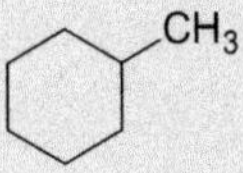	Cyclohexane

TABLE 2.9 Properties of Selected *n*-Alkanes (Straight-Chain Hydrocarbons), or Alkanes

Name	Molecular Formula	Melting Point (°C)	Boiling Point (°C)	State at 25°C
Methane	CH_4	−182.5	−164	Gas
Ethane	C_2H_6	−183.3	−88.6	Gas
Propane	C_3H_8	−189.7	−42.1	Gas
Butane	C_4H_{10}	−138.4	−0.5	Gas
Pentane	C_5H_{12}	−129.7	36.1	Liquid
Hexane	C_6H_{14}	−95	68.9	Liquid
Heptane	C_7H_{16}	−90.6	98.4	Liquid
Octane	C_8H_{18}	−56.8	124.7	Liquid
Nonane	C_9H_{20}	−51	150.8	Liquid
Decane	$C_{10}H_{22}$	−29.7	174.1	Liquid
Undecane	$C_{11}H_{24}$	−24.6	195.9	Liquid
Dodecane	$C_{12}H_{26}$	−9.6	216.3	Liquid
Eicosane	$C_{20}H_{42}$	36.8	343	Solid
Triacontane	$C_{30}H_{62}$	65.8	449.7	Solid

Branched alkanes are named by indicating the branch attached to the principal chain. Branches, known as substituents, are named by taking the name of an alkane and replacing the suffix with -*yl*: for example, methyl, ethyl, with similar relevant names for the high-molecular-weight alkanes. The general term for an alkane which functions as a substituent is alkyl. Also, to add to the compilation of the alkane series, for any specific alkane, the properties of the branched chain isomers vary with the structure insofar as the three-dimensional structure of the hydrocarbon will influence properties and behavior.

Finally, the complexity of the alkane series of hydrocarbons is reflected in the number of possible isomers as the number of carbon atoms in the molecule increases (Table 2.10). In addition, the degree of branching affects characteristics, such as the ability of the compound to degrade (by the action of aerial oxygen or by the action of bacteria) or survive in the environment and whether or not the compound can be leached form soil.

2.1.2   Cycloaliphatic Hydrocarbons

Cycloaliphatic hydrocarbons (also called *cycloalkanes, alicyclic hydrocarbons, naphthenes*) are saturated hydrocarbons containing one or more rings,

TABLE 2.10 Boiling Point of the *n*-Isomers of the Various Paraffins and the Number of Possible Isomers Associated With Each Carbon Number

Carbon Atoms in Molecule	Boiling Point of *n*-Isomer (°C)	Boiling Point of *n*-Isomer (°F)	Number of Possible Isomers
5	36	97	3
10	174	345	75
15	271	519	4347
20	344	651	366,319
25	402	755	36,797,588
30	450	841	4,111,846,763
40	525	977	62,491,178,805,831

each of which may have one or more paraffin (alkyl) side chains. Cycloalkanes are alkanes joined in a closed loop to form a ring-shaped molecule, and they are named by using the alkane names above, with cyclo-as a prefix. These start with propane, or rather cyclopropane, which has the minimum number of carbon atoms to form a closed shape: three atoms, forming a triangle.

Thus, cycloalkane rings may be built up of a varying number of carbon atoms, and among the synthesized hydrocarbons, there are individual constituents with rings of the three-, four-, five-, six-, seven-, and eight carbon atoms. Saturated cyclic compounds contain single bonds only, whereas aromatic rings have an alternating (or conjugated) double bond. Cycloalkanes do not contain multiple bonds, whereas the cycloalkene derivatives and the cycloalkyne derivatives do contain the olefin bond ($>C=C<$) and the alkyne bond ($-C\equiv C-$), respectively. However, thermodynamic studies show that ring systems that contain with five or six carbon atoms are the most stable ring systems. The smallest cycloalkane family is the three-membered cyclopropane and is reactive because of the strain within the ring system.

2.1.3 Aromatic compounds

Aromatic hydrocarbon compounds are hydrocarbons containing one or more aromatic rings, such as the single-ring benzene as well as the multiring systems: naphthalene, anthracene, and phenanthrene ring systems, which may be linked up with (substituted) naphthene rings and/or paraffinic side chains.

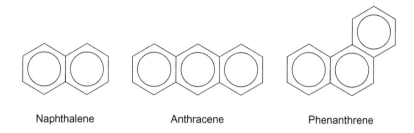

Naphthalene   Anthracene   Phenanthrene

Benzene is the simplest and most stable aromatics and is a single-ring system that contains conjugated double bonds:

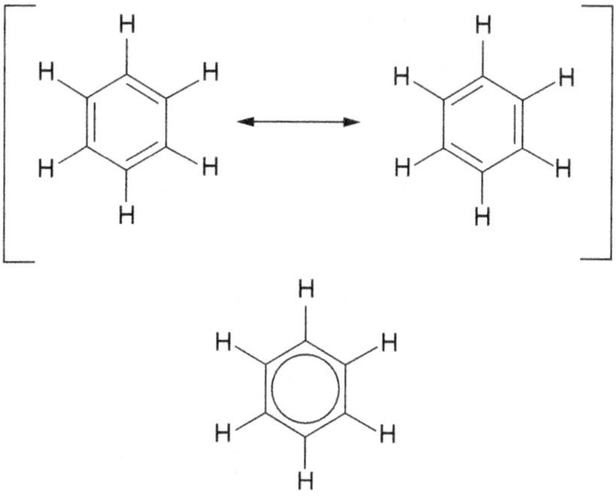

Thus, every carbon atom in the ring is sp^2 hybridized, allowing for added stability of the ring system. After benzene, the polynuclear aromatic system increases in number of rings in the system with corresponding variations in properties. Furthermore, the naphthene-aromatic hydrocarbons are classified according to the number of aromatic rings in the molecule. The concept of the occurrence of identifiable (and different) cyclic systems in nature is a reality and well documented, especially in natural products such as lignin, coal, and crude oil (Table 2.11) (Sakarnen and Ludwig, 1971; Durand, 1980; Weiss and Edwards, 1980; Olah, 2005; Speight, 2013, 2014).

Polynuclear aromatic hydrocarbons are a class of organic compounds produced by incomplete combustion or high-pressure processes and are a large group of organic compounds with two or more fused aromatic rings. Polynuclear aromatic hydrocarbons (PNAs) are known by several names: (1) polynuclear aromatic hydrocarbons, PAHs; (2) polynuclear hydrocarbons, PHs; and (3) polycyclic organic matter, POM (Bjorseth, 1983; Neilson, 1998; Fetzer, 2000). Polynuclear aromatic hydrocarbons are solids with low (to no) volatility at room temperature and have relatively high molecular

TABLE 2.11 Varying Types of Natural Products

Base Units	Amino Acids	Fatty Acids	Carbohydrates	Purine Bases
	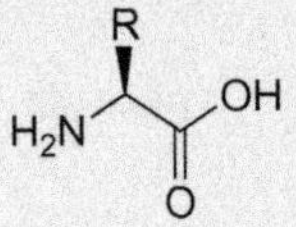 (amino acid structure with R, H_2N, OH, O)	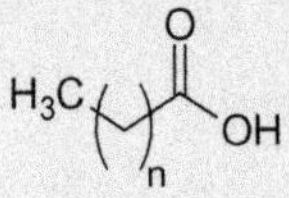 (fatty acid structure H_3C, O, OH, n)	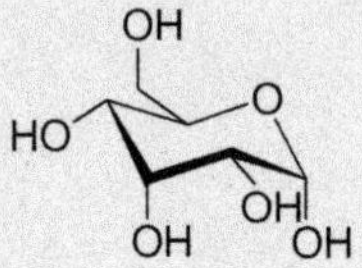 (carbohydrate structure with OH, HO, O, OH)	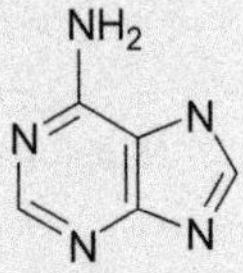 (purine base structure with NH_2, N)
Products	Polypeptides	Phospholipids	Monosaccharides	Nucleotides
	Proteins		Polysaccharides	Ribonucleic acid (RNA)
				Deoxyribonucleic acid (DNA)

weights. They are soluble (sometimes with difficulty) in many organic solvents but are relatively insoluble in water. More pertinent to their presence in the environment, most are susceptible to photooxidation which can have serious consequences for the properties of these compounds and the ensuing effects on an ecosystem.

Polynuclear aromatic hydrocarbons are typically produced from incomplete combustion. Generally, the polynuclear aromatic hydrocarbons of interest to the environmental scientists and environmental engineers consist of three or more fused benzene rings containing only carbon and hydrogen. The more common polynuclear aromatic hydrocarbons (because of the serious effects when released into the environment) include the multiring compounds:

- benzo(a)anthracene,
- benzo(a)pyrene,
- benzo(e)pyrene,
- benzo(g,h,i)perylene,
- benzo(k)fluoranthene,
- chrysene,
- coronene,
- dibenz(a,h)anthracene, and
- pyrene

Most of the polynuclear aromatic hydrocarbons with low vapor pressure in the air are adsorbed on particles. When dissolved in water or adsorbed on particulate matter, polynuclear aromatic hydrocarbons can undergo photodecomposition when exposed to ultraviolet light from solar radiation. Furthermore, in the atmosphere, polynuclear aromatic hydrocarbons can react with pollutants such as ozone, nitrogen oxides, and sulfur dioxide, yielding diones, nitro- and dinitro-polynuclear aromatic hydrocarbons, and sulfonic acids, respectively. Polynuclear aromatic hydrocarbons may also be degraded by some microorganisms in the soil (Speight and Arjoon, 2012).

2.2 Heterocyclic Compounds

The previous sections present some indication of the types and nomenclature of the organic hydrocarbons. Inclusion within the hydrocarbon structures of atoms of nitrogen or oxygen or sulfur or more than one of these heteroatoms adds a further level of complexity to the original hydrocarbon structures and the chemical and physical properties of the (original) cyclic hydrocarbons are again altered if heteroatoms are present (Table 2.3–2.5). The heteroatoms (nitrogen or oxygen or sulfur) can exist as either substituents attached externally to the ring (exocyclic) or as a member of the ring itself (endocyclic). Pyridine and furan are examples of aromatic heterocycles, while piperidine and tetrahydrofuran are the corresponding alicyclic heterocyclic derivatives. The heteroatom of heterocyclic molecules is generally oxygen, sulfur, or nitrogen, with the latter being particularly common in biochemical systems. Heterocycles are commonly found in a wide range of products including aniline dyes and medicines.

3 FUNCTIONAL GROUPS

A functional group is the part of an organic molecule which contains atoms other than carbon and hydrogen, or which contain bonds other than a carbon-carbon single bond (C—C) and a carbon-hydrogen bond (C—H). Some of the most common functional groups in organic chemistry are: alkenes, alkynes, aromatics, nitriles, amines, amides, nitro compounds, alcohols, phenols, ethers, aldehydes, ketones, carboxylic acids, acid chlorides, acid anhydrides, esters, alkyl halides, thiols, and thioethers.

Thus, a functional group is any atom or collection of atoms that is capable or reacting with a reactive species to produce a product. A functional group is also capable of affecting the properties of the original nonfunctional molecule in which the group occurs. Some of the most common functional groups in organic chemicals are: (1) the double bond of alkene derivatives ($R_1C\!\!=\!\!CR_2$), (2) the triple bond of alkyne derivatives ($R_1C\!\!\equiv\!\!CR_2$), (3) the aromatic ring in aromatic derivatives such as in benzene derivatives and including the ring systems in condensed-ring aromatic derivative, (4) the carbon-nitrogen function in nitrile derivatives ($RC\!\!\equiv\!\!N$), (5) the carbon-nitrogen function in amine derivatives (RNH_2, R_1NHR_2, $R_1NH(R_3)R_2$), (6) the oxygen-carbon-nitrogen function in amide derivatives ($RCONH_2$), (7) the nitrogen-oxygen function in nitro compounds (RNO_2), (8) the oxygen-hydrogen function in alcohol derivatives (ROH), (9) the oxygen-hydrogen function in phenol derivatives ($ArOH$), (10) the carbon-oxygen-carbon function in ether derivatives (R_1OR_2), (11) the carbon-oxygen double bond in aldehyde derivatives ($RCH\!\!=\!\!O$), (12) the carbon-oxygen double bond in ketone derivatives (R_1COR_2), (13) the carbon-oxygen double bond and oxygen-hydrogen bond in carboxylic acid derivatives (RCO_2H), (14) the carbon-oxygen double bond and oxygen-chloride bond in acid chloride derivatives (RCOCl), (15) the carbon-oxygen-carbon bond in acid anhydride derivatives (R_1CO—O—COR_2), (16) the carbon-oxygen-carbon bond in ester derivatives (R_1COOR_2), and the carbon-halogen bond in alkyl halide derivatives, RX, where X is a halogen, (17) the sulfur-hydrogen bond in thiol derivatives (RSH), and (18) the carbon-sulfur-carbon bond in thioether derivatives R_1SR_2. In each of the aforementioned examples, cases R_1 and R_2 are alkyl groups (the same or different) and Ar is a phenyl group or other aromatic group.

Functional groups are classed as aliphatic if there is no aromatic ring directly attached to the group, but it is possible to have an aromatic molecule containing an aliphatic functional group if the aromatic ring is not directly attached to the functional group—benzyl alcohol ($C_6H_5CH_2OH$) is an example. On the other hand, functional groups are classed as aromatic if they have an aromatic ring directly attached to them—phenol (C_6H_5OH) is an example. In the case of ester derivatives and amide derivatives, the aromatic ring must be attached to the carbon atom of the acid function—methyl benzoate ($C_6H_5CO_2CH_3$) is an example—but if the aromatic ring is attached to the oxygen atom of the carboxylic acid function, the functional groups are classed as aliphatic ($CH_3CO_2C_6H_5$) phenyl acetate in an example. Generally, the carbon-hydrogen bond in most locations is not classed as a functional group but must be recognized as having the potential of functionality.

Early in the study of organic compounds, it was observed that certain groups of atoms and associated bonds (functional groups) confer specific reactivity patterns on the organic molecules of which they were a part. Although the properties of each of the several million organic molecules whose structure is known are unique in some way, all molecules that contain the same functional group have a similar pattern of reactivity at the functional group site. The exception is when the functional group is sterically hindered because of the structure of the molecule. An example of steric hindrance (or steric interference) with the reactivity occurs in the case of dimethyldibenzothiophene derivatives:

H_3C CH_3

Dimethyldibenzothiophene

In this molecule, hydrogen (removal of the sulfur atom as hydrogen sulfide, H_2S) is much slower that the desulfurization of dibenzothiophene because of the protection of the carbon-sulfur-carbon function by the two methyl groups (Macaud et al., 2000; Xu et al., 2004; Speight, 2014). Thus, functional groups are a key organizing feature of organic chemistry. By focusing on the functional groups present in a molecule (some molecules have more than one functional group, especially pharmaceutical molecules), several of the reactions that the molecule will undergo can not only be understood but are also predictable.

Further to whether or not a carbon-carbon bond and/or carbon-hydrogen bonds can be classed as a functional group, carbon-carbon and carbon-hydrogen bonds are extremely strong and the charge of the electrons in these covalent bonds is spread more or less evenly over the bonded atoms and, therefore, hydrocarbons that contain only single bonds of these two types are not very reactive—the name *paraffin* is derived from the Latin *parum* and *affinis*, which mean meaning *lacking affinity* or *lacking reactivity*, referring to unreactive nature of the paraffin hydrocarbons.

However, the reactivity of a paraffin molecule increases if it contains one or more weak bonds or bonds that have an unequal distribution of electrons between the two atoms. If the two electrons of a covalent bond are drawn more closely to one of the bonded atoms, that atom will develop a partial negative charge (a region of high electron density and indicated by δ^-) and the atom to which it is bonded will develop a partial positive charge (indicated by δ^+). A covalent bond in which the electron pair linking the atoms is shared unequally becomes a polar (reactive) bond which has unique electronic properties that confer the potential for chemical reaction on the molecule in which they are present. The presence of a partial negative charge (δ^-) is attracted to other atoms or groups of atoms that are deficient in electron density (δ^+). This initiates the process of bond breaking that is a prerequisite for a chemical reaction and molecules with regions of increased electron density or decreased electron density are especially important for chemical change.

Thus, there are two major bonding features that generate the reactive sites of functional groups. The first type of reactive site is a site of the presence of multiple bonds—both double and triple bonds (designated as functional groups) have regions of high electron density lying outside the atom-to-atom bond axis. A second type of reactive site occurs when an atom other than carbon or hydrogen (a *heteroatom* such as nitrogen, oxygen, or sulfur) is bonded to a carbon atom. Since all heteroatoms have a greater or lesser attraction for electrons than the carbon atom, each bond between a carbon and a heteroatom is polar, and the degree of polarity depends on the difference between the electron-attracting properties of the carbon atom and the heteroatom. Molecules with more than one functional group (polyfunctional compounds) may have more complicated properties that result from the identity and intermolecular relations of the multiple functional groups. Many natural products and pharmaceutical products contain more than one functional group located at specific sites within a large, complicated, three-dimensional structure.

The concept of functional groups requires that, in order to be classed as a functional group, a bonded atom or a collection of bonded atoms must confer chemical reactivity upon the molecule in which the bonded atom or the collection of bonded atoms occurs. In this section, the various functional groups will be presented: (1) alkanes, (2) alkenes, (3) alkynes, (4) aromatic hydrocarbons, (5) alcohols and phenols, (6) halides, (7) ethers and epoxides, (8) thiols, (9) amines, (10) aldehydes and ketones, (11) carboxylic acids, (12) polyfunctional compounds, (13) petrochemicals, and (14) polymers. In each case, descriptions of the qualities of the functional groups will be presented.

Although, there is some question as to whether or not the alkanes should be considered as molecules that contain a functional group, alkanes do (in spite of the derivation of the name of this group of hydrocarbons; see earlier) are included because, in certain instances, a carbon-hydrogen bond or carbon-carbon bond will react to produce a product. In addition, when considering the reaction of a functional group in an otherwise hydrocarbon molecule, using the corresponding alkane as the base case for reactivity is always a worthwhile comparison.

3.1   Alkanes

Alkanes are compounds that consist entirely of atoms of carbon and hydrogen bonded to one another by carbon-carbon and carbon-hydrogen single bonds. The alkanes are also known as saturated hydrocarbons because all the bonds not used to form the molecule are used to the maximum single bond capacity—that is, the valence bonding capacity of each carbon atom is filled by either bonding to another carbon atom or bonding to hydrogen atoms and the molecule is classed as *saturated*. As a result, the formula for any (noncyclic) alkane is C_nH_{2n+2}, where n is the number of carbon atoms. In the case of a linear, unbranched alkane, every carbon atom has two hydrogen atoms attached, but the two end carbon atoms each have an extra hydrogen, as in the example for *n*-decane:

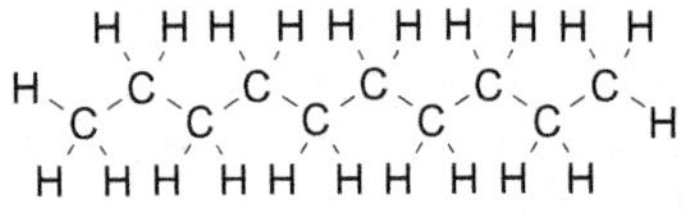

n-Decane

The shared electron pair in each of these single bonds occupies space directly between the two atoms, and the bond generated by this shared pair is a sigma (σ) bond. Both carbon-carbon and carbon-hydrogen sigma bonds are single strong, nonpolar covalent bonds that are typically the least reactive bonds in organic molecules.

What follows are the names and formulas for the first 10 normal alkanes (or unbranched alkanes). Note that the first four of these received common names before their structures were known; from C_5 onward, however, they were given names with Greek or Latin roots indicating the number of carbon atoms (e.g., decane, a reference to *eight*):

- Methane (CH_4)
- Ethane (C_2H_6)
- Propane (C_3H_8)
- Butane (C_4H_{10})
- Pentane (C_5H_{12})
- Hexane (C_6H_{14})
- Heptane (C_7H_{16})
- Octane (C_8H_{18})
- Nonane (C_9H_{20})
- Decane ($C_{10}H_{22}$)

Alkane sequences form the inert framework of most organic compounds. For this reason, alkanes are not formally considered a functional group. In terms of occurrence and use, the simplest examples of alkanes are methane (CH_4, the principal constituent of natural gas), ethane (C_2H_6), propane (C_3H_8, widely used as a gaseous fuel), and butane (C_4H_{10}, the liquid fuel in pocket lighters). Hydrocarbon chains commonly occur in cyclic forms, or rings; the most common example is cyclohexane (C_6H_{12}).

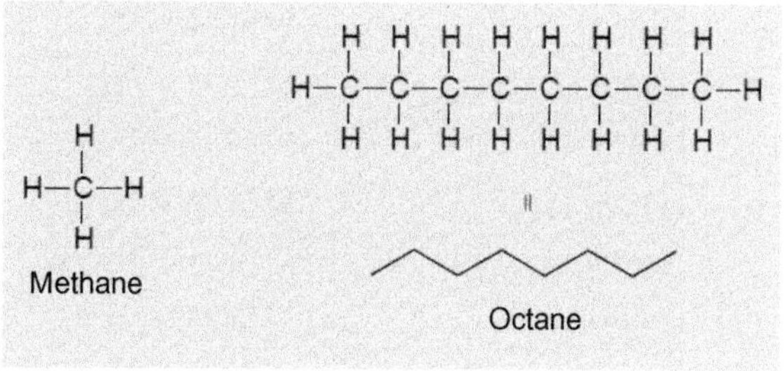

Methane

Octane

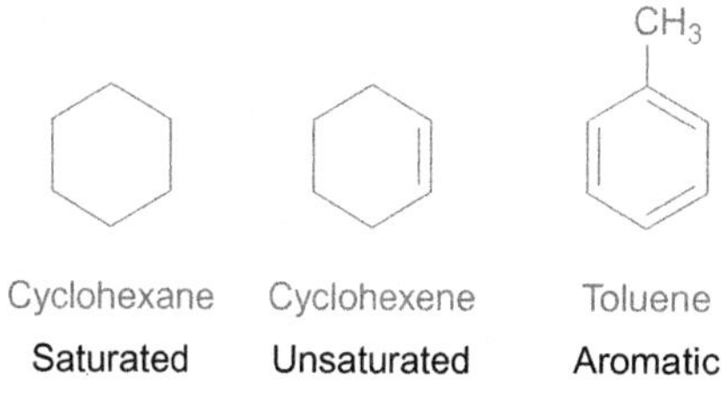

Cyclohexane	Cyclohexene	Toluene
Saturated	Unsaturated	Aromatic

Cyclohexene and toluene are included here for structural comparison.

When a hydrocarbon chain is connected as a substituent to a more fundamental structural unit, it is termed an alkyl group: An alkyl group is formed by removing one hydrogen from the alkane chain and is described by the formula C_nH_{2n+1}. The removal of this hydrogen results in a stem change from -ane to -yl. For example:

$$CH_4 \longrightarrow CH_3^-$$
Methane   Methyl

$$CH_3CH_2CH_3 \longrightarrow CH_3CH_2CH_2^-$$
Propane   Propyl

The same concept can be applied to any of the straight-chain alkane names (Table 2.12).

TABLE 2.12 Formula and Common Names of the First Twenty Alkanes

Name	Molecular Formula	Condensed Structural Formula
Methane	CH_4	CH_4
Ethane	C_2H_6	CH_3CH_3
Propane	C_3H_8	$CH_3CH_2CH_3$
Butane	C_4H_{10}	$CH_3(CH_2)_2CH_3$
Pentane	C_5H_{12}	$CH_3(CH_2)_3CH_3$
Hexane	C_6H_{14}	$CH_3(CH_2)_4CH_3$
Heptane	C_7H_{16}	$CH_3(CH_2)_5CH_3$
Octane	C_8H_{18}	$CH_3(CH_2)_6CH_3$
Nonane	C_9H_{20}	$CH_3(CH_2)_7CH_3$
Decane	$C_{10}H_{22}$	$CH_3(CH_2)_8CH_3$
Undecane	$C_{11}H_{24}$	$CH_3(CH_2)_9CH_3$
Dodecane	$C_{12}H_{26}$	$CH_3(CH_2)_{10}CH_3$
Tridecane	$C_{13}H_{28}$	$CH_3(CH_2)_{11}CH_3$
Tetradecane	$C_{14}H_{30}$	$CH_3(CH_2)_{12}CH_3$
Pentadecane	$C_{15}H_{32}$	$CH_3(CH_2)_{13}CH_3$
Hexadecane	$C_{16}H_{34}$	$CH_3(CH_2)_{14}CH_3$
Heptadecane	$C_{17}H_{36}$	$CH_3(CH_2)_{15}CH_3$
Octadecane	$C_{18}H_{38}$	$CH_3(CH_2)_{16}CH_3$
Nonadecane	$C_{19}H_{40}$	$CH_3(CH_2)_{17}CH_3$
Eicosane	$C_{20}H_{42}$	$CH_3(CH_2)_{18}CH_3$

3.2 Alkenes

Organic compounds are termed alkenes if the compound contains a carbon-carbon double bond ($>C=C<$). The shared pair of electrons of one of the bonds is a σ bond, while the second pair of electrons occupies space on both sides of the σ bond, and this shared pair of electrons constitutes a pi (π) bond. The π-bond forms a region of increased electron density (δ^-) because the electron pair is more distant from the positively charged carbon nuclei than is the electron pair of the σ bond. Even though a carbon-carbon double bond is a strong bond, a π bond will draw to itself atoms or atomic groupings that are electron-deficient, thereby initiating a process of bond breaking that can lead to rupture of the π bond and formation of new σ bonds during a reaction.

The names of the alkenes, hydrocarbons that contain one or more double bonds per molecule, are parallel to those of the alkanes, but the family ending is -*ene*. Likewise, the alkenes have a common formula: C_nH_{2n}. Both alkenes and alkynes (the latter hydrocarbons are discussed later) are unsaturated—not all of the carbon atoms are saturated by binding to hydrogen atoms, and these carbon atoms in them are free to form other bonds. Alkenes with more than one double bond can be referred to as dienes (two double bonds in a molecule), trienes (three double bonds in a molecule), or even polyunsaturated in which there can be four or more double bonds in a molecule.

A simple example of an alkene reaction, which illustrates the way in which the electronic properties of a functional group determine its reactivity, is the addition of molecular hydrogen to form alkanes, which contain only σ bonds.

$$
\underset{\text{Alkene}}{\overset{R}{\underset{R}{>}}C=C\overset{R}{\underset{R}{<}}} \;+\; \underset{\text{Hydrogen}}{H—H} \;\longrightarrow\; \underset{\text{Alkane}}{R—\overset{\overset{H}{|}}{\underset{\underset{R}{|}}{C}}—\overset{\overset{H}{|}}{\underset{\underset{R}{|}}{C}}—R}
$$

Such reactions, in which the π bond of an alkene reacts to form two new σ bonds, are energetically favorable because the new bonds formed (two carbon-hydrogen σ bonds) are stronger than the bonds broken (one carbon-carbon π bond and one hydrogen-hydrogen σ bond). Because the addition of atoms to the π bond of alkenes to form new σ bonds is a general and characteristic reaction of alkenes, alkenes are said to belong to the class of hydrocarbons known as *unsaturated hydrocarbons*. On the other hand, the product alkane, which cannot be transformed by addition reactions into molecules with a greater number of σ bonds, belongs to the class of hydrocarbons known as *saturated hydrocarbons* (see earlier).

The alkene functional group is an important one in chemistry and is widespread in nature with the exception of the fossil fuels: coal, crude oil, and natural gas. The maturation process for each of these fossil fuel is not conducive to the prolonged existence of alkene hydrocarbon (such as ethylene derivatives, $>C=C<$) or alkyne hydrocarbons (such as acetylene derivatives, $-C\equiv C-$).

Some common examples of the alkene series of hydrocarbons (shown here) include ethylene (used to make polyethylene) and 2-methyl-1,3-butadiene (isoprene, used to make synthetic rubber).

$$CH_2 = CH_2 \quad CH_2 = C(CH_3)CH = CH_2$$
Ethylene   Isoprene

For ethylene, both of the carbon atoms of an alkene and the four atoms connected to the double bond lie in a single plane.

The IUPAC rules for naming the alkene hydrocarbons are somewhat more complex that the rule for naming the alkanes but, are nevertheless, fairly easy to follow. Generally, the suffix -*ene* is used to indicate and alkene (Table 2.13) or a

TABLE 2.13 The Systemic Name for the First Twenty Straight-Chain Alkenes

Name	Molecular Formula
Ethene	C_2H_4
Propene	C_3H_6
Butene	C_4H_8
Pentene	C_5H_{10}
Hexene	C_6H_{12}
Heptene	C_7H_{14}
Octene	C_8H_{16}
Nonene	C_9H_{18}
Decene	$C_{10}H_{20}$
Undecene	$C_{11}H_{22}$
Dodecene	$C_{12}H_{24}$
Tridecene	$C_{13}H_{26}$
Tetradecene	$C_{14}H_{28}$
Pentadecene	$C_{15}H_{30}$
Hexadecene	$C_{16}H_{32}$
Heptadecene	$C_{17}H_{34}$
Octadecene	$C_{18}H_{36}$
Nonadecene	$C_{19}H_{38}$
Eicosene	$C_{20}H_{40}$

cycloalkene and the longest chain chosen for the root name must include both carbon atoms of the double bond. The root chain must be numbered from the end nearest a double bond carbon atom. If the double bond is in the center of the chain, the nearest substituent rule is used to determine the end where numbering starts. The smaller of the two numbers designating the carbon atoms of the double bond is used as the double bond locator. If more than one double bond is present, the compound is named as a diene, triene, or equivalent prefix indicating the number of double bonds, and each double bond is assigned a locator number. In the cycloalkene hydrocarbon series, the double bond carbons are assigned ring locations Number 1 and Number 2—the carbon atom designated as Number 1 is determined by the nearest substituent rule. Finally, substituent groups containing double bonds are the vinyl group (H_2C=CH—) and the allyl group (H_2C=$CHCH_2$-).

3.3   Alkynes

Molecules that contain a triple bond between two carbon atoms (—C≡C—) are known as alkynes. The triple bond is made up of one σ bond and two π bonds. As in alkenes (>C=C<), the π bonds constitute regions of increased electron density lying parallel to the carbon-carbon bond axis. Carbon-carbon triple bonds are strong bonds but are still capable of participating in reactions that break the π bonds to form stronger σ bonds.

The most common example of an alkyne is acetylene (CH≡CH, also known as ethyne) which is used as a fuel for oxyacetylene torches in welding applications. Alkynes are not abundant in nature due to the reactivity of the triple bond, but the fungicide capillan contains two alkyne functional groups:

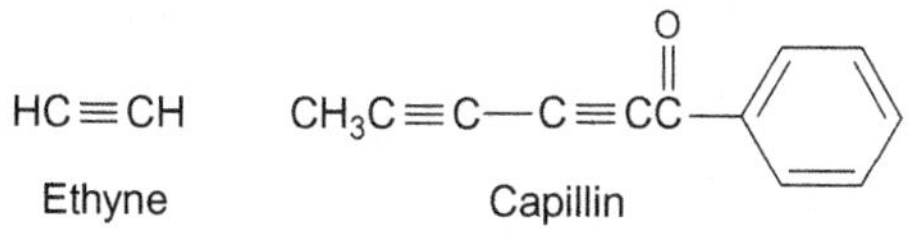

HC≡CH   Ethyne   CH₃C≡C—C≡CC   Capillin

As with the alkenes, the names of alkynes (hydrocarbons containing one or more triple bonds per molecule) are parallel to those of the alkanes (Table 2.12), only with the replacement of the suffix -*yne* in place of -*ane* (Table 2.14). The formula for alkenes is C_nH_{2n-2}. Among the members of this group is acetylene, or C_2H_2, used for welding steel.

3.4   Aromatic Hydrocarbons

Aromatic hydrocarbons (also called *arenes*), despite their name, often (but not necessarily always) have distinctive aromas (odors). In fact, the name is a traditional one, and these organic compounds are characterized by the presence of one or more benzene rings in the molecule. Benzene, and all

TABLE 2.14 Molecular Formulas and Names of the First Ten Straight-Chain Alkynes

Name	Molecular Formula
Acetylene[a]	C_2H_2
Propyne	C_3H_4
1-Butyne	C_4H_6
1-Pentyne	C_5H_8
1-Hexyne	C_6H_{10}
1-Heptyne	C_7H_{12}
1-Octyne	C_8H_{14}
1-Nonyne	C_9H_{16}
1-Decyne	$C_{10}H_{18}$

[a]Acetylene is the more commonly used name for ethyne; in the remaining alkynes, the number 1 indicates that that tripe bond is between the number 1 and number 2 carbon atoms in the carbon chain, e.g., 1-butyne: $CH_3CH_2C \equiv CH$.

the larger arenes, has a characteristic planar structure forced on them by the electronic requirements of the six (or more) π electrons. When named as substituents on other structural units, the aromatic units are referred to as *aryl* substituents.

The aromatic hydrocarbon group contains not only benzene but also derivatives such as toluene ($C_6H_5CH_3$), the isomeric dimethyl benzene derivatives ($CH_3C_6H_4CH_3$), which are used as solvents, as well as in the synthesis of drugs, dyes, and plastics.

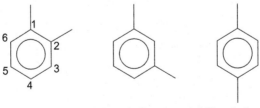

1,2-Dimethylbenzene   1,3-Dimethylbenzene   1,4-Dimethylbenzene
(*ortho*-xylene)   (*meta*-xylene)   (*para*-xylene)

The isomeric xylene derivatives: the line projections for the rings indicate the relative positions of the two methyl groups on each benzene ring.

One of the more famous (or infamous) products in the single-ring aromatic hydrocarbon group is trinitrotoluene (or TNT):

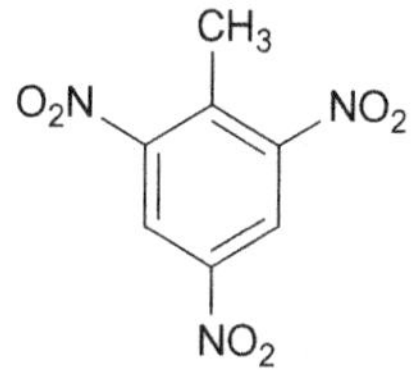

Trinitrotoluene

Naphthalene, the fused two-ring condensed aromatic hydrocarbon, is derived from coal tar and used in the synthesis of other compounds. A crystalline solid with a powerful odor is found in mothballs and various deodorant-disinfectants. Benzo(a)pyrene, an aromatic hydrocarbon produced in small amounts by the combustion of organic substances, contains five fused benzene rings. Like several other polycyclic aromatic hydrocarbons, it is carcinogenic. Aromatic compounds are widely distributed in nature. Benzaldehyde, anisole, and vanillin, for example, have pleasant aromas.

Benzene 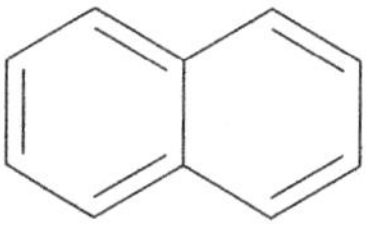   Naphthalene 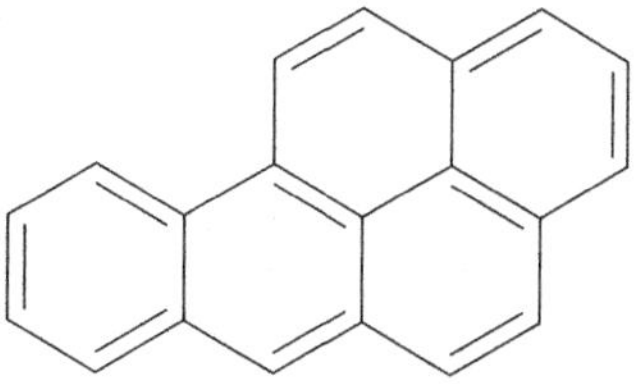   Benzo[a]pyrene

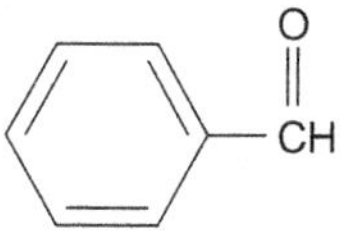

Benzaldehyde 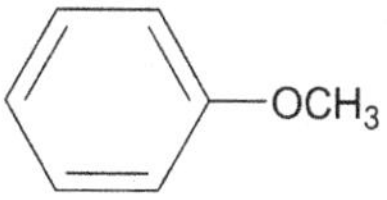   Anisole 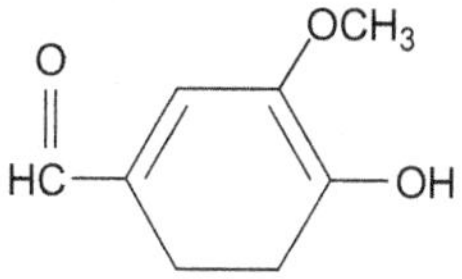   Vanillin

3.5   Alcohols and Phenols

An oxygen atom is much more electronegative than carbon or hydrogen atoms, so both carbon-oxygen and hydrogen-oxygen bonds are polar. The oxygen atom is slightly negatively charged, and the carbon and hydrogen atoms are slightly positively charged. The polar bonds of the hydroxyl group are responsible for the major reaction characteristics of alcohols and phenols. In general, these reactions are initiated by reaction of electron-deficient groups with the negatively charged oxygen atom or by reaction of electron-rich groups with the positively charged atoms—namely, carbon or hydrogen—bonded to oxygen.

Alcohols are the oxygen-hydrogen functional group (—OH) within a hydrocarbon molecule. The two most important commercial types of alcohol are methanol (CH_3OH, also called methyl alcohol or wood alcohol) and ethanol (C_2H_5OH), which needs very little introduction since it is found in alcoholic beverages, such as beer, wine, whiskey, and many other such beverages. Though methanol is still known as *wood alcohol*, it is no longer obtained by heating wood, but rather by the industrial hydrogenation of carbon monoxide:

$$CO + H_2 \rightarrow CH_3OH$$
Syngas

Methanol is used in adhesives, fibers, and plastics; it can also be used as a fuel. Ethanol, too, can be burned in an internal-combustion engine when combined with gasoline (a product known as *gasohol*).

Ethyl alcohol (C_2H_5OH) is an example of replacing on hydrogen atom of ethane with a hydroxyl function. Ethylene glycol is an example of replacing a hydrogen atom on each carbon atom of ethane with hydroxyl function, and phenol is an example of replacing one hydrogen atom in benzene with a hydroxyl function. When the hydroxyl group is *directly* attached to a benzene ring, the resulting compound is phenol, which can lead to a variety of phenol derivatives (Fig. 2.1). Both alcohols and phenols are widespread in nature, with alcohols being especially ubiquitous.

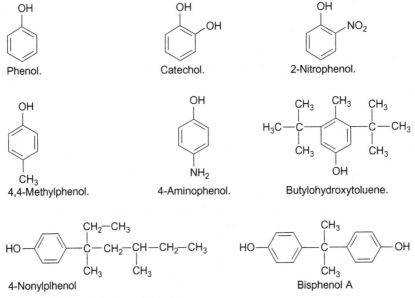

FIG. 2.1 Various derivatives of phenol.

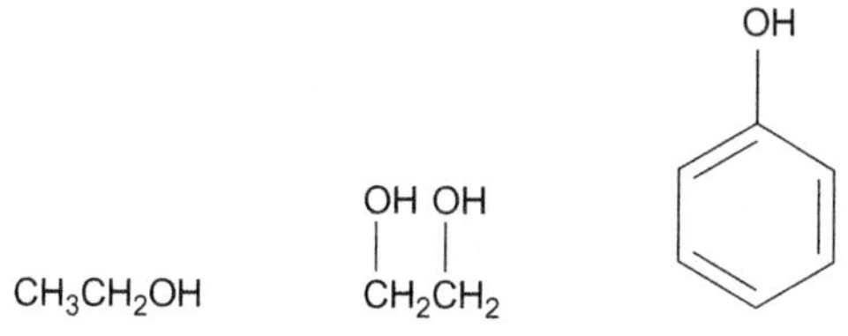

OH

OH OH
| |
CH₃CH₂OH   CH₂CH₂

Ethanol   Ethylene   Phenol
             glycol

All alcohols can be classified as primary, secondary, or tertiary alcohols. In a primary alcohol, the carbon bonded to the hydroxyl group is also bonded to only one other carbon. In a secondary alcohol and tertiary alcohol, the carbon is bonded to two or three other carbons, respectively. The sulfur analog of an alcohol (—SH) is a *thiol* (often referred to as a *mercaptan*).

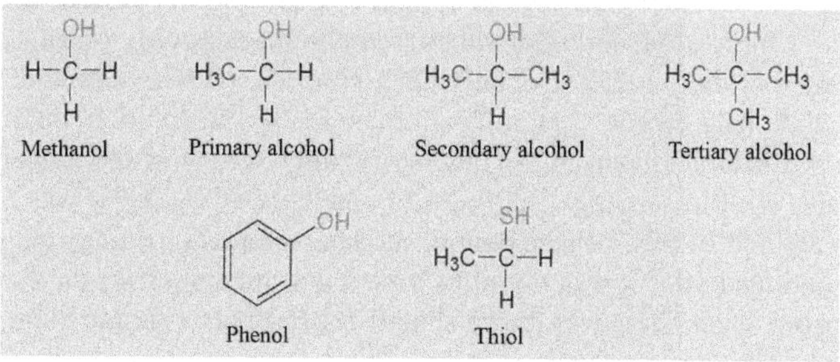

Phenol derivatives of anthropogenic origin, which drain from municipal sewage systems or industrial sewage systems, penetrate ecosystems such as surface water. Moreover, the occurrence of phenol derivatives in the environment also arises from the production and use of numerous pesticides, in particular phenoxy-herbicides such as 2,4-dichlorophenoxyacetic acid (2,4-D) or 4-chloro-2-methylphenoxyacetic acid (MCPA) and also from phenolic biocides such as like pentachlorophenol (PCP).

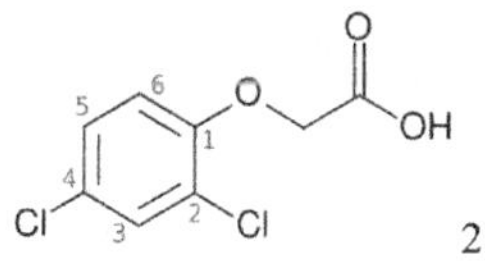

2,4-Dichlorophenoxyacetic acid (2,4-D)

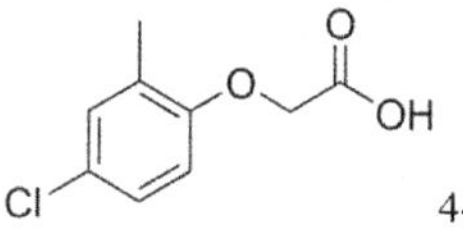

4-Chloro-2-methylphenoxyacetic acid (MCPA)

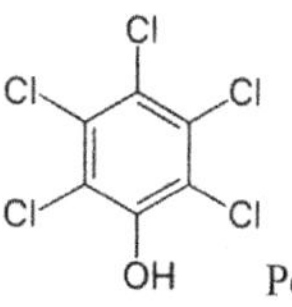

Pentachlorophenol (PCP)

In addition, phenol derivatives may be formed as a result of natural processes as, for example, during the formation of phenol and *p*-cresol as a result of the decomposition of organic matter or synthesis of chlorinated phenol derivatives by fungi and plants (Laine and Jorgensen, 1996; Swarts et al., 1998; Jaromir et al, 2005; Michałowicz and Duda, 2007).

3.6   Halides

When the carbon atom of an alkane is bonded to one or more halogen atoms, the compound is referred to as an alkyl halide or haloalkane. Halide derivatives, or organo-halide derivatives, are compounds that contain a halogen atom (fluorine, F; chlorine, Cl; bromine, Br; or iodine, I) bonded to a carbon atom by a polar bond. The slightly positive charge that exists on the carbon atom in carbon-halogen bonds is the source of reactivity of halides.

Chloroethane (ethyl chloride) is a volatile liquid that is used as a topical anesthetic. Chloroethylene (chloroethene, vinyl chloride, $CH_2=CHCl$) is the monomeric building block for polyvinyl chloride (PVC), and the mixed organo-halide halo-ethane is an inhalation anesthetic. In addition, chloroform ($CHCl_3$) is a useful solvent in the laboratory and was one of the earlier anesthetic drugs used in surgery. Chlorodifluoromethane ($CH_2Cl_2F_2$) was used as a refrigerant and in aerosol sprays until the late twentieth century, but its use was discontinued after it was found to have a detrimental effect on the atmospheric ozone layer. However, bromoethane (C_2H_5Br) is a simple alkyl halide often used in organic synthesis.

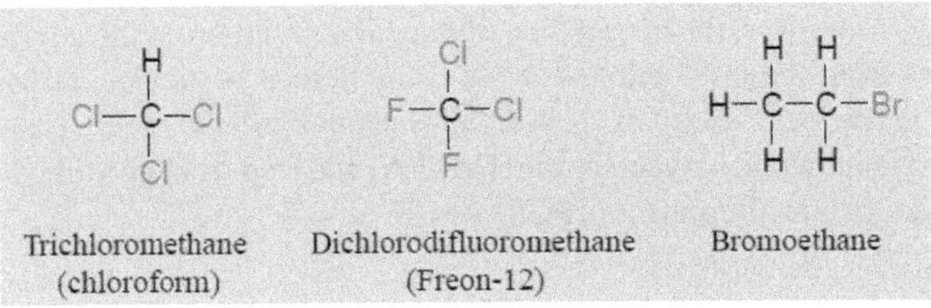

Trichloromethane (chloroform)   Dichlorodifluoromethane (Freon-12)   Bromoethane

Alkyl halide derivatives do occur in biomolecules (naturally occurring organic compounds), but the origin of such compounds is subject to debate—natural vs anthropogenic. For example, a variety of organohalides have been discovered in marine organisms, and several simple halide compounds have important commercial applications. The compound epibatidine is a naturally occurring organic chloride:

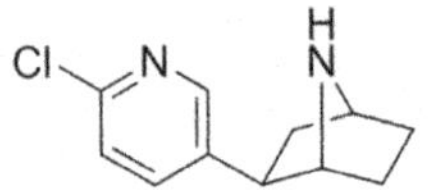

Epibatidine is isolated from the glands on the back of an Ecuadorian poison frog, and the compound has been found to be an especially potent painkiller but only in the correct medically supervised dosage, otherwise serious incapacitation or death may result.

3.7 Ethers and Epoxides

An organic molecule in which an oxygen atom is bonded to two carbon atoms through two sigma bonds is known as an *ether*, and ether compounds occur widely in nature. Although ethers contain two polar carbon-oxygen bonds, they are much less reactive than alcohols or phenols.

Diethyl ether was once widely used as an anesthetic and aromatic ether—2-ethoxynaphthalene (commercially known as Nerolin II) is used in perfumes to impart the scent of orange blossoms. Cyclic ethers, such as tetrahydrofuran, are commonly used as organic solvents.

$CH_3CH_2OCH_2CH_3$

Diethyl ether

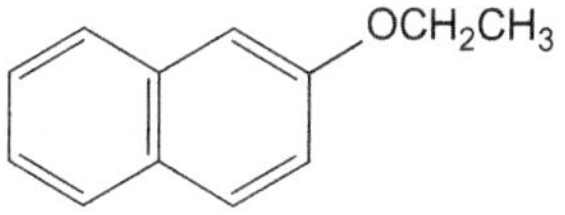

OCH_2CH_3

Nerolin II

O

Tetrahydrofuran

Epoxides are cyclic ethers that contain a three-membered ring of which the simplest derivative is oxirane (ethylene oxide). An epoxide is one of the functional groups in the insect hormone known as juvenile hormone.

3.8 Thiols

A thiol (RSH) is structurally similar to an alcohol (ROH) but contains a sulfur atom in place of the oxygen atom normally found in an alcohol. The outstanding feature of thiol derivatives is the foul smell. The simplest thiol is hydrogen sulfide (H_2S, HSH) the sulfur analog of water (H_2O). It can be detected by the human nose at a concentration of a few parts per billion and is readily identifiable as having the odor of rotten eggs. Ethane thiol (CH_3CH_2SH) is added in trace amounts to natural gas to give it a detectable odor, and the North American skunk deters predators by releasing a liquid spray containing 3-methyl-1-butanethiol (($CH_3)_2CHCH_2CH_2SH$; also called *iso*-amyl thiol) (Wood, 2000):

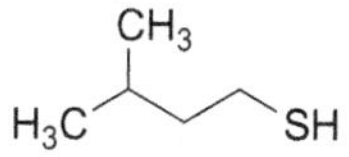

CH_3

H_3C SH

Iso-amyl thiol

When present as a substituent on another structural unit, the thiol (—SH) group is commonly termed *mercapto*, as in 2-mercaptoethanol ($HSCH_2CH_2OH$).

3.9 Amines

Amines are functional group compounds that contain at least one nitrogen atom bonded to hydrogen atoms or to alkyl or aryl groups. If the substituents (other than hydrogen atoms) are alkyl groups, the resulting compounds are termed *alkyl amines*. If one or more substituents are an aryl group, the compounds are termed *aryl amines*. Amines are commonly categorized as primary (—NH$_2$), secondary (>NH), or tertiary (≡N), depending on whether the nitrogen atom is bonded to one, two, or three alkyl groups or three aryl groups, respectively. The nitrogen atom is bonded to the hydrogen atom(s) and to the alkyl groups (or aryl groups) by sigma bonds (σ-bonds), but the nitrogen atom also bears a nonbonded electron pair. The three σ bonds and nonbonded electron pair are oriented around the nitrogen atom in a distorted form of tetrahedral geometry. In some compounds, the nonbonded electron pair on the nitrogen atom is replaced by a fourth σ-bond to a hydrogen atom or to an alkyl or aryl group. The resulting compound, a quaternary ammonium salt, has a positive charge on the nitrogen atom and a tetrahedral arrangement of groups around the nitrogen atom:

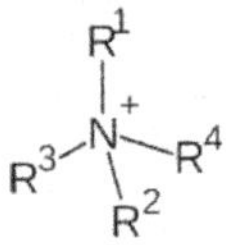

$$R^3 - \overset{\overset{\displaystyle R^1}{|}}{\underset{\underset{\displaystyle R^2}{|}}{\overset{+}{N}}} - R^4$$

A quaternary ammonium ion; the R designations represent alkyl groups or aryl groups.

Amines are particularly valuable because of their ability to act as bases, a property that is a consequence of the ability of amines to accept hydrogen atoms from acidic molecules. One of the most important properties of amines is that they are basic and react with acids to form cations (positively changed organic species).

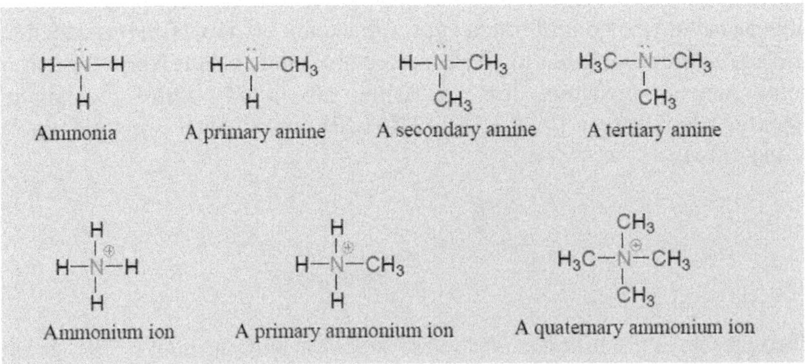

H—N̈—H | H	H—N̈—CH$_3$ | H	H—N̈—CH$_3$ | CH$_3$	H$_3$C—N̈—CH$_3$ | CH$_3$
Ammonia	A primary amine	A secondary amine	A tertiary amine

H |⊕ H—N—H | H	H |⊕ H—N—CH$_3$ | H	CH$_3$ |⊕ H$_3$C—N—CH$_3$ | CH$_3$
Ammonium ion	A primary ammonium ion	A quaternary ammonium ion

3.10   Aldehydes and Ketones

When an oxygen atom forms a double bond to a carbon atom, a carbonyl functional group ($>C=O$) is obtained. The carbon atom of a carbonyl group is bonded to two other atoms in addition to the oxygen atom. A wide range of functional groups are produced by the presence of different atomic groupings on the carbon of the carbonyl group. Two of the most important are aldehyde derivatives and ketone derivatives. In a ketone, both atoms bonded to the carbonyl carbon are other carbon atoms, and, in an aldehyde, at least one atom on the carbonyl carbon is a hydrogen. Similar to the double bond of alkenes, the carbon-oxygen double bond is made up of an σ bond in which the electron pair lies between the bonded atoms, and a π bond, whose electron pair occupies space on both sides of the σ bond.

One prominent example of a ketone is acetone, used in nail polish remover. Acetone is a manufactured chemical that is also found naturally in the environment. It is a colorless liquid with a distinct smell and taste. It evaporates easily, is flammable, and dissolves in water. It is also called dimethyl ketone, 2-propanone, and beta-ketopropane. Acetone is used to make plastic, fibers, drugs, and other chemicals. It is also used to dissolve other substances. It occurs naturally in plants, trees, volcanic gases, forest fires, and as a product of the breakdown of body fat. It is present in vehicle exhaust, tobacco smoke, and landfill sites. Industrial processes contribute more acetone to the environment than natural processes.

Aldehydes often appear in nature—for instance, as vanillin, which gives vanilla beans their pleasing aroma. The ketones carvone and camphor impart the characteristic flavors of spearmint leaves and caraway seeds.

The carbonyl group has a wide variety of possible reaction paths—because of the π bond, the carbonyl group undergoes addition reactions similar to those that occur with alkenes but with a few important differences. Whereas carbon-carbon double bonds are nonpolar, carbon-oxygen double bonds are polar. Species that add to a carbonyl group to form new σ bonds react in such a way that electrophilic (electron-seeking) groups attack the oxygen atom and nucleophilic groups (those seeking positively charged centers) attack the carbon atom. Furthermore, addition to a carbonyl group results in the breaking of a strong π bond. The energy relationships of carbonyl addition reactions are consequently very different from those of alkene addition reactions. Other reaction possibilities of carbonyl compounds depend on the nature of the atomic groupings, termed *substituents*, attached to the carbonyl carbon. When both substituents are unreactive alkane fragments, as in ketones, there are few reactions other than carbonyl additions. When one of the substituents is not an alkane fragment, different possibilities emerge. In aldehydes, the carbonyl carbon is bonded to a hydrogen atom, and reactions that involve this hydrogen atom distinguish the reactions of aldehydes from those of ketones.

Many aldehydes and ketones have pleasant, fruity aromas, and these compounds are frequently responsible for the flavor and smell of fruits and vegetables. A 40% (v/v) solution of formaldehyde (HCH=O) in water is known as *formalin*, a liquid used for preserving biological specimens. Benzaldehyde ($C_6H_5CH=O$) is an aromatic aldehyde and imparts much of the aroma to cherries and almonds. Butanedione, a ketone with two carbonyl groups, is partially responsible for the odor of cheeses. Civetone, a large cyclic ketone, is secreted by the civet cat and is a key component of many expensive perfumes. Also, but perhaps more difficult to accept and less palatable, *kopi luwak*, which is also known as *caphe cut chon* (fox-dung coffee) in Vietnam and *kape alamid* in the Philippines, is coffee that is prepared using coffee beans that have been eaten, partially digested by the Asian palm civet, and then harvested from its fecal matter.

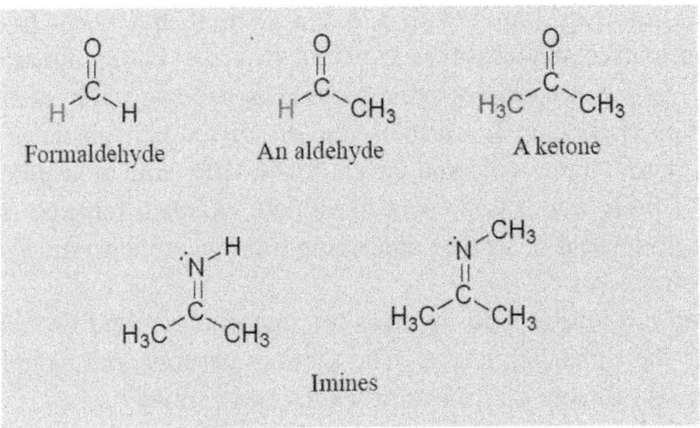

A group with a carbon-nitrogen double bond is an imine.

3.11 Carboxylic Acids

The structural unit containing an alkyl group bonded to a carbonyl group is known as an acyl group. A family of functional groups, known as carboxylic acid derivatives, contains the acyl group bonded to different substituents:

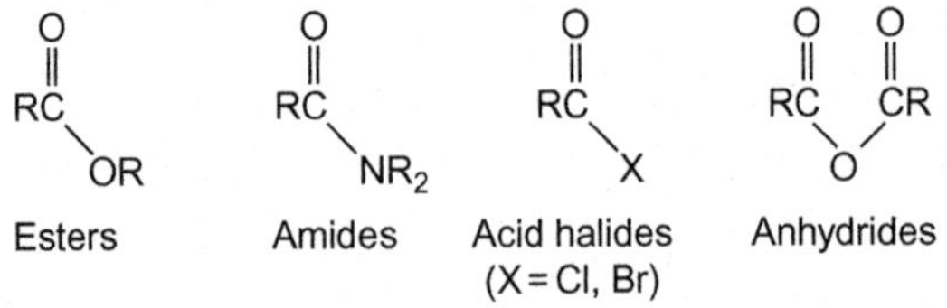

Examples of acyl groups.

The conjunction of a carbonyl (acyl) group and a hydroxyl group forms a functional group known as a carboxylic acid group and forms the functional group known as carboxylic acids. Carboxylic acids all have in common what is known as a carboxyl group, designated by the symbol —COOH. This consists of a carbon atom with a double bond to an oxygen atom, and a single bond to another oxygen atom that is, in turn, wedded to a hydrogen. All carboxylic acids can be generally symbolized by RCOOH, with R as the standard designation of any hydrocarbon. The hydrogen of a carboxyl acid group can be removed (to form a negatively charged carboxylate ion, $-CO_2^-$), and, thus, molecules containing the carboxyl acid group have acidic properties.

Esters have an alkoxy (—OR) fragment attached to the acyl group; amides have attached amino groups ($-NR_2$); acyl halides (RCOCl) have an attached halogen, typically a chlorine atom; and anhydrides have an attached carboxyl group. Each type of acid derivative has a set of characteristic reactions that qualifies it as a unique functional group, but all acid derivatives can be readily converted to a carboxylic acid under appropriate reaction conditions. Many simple esters are responsible for the pleasant odors of fruits and flowers. Methyl butanoate, for example, is present in pineapples. Urea, the major organic constituent of urine and a widely used fertilizer, is a double amide of carbonic acid (carbonic acid: H_2CO_3 or HO—CO—OH; urea: H_2NCONH_2). Acyl chlorides and anhydrides are the most reactive carboxylic acid derivatives and are useful chemical reagents although they are not important functional groups in natural substances.

As examples of carboxylic acids, vinegar is a 5% (v/v) solution of acetic acid (CH_3CO_2H) in water, and its sharp acidic taste is due to the carboxylic acid (acetic acid) present. Lactic acid provides much of the sour taste of pickles and sauerkraut and is produced by contracting muscles. Lactic acid is also generated by the human body when a person overexerts, the muscles generate lactic acid, resulting in a feeling of fatigue until the body converts the acid to water and carbon dioxide. Another example of a carboxylic acid is butyric acid, responsible in part for the smells of rancid butter and human sweat. Citric acid is a major flavor component of citrus fruits, such as lemons, grapefruits, and oranges.

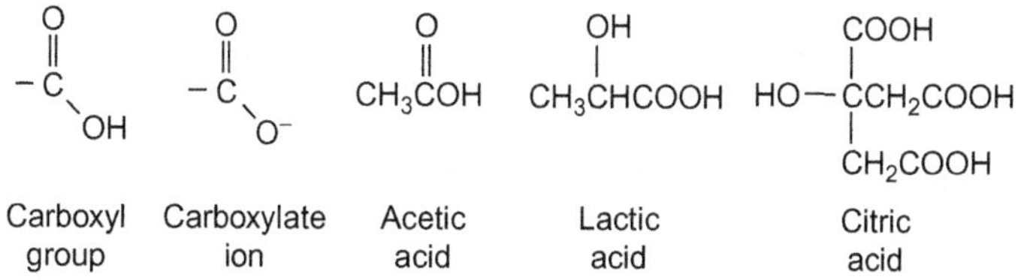

| Carboxyl group | Carboxylate ion | Acetic acid | Lactic acid | Citric acid |

Ibuprofen, an effective analgesic and antiinflammatory agent, contains a carboxyl group.

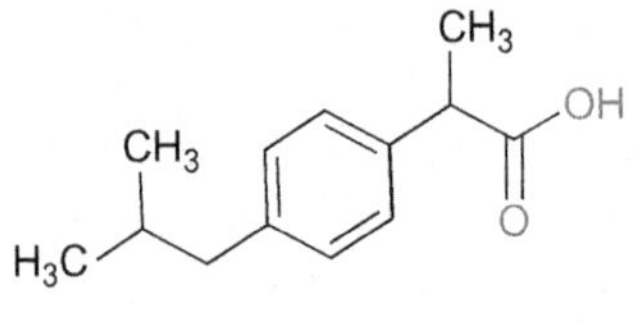

Ibuprofen [2-(4-Isobutylphenyl)propanoic acid]

When a carboxylic acid reacts with an alcohol, it forms an ester:

$$R^1CO_2H + R^2OH \rightarrow R^1CO_2R^2 + H_2O$$

An ester has a structure similar to that described for a carboxylic acid, but with prominent differences. In addition to the bonds (one double, one single) with the oxygen atoms, the carbon atom is also attached to a hydrocarbon, which originates from the carboxylic acid. Furthermore, the single-bonded oxygen atom is attached not to a hydrogen, but to a second hydrocarbon moiety from the alcohol. One well-known (and well-used) ester is acetylsalicylic acid—better known as Aspirin:

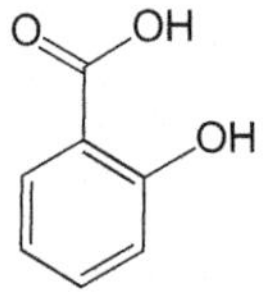

Salicylic acid

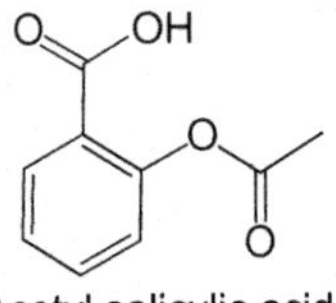

Acetyl salicylic acid (Aspirin).

Esters, which are a key factor in the aroma of various types of fruit, are often noted for their pleasant smell.

3.12 Polyfunctional Compounds

Although each of the functional groups introduced above has a characteristic set of preferred reactions, it is not always possible to predict the properties of organic compounds that contain several different functional groups. For example, the functional groups may interact within the host molecule (intramolecular reaction) and confer unique reactivity as well as physical and chemical properties to the host compound. On the other hand, the functional groups may interact with the functional group(s) of another compound (intermolecular reaction), thereby conferring unique reactivity as well as physical and chemical properties to the nonhost organic compound. In the polyfunctional compounds, one functional group may dictate the reactivity and properties, depending upon the position of the group in the molecule and the three-dimensional structure of the molecule.

3.13   Petroleum Products and Petrochemicals

Petroleum and the equivalent term *crude oil* cover a wide assortment of materials consisting of mixtures of hydrocarbons and other compounds containing variable amounts of sulfur, nitrogen, and oxygen, which may vary widely in volatility, specific gravity, and viscosity. Metal-containing constituents, notably those compounds that contain vanadium and nickel, usually occur in the more viscous crude oils in amounts up to several thousand parts per million and can have serious consequences during processing of these feedstocks. Because petroleum is a mixture of widely varying constituents and proportions, its physical properties also vary widely and the color from colorless to black. Chemically, petroleum is a mixture of gaseous, liquid, and solid hydrocarbon compounds that occur in sedimentary rock deposits throughout the world and also contains small quantities of nitrogen-, oxygen-, and sulfur-containing compounds as well as trace amounts of metallic constituents (Speight, 2014).

Crude petroleum can be separated into a variety of different generic fractions by distillation—the terminology of these fractions has been bound by utility and often bears little relationship to composition. Furthermore, there is a wide variation in the properties of crude petroleum because the proportions in which the different constituents occur vary with origin. Thus, some crude oils have higher proportions of the lower boiling components, and others (such as heavy oil and bitumen) have higher proportions of higher boiling components (asphaltic components and residuum) (Speight, 2014). These fractions are then refined further to produce a variety of petroleum products.

Petroleum products, in contrast to *petrochemicals*, are those bulk fractions that are derived from petroleum and have commercial value as a bulk product. In the strictest sense, petrochemicals are also petroleum products, but they are individual chemicals that are used as the basic building blocks of the chemical industry. Furthermore, *petroleum products* (bulk products, bulk fractions) are among the products derived from the fractional distillation of petroleum are the following, listed from the lowest boiling range fraction (that is, the first material to be separated) to the highest boiling range fraction: (1) natural gas; (2) naphtha, which is used as a precursor to gasoline and as a solvent; (3) kerosene, which is used as a precursor to diesel fuel as well as a fuel for heating; (4) atmospheric gas oil and vacuum gas oil, which are used for manufacture of fuel oil, lubricating oil, petroleum jelly; paraffin wax, and residuum, which can be used for asphalt manufacture or as feedstock for a coking unit.

By-products from many of the refining units include alkanes gases and olefin gases which are sent to the petrochemical section of the refinery as the basis for production of a host of other organic chemicals, including various drugs, plastics, paints, adhesives, fibers, detergents, synthetic rubber, and agricultural chemicals.

On the other hand, for the purposes of this text, a *petrochemical* is any organic chemical (as distinct from fuels and petroleum products) manufactured from petroleum (and natural gas) and used for a variety of commercial purposes. The definition, however, has been broadened to include the whole range of aliphatic, aromatic, and naphthenic organic chemicals, as well as carbon black and such inorganic materials as sulfur and ammonia. Petroleum and natural gas are made up of hydrocarbon molecules, which are comprised of one or more carbon atoms, to which hydrogen atoms are attached. Currently, through a variety of intermediates petroleum and natural gas are the main sources of the raw materials because they are the least expensive, most readily available, and can be processed most easily into the primary petrochemicals. An aromatic petrochemical is also an organic chemical compound but one that contains, or is derived from, the basic benzene ring system.

Primary petrochemicals include: olefins (ethylene, propylene, and butadiene), aromatics (benzene, toluene, and the isomers of xylene),;and methanol that are produced from a variety of feedstocks (Table 2.15). Thus, petrochemical feedstocks can be classified into three general groups: olefins, aromatics, and methanol; a fourth group includes inorganic compounds and synthesis gas (mixtures of carbon monoxide and hydrogen). In many instances, a specific chemical included among the petrochemicals may also be obtained from other

TABLE 2.15 Examples of the Production of Petrochemicals from Various Feedstocks

Feedstocks	Primary Petrochemicals	Derived Petrochemicals
Natural gas	Methane, CH_4	Methanol, CH_3OH
	Ethane, CH_3CH_3	Ethylene, $CH_2{=}CH_2$
	Propane, $CH_3CH_2CH_3$	Propylene, $CH_2{=}CH_2$
	Butane, $CH_3CH_2CH_3$	1-Butylene, $CH_3CH_2CH_2{=}CH_2$
		2-Butylene, $CH_3CH{=}CHCH_3$
		Butadiene, $CH_2{=}CHCH{=}CH_2$
Crude oil	Naphtha	Benzene, C_6H_6
		Toluene, $C_6H_5CH_3$
		Xylene isomers, $CH_3C_6H_4CH_3$

sources, such as coal, coke, or vegetable products. For example, materials such as benzene and naphthalene can be made from either petroleum or coal, while ethyl alcohol may be manufactured as a petrochemical or from feedstocks of vegetable origin (Speight, 2011). Petrochemical intermediates are generally produced by chemical conversion of primary petrochemicals to form more complicated derivative products.

Another group of refining operations that contributes to gas production is that of the *catalytic cracking processes* (Speight, 2014, 2016). Both catalytic and thermal cracking processes result in the formation of unsaturated hydrocarbons, particularly ethylene ($CH_2{=}CH_2$), but also propylene (propene, $CH_3CH{=}CH_2$), *iso*-butylene [*iso*-butene, $(CH_3)_2C{=}CH_2$], and the *n*-butenes ($CH_3CH_2CH{=}CH_2$, and $CH_3CH{=}CHCH_3$) in addition to hydrogen (H_2), methane (CH_4) and smaller quantities of ethane (CH_3CH_3), propane ($CH_3CH_2CH_3$), and butanes [$CH_3CH_2CH_2CH_3$, $(CH_3)_3CH$]. Diolefins such as butadiene ($CH_2{=}CHCH{=}CH_2$) are also present. A further source of refinery gas is *hydrocracking*, a catalytic high-pressure pyrolysis process in the presence of fresh and recycled hydrogen. The feedstock is again heavy gas oil or residual fuel oil, and the process is mainly directed at the production of additional middle distillates and gasoline. Since hydrogen is to be recycled, the gases produced in this process again have to be separated into lighter and heavier streams; any surplus recycle gas and the liquefied petroleum gas from the hydrocracking process are both saturated.

Petrochemical derivative products can be made in a variety of ways: (1) directly from primary petrochemicals; (2) from intermediate products which still contain only carbon and hydrogen; and (3) from intermediates which incorporate chlorine, nitrogen, or oxygen in the finished derivative. In some cases, they are finished products; in others, more steps are needed to arrive at the desired composition (Speight, 2014). Of all the processes used, one of the most important is polymerization. It is used in the production of plastics, fibers, and synthetic rubber, the main finished petrochemical derivatives. Some typical petrochemical intermediates are: vinyl acetate for paint, paper, and textile coatings, vinyl chloride for PVC, resin manufacture, ethylene glycol for polyester textile fibers, styrene which is important in rubber and plastic manufacturing. The end products number in the thousands, some going on as inputs into the chemical industry for further processing. The more common products made from petrochemicals include adhesives, plastics, soaps, detergents, solvents, paints, drugs, fertilizer, pesticides, insecticides, explosives, synthetic fibers, synthetic rubber, and flooring and insulating materials.

3.14 Polymers

Polymers are long-chain molecules produced from monomers (single molecules, such as ethylene, $CH_2{=}CH_2$) by a variety of processes. The structure of even the simplest polymer, polyethylene—represented simply as

[H—(CH$_2$CH$_2$)—H]—is in reality a complicated structure with properties that vary depending upon the stature and molecular size. For example, polyethylene is the plastic used in garbage bags, electrical insulation, bottles, and a host of other applications. A variation on polyethylene is Teflon [poly(tetrafluoroethylene), represented simply as H—(CF$_2$CF$_2$)—H], used not only in nonstick cookware but also in a number of other devices, such as bearings for low-temperature use. Polymers of various kinds are found in siding for houses, tire tread, toys, carpets and fabrics, and a variety of other products.

4 BONDING

A chemical bond is a lasting attraction between atoms and which contributes to the formation (in the current context) of organic chemical compounds. The bond may result by the sharing of electrons as in the formation of covalent bonds, such as the typical carbon-carbon bond or the typical carbon-hydrogen bond or from the electrostatic force of attraction between atoms with opposite charges. The strength of a chemical bond varies considerably—there are the so-called relatively *strong bonds* such as covalent or ionic bonds and the relatively *weak bonds* such as (1) hydrogen bonds, (2) bonds formed by, and (3) van der Waals bonds (Table 2.16).

Intermolecular bonding takes place between different molecules. This can take the form of ionic bonding, hydrogen bonding, dipole-dipole interactions, and van der Waals interactions. The type of bonding involved in organic compounds depends on the functional groups that are present in the molecule. On the other hand, *intramolecular bonding* occurs within the same molecule and also depends on the functional groups that are present in the molecule.

Ionic bonds are possible between ionized functional groups such as carboxylic acids and amines. Ionic bonding takes place between molecules having opposite charges and involves an electrostatic interaction between the two opposite charges. The functional groups which most easily ionize are amines and carboxylic acids, such as the reaction of ammonia (NH$_3$) with a carboxylic acid (carboxylate ion). Some important naturally occurring molecules contain both groups—the amino acids.

Intermolecular hydrogen bonding is possible functional group such as alcohol derivatives (ROH), carboxylic acid derivatives (RCO$_2$H), amide derivatives (RCONH$_2$), amine derivatives (RNH$_2$), and phenol derivatives (C$_6$H$_5$OH) which contain a hydrogen atom bonded to nitrogen or to oxygen. Hydrogen bonding involves the interaction of the partially positive hydrogen (δ^+) on one molecule which carries a fractional positive charge and the partially negative heteroatom which carries a fractional negative charge (δ^-) on another molecule.

Dipole-dipole interactions are possible between molecules having polarizable bonds, in particular the carbonyl group (C=O). Such bonds have a dipole

TABLE 2.16 Different Types of Bond Arrangements

Covalent bond	A bond in which one or more electrons (often a pair of electrons) are drawn into the space between the two atomic nuclei. These bonds exist between two particular identifiable atoms and have a direction in space, allowing them to be shown as single connecting lines between atoms in drawings, or modeled as sticks between spheres in models
Ionic bond	Occurs between ionized functional groups such as carboxylic acids and amines
Hydrogen bond	Occurs between alcohol derivatives, carboxylic acid derivatives, amide derivatives, amine derivatives, and phenol derivatives. Hydrogen bonding involves the interaction of the partially positive hydrogen on one molecule and the partially negative heteroatom on another molecule. Hydrogen bonding is also possible with elements other than nitrogen or oxygen and can occur intermolecularly or intramolecularly
Dipole-dipole interaction	Possible between molecules having polarizable bonds, in particular the carbonyl group ($C=O$) which have a dipole moment and molecules can align themselves such that their dipole moments are parallel and in opposite directions. Ketones and aldehydes are capable of interacting through dipole-dipole interactions
Van der Waals interaction	Weak intermolecular bonds between regions of different molecules bearing transient positive and negative charges which are caused by the movement of electrons. Alkanes, alkenes, alkynes, and aromatic rings interact through van der Waals interactions
Intermolecular bond	Occurs between different molecules and can take the form of ionic bonding, hydrogen bonding, dipole-dipole interactions, and van der Waals interactions
Intramolecular bond	Occurs within a molecule and can take the form of ionic bonding, hydrogen bonding, dipole-dipole interactions, and van der Waals interactions

moment, and molecules can align themselves such that their dipole moments are parallel and in opposite directions. Ketones and aldehydes are capable of interacting through dipole-dipole interactions. Dipole-dipole interactions are possible between polarized bonds other than N—H interactions or O—H bonds, and the most likely functional groups which can interact in this way are those containing a carbonyl group ($>C=O$). The electrons in the carbonyl bond are polarized toward the more electronegative oxygen a slight negative charge and the carbon gains a slight positive charge such that the oxygen gains a slight negative charge and the carbon gains a slight positive charge. Molecules containing dipole moments can align themselves with each other such that the dipole moments are pointing in opposite directions.

Van der Waals interactions are weak intermolecular bonds between regions of different molecules bearing transient positive and negative charges. These transient charges are caused by the random fluctuation of electrons. Alkane hydrocarbons, alkene hydrocarbons, alkyne hydrocarbons, and aromatic rings interact through van der Waals interactions. Van der Waals interactions are the weakest of the intermolecular bonding forces interactions and involve the transient existence of partial charges in a molecule. At any moment of time, there may be a slight excess of electrons (δ^-) in one part of the molecule and a slight deficit of electrons (δ^+) in another part of the molecule, and although these charges are very weak and fluctuate within the molecule, they are sufficiently strong to allow a weak interaction between molecules, where regions of opposite charge in different molecules attract each other. Alkane molecules can interact in this way, and the strength of the interaction increases with the size of the alkane molecule. Van der Waals interactions are also important for alkene derivatives, alkyne derivatives, and aromatic rings.

The presence of functional groups affects the chemical and physical properties of molecules, especially properties such as (1) melting point, (2) boiling point, (3) polarity, (4) dipole moment, and (5) solubility (Patrick, 2004). For example, a molecule with a strongly polar functional group can be predicted with a degree of certainty to have a higher melting point and a higher boiling point than a molecule with a nonpolar functional group, and, in addition, a polar molecule would preferentially dissolve in a polar solvent rather than in a nonpolar solvent. Thus, the types of reactions in which organic compounds participate are determined by the types of functional groups that are present in the molecule. Functional groups do undergo characteristic reactions, but the rates of these reactions are affected by steric factors.

In most cases, the presence of a polar functional group will determine the physical properties of the molecule. However, this is not always true. For example, if a molecule has a polar group such as a carboxylic acid, but has a long hydrophobic alkane chain, the effects of the alkyl chain can overcome the effects of the polar function rendering the hydrophobic rather than hydrophilic.

Finally, as a result of the various types of binding arrangements that can occur because of the presence of functional groups, properties such as melting point, boiling point, polarity, and solubility are affected. For example, a molecule with a strongly polar functional group will typically have higher melting points and boiling point than a similar molecule with a nonpolar or less polar functional group—the solubility of the two molecules in different solvents will also be affected by the relative polarity of the functional group. In addition, the types of chemical reactions which compounds undergo are also determined by the types of functional group that is present. Functional groups undergo characteristic reactions, but the rates of these reactions are affected by the stereochemical structure.

5   REACTIONS

The vast majority of the reactions of organic compounds take place at site of the functional groups and are characteristic of that functional group. However, the reactivity of the functional group is affected by stereoelectronic effects. For example, a functional group may be surrounded by bulky groups which hinder the approach of a reagent and slow down the rate of reaction—this is referred to as steric shielding. Electronic effects can also influence the rate of a reaction—an example is the presence of neighboring groups (in polyfunctional molecules) that can influence the reactivity of a functional group if the neighboring group is an electron-withdrawing group or an electron-donating group, both of which influence the electronic density within the functional (reacting) group. Conjugation of any electronic effects (due to the presence of any neighboring groups) and aromaticity also have an important effect on the reactivity of functional groups. For example, an aromatic ketone reacts at a different rate from an aliphatic ketone because the aromatic ring is in conjugation with the carbonyl group (a conjugating effect) which serves to increase the stability of the overall system, thereby reducing the reactivity of the molecule.

5.1   Nucleophilic Reactions

Most reactions of organic compounds involve the reaction between a molecule which is relatively rich in electrons and a molecule which is relatively deficient in electrons. The reaction involves the formation of a new bond where the electrons are provided by the electron-rich molecule. Electron-rich molecules are called nucleophiles (meaning nucleus loving), and the easiest nucleophilic species to identify is the negatively charged ions which has a lone pair of electrons (an example is the hydroxide ion, OH^-), but neutral molecules can also act as nucleophiles if they contain electron-rich functional groups (e.g., an amine).

Nucleophilic species also have a specific atom or region of the molecule which is electron-rich—the nucleophilic center of an ion is the atom that carries a lone pair of electrons and, hence, a negative charge. On the other hand, the nucleophilic center of a neutral molecule is usually an atom with a lone pair of electrons (such as a nitrogen atom or an oxygen atom), or a multiple bond (such as an alkene molecule, an alkyne molecule, or an aromatic ring).

5.2   Electrophilic Reactions

Electron-deficient molecules (electrophiles, electron-loving species) react with nucleophilic species. Positively charged ions can easily be identified as electrophiles (e.g., a carbocation—and ion with a positively charged carbon atom and until the early 1970s, all carbocations were called carbonium ions), but neutral molecules can also act as electrophiles if they contain certain types of functional groups (such as carbonyl groups or alkyl halides). Electrophiles have a specific atom or region of the molecule which is electron center deficient (the *electrophilic center*). In a positively charged ion, the electrophilic center is the atom bearing the positive charge (i.e., the carbon atom of a carbocation). In a neutral molecule, the electrophilic center is an electron-deficient atom within a functional group (such as a carbon atom or a hydrogen atom linked to an electronegative atom such as oxygen atom or a nitrogen atom).

REFERENCES

Bjorseth, A., 1983. Handbook of Polycyclic Aromatic Hydrocarbons. Marcel Dekker, New York.

Durand, B., 1980. Kerogen: Insoluble Organic Matter From Sedimentary Rocks. Editions Technip, Paris.

Fetzer, J.C., 2000. The chemistry and analysis of the large polycyclic aromatic hydrocarbons. Polycycl. Aromat. Compd. 27 (2), 143.

Jaromir, M., Ożadowicz, R., Duda, W., 2005. Analysis of chlorophenols, chlorocatechols, chlorinated methoxyphenols and monoterpenes in communal sewage of Łódź and in the Ner River in 1999-2000. Water Air Soil Pollut. 16, 205–222.

Laine, M., Jorgensen, K., 1996. Straw compost and bioremediated soil as inocula for the bioremediation of chlorophenol-contaminated soil. Appl. Environ. Microbiol. 54, 1507.

Macaud, M., Milenkovic, A., Schulz, E., 2000. Hydrodesulfurization of alkyldibenzothiophenes. Evidence of highly unreactive aromatic sulfur compounds. J. Catal. 193, 255–263.

Michałowicz, J., Duda, W., 2007. Phenols—sources and toxicity. Pol. J. Environ. Stud. 16 (3), 347–362.

Neilson, A.H., 1998. PAHs and Related Compounds. In: Hutzinger, O. (Ed.), Handbook of Environmental Chemistry. In: Vol. 3. Springer-Verlag, Berlin.

Olah, G.A., 2005. Beyond oil and gas: the methanol economy. Angew. Chem. Int. Ed., 2636–2639.

Patrick, G.L., 2004. Instant Notes: Organic Chemistry: Instant Notes. Garland Science/BIOS Scientific Publishers, Abingdon, Oxfordshire.

Sakarnen, K.V., Ludwig, C.H., 1971. Lignins: Occurrence, Formation, Structure and Reactions. John Wiley & Sons, New York.

Speight, J.G. (Ed.), 2011. The Biofuels Handbook. Royal Society of Chemistry, London.

Speight, J.G., 2013. The Chemistry and Technology of Coal, third ed. CRC Press, Taylor and Francis Group, Boca Raton, FL.

Speight, J.G., 2014. The Chemistry and Technology of Petroleum, fifth ed. CRC Press, Taylor and Francis Group, Boca Raton, FL.

Speight, J.G., 2016. Handbook of Refining Processes. CRC Press, Taylor and Francis Group, Boca Raton, FL.

Speight, J.G., Arjoon, K.K., 2012. Bioremediation of Petroleum and Petroleum Products. Scrivener, Salem, MA.

Swarts, M., Verhagen, F., Field, J., Wijnberg, J., 1998. Trichlorinated phenols from *Hypholoma elongatum*. Phytochemistry 49, 203.

Weiss, V., Edwards, J.M., 1980. The Biosynthesis of Aromatic Compounds. John Wiley & Sons, New York.

Wood, W.F., 2000. The history of skunk defensive secretion research. Chem. Educ. 5 (3), S1430–S4171.

Xu, Y., Shang, H., Liu, C., 2004. The study of hydrodesulfurization of 4,6-dimethyldibenzothiophene over phosphorus-added NiMo/γ-Al2O3. Prepr. Pap.—Am. Chem. Soc., Div. Fuel Chem. 49 (2), 987.

Chapter 3

Industrial Organic Chemistry

1 INTRODUCTION

The chemical process industries play an important role in the development of a country by providing a wide variety of products and use raw material derived from petroleum and natural gas, salt, oil and fats, biomass and energy from coal, natural gas and a small percentage from renewable energy resources. Although manufacture of organic chemicals initially started with coal and alcohol from fermentation industry, later due to availability of petroleum and natural gas dominated the scene and now more than 90% of organic chemicals are produced from petroleum and natural gas routes. However, rising cost of petroleum and natural gas and continuous decrease in the reserves has spurred the chemical industry for alternative feedstocks like coal, biomass, coal bed methane, shale gas, and sand oil as an alternate source of fuel and chemical feedstock.

However, scientific and engineering interests have not always been aligned and, therefore, maintaining a partnership between chemists and engineers with the focus on industrial processes and the production of organic chemicals is an opportunity for both disciplines that and can create a true symbiosis. This type of collaboration is engendered by the desire of industrial operations to access the chemical and engineering expertise that is required for the development of new technologies at the forefront of process innovations. From both, the chemical standpoint and the engineering standpoint, the symbiotic relationship represents the perfect occasion to apply scientific and engineering research concepts to solve problems that benefit a variety of processes.

Furthermore, the umbrella term *the industrial organic chemicals industry* includes thousands of chemicals and hundreds of processes. In general, a set of building blocks (feedstocks) is combined in a series of reaction steps to produce both intermediates and end-products. Moreover, organic chemicals, particularly petrochemicals, play an indispensable role in the modern world. They are essential ingredients in plastics, synthetic fibers, rubber, fertilizers, and chemical intermediates, which are converted into a wide range of industrial products. They are the primary building blocks of important materials supporting the health, food, transportation, and communication industries. Organic substances also have made possible many important specialty items, such as protective clothing and materials used for space exploration. The primary organic chemical building blocks (generated principally from petroleum

Environmental Organic Chemistry for Engineers. http://dx.doi.org/10.1016/B978-0-12-804492-6.00003-4

87

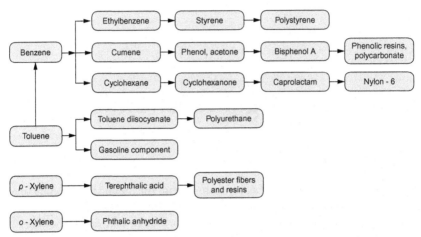

FIG. 3.1 Selected examples (benzene, toluene, and the xylenes) of the organic chemical building blocks.

refining)—using benzene, toluene, and the xylene isomers as examples (Fig. 3.1)—are a key subset of the large volume secondary building blocks and a set of large volume tertiary building blocks that participate in a variety of reaction types and a large variety of processes that are used in manufacture of organic chemicals.

The industrial organic chemical sector produces organic chemicals (chemicals that contain carbon) used as either chemical intermediates or end-products. The industrial organic chemical industry uses feedstocks derived from petroleum and natural gas (about 90%) and from recovered coal tar condensates generated by coke production (about 10%). The chemical industry produces raw materials and intermediates, as well as a wide variety of finished products for industry, business, and individual consumers. Important classes of chemicals produced by organic chemical industry facilities include: (1) noncyclic organic chemicals such as acetic, chloroacetic, adipic, formic, oxalic acids and their metallic salts, chloral, formaldehyde, and methylamine; (2) solvents such as amyl, butyl and ethyl alcohols; methanol; amyl, butyl, and ethyl acetates; ethyl ether, ethylene glycol ether and diethylene glycol ether; acetone, carbon disulfide, and chlorinated solvents such as carbon tetrachloride, tetrachloroethylene, and trichloroethylene; (3) polyhydric alcohols such as ethylene glycol, sorbitol, pentaerythritol, and synthetic glycerin; (4) synthetic perfumes and flavoring materials such as coumarin, methyl salicylate, saccharin, citral, citronellal, synthetic geraniol, ionone, terpineol, and synthetic vanillin; (5) rubber processing chemicals such as accelerators and antioxidants, both cyclic and acyclic; (6) plasticizers, both cyclic and acyclic, such as esters of phosphoric acid, phthalic anhydride, adipic acid, lauric acid, oleic acid, sebacic acid, and stearic acid; (7) synthetic tanning agents such as sulfonic acid condensates; and (8) esters and amines of polyhydric alcohols and fatty and other acids.

Relatively few organic chemical manufacturing facilities are single-product (or single-process) plants and many process units are designed to be flexible (with options) so that production levels of related products (or by-products) can be varied over wide ranges. This flexibility is required to accommodate variations in feedstock purity properties which can change the production rate and processes used, even on a short-term (less than a year) basis. Furthermore, the process by-products are also valuable saleable products and the value of these by-products can change the economics of the process.

The typical chemical synthesis process involves combining multiple feedstocks in a series of unit operations—the petroleum refining industry is the best example of the integration of a series of unit processes to produce the desired products (Speight, 2014a, 2016). This is not always the case in the organic chemicals industry where the first unit operation is typically a chemical reaction to produce the product or an intermediate product that leads to the desired chemical. In addition, there is some differentiation between the chemicals to be produced—commodity chemicals tend to be synthesized in a continuous reactor while specialty chemicals usually are typically produced in a batch reactor. Many, but not all, reactions (1) take place at high temperatures, (2) involve metal catalysts, and (3) include one or two additional reaction components. The yield of the organic chemical will/may determine the quantity, type and by-products, including gaseous emissions. The production of many organic specialty chemicals often requires a series of two or more reaction steps and once the reaction is complete, the desired product must be separated from the by-products by a second unit operation. In the separation stage, a number of separation techniques such as settling, distillation or refrigeration may be used and the final product may require further processing (for example, by spray drying or pelletizing) to produce the saleable item. The separation technology employed depends on many factors including the phases of the substances being separated, the number of components in the mixture, and whether recovery of by-products is important. Numerous techniques such as distillation, extraction, filtration, and settling can be used singly or in combination to accomplish the separation.

Finally, regulatory laws regarding hazardous emissions generated during the production of organic chemicals are important dynamics that shape the industry. To minimize the detrimental effects of chemical industry pollutants, multiple local, state, and federal laws govern producers. For example, the federal Emergency Planning and Community Right-to-Know Act requires many manufacturers to submit details of any emissions data to the United States Environmental Protection Agency (US EPA). Similarly, the Pollution Prevention Act requires those same companies to report their waste management and pollution reduction activities. Other federal regulations impacting producers include the Safe Drinking Water Act, the Clean Air Act and Amendments, and other laws that restrict hazardous wastes. In addition to legal restrictions, both the US EPA and the Chemical Manufacturers Association (CMA) sponsor successful voluntary pollution reduction programs that encourage environmental sensitivity. The US EPA has continued to monitor the industry and in the light of current strong emphasis on chemical safety and pollution controls, it is likely that regulations will continue to be added and modified.

There are also voluntary programs where the member companies work with the public to address such issues as chemical safety. The mechanism for these programs involves a combination of soliciting information from the public about the various concerns which are addressed and the progress is reported back to the public. While increasing federal and state regulations pose an ongoing challenge to chemical industry participants, positive signs have indicated that the organic chemicals industry has been successful in clearing these hurdles. Overall chemical industry has managed to reduce the various emissions of waste that appear on the Toxics Release Inventory.

The previous chapter (Chapter 2) has introduced the reader to the varied, but fundamental, aspects of organic chemistry. However, organic chemistry as practiced on the industrial stage is not so simple and/or straightforward. Industrial organic chemistry is an extremely comprehensive and practical discipline and, although work there benefits from understanding the basic organic chemical science, there is still the need to gain a valuable insight into chemical technology. Basic organic chemistry does provide but other chemicals used and produced in industrial processes offer a considerable (but valuable) challenge to (1) an understanding of the processes, (2) the process parameters, (3) the properties of the feedstocks, (4) the properties of the products, (5) the properties of the by-products, and (6) the influence of these various chemicals on the environment. These effects are difficult to understand on the basis of laboratory chemical studies alone and it is for this reason that this chapter is included in the book.

Thus, this chapter deals with the foundations of organic chemistry from an industrial perspective and presents the various processes that are used on a regular (almost a day-to-day) basis. This will give the reader the ability to understand the necessary links between laboratory organic chemistry and industrial process chemistry that is a necessary and growing phenomenon within the chemistry community. These chemistry and engineering sectors of industry have long held strong ties since chemistry points the way to synthetic pathways and engineering points that way by which these pathways might be achieved on a commercial scale.

2 PRODUCTION OF ORGANIC CHEMICALS

The typical organic chemical synthesis process involves combining one or more feedstocks in a series of unit process operations. Commodity chemicals tend to be synthesized in a continuous reactor while specialty chemicals usually are produced in batches. Most reactions take place at high temperatures, involve metal catalysts, and include one or two additional reaction components. The yield of the organic chemical will partially determine the kind and quantity of by-products and releases. In fact, many specialty organic chemicals require a series of two or three reaction steps, each involving a different reactor system and each capable of producing by-products. Once the reaction is complete, the desired product must be separated from the by-products by a second unit operation. A number of separation techniques such as settling, distillation or refrigeration may be used. The final product may be further processed, such as by spray drying or pelletizing, to produce the saleable item. Frequently, by-products are also sold and their value can influence the economics of the process.

Feedstock costs are the highest variable cost in production of the organic commodity chemicals. The larger producers integrate feedstocks and derivatives production in order to minimize production costs and price fluctuations. Smaller firms do not possess this integration flexibility, making them more susceptible to variations in feedstock price swings. When feedstock prices rise, manufacturers often lower operating rates or suspend production if price increases are not possible. Some producers have the ability to switch feedstocks in order to obtain better feedstocks which lead to better market prices. Moreover, the type of reaction process used to manufacture chemicals depends on the intended product.

Because of the heightened interest in alternative fuels production that has expanded over the past three decades, the use of alternative feedstocks for production of the high-volume commodity chemicals has been, and continues to be, evaluated. Petroleum is still a major feedstock for the production of organic chemicals. However, the currently high prices for petroleum and natural gas have spurred the US chemical industry to evaluate alternative feedstocks for the production of commodity chemicals. These feedstocks include unconventional processing technologies, such as (1) the increased use of natural gas, (2) coal, which includes: coal gasification and coal liquefaction, (3) heavy oil, (4) bitumen from tar sand formations, (5) liquids from oil shale processing, and (6) novel resources such as the various types of biomass. Thus, as part of this increased interest, new pathways for commodity chemicals manufacture are continually being developed using feedstocks that offer alternatives to petroleum as well as the development of the varying process options from production of the products.

Finally, there is the need to realize that crude oil has become integral to the necessities of the modern world. However, although the era of crude oil may be drawing to a close within the next 50 years or so (Hirsch et al., 2006; Speight, 2011a,b; Speight and Islam, 2016), petrochemicals have remained the mainstay of the chemical and energy industries and will continue to do so during the next five decades. The infrastructure of the chemicals industry remains focused on crude oil and natural gas and any alternative must work within that system and there is a much better chance of making the transition to biological sources if new processes are integrated into existing technology. However, despite the efforts to facilitate biomass conversion, there have been tangible successes but renewable feedstocks are not yet fulfilling their potential to replace crude oil and natural gas. In addition, biomass derivatives rarely have the purity and homogeneity that crude oil and natural gas feedstocks offer for production of organic chemicals and, in addition, purification processes for bioderived streams may be particularly difficult to process. The challenge is to transform low cost bioderived feedstocks into high-value derivatives with commercial applications. Currently, crude oil, natural gas liquids including gas condensate, and natural gas account for the majority of the feedstock materials used by the organic chemicals industry. Natural gas is predominately used to manufacture methanol and ammonia, and the majority of the olefin productions (particularly ethylene) production is based on natural gas liquids. Coal, once the prime feedstock for the chemicals industry became a lesser feedstock for chemical production after World War II when crude oil and natural gas, being more easy to convent to organic chemicals, became the leading feedstocks for the chemical industry.

The price volatility of crude oil and natural gas and the possible dearth of these feedstocks in the distant future (Speight, 2011a,b; Speight and Islam, 2016) have spurred the chemical industry to examine alternative feedstocks for the production of commodity chemicals. Thus, over the last 30 years, alternatives to conventional petroleum and natural gas feedstocks have been developed, but have limited, if any, commercial implementation. Alternative feedstocks under consideration include coal from known processing technologies, such as gasification and liquefaction, novel resources such as biomass, heavy oil, and bitumen from tar sands or oil shale.

Sources of organic compounds, such as ethanol from sugar fermentation and bitumen-derived heavy crude oil are now being primarily exploited for fuels, rather than for chemical feedstocks. In fact, over the last 50 years, there has been much activity in the development of alternative feedstocks, but little activity in bringing technologies to market—the delay in implementation of any such technologies rests on the economics of the process. The economic competitiveness of technologies such as integrated gasification combined cycle (IGCC) and gas-to-liquids (GTL) depend on the current and predicted prices of crude oil and alternative feedstocks, and the costs for transportation and storage.

Most of the work into alternative feedstocks has focused on energy production, either for electricity, liquid fuels from synthesis gas (syngas, a mixture of carbon monoxide and hydrogen), or bioethanol (i.e., ethanol from biomass). Some technologies for chemicals production are mature, such as coal gasification, and are ready for implementation if economically feasible. Other unconventional sources of organic compounds, such as ethanol from sugar fermentation, are now being exploited for fuels, rather than for chemical feedstocks.

This section presents a review of the various types of feedstocks that are used for production of organic chemicals (single chemicals and mixtures of chemicals such as petroleum products) and the means by which these feedstocks are used to produce chemical products and how they fit into the current future production of organic chemicals.

2.1 Chemicals From Petroleum

The organic commodity chemicals are a group of crude oil-derivative chemicals (also known as petrochemicals) used as intermediates to produce other chemicals, which, in turn, are used to manufacture a wide variety of end-use products, including construction materials, apparel, adhesives, plastics, and tires. The majority of the organic commodity chemicals are derived from benzene, a chemical derived from crude oil refining. Examples of specific compounds in this hydrocarbon group include ethylbenzene ($C_6H_5C_2H_5$), styrene ($C_6H_5C{=}H_2$), cumene [$C_6H_5CH(CH_3)_2$], *ortho*-xylene (0-1,2-$CH_3C_6H_4CH_3$), *meta*-xylene (1,3-$CH_3C_6H_4CH_3$), and *para*-xylene (1,4-$CH_3C_6H_4CH_3$) while (C_6H_5OH) and aniline ($C_6H_5NH_2$) are products that contain the oxygen and nitrogen heteroatoms, respectively.

As commodities, the chemicals produced by one manufacturer should be indistinguishable from those same chemicals produced by another manufacturer, given the same levels of purity. This similarity of products allows consumers to purchase similar product from a wide variety of suppliers, making price the dominant economic factor in purchasing decisions. But it should be noted that some manufactures sell chemical products that may not be 100% pure, and the level of purity must be stated on the packaging.

Chemicals from crude oil usually take the form of products that are mixtures or petrochemicals in which the product is typically an identifiable single organic compound. *Petroleum products* in contrast to *petrochemicals*, are those bulk fractions that are derived from petroleum and have commercial value as a bulk product. In the strictest sense, petrochemicals are also petroleum products but they are individual chemicals that are used as the basic building blocks of the chemical industry (Speight, 2011a,b, 2014a).

The constant demand for products, such as liquid fuels, is the main driving force behind the petroleum industry. Other products, such as lubricating oil, wax, and asphalt, have also added to the popularity of petroleum as a national resource. Indeed, fuel products that are derived from petroleum supply more than half of the total supply of energy use on a worldwide basis. Gasoline, kerosene, and diesel oil provide fuel for automobiles, tractors, trucks, aircraft, and ships. Fuel oil is used to heat homes and commercial buildings, as well as to generate electricity. Furthermore, petroleum products are the basic materials used for the manufacture of synthetic fibers for clothing and in plastics, paints, fertilizers, insecticides, soaps, and synthetic rubber. The uses of crude oil as a source of raw material in manufacturing are central to the functioning of modern industry.

Unlike processes, products are more difficult to be placed on an individual evolutionary scale. Processes changed and evolved to accommodate the demand for, say, higher-octane fuels, longer-lasting asphalt, or lower sulfur coke. Another consideration that must be acknowledged is the change in character and composition of the original petroleum feedstock. In the early days of the petroleum industry several products were obtained by distillation and could be used without any further treatment. In the modern refinery, the different character and composition of the petroleum dictates that any liquids obtained by distillation must go through one or more of the several available product improvement processes. Such changes in feedstock character and composition have caused the refining industry to evolve in a direction such that changes in the petroleum can be accommodated.

There is a myriad of products that have evolved through the life of the petroleum industry and the complexities of product composition have matched the evolution of the products. In fact, it is the complexity of product composition that has served the industry well and, at the same time, had an adverse effect on product use. Product complexity has made the industry unique among industries. Product complexity, and the means by which the product is evaluated, has made the industry unique among all industries. But product complexity has also brought to the fore issues such as instability and incompatibility. In order to understand the evolution of the products it is essential to have an understanding of the composition of the various products.

In the simplest sense, naphtha contains varying amounts of paraffins, olefins, naphthene constituents, and aromatics and olefins in different proportions in addition to potential isomers of paraffin that exist in naphtha boiling range. As a result, naphtha is divided predominantly into two main types: (1) aliphatic naphtha and (2) aromatic (naphtha). The two types differ in two ways: first, in the kind of hydrocarbons making up the solvent, and second, in the methods used for their manufacture. Aliphatic solvents are composed of paraffinic hydrocarbons and cycloparaffins (naphthenes), and may be obtained directly from crude petroleum by distillation. The second type of naphtha contains aromatics, usually alkyl-substituted benzene, and is very rarely, if at all, obtained from petroleum as straight-run materials. The products that are higher boiling than naphtha contain higher molecular weight constituents that vary in molecular type.

The high-boiling and more complex lubricating oil is distinguished from other fractions of crude oil by their usually high (>400°C, >750°F) boiling point, as well as their high viscosity. Materials suitable for the production of lubricating oils are comprised principally of hydrocarbons containing from 25 to 35 or even 40 carbon atoms per molecule, whereas residual stocks may contain hydrocarbons with 50 or more (up to 80 or so) carbon atoms per molecule. The composition of lubricating oil may be substantially different from the lubricant fraction from which it was derived, since wax (normal paraffins) is removed by distillation or refining by solvent extraction and adsorption preferentially removes nonhydrocarbon constituents as well as polynuclear aromatic compounds and the multiring cycloparaffins (Speight, 2014a, 2016).

There are general indications that the lubricant fraction contains a greater proportion of normal and branched paraffins than the lower boiling portions of petroleum. For the polycycloparaffin derivatives, a good proportion of the rings appear to be in condensed structures, and both cyclopentyl and cyclohexyl nuclei are present. The methylene groups appear principally in unsubstituted chains at least four carbon atoms in length, but the cycloparaffin rings are highly substituted with relatively short side chains. Mono-, di-, and trinuclear aromatic compounds appear to be the main constituents of the aromatic portion, but material with more aromatic nuclei per molecule may also be present. For the binuclear aromatics, most of the material consists of naphthalene types. For the trinuclear aromatics, the phenanthrene type of structure predominates over the anthracene type. There are also indications that the greater part of the aromatic compounds occur as mixed aromatic-cycloparaffin compounds.

After lubricating oil, there is paraffin wax which is a solid crystalline mixture of straight-chain (normal) hydrocarbons ranging from C_{20} to C_{30} and possibly higher, that is, $CH_3(CH_2)_nCH_3$ where $n \geq 18$. Wax is distinguished by its solid state at ordinary temperatures (25°C, 77°F) and low viscosity when melted. However, in contrast to petroleum wax, petrolatum (*petroleum jelly*), although solid at ordinary temperatures, does in fact contain both solid and liquid hydrocarbons. It is essentially a low-melting, ductile, microcrystalline wax.

Another product, asphalt which is used as a mastic in various applications (such as road construction and repair), is the residue of mixed-base and asphalt-base crude oils. It cannot be distilled even under the highest vacuum, because the temperatures required to do this promote formation of coke. Asphalt have complex chemical and physical compositions that usually vary with the source of the crude oil.

The final product, coke is the residue left by the destructive distillation of petroleum residua. That formed in catalytic cracking operations is usually nonrecoverable, as it is often employed as fuel for the process. The composition of petroleum coke varies with the source of the crude oil, but in general, large amounts of high-molecular-weight complex hydrocarbons (rich in carbon but correspondingly poor in hydrogen) make up a high proportion. The solubility of petroleum *coke* in carbon disulfide has been reported to be as high as 50–80%, but this is in fact a misnomer, since the coke is the insoluble, honeycomb material that is the end product of thermal processes.

2.2 Chemicals From Natural Gas

The principal constituent of natural gas is methane (CH_4). Other constituents are paraffinic hydrocarbons such as ethane (CH_3CH_3), propane ($CH_3CH_2CH_3$), and the butanes [$CH_3CH_2CH_2CH_3$ and/or $(CH_3)_3CH$]. Many natural gases contain nitrogen (N_2) as well as carbon dioxide (CO_2) and hydrogen sulfide (H_2S). Trace quantities of argon, hydrogen, and helium may also be present. Generally, the hydrocarbons having a higher molecular weight than methane, carbon dioxide, and hydrogen sulfide are removed from natural gas prior to its use as a fuel. Gases produced in a refinery contain methane, ethane, ethylene, propylene, hydrogen, carbon monoxide, carbon dioxide, and nitrogen, with low concentrations of water vapor, oxygen, and other gases.

Transportation of natural gas from isolated sources could best be effected by converting the light hydrocarbons to easily transportable liquids, such as methanol, in situ. Natural gas is used to manufacture methanol offshore, but production in the United States has been curtailed because of the high cost of natural gas. Methanol is a key building block chemical, and a methanol-based economy has been touted as an alternative to a hydrogen economy. Technologies exist for methanol-to-olefin production that are ready for implementation. These processes will become viable depending on the prices of the feedstocks and oil, all of which are highly variable, as well as costs for transportation and storage.

Unconventional or stranded natural gas is being evaluated as an alternative to alleviate current shortages in natural gas supply. Stranded, or unconventional, natural gas is that which is not easily transported from source to end use by pipeline, or is uneconomical to transport as a gas. Unconventional natural gas may come from a variety of sources: natural gas hydrates, stranded or geographically remote methane, such as that from sites in Alaska or the Rocky Mountains, coal bed methane, and methane from anaerobic fermentation such as occurs in landfills.

Gas from unconventional sources must be controlled, concentrated, or converted to a liquid form or stabilized in some manner, before being transported to the chemical manufacturer, power plant, or end user. The conversion of methane to a liquid product, such as methanol (CH_3OH), will allow easier transportation of materials from remote sites.

Liquid products can be made from natural gas through a process involving the conversion of synthesis gas (syngas). Some synthesis routes, such as the production of diesel fuel, are well understood (Chadeesingh, 2011) and one example of gas to liquids is the Fischer-Tropsch production of liquid fuels or methanol to gasoline (MTG). Methanol can give rise to a number of building block organic chemicals such as: acetic acid (CH_3CO_2H), dimethyl ether (CH_3OCH_3), formaldehyde (HCHO), ethane (C_2H_6), and propane ($CH_3CH_2CH_3$).

2.3 Chemicals From Coal

Substantial worldwide coal reserves make coal an attractive alternative to natural gas and petroleum. Historically, research into coal gasification has been focused on energy fuels, and more recently on power production, with less emphasis on commodity chemicals production. Efforts to develop pathways for chemicals production using optimal catalysts and for process scale-up are needed to replace conventional petroleum. The Fischer-Tropsch synthesis hydrocarbons (Chadeesingh, 2011) is particularly useful as a means of converting coal (via the production of synthesis gas) to commodity chemicals. While this pathway to organic chemical is different to the pathways used for crude oil and crude oil products, they are an option for use of a plentiful resource (coal). In fact, there have been several assessments of the use of coal for the generation of synthesis gas and liquid fuels through Fischer-Tropsch process.

2.3.1 Gasification

Organic chemicals can be produced through the gasification of coal. Because of the large domestic reserves of coal on a worldwide basis, this feedstock option is one of the strongest for chemical production in the long term. The technology associated with the gasification of coal [and other carbonaceous feedstocks is well understood (Chadeesingh, 2011)], having been used in Germany in WWII and in South Africa to produce liquid fuels in combination with Fischer-Tropsch processing.

The typical gasification process starts with the production of synthesis gas (a mixture of carbon monoxide and hydrogen) in the gasifier:

$$C_{coal} + H_2O \rightarrow H_2 + CO$$
$$CO + H_2O \rightarrow H_2 + CO_2$$

Water-gas shift reaction:

$$CO + H_2O \rightarrow CO_2 + H_2$$

A generic oxygen gasification system comprises the following main steps: (1) reaction of the coal with oxygen or steam at 1000+psi, (2) quench with water to remove particles and cool the synthesis gas, (3) application of the water-gas-shift reaction to produce hydrogen and carbon dioxide, and (4) application of gas cleaning technology by the use of physical solvents for the simultaneous removal of hydrogen sulfide and carbon dioxide (Mokhatab et al., 2006; Speight, 2007, 2013, 2014a,b). The solvents are recovered by depressurization and the hydrogen sulfide is converted to sulfur in a two-step process by first heating in oxygen to produce sulfur dioxide which then reacts further with hydrogen sulfide to produce sulfur and steam. Further catalysis increases the production of hydrogen by the reaction of carbon monoxide to carbon dioxide in the presence of water to produce additional hydrogen. The Fischer-Tropsch process can be used to produce alkanes, which are building blocks for many large-volume chemicals.

In addition to the production of synthesis gas, other products of coal gasification which may find use in the organic chemicals industry are (1) *producer gas*, which is a low Btu gas obtained from a coal gasifier (fixed-bed) upon introduction of air instead of oxygen into the fuel bed—the composition of the producer gas is approximately 28% v/v carbon monoxide, 55% v/v nitrogen, 12% v/v hydrogen, and 5% v/v methane with some carbon dioxide, (2) *water-gas*, which is a medium Btu gas that is produced by the introduction of steam into the hot fuel bed of the gasifier—the composition of the gas is approximately 50% v/v hydrogen and 40% v/v carbon monoxide with small amounts of nitrogen and carbon dioxide, (3) *town gas*, which is a medium Btu gas that is produced in the coke ovens and has the approximate composition: 55% v/v hydrogen, 27% v/v methane, 6% v/v carbon monoxide, 10% v/v nitrogen, and 2% v/v carbon dioxide—carbon monoxide can be removed from the gas by catalytic treatment with steam to produce carbon dioxide and hydrogen, and (4) *synthetic natural gas* (SNG), which is methane obtained from the reaction of carbon monoxide or carbon with hydrogen—depending on the methane concentration, the heating value can be in the range of high-Btu gases.

Because any organic carbonaceous material can be gasified (Speight, 2013, 2014b), existing gasifier designs can be adapted to use any type of coal as gasifier feed. Thus, coal characteristics (and other feedstock characteristics) do not offer insurmountable obstacles to its use for the production of synthesis gas as a first step to chemicals or fuel production (Speight, 2014b).

2.3.2   Liquefaction and Carbonization

Liquefaction

Coal can also be liquefied directly, without going through the production of synthesis gas. This process is termed coal-to-liquid (CTL) and is a reasonably mature technology. The process typically uses the technique of heating under pressure (up to 470°C, 200 bar) and hydrogenation where hydrogen is added to a coal-water slurry. The slurry increases the H/C ratio to a crude oil level and removes impurities such as sulfur, nitrogen, and oxygen. Coal liquefaction has been reviewed by United Kingdom Department of Trade and Industry (1999), in a paper that gives advantages and disadvantages of the technology.

Recent developments in coal liquefaction are mostly based on the pre-World War II technologies, with the exception being a process developed by Conoco based on coal dissolution in molten zinc chloride. As a response to the crude oil crisis in the 1970s, two-stage liquefaction was developed, which separated coal dissolution and hydrogenation. Experience from coal liquefaction has indicated that the areas that pose the most risk to implementation on a large scale include the handling of liquids with a high load of solids that can cause mechanical difficulties in the plant. Waste handling is also an issue because of emissions of trace contaminants, such as mercury, that need to be quantified and regulated, handling of reduced sulfur by-products, and polyaromatic hydrocarbon residues that are carcinogenic and so cannot be disposed of easily.

An advantage of coal liquefaction over gasification is the thermal efficiency of fuel production, with the former being 60–70% and the latter no higher than 55% (Research Reports International 2006). Product compositions are also different, with the direct process being skewed towards lower cetane fuels, and a higher aromatic content. The use of coal liquefaction for production of chemical feedstocks has not been discussed much in the literature, although the production of lower specific gravity hydrocarbons generated from coal gasification should be more easily be adapted to prevalent petrochemical-reaction pathways to commodity chemical production.

Coal liquefaction and the upgrading of the coal liquids are being considered as a future alternative of petroleum to produce synthetic liquid fuels due to the declining crude oil reserves and the high dependence on the foreign oil supplies. Since the late 1970s, coal liquefaction processes have been developed into integrated two-stage processes, in which coal is liquefied in the presence of hydrogen in the first stage and the products are upgraded in the second stage. Upgrading of the coal liquids is an important aspect of this approach and may determine whether such liquefaction can be economically feasible.

However, coal liquids have remained largely unacceptable as refinery feedstocks because of their high concentrations of aromatic compounds and high heteroatom and metals content. Successful upgrading process will have to achieve significant reductions in the content of the aromatic components. However, the hydrogenation of coal liquids with multiring, aromatic hydrocarbons with hydrogen is a difficult process from the technological point of view due to the stable structures of the aromatic compounds and the poor dynamic yields at low pressure and low temperature. However, liquid products from coal are generally different from those produced by petroleum refining, particularly as they can contain substantial amounts of phenols. Therefore, there will always be some question about the place of coal liquids in refining operations as well as the use of the liquids as suitable feedstocks for the production of organic chemicals.

Coprocessing of coal and a noncoal liquid hydrocarbon has also been considered, in particular for heavy oil. Coprocessed feed has the advantage that recycled solvents are not required, and is a method of facilitating refining of heavy oil.

Carbonization

Carbonization (which includes pyrolysis) is the oldest direct methods of production of liquids from coal, involving heating of the coal and capture of volatilized liquids leaving a char reduced in hydrogen. The amount of liquid generated is small, less than 20%, and the quality is poor being a complex blend of chemicals with water contamination. However, carbonization is a process predominantly for the production of a carbonaceous residue (coke) by the thermal decomposition (with simultaneous removal of distillate) of organic substances (Wilson and Wells, 1950; McNeil, 1966; Gibson and Gregory, 1971). The process, which is also referred to as destructive distillation, has been applied to a whole range of organic (carbon-containing) materials particularly natural products such as wood, sugar, and vegetable matter to produce charcoal. In this present context, the carbonaceous residue from the thermal decomposition of coal is usually referred to as "coke" (which is physically dissimilar from charcoal) and has the more familiar honeycomb-type structure. But, coal carbonization is not a process that has been designed for the production of liquids as the major products.

Carbonization is essentially a process for the production of a carbonaceous residue by thermal decomposition (with simultaneous removal of distillate) of organic substances:

$$C_{organic} \rightarrow C_{coke/char/carbon} + liquids + gases$$

The process may also be referred to as destructive distillation and has been applied to a whole range of organic materials, but more particularly to natural products such as wood, sugar, and vegetable matter to produce charcoal. Coal usually yields "coke" which is physically dissimilar from charcoal and has the more familiar honeycomb-type structure.

Coal tar is the volatile material that is released during the thermal decomposition of coal and which condenses at room temperature. The tar may be composed of solid material (pitch) and liquid or semisolid materials (coal tar). The carbonization of coal to produce *coal gas* for street and house lighting in the closing years of the 18th century produced substantial quantities of tar which (during the following 50 years) were mostly discarded as a troublesome and unnecessary by-product. However, the development of a western European chemical industry brought increasing importance to coal tar as a source of the precursors that were to be used for the synthesis of dyes as well as raw materials for the production of solvents, pharmaceutical products, synthetic fibers, and plastics (Speight, 2013, 2014a, 2016).

Coal tar can also be upgraded to gasoline and other liquid fuels. In fact, in the manner of crude petroleum, high temperature tars can be fractionated by distillation into (1) light oil, (2) middle (or tar acid) oil, and (3) heavy (or anthracene) oil. This primary separation is carried out by means of batch stills (vertical or horizontal; 3000–8000 US gallon, $11–30 \times 10^3$ L, capacity) or by means of continuous "pipe" stills in which the tar is heated to a predetermined temperature before injection into a fractionating tower.

The light oil fraction (b.p. 220°C; 430°F, c.f. petroleum naphtha b.p. 205°C; 400°F) consists mostly of benzene (45–72% w/w), toluene (11–19% w/w), xylene (3–8% w/w), styrene (1–1.5% w/w), and indene (1–1.5% w/w) and is processed either into gasoline and aviation fuel components or is fractionated further to provide solvents and petrochemical feedstocks. In either case, upgrading involves removal of sulfur compounds, nitrogen compounds, and unsaturated materials. This is usually accomplished by acid-washing in batch agitators or by hydrogenation over a suitable catalyst (e.g., cobalt-molybdenum or nickel-tungsten on a support). Thus, in the acid wash, the crude material is mixed with strong sulfuric acid, neutralized (with ammonical liquor or caustic soda), and after separation of the aqueous phase, steam-distilled or stripped of higher molecular weight material by centrifuging. The hydrogenation process conditions vary with the nature of the material to be removed (such as removal of sulfur or removal of olefin products by hydrogenation), but could typically be 300–400°C (570–750°F) and 500–1500 psi hydrogen. The middle oil typically boils over the range 220–375°C (430–710°F) and after extraction of the tar acids, tar bases, and naphthalene, can be processed to obtain diesel fuel, kerosene, or creosote.

Coal tar creosote consists of aromatic hydrocarbons, anthracene, naphthalene, and phenanthrene derivatives. At least 75% of the coal tar creosote mixture is polycyclic aromatic hydrocarbons (PAHs). Unlike the coal tars and coal tar creosotes, coal tar pitch is a residue produced during the distillation of coal tar. The pitch is a shiny, dark brown to black residue which contains PAHs and their methyl and polymethyl derivatives, as well as heteronuclear compounds.

Coal tar pitch is the tar distillation residue produced during coking operations. The grade of pitch thus produced is dependent on distillation conditions, including time and temperature. The fraction consists primarily of condensed ring aromatics, including 2–6 ring systems, with minor amounts of phenolic compounds and aromatic nitrogen bases. The number of constituents in coal tar pitch is estimated to be in the thousands.

In this context, it should be noted that the tar acids, which are mostly phenol derivative, including cresol derivatives (cresol is $CH_3C_6H_4OH$) and xylenol derivatives [xylenol is $(CH_3)_2C_6H_3OH$] which can be recovered by mixing the crude middle oils with a dilute solution of caustic soda, separating the aqueous layer, and passing steam through it to remove residual hydrocarbons. The acidic products—the phenol derivative, including cresol derivatives (cresol is $CH_3C_6H_4OH$) and xylenol derivatives—are then recovered by treatment of the aqueous extract with carbon dioxide or with dilute sulfuric acid and are then fractionated by vacuum distillation. Tar bases are isolated by treating the acid-free oil with dilute sulfuric acid and the bases are regenerated from the acid solution by addition of an excess of alkali (e.g., caustic soda or lime slurry). The mixture is then fractionated to produce pyridine, quinoline, and isoquinoline as well as other nitrogen-containing products.

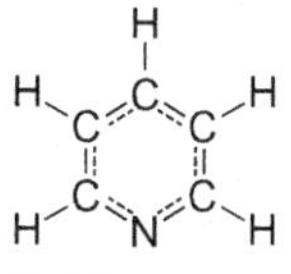

Pyridine

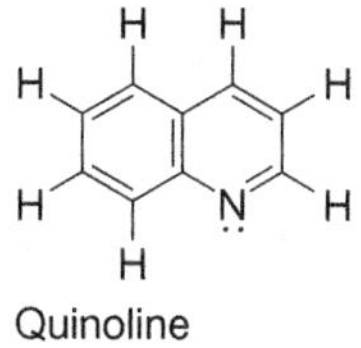

Quinoline

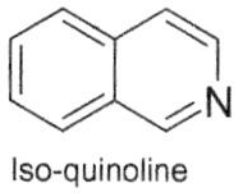

Iso-quinoline

The temperature to which the distillation of the heavy oil fraction is taken depends on the type of residue pitch; that is desired but usually lies within the range 450–550°C (840–1020°F). In all cases, the distillate is an excellent source of hydrocarbons such as anthracene, phenanthrene, acenaphthene, fluorene, and chrysene. The residual coal tar pitches are complex mixtures that contain several thousand compounds (mostly condensed aromatic compounds) and may, by analogy, be likened to the vacuum residua that are produced in a petroleum refinery and represent the materials in petroleum which have boiling points in excess of 565°C (1050°F).

One important aspect of coal tar chemistry relates to the presence and structure of the *nitrogen species* in the tar. Initial work on tar nitrogen chemistry indicated that the nitrogen structure of tar is similar to the nitrogen structures in the parent coal. This was due to the belief that nitrogen in coal exists in tightly bound compounds and hence the most thermally stable structures in the coal during devolatilization. Furthermore, during the thermal reactions that led to tar formation these nitrogen compounds were released without rupture as part of the tar.

2.4 Chemicals From Tar Sand Bitumen

Tar sand bitumen, derived from unconventional sources such as tar sands in Canada and Venezuela has a higher aromatic content than conventional crude. The key aspect that needs to be addressed in the use of heavy oil for chemicals is the development of a ring opening catalyst to break down the polyaromatic tar compounds into smaller molecules. Oil shale in Colorado has the potential to be a large scale domestic supply of petroleum. Extraction technologies are currently being developed and tested, with the product of leading processes predicted to be sweet crude.

Unconventional sources of petroleum include the Canadian Athabasca oil sand, Venezuelan heavy oil, and oil shale in the Western United States. Historically, the relatively high cost of extraction of these hydrocarbons has been a major detriment to the use of heavy oil. These feedstocks are now more competitive due to the lowering costs of the production of synthetic crude oil and the volatility of the price of conventional crude oil.

The extraction of bitumen-derived crude oil follows several steps. The first step is the extraction of the heavy oil from the rock. Tar sands (7% bitumen by weight from the tar sands) and oil shale contain relatively low concentrations of hydrocarbons. Some loss (<10% of bitumen) occurs in the primary extraction process. Increased mechanical breakup or froth treatment can increase the yield. Current research is focused on making the extraction of unconventional crude more sustainable with less of an environmental impact, including better handling of mine tailings and the water entrained in the bitumen as it is extracted from the tar sands.

To date there has been little published on the use of heavy oil, oil shale, or tar sands for chemicals production because these sources of oil have only recently (within the past two decades) become cost competitive with natural gas and with crude oil. In addition, the sources of unconventional crude oil are far from current petrochemical plants. Synthetic crude oil, derived from tar sands in Canada and from oil shale in Colorado, has a higher aromatic content than conventional crude oil. The key research aspect that needs to be addressed in the use of synthetic crude oil for the production of organic chemicals is the development of a ring opening catalyst to break down the polynuclear aromatic tar compounds into lower molecular weight less complex products.

2.5   Chemicals From Biomass

Thousands of years of innovations in agriculture have optimized crops for food and fiber production, but not for energy production. In general, the fermentation of sugar from crops such as corn and sugarcane will provide oxygenated organics, but these are often small volume niche chemicals with limited potential for large scale manufacture (Chang and Holtzapple, 2000; Chang et al., 2001a,b; Hammerschlag, 2006). However, some biobased chemicals that have potential for large scale manufacture include the carboxylic acid derivatives and glycol derivatives (Table 3.1).

Projects that can lead to the production of organic chemicals from various sources include: (1) biomass gasification, (2) fermentation of sugars, (3) decomposition of cellulose, (4) separation of lignin and other plant components, (5) high temperature pyrolysis, and (6) biorefining of wood and waste materials (Speight, 2011c). Issues in the replacement of petroleum by biomass feedstocks

TABLE 3.1 Conversion of Biomass to Chemicals

Process	Primary Products	Secondary Products
Gasification	Hydrogen	
	Alcohol derivatives	
	Naphtha	Gasoline
	Kerosene	Diesel fuel
	Olefin derivatives	Polymers
	Oxo chemicals	
	Synthetic natural gas	Hydrogen
		Synthesis gas
Pyrolysis	Hydrogen	
	Biooil	Various chemicals
Anaerobic digestion	Biogas	Hydrogen
Fermentation	Ethanol	Ethanol derivatives
		Hydrogen

include impurities, variabilities of feedstock composition, distributed supply, scalability, and pathways for breakdown of cellulose. Although some large-scale chemicals production occurs as a by-product of fuel production, widespread use of biomass feedstocks for commodity chemical manufacture will require sustained research and development in a variety of fields such as plant science, microbiology, genomics, catalysis, and chemical separation technologies.

Chemicals can be manufactured from biomass through gasification, pyrolysis, and fermentation in dedicated plants or in biorefineries. In general, the fermentation of sugar from crops such as corn and sugarcane will provide oxygenated organics, but these are often small volume niche chemicals with limited potential for large scale manufacture. By-products of transportation fuel production, such as biodiesel, may be used in bulk chemical manufacture—one example being glycerin (also called glycerol, glycerine, and propane triol):

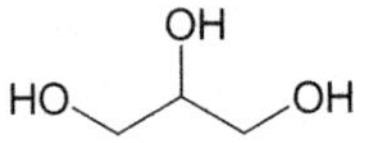

OH

HO ⋀ OH

Glycerin

Efficient utilization of intermediate reagents such as these will eventually lead to the development of new synthesis pathways, or optimization of synthetic routes that are currently only being done in the laboratory.

Use of biomass to make biofuels such as ethanol and biodiesel, is of great interest in the United States and elsewhere (Speight, 2011c). Biofuels provide energy security, support local agriculture, and reduce net emissions of carbon dioxide to the atmosphere. The use of biomass to make chemicals has received less public attention to date. Chemical manufacturers currently use biomass to make organic acids, textile fibers, polymers, adhesives, lubricants and greases, and soy-based inks.

Biorefining is the name given to the use of biologically derived feedstocks for chemicals manufacture. The biorefinery concept generally involves feeding biomaterials, along with waste oils and other carbon-based materials, into steam or catalyst crackers to make chemicals (Speight, 2011c). Alternatively, these feedstocks may be hydroprocessed directly. Mills are used to process biomass (corn, pulp) to produce carbohydrates, oils, lignin, and fuel compounds. Once broken down, fermentation will produce alcohols from sugars and starch. Biorefining feedstocks are crops, waste plant or animal material, and recycled fibers. Eventually, biorefining will not only utilize the starch or sugar component of biomaterials, but also consume lignin, hemicellulose, and cellulose in value-added processes beyond the current practice of burning these materials for fuel. An example is bioconversion of sugar derivatives to produce polyols (polyhydroxy compounds which are *building block* chemicals). In addition, another example of biorefining is the conversion of vegetable oils to lubricants, hydraulic fluids, and other chemical products that can serve as monomer for polymerization processes or building block chemicals.

Bioprocessing of corn to produce ethanol can either be done by wet-milling or dry-milling—the wet-milling process produces gluten feed, gluten meal, and corn oil, as well as ethanol while the dry-milling process gives ethanol only. In spite of additional expenses incurred for separations equipment and operation, these by-products can be sold to improve the economics of ethanol production. Wet mills are large, and generally require a coal-fired plant to operate, whereas dry mills (~40 M gal) are smaller, and use natural gas. The by-product of sugarcane processing is bagasse, which can be burned for fuel, providing sufficient energy to run the processing plant. Also, the production of biodiesel (fatty acid methyl esters) from various feedstocks as well as from biomass residues is of interest in many countries, particularly in the developing world. The processes for manufacturing diesel is an alternative to fermentation and the use of the reactive distillation concept to produce esters of dibasic acids takes advantage of differences in volatility to give high purity products. The reactive distillation is a process where the still pot is the chemical reactor and separation of the product from the reaction mixture does not need a separate distillation step, thereby saving process energy (for heating) as well as the need from an extra piece of equipment (the reactor). Thus, production of biofuels will have a direct impact on the chemicals industry. For example, glycerine produced as by-product of biodiesel manufacture, will supplant propylene as a feedstock in the production of epichlorohydrin which can serve as an intermediate organic chemical for the production of epoxy resins.

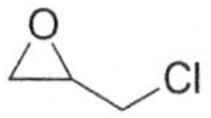

Epichlorohydrin

Epichlorohydrin (ECH) is an organochlorine compound and an epoxide (a three-membered ring that contains two carbon atoms and an oxygen atom) and, despite the name, it is not an acid chloride (like acetyl chloride, CH_3COCl):

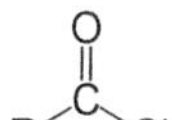

An acid chloride; when R=CH$_3$, the acid chloride is acetyl chloride.

2.5.1 Thermochemical Gasification

Thermochemical gasification is one of several thermochemical conversion processes which also include combustion and pyrolysis and serve to convert carbonaceous feedstock to gases, liquids, and solid products.

Combustion is the complete conversion (in the presence of oxygen) to carbon dioxide, water, and energy. Combustion is also a means of converting waste to energy and involves oxidation of the fuel for the production of heat at elevated temperatures without generating useful intermediate fuel gases, liquids, or solids. Combustion typically employs an excess of the oxidizer (air) to ensure maximum fuel conversion and the products of combustion processes include heat, oxidized species (such as carbon dioxide and water), products of incomplete combustion (such as carbon monoxide and hydrocarbons), other reaction products (most as pollutants), and ash from any inorganic mineral matter (or organo-minerals) in the feedstock.

Pyrolysis is the thermal degradation of a material usually without the addition of any air or oxygen. The process is similar to gasification but generally optimized for the production of fuel liquids or pyrolysis oils (sometimes called biooils if biomass feedstock is used). Pyrolysis also produces gases and a solid char product. Pyrolysis liquids can be used directly (e.g., as boiler fuel and in some stationary engines) or refined for higher quality uses such as organic chemicals, gasoline, diesel fuel, and other products.

Gasification typically refers to conversion in an oxygen- or air-deficient environment to produce fuel gases (such as synthesis gas). The fuel gases are principally carbon monoxide, hydrogen, methane, and lighter hydrocarbons, but depending on the process used, can contain significant amounts of carbon dioxide and nitrogen, the latter mostly from air. Gasification processes also produce liquid products (tars, oils, and other condensates) and solids (char, ash) from solid feedstocks. The combustion of gasification-derived fuel gases generates the same categories of products as direct combustion of solids, but pollution control and conversion efficiencies may be improved. Synthesis gases can produce fuel products and other chemicals by chemical reactions such as Fischer-Tropsch synthesis (Chadeesingh, 2011).

Synthesis gas for commodity chemical production can be derived from any carbonaceous feedstock, such as coal, petroleum residua, or biomass. Issues such as the production of clean syngas via biomass thermochemical processing are similar to issues associated with coal gasification. Gasification of biomass can take place under slightly milder conditions than coal gasification (800–1000°C at 20–30 bar instead of 1400°C at 20–70 bar). Biomass (feedlot and chicken litter) can also be combined with a coal-syngas feedstock (Priyadarsan et al., 2004, 2005).

However, using biomass for thermochemical conversion raised several issues and some pretreatment is necessary and is unique to gasification of biomass. Biomass has a large water content that must be removed before gasification. Also, biomass components (alkali metals, halides, sulfur compounds, and tars) have a significant potential to poison downstream noble metal catalysts used in production of syngas and chemicals (Ragauskas et al., 2006). Technologies have been developed to handle these impurities, but they add to the complexity and cost of the gasification process. In addition, because production of biomass requires a large land base, feedstocks are diffuse, and manufacturing is distributed (e.g., forest pulp mills). Hence, additional methods as required by the character of the feedstock, such as feedstock densification or on-site drying prior to shipping, will be required to achieve economies of scale.

2.5.2 Sugar Fermentation

Fermentation is a metabolic process that converts sugar to acid derivatives, various gases, or alcohol. Another definition is that fermentation is the chemical process by which molecules such as glucose ($C_6H_{12}O_6$) are decomposed anaerobically. The process can involve complete decomposition of the glucose to carbon dioxide and water (+energy) or can be adapted to produce ethanol (ethyl alcohol + energy):

$$C_6H_{12}O_6 \rightarrow [\,fermentation\,] \rightarrow CO_2 + H_2O + energy$$
$$C_6H_{12}O_6 \rightarrow [\,fermentation\,] \rightarrow C_2H_5OH + energy$$

The presence of oxygen during the fermentation offers routes to a suite of other chemical products.

Thus, fermentation offers a pathway for the use of biomass as a means of producing organic chemicals either directly or organic chemicals as feedstocks for the production of other chemicals. In fact, the fermentation of sugars has been the focus of much recent investigation because of the chemical character and properties of the products, which include: (1) the presence of functional groups that give rise to building block chemicals, (2) the facile adaptation to petrochemical pathways, and (3) the performance of the production process, which is a well-tested operation.

Chemical transformation of sugars can take place by oxidation (such as during the decomposition of polysaccharides of which starch is an example), oxidative dehydration (of the six-carbon sugars), hydrogenation (sometimes acid-catalyzed) of cellulose derivatives and sugar derivatives, acid amination, and esterification of oils. These products can be further converted by chemical means to derivatives by: (1) oxidation, which tends to be less important for biomass as these compounds are already oxidized, (2) by hydrogenation, (3) by dehydration, (4) by bond cleavage, and (5) by direct polymerization. Biological reactions to derivatives can also be achieved with the advantages that they are generally enzymatic and so very selective, and may go directly from sugar to the end product.

2.5.3 Nonsugar Fermentation

Annual crops such as corn that have a high sugar yield are typically difficult to grow, and need fertilization, irrigation, herbicides, and pesticides. In addition, these crops as well as waste biomass have been engineered for food production, rather than for efficiency of photosynthesis, or energy, or chemicals production (Ragauskas et al., 2006). Use of cellulose from waste biomass would allow raw material to come from corn stalks (Thanakoses et al., 2003a), wood, bagasse from sugarcane (Thanakoses et al., 2003b), and even animal products (if proteins can be used) as well as various types of waste (Chan and Holtzapple, 2003; Aiello-Mazzarri et al., 2005, 2006).

Conversion of cellulose to fuel and hydrocarbons is a multistage process. The biomass must first be physically or chemically broken down, to separate the cellulose from other components, such as lignin. Pretreatment issues dominate cellulose and lignin processing, and often involve acid- or base-catalyzed hydrolysis to facilitate enzymatic breakdown. The costs of pretreatment are high, but some suggest that few major technical improvements in the chemical or acid processing are possible. However, pretreatment is an active area of research, one example being investigation of improvements in lime pretreatment (e.g., Chang and Holtzapple, 2000; Chang et al., 2001a,b).

Once isolated, the next step is the breakdown of cellulose to form sugars. The natural rotting process, facilitated by bacteria or fungi, is slow. Enzymatic hydrolysis increases the rate of the process, but the production of by-products is a problem when the processes of oxidation produce aldehydes and acids as well as sugars. Once produced, these sugars can be fermented and processed by conventional means. Hence, the key needs for fermentation of biomass include bacteria that break down cellulose quickly. Other process goals include the use of different conversion pathways (Lynd, 1996). In addition, separation processes are essential in handling of the varied and variable biofeedstocks which includes separation of cellulose from lignin and other plant materials and separation of the by-products from the product after fermentation is complete. A separation step in itself may allow production of a value-added chemical, such as the production of xylitol [$CH_2OH(CHOH)_3CH_2OH$, a polyalcohol or sugar alcohol] from the pretreatment of cellulose.

2.5.4 Pyrolysis

A third means of extracting chemicals from biomass feedstocks is by means of pyrolysis (Speight, 2011c). The biomass feedstocks can be wood wastes, bark, or other forest products and the resulting products are biooils which are comprised of oxygenated organic compounds, and water. Pyrolysis is complex, incorporating both evaporation and combustion of a chemically unknown fuel. The process involves (1) heating and vaporization of water, (2) separation of the volatile components, and (3) formation of a porous char or cenosphere from the high-boiling nonvolatile components. Unlike direct combustion, pyrolysis occurs at high temperatures, on the order of 400°C (750°F).

High temperature pyrolysis has been suggested as a more effective method of black liquor gasification than traditional lower temperature approaches because tarry residues are less of a problem. Black liquor is the waste product from the Kraft process when digesting pulpwood into paper pulp, removing lignin, hemicellulose derivatives, and other extractives from the wood to free the cellulose fibers. At temperatures in the range of 700°C and 1000°C

(1290–1830°F) the process yields tar products, semivolatile products, and non-volatile char that were identified from the heating of black liquor (Sricharoenchaikul et al., 2002). In addition, in organic chemicals such as single benzene ring derivatives up to pyrene and fluoranthene (4-ring compounds) derivatives are produced.

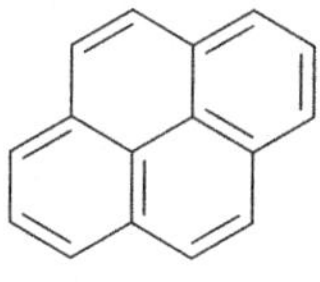

Pyrene

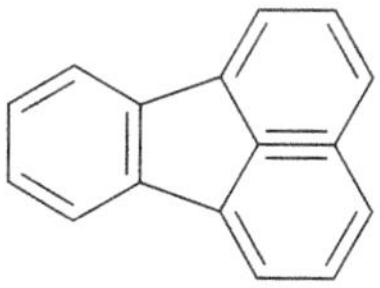

Fluoranthene

3 PROCESS CHEMISTRY

Industrial organic chemicals are those organic compounds produced in the United States and other countries as single compounds or as the more complex mixtures, such as fuels that are composed of organic compounds (Tables 3.2 and 3.3) (Kroschwitz et al., 1991; Chenier, 1992; Speight, 2002, 2014a,b; Wittkoff et al., 2012). Historically, during the 19th century and early 20th century most organic chemicals had been obtained as by-products from the coking of coal, e.g., from coal oil. During the 20th century, however, oil and natural gas became the dominant sources of the world's industrial organic chemicals. By 1950, a substantial portion of the industrial organic chemicals were being produced from crude oil and natural gas, and by 2000 more than 90% of the organic chemical industry was based on crude oil and natural gas as the prime feedstocks. As a result, the term *petrochemicals* has almost become synonymous with industrial organic chemistry and the chemicals industry.

Nevertheless, terminology aside, there are several types of process reactions common to the organic chemicals manufacturing industry: (1) polymerization, (2) oxidation, and (3) addition. Polymerization is a chemical reaction usually carried out with a catalyst, heat, or light (often under high pressure) in which a large number of relatively simple olefin derivatives (such as ethylene, $CH_2=CH_2$) combine to form a high-molecular weight product [a macromolecule, such as polyethylene, $H-(CH_2CH_2)_n-H$]. Oxidation, in the strictest chemical sense, is the combination of oxygen chemically with another

TABLE 3.2 Example of Organic Chemicals Manufactured in the United States

Chemical	Volume of Production 10^6 tons (US, 1997)	Alternative Feedstock or Precursor	Industry/Product
Olefins			
Ethylene	24.125	Gasification of coal, biomass. DME	Gas to polymers
Propylene	14.350		
Butadiene (1,3-)	2.038	Methanol (MTO. MTP)	
Ammonia	14.204	Gasification of coal, biomass	Fertilizer, reagent, explosives
		Natural gas, Haber cycle	
Chlorinated organics			
Ethylene dichloride	10.088	Chlorination of olefins	Polyurethanes, solvents, pulp and paper solvents
Vinyl chloride	8.753		
Methyl chloride	0.563		
MTBE	9.038	CH_3OH + isobutylene	Fuel additive
Aromatics			
Benzene	7.463	Gasification of coal, biomass	Polymerization/ surfactants
Ethylbenzene	6.950	From methanol from GTL	
Toluene	4.138		
p-Xylene	3.963	Friedel-Crafts alleviation	
Cumene	2.913		
Methanol	6.013	Gasification of coal, biomass	Building block
		FT	
Urea	5.918	Ammonia, CO_2;	Fertilizer, resins, adhesives
Styrene	5.700	Ethylbenzene	Polystyrene

TABLE 3.2 Example of Organic Chemicals Manufactured in the United States—cont'd

Chemical	Volume of Production 10^6 tons (US, 1997)	Alternative Feedstock or Precursor	Industry/Product
Terephthalic acid	5.000	p-Xylene	Intermediate
Aldehydes			
Formaldehyde	4.188	From methanol (natural gas)	Building block for olefins
Oxygenated organics			
Ethylene oxide	3.550	Gasification of coal, biomass, through ethylene oxide	Building block, reagent
Propylene oxide	1.963		
Ethylene glycol	2.813	Hydration of ethylene oxide	Polyester
Propylene glycol	0.538		
Carboxylic acids			
Acetic acid and anhydride	2.425	Gasification of coal, biomass Methanol from syngas	Adds C_2, reagents and intermediates
Phenol	2.175	Cumene	Resins, paints, adhesives, coatings, solvents, polycarbonate
Bis phenol A	0.863		
Acrylonitrile	1.663	NH_3+propylene, propane	Acrylic fibers
Esters			
Vinyl acetate	1.500	From ethylene	Intermediate coatings, plastics
Methyl methacrylate	0.313		

Continued

TABLE 3.2 Example of Organic Chemicals Manufactured in the United States—cont'd

Chemical	Volume of Production 10^6 tons (US, 1997)	Alternative Feedstock or Precursor	Industry/Product
Acetone	1.463	Gasification of coal, biomass, BTEX	Intermediate
Cyclohexane	1.100	Hydrogenation of benzene	Intermediate
Caprolactam	0.825	Oxidation of cyclohexane	Nylon
Aniline	0.713	Nitro or chlorobenzene	Intermediate
Isopropyl alcohol Butanol	0.700	Gasification of coal, biomass, propylene	Intermediate

TABLE 3.3 Manufacture of Fuels

Paraffins: Ethane Propane Butanes	Gasification of coal, biomass Methanol dehydrogenation FT liquid	Fuel building block
Ethanol	Fermentation of biomass Gasification of coal, biomass	Fuel
Biodiesel, Glycerol	Enzymatic transformation of biomass Transesterification of oils	Fuel Building block Polyurethane, glycol, 1-propane diol, biodiesel
Dimethyl ether		Replacement for propane

substance although this name is also applied to reactions where electrons are transferred (Chapter 2). The addition reaction covers a wide range of reactions where a double bond ($>C=C<$) or a triple bond ($-C\equiv C-$) is broken and a component molecule is added to the reactant molecule. The alkylation reaction (often considered to be an addition reaction as practiced in the petroleum refining industry) in which a hydrogen atom is converted to an alky group can be considered an addition reaction:

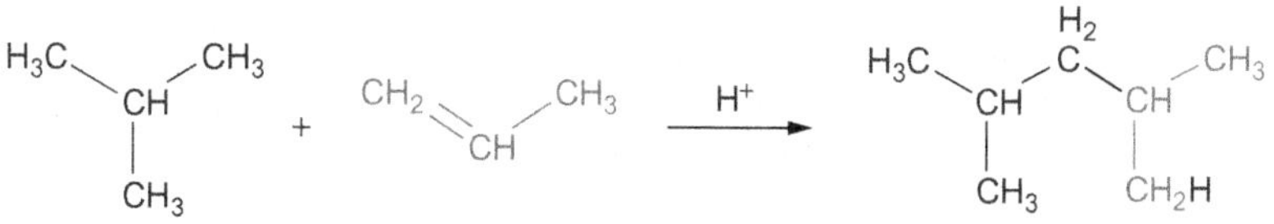

Alkylation by replacement of an active hydrogen by an alkyl group.

However, an important aspect of process chemistry is that most industrial organic chemistry is considered to fall into one of the following categories: (1) C-1 chemistry, based on synthesis gas, also called syngas which is a variable mixture of carbon monoxide (CO) and hydrogen (H_2) produced by the high temperature reaction of steam water with crude oil residua, coal, oil, or natural gas, or for that matter any carbonaceous material, including biomass (Speight, 2013, 2014a,b), (2) C-2 chemistry, based on the chemistry of ethylene, (3) C-3 chemistry, based on the chemistry of propylene, (4) C-4 chemistry, based on the chemistry of the butane isomers and the butene isomers, and (5) BTX chemistry, based on the chemistry of benzene, toluene, and the xylene isomers.

Alkenes or olefins (ethylene, propylene, butene isomers, and butadiene) are mainly produced via thermal steam cracking as used in the petroleum refining industry (Speight, 2014a, 2016). In the steam cracking process, a fraction of crude oil (usually isolated by distillation without any further purification, unless a catalyst is involved) is mixed with water and heated using a short residence time (usually on the order of 1 s) at 800–900°C (1472–1652°F). These process parameters cause rupture of the carbon-carbon bonds to yield lower molecular weight products which also are reduced in hydrogen content compared to the feedstock (as deduced by comparing the atomic hydrogen-to-carbon ratio of the feedstock and products) with the formation of double bonds. On the other hand, the BTX compounds (benzene, toluene, and xylene isomers) the simplest aromatics, are largely produced during catalytic reforming (the Platforming process). In this process a naphtha fraction that is rich in the pentane *t*-nonane alkane hydrocarbons (C_5 to C_9 alkanes) is reacted at temperatures on the order of 450°C (842°F) and approximately 300–400 psi, over a platinum (Pt/SiO_2) catalyst, to yield a reaction product that contains approximately 60% aromatic hydrocarbons. Typically, the product mixture might contain approximately 3% v/v benzene, 12% v/v toluene, 18% v/v xylene isomers, and 27% v/v C_9 alkylbenzenes. The product, because of the presence of the aromatic components has a high octane number and can be blended (as a mixture) as individual aromatic constituents as a blend-stock for gasoline production. Because benzene is much more in demand for industrial purposes than toluene, the methyl group of toluene is often removed by hydrogenation.

$$C_6H_5CH_3 + H_2 \rightarrow C_6H_6 + CH_4$$
$$\text{Toluene} \qquad\qquad \text{Benzene}$$

Thus, the following sections relate to the production of high-volume organic chemicals which also serve to illustrate three key points, which are: (1) primary building blocks are typically used in more reactions than the chemical building blocks in the later stages of chemical production, (2) most feedstocks for chemicals production can participate in more than one reaction, and (3) there is typically more than one reaction route to an end-product. Furthermore, the end-products of all of these chemical processes can be used in several commercial applications.

3.1 C-1 Chemistry

C-1 chemistry refers to the conversion of carbon-containing organic chemicals materials that contain one carbon atom per molecule into more valuable products. The feedstocks for C-1 chemistry include natural gas (predominantly methane), carbon dioxide, carbon monoxide, methanol, and synthesis gas (syngas, a mixture of carbon monoxide and hydrogen). Synthesis gas is produced primarily by the reaction of natural gas, which is principally methane, with steam as well as by the reaction of any high-molecular weight carbonaceous feedstock (such as coal, petroleum residua, petroleum coke, or biomass) (Speight, 2013, 2014a,b). The availability of synthesis gas from such feedstocks coal gasification is expected to increase significantly in the future because of increasing development of IGCC power generation and it is anticipated that the refinery of the future will also have a gasifier integrated into the refinery unit processes (Speight, 2011b).

Synthesis gas is also the feedstock for all methanol and Fischer-Tropsch plants. In fact, many important organic chemicals can be produced from synthesis gas (syngas)—the carbon monoxide (CO) and hydrogen (H_2) mixture produced as a result of the gasification of a variety of carbonaceous feedstocks either as a single feedstock or as a mixture of feedstocks (Speight, 2014b). They range from simple molecules, such as methanol, to high-grade synthetic crude oil. The basic reaction for conversion of synthesis gas to mixtures of hydrocarbons (Fischer-Tropsch reaction) was used in Germany during World War II to produce fuel mixtures for the military diesel and gasoline engines. Since the 1950s South Africa has also used this reaction, and currently there is much interest in using it to convert natural gas (methane) to more easily transported liquids as well as for the end use of these liquids.

Ammonia (NH_3), although it is not an organic chemical, is often considered as part of C-1 chemistry, since it is produced via a reaction that uses hydrogen gas obtained from methane. It is made by the Haber process. Ammonia and its derivatives, HNO_3, NH_4NO_3, and $CO(NH_2)_2$, are key fertilizers and ingredients for explosives, and their production consumes nearly 5% v/v of the worldwide natural gas.

Methanol (methyl alcohol, CH_3OH), an important C-1 chemical that is used as a solvent as well as a precursor for many organic chemicals, is made by the hydrogenation of carbon monoxide—a process developed in the 1920s:

$$CO + 2H_2 \rightarrow CH_3OH$$

For example, as a precursor to other chemicals, a major use of methanol is the production of acetic acid. Acetic acid (ethanoic acid, CH_3CO_2H) was for many years made by the oxidation of ethanol:

$$C_2H_5OH + O_2 \rightarrow CH_3COOH + H_2O$$

However, there are other processes by which acetic acid can be manufactured including the carbonylation of methanol:

$$CH_3OH + CO \rightarrow CH_3CO_2\,H$$

C-1 chemistry is expected to become a major means of chemical production (or, at least, an initial step in the production of many organic chemicals) for the production of chemicals and transportation fuels in the near future (Speight, 2011b). In addition, synthesis gas is a major source of refinery hydrogen for use in a variety of refining processes (i.e., processes that require a hydrogen presence in the reactor) as well as for a variety of chemical and petrochemical operations.

3.2   C-2 Chemistry

C-2 chemistry usually refers to the processes that use ethylene as the starting organic chemical and it is one of the largest volume organic compounds used worldwide. Ethylene is produced in the petrochemical industry by steam cracking. In this process, gaseous or light liquid hydrocarbons are heated to 750–950°C (1380–1740°F). The products are quenched to prevent further reactions (often leading to the formation of tar-like products) and ethylene is separated from the resulting mixture by repeated compression and distillation. In a related process used in oil refineries, high-molecular weight feedstocks are cracked over zeolite catalysts. Feedstocks such as naphtha and gas oil require at least two quench towers downstream of the cracking furnaces to recirculate pyrolysis-derived gasoline and process water. When cracking a mixture of ethane and propane, only one water quench tower is required.

The major uses for ethylene are in the synthesis of polymers (such as polyethylene) and for the production of ethylene dichloride, a precursor to vinyl chloride (Tables 3.4 and 3.5). Other important products are ethylene oxide (a precursor to ethylene glycol) and ethylbenzene (a precursor to styrene).

TABLE 3.4 Industrial Uses of Ethylene

Process	Target Product	Process Conditions			Reaction Components	Other Characteristics
		Pressure (MPa)	Temperature (°C)	Catalyst		
Polymerization	Low-density polyethylene (LDPE)	60–350	350		Oxygen or peroxide	
	High-density polyethylene	0.1–20	50–300	Molybdenum Chromium oxide		
	Polyethylene	Low		Aluminum alkyls Titanium oxide		
Oxidation	Ethylene oxide	1–2	250–300	Silver	1,2-Dichloroethane, oxygen	60% is converted to ethylene glycol using an acid catalyst
	Acetaldehyde	0.3	120–130	Copper chloride/ palladium chloride	Oxygen	Vapor phase
	Vinyl acetate	0.4–1	170–200	Palladium	Acetic acid	

Addition

Halogenation \hydrohalogenation	Ethylene dichloride	60		Iron, aluminum, copper, or antimony chlorides	Chlorine	Feedstock for vinyl chloride and trichloroethylene and tetrachloroethylene
	Ethyl chloride		0.3–0.5	Aluminum or iron chlorides	HCl	Precursor of styrene
Alkylation	Ethyl benzene			Aluminum, iron, and boron chlorides	Benzene	
Hydroformation	Propionaldehyde	60–200	4–35	Cobalt	Synthesis gas (carbon monoxide and hydrogen)	

TABLE 3.5 Industrial Uses of Vinyl Chloride

Process	Target Product	Process Conditions			Reaction Components	Other Characteristics
		Pressure (MPa)	Temperature (°C)	Catalyst		
Polymerization	Polyvinylchloride		50	Peroxides		
Substitution at the carbon–chlorine bond	Vinyl acetates, alcholates, vinyl esters and vinyl ethers			Palladium	Alkyl halides	
Addition	Various halogen addition products					

Ethylene oxide (C_2H_4O, also called oxirane) is a cyclic ether that is a colorless flammable gas at room temperature, with a faintly sweet odor. It is the simplest epoxide—a three-membered ring consisting of one oxygen atom and two carbon atoms:

Ethylene oxide

Because of this molecular structure, ethylene oxide easily participates in addition reactions, such as ring opening followed by further reactions and polymerization. This chemical reactivity has made ethylene oxide a key industrial chemical and is usually produced by direct oxidation of ethylene in the presence of a catalyst. It is extremely flammable and explosive and is used as a main component of thermobaric weapons—therefore, it is commonly handled and shipped as a refrigerated liquid. While ethylene itself is not generally considered a health threat, ethylene oxide has been shown to cause cancer.

Most ethylene oxide (about 60% v/v) is converted to ethylene glycol ($HOCH_2CH_2OH$) via acid-catalyzed hydrolysis:

$$CH_2{=}CH_2 + 2OH^- \rightarrow HOCH_2CH_2OH$$

Caution is always advised when handling a toxic chemical such as ethylene glycol. It is a toxic chemical that is used as an antifreeze, heat transfer agent, and also as a solvent in industrial organic chemical facilities. Long-term inhalation exposure to low levels of ethylene glycol may cause throat irritation, headache, and backache while exposure to high concentrations can lead to loss of consciousness. Liquid ethylene glycol is irritating to the eyes and skin and toxic effects from ingestion of ethylene glycol include damage to the central nervous system and kidneys, intoxication, conjunctivitis, nausea and vomiting, abdominal pain, weakness, low blood oxygen, tremors, convulsions, respiratory failure, and coma. Ethylene glycol readily biodegrades in water and in soils—biodegradation is probably the dominant removal mechanism.

Vinyl chloride ($CH_2{=}CHCl$) is the second-largest-volume chemical made from ethylene and is produced by adding chlorine to ethylene (to produce ethylene dichloride, $ClCH_2CH_2Cl$) and then thermal treatment to remove hydrogen chloride from the intermediate ethylene dichloride.

$$CH_2{=}CH_2 + Cl_2 \rightarrow ClCH_2CH_2Cl$$
$$ClCH_2CH_2Cl \rightarrow CH_2{=}CHCl + HCl$$

The uses of vinyl chloride include polymerization to polyvinyl chloride (PVC) which is used to make pipe, floor covering, wire coating, house siding, imitation leather, and many other products (Table 3.5). Vinyl chloride is one of the largest commodity chemicals but is also considered a human carcinogen by the US EPA. Vinyl chloride polymers are the primary end use but various vinyl ethers, esters, and halogen products can also be made as shown in the table below.

Styrene (phenylethylene, vinyl benzene, $C_6H_5CH=CH_2$) is made from ethylene by reaction with benzene to form ethylbenzene, followed by dehydrogenation.

$$C_6H_6 + CH_2=CH_2 \rightarrow C_6H_5CH_2CH_3$$
$$C_6H_5CH_2CH_3 \rightarrow C_6H_5CH=CH_2 + H_2$$

Over 50% of manufactured styrene is polymerized to polystyrene for toys, cups, containers, and foamed materials used for insulation and packing. The rest is used to make styrene copolymers, such as styrene-butadiene rubber (SBR).

At the high-molecular weight end of the uses of ethylene, polyethylene comes in two basic types: high density and low density. The original polymer was a highly flexible branched product, first prepared in 1932 by a process that required high temperatures and ultrahigh pressures. It is now known as low-density polyethylene (LDPE), to differentiate it from a linear polymer discovered later and known as high-density polyethylene (HDPE). For many applications the original branched LDPE has now been replaced by linear low-density polyethylene (LLDPE). HDPE is more rigid and less translucent than LDPE or LLDPE, and it has a higher softening point and tensile strength. HDPE is used to make bottles, toys, kitchenware, and so on, whereas LDPE and LLDPE are predominantly used for film used in packaging (such as plastic bags).

3.3 C-3 Chemistry

C-3 chemistry is based on the chemistry of propylene ($CH_3CH=CH_2$) from which polypropylene is produced. At room temperature and atmospheric pressure, propylene is a gas, and as with many other alkenes, it is also colorless. Propylene is produced in nature as a by-product of vegetation and fermentation processes. On the industrial scale, propylene is produced from crude oil and natural gas—during crude oil refining ethylene, propylene, and other compounds are produced as a result of the thermal decomposition (cracking) of higher molecular weight larger hydrocarbons (Speight, 2014a, 2016). Thus, a major source of propylene is from naphtha cracking which is a major process for ethylene production, but propylene is also a product from refinery cracking processes that produce other products. Propene can be separated from product mixtures by fractional distillation.

Propylene is also produced by olefin disproportionation—a reversible reaction between ethylene and butenes in which double bonds are broken and then reformed to form propene, for example in the simplest chemical sense the reaction can be represented as:

$$CH_2=CH_2 + CH_3CH_2CH=CH_2 \leftrightarrow 2CH_3CH=CH_2$$

Also, propylene is produced by dehydrogenation of propane converts propane into propene and by-product hydrogen:

$$CH_3CH_2CH_3 \rightarrow CH_3CH{=}CH_2 + H_2$$

In addition, olefins cracking processes include a broad range of technologies that catalytically convert higher molecular weight olefins (butene olefins to octene olefins, C_4 to C_8) into mostly propene (predominantly) and ethylene (lesser amounts).

The primary products manufactured from propylene are polypropylene, acrylonitrile, propylene oxide, and isopropyl alcohol (Table 3.6). Acrylonitrile and propylene oxide have both been shown to cause cancer, while propylene itself is not generally considered a health threat. Polypropylene is used to make injection-molded articles, such as automotive battery cases, steering wheels, outdoor chairs, toys, and luggage as well as fibers for upholstery, carpets, and special sports clothing. Oligomers (dimers, trimers, and tetramers) of propylene, which are made by acid-catalyzed polymerization, form mixtures known as polygas, used as high-octane motor fuel.

Propylene oxide is made via several methods. The classical one involves treating propylene with chlorine water to produce propylene chlorohydrin, and then using base to split out HCl. The primary use for propylene oxide is its oligomerization (to polypropylene glycols). These products combine with diisocyanates to produce high-molecular weight polyurethane foams, which make very good padding for furniture and vehicle seats. Manufacture of propylene glycol ($CH_3CHOHCH_2OH$) consumes about 30% of the propylene oxide produced. Like ethylene oxide, propylene oxide undergoes hydrolysis to yield the corresponding glycol. Propylene glycol is mainly used to make polyester resins, but it is also used in foods, pharmaceuticals, and cosmetics.

Another chemical, acrylonitrile ($CH_2{=}CHC{\equiv}N$) is made by direct ammoxidation of propylene:

$$2CH_3CH{=}CH_2 + 2NH_3 + 3O_2 \rightarrow 2CH_2{=}CHCN + 6H_2O$$

The major use is in making polyacrylonitrile, which is mainly converted to fibers (Orlon). It is also copolymerized with butadiene and styrene to produce high impact plastics.

3.4 C-4 Chemistry

C-4 Chemistry is the chemistry of butane isomers, butylene isomers, and butadiene. Butane (C_4H_{10}) is an alkane which is a gas at room temperature and atmospheric pressure and the term includes either of two structural isomers: *n*-butane or isobutane (or 2-methylpropane), or to a mixture of these isomers. However, using the correct nomenclature, butane refers only to the *n*-butane isomer:

TABLE 3.6 Industrial Uses of Propylene

Process	Target Product	Process Conditions			Reaction Components	Other Characteristics
		Pressure (MPa)	Temperature (°C)	Catalyst		
Polymerization	Polypropylene			Aluminum alkyls/ Titanium oxide		
Oxidation	Acrylonitrile		400	Phosphomolybdate	Ammonia Oxygen	Commercially valuable by-products are acetonitrile and hydrogen cyanide
	Propylene oxide				Oxygen Ethylbenzene	Commercially valuable by-product is *tert*-butyl alcohol
Addition						
Chlorohydrination	Propylene oxide	25	37	Tungsten	Hypochlorous acid	
Hydrolysis	Isopropyl alcohol		267		Water	

Common name	Normal butane, *n*-butane	Iso-butane, *i*-butane
IUPAC name	Butane	2-Methylpropane
Molecular structure	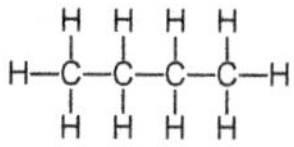	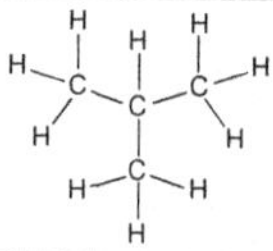
Skeletal structure		

Whatever the structure, butane isomers are highly flammable, colorless, and easily liquefiable gases.

Butene (C_4H_8) is a colorless gas that is present in crude oil as a minor constituent in quantities that are too small for viable extraction. Typically, butene is obtained by catalytic cracking of higher molecular weight hydrocarbons produced during crude oil refining. The cracking process yields a mixture of products, and the butene is extracted from this by fractional distillation. Butene can be used as the monomer for polybutene but polybutene is therefore commonly used as a copolymer (mixed with another polymer, either during or after reaction), such as in hot-melt adhesives.

Among the molecules which have the chemical formula C_4H_8 four isomers are alkenes which have four carbon atoms and one double bond in the structure but have different chemical structures. The IUPAC names and the common names of the butene isomers are:

IUPAC name	Common name	Molecular structure	Skeletal structure
1-Butene	α-Butylene But-1-ene Butene-1	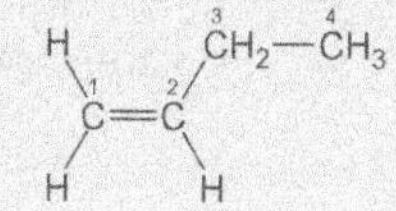	
Cis-but-2-ene[a]	*cis*-β-butylene *cis*-But-2-ene *cis*-Butene-2	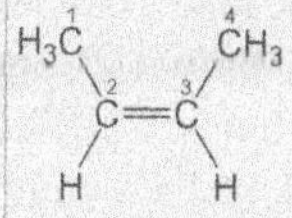	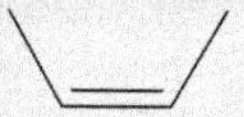
Trans-but-2-ene[a]	*trans*-β-Butylene *trans*-But-2-ene *trans*-Butene-2	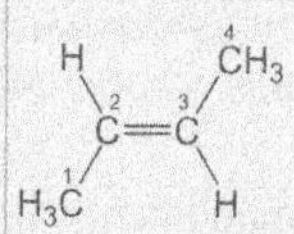	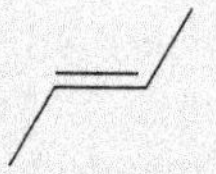
2-Methylpropene 2-Methylpropylene	Isobutylene	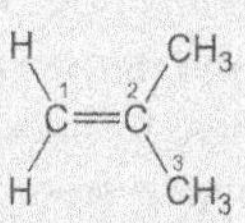	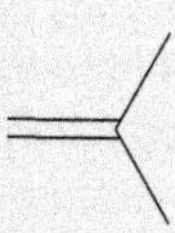

[a]Refers to the relative position of the methyl groups: *cis*—both methyl groups on the same side of the molecule and *trans* refers to both methyl groups on the opposite side of the molecule.

All four of these isomers are gases at room temperature and pressure but can be liquefied by lowering the temperature or raising the pressure. These gases are colorless, but do have distinct odors, and are highly flammable. Although not naturally present in crude oil in high percentages, they can be produced in a refinery by catalytic cracking or from petrochemical intermediates. Although they are stable compounds, the carbon-carbon double bonds make them more reactive than the respective alkane hydrocarbons.

Because of the double bonds, these 4-carbon alkenes can act as monomers in the formation of polymers, as well as having other uses as petrochemical feedstocks. They are used in the production of synthetic rubber. But-1-ene is a linear or normal alpha-olefin and isobutylene is a branched alpha-olefin. In a rather low percentage, but-1-ene is used as one of the comonomers, along with other alpha-olefins, in the production of HDPE and LLDPE. Butyl rubber is made by polymerization of isobutylene with isoprene 2–7% by weight. Isobutylene is also used for the production of methyl *tert*-butyl ether (MTBE) and isooctane, both of which improve the combustion performance of gasoline.

1,3-Butadiene is a simple conjugated diene with the formula C_4H_6 and is an important industrial chemical that is used as a monomer in the production of synthetic rubber. In the simplest sense, the molecule is two vinyl groups (CH_2=CH—) joined together and the term *butadiene* typically refers to 1,3-butadiene which has the structure H_2C=CH—CH=CH_2:

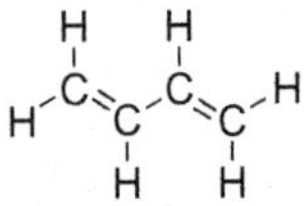

1,3-Butadiene

The name butadiene can also refer to the isomer, 1,2-butadiene (H_2C=C=CH—CH_3, also called buta-1,2-diene and methyl-allene) which is a *cumulated diene* but this allene is difficult to prepare and has no industrial significance. Briefly, an allene is a compound where one carbon atom has double bonds with each of its two adjacent carbon neighbors—allene derivatives are classified as polyenes—and will not undergo the same reactions as 1,3-butadiene.

Maleic anhydride is the main chemical made from *n*-butane. A complex catalyst is used for the oxidation reaction. The major uses for maleic anhydride are the making of unsaturated polyester resins (by reaction with glycol and phthalic anhydride) and tetrahydrofuran (by hydrogenation). MTBE is one of the leading chemicals currently being made from isobutylene (methyl propene) via the acid-catalyzed addition of methyl alcohol. MTBE has been added to gasoline as a required oxygenate. However, it is under attack as a groundwater contaminant and has been phased out of general use.

Polyisobutylene derivatives are easily made via the acid-catalyzed polymerization of isobutylene. The low molecular weight polymers are used as additives for gasoline and lubricating oils, whereas higher molecular weight polymers are used as adhesives, sealants, caulks, and protective insulation. Butyl rubber is made by polymerizing isobutylene with a small quantity of isoprene. Its main uses are in the making of truck tire inner tubes, inner coatings for tubeless tires, and automobile motor mounts. Hexamethylenediamine [HMDA, H_2H $(CH_2)_6NH_2$] is the principal industrial chemical made from butadiene. HMDA is polymerized with adipic acid to make a kind of nylon.

SBR accounts for about 40% of the total consumption of butadiene. SBR is the material used to make most automobile tires. Other synthetic rubbers, such as polybutadiene and polychloroprene (neoprene), make up another 25% of the butadiene market. The acrylonitrile-butadiene-styrene resin (ABS resin) is a widely used terpolymer that accounts for about 8% of the butadiene market.

3.5 BTX Chemistry

The heavy chemical industry, in its classical form, was based on inorganic chemistry, concerned with all the elements except carbon and their compounds, but including, as has been seen, the carbonates. Similarly, the light chemical industry uses organic chemistry, concerned with certain compounds of carbon such as the hydrocarbons, combinations of hydrogen and carbon. In the late 1960s the phrase "heavy organic chemicals" came into use for compounds such as benzene, phenol, ethylene, and vinyl chloride. Benzene and phenol are related chemically, and they are also related to toluene and the xylenes, which can be considered together as part of the aromatic group of organic chemicals, the aromatic compounds being most easily defined as those with chemical properties like that of benzene, toluene, and the xylene isomers.

In the crude oil refining and petrochemical industries, the acronym BTX refers to mixtures of benzene, toluene, and the three xylene (1,2-dimethylbenzene, 1,3-dimethylbenzene, and 1,4-dimethylbenzene), all of which are aromatic hydrocarbons.

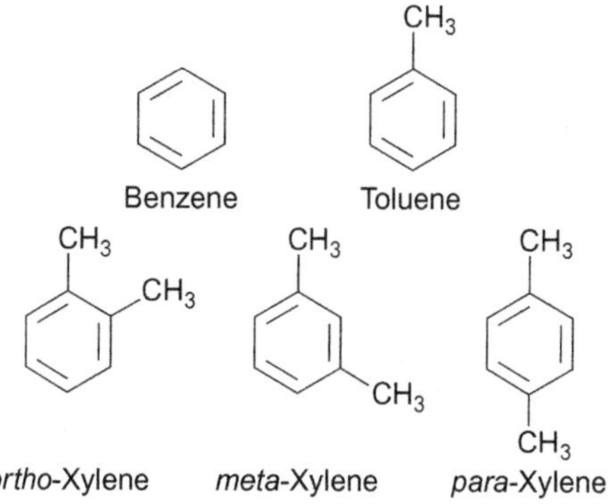

Benzene, toluene, and the isomeric xylenes.

If ethylbenzene ($C_6H_5C_2H_5$) is included in the mixture, its presence in the mixture is often formally acknowledged by the acronym BTEX.

	Benzene	Toluene	Ethyl-benzene	p-Xylene	m-Xylene	o-Xylene
Molecular formula	C_6H_6	C_7H_8	C_8H_{10}	C_8H_{10}	C_8H_{10}	C_8H_{10}
Molecular mass	78.12	92.15	106.17	106.17	106.17	106.17
Boiling point, °C	80.1	110.6	136.2	138.4	139.1	144.4
Melting point, °C	5.5	−95.0	−95.0	13.3	−47.9	−25.2

The three isomeric xylenes ($1,2$-$CH_3C_6H_4CH_3$, $1,3$-$CH_3C_6H_4CH_3$, and $1,4$-$CH_3C_6H_4CH_3$) and another isomer, ethylbenzene ($C_6H_5C_2H_5$) can be separated only with difficulty, but numerous separation methods have been worked out. The small letters o-, m-, and p- (standing for ortho-, meta-, and para-) preceding the name xylene are used to identify the three different isomers that vary in the ways the two methyl groups displace the hydrogen atoms of benzene. Ortho-xylene is used mostly to produce phthalic anhydride, an important intermediate that leads principally to various coatings and plastics. The least valued of the isomers is meta-xylene, but it has uses in the manufacture of coatings and plastics. Para-xylene leads to polyesters, which reach the ultimate consumer as polyester fibers under various trademarked names.

BTX chemistry focuses on the chemistry of benzene, toluene, and the xylene isomers (Fig. 3.1). Styrene, discussed under C-2 chemistry, is one of the main industrial chemicals made from benzene. Most benzene is alkylated with ethylene to form ethylbenzene, which is dehydrogenated to styrene.

$$C_6H_6 \rightarrow C_6H_5C_2H_5 \rightarrow C_6H_5CH{=}CH_2$$

Benzene is an important intermediate in the manufacture of industrial chemicals (Table 3.7) and the products from benzene are frequently feedstocks for the synthesis of additional organic chemicals. Chemically benzene, which forms the basis of the aromatics and considered a human carcinogen by the US EPA, is a closed, six-sided ring structure of carbon atoms with a hydrogen atom at each corner of the hexagonal structure. Thus, a benzene molecule is made up of six carbon (C) atoms and six hydrogen (H) atoms and has the chemical formula C_6H_6. Benzene is the simplest and most stable aromatics and is a single-ring system that contains conjugated double bonds (Chapter 2):

TABLE 3.7 Industrial Uses of Benzene

Process	Target Product	Process Conditions			Reaction Components	Other Characteristics
		Pressure (MPa)	Temperature (°C)	Catalyst		
Oxidation	Phenol	0.6	90–100		Cumene, oxygen	Most important phenol synthesis
	Maleic anhydride	0.1–0.2	350–400	Vanadium oxide	Butane oxygen	
	Styrene	0.1	580–590	Iron oxide	Ethylene benzene	
Addition						
Alkylation	Ethylbenzene	0.2–0.4	125–140	Aluminum chloride	Benzene, ethylene	Precursor to styrene
	Ethylbenzene	2.0	420–430	Zeolite	Benzene, ethylene	Precursor to styrene
	Cumene	0.3–1.0	250–350	Phosphoric acid silicate	Benzene, propylene	
	2.6-Xylenol	0.1–0.2	300–400	Aluminum oxide	Phenol, methanol	
Hydrogenation	Cyclohexanone	0.1	140–170	Palladium	Phenol, hydrogen	
	Cyclohexanol	1.0–2.0	120–200	Nickel/silicon oxide and aluminum oxide	Phenol, hydrogen	

Continued

TABLE 3.7 Industrial Uses of Benzene—cont'd

| Process | Target Product | Process Conditions | | | Reaction Components | Other Characteristics |
		Pressure (MPa)	Temperature (°C)	Catalyst		
	Cyclohexane	2.0–5.0	150–200	Nickel	Benzene, hydrogen	
	Aniline	0.18	270	Copper	Nitrobenzene, hydrogen	
Nitration	Nitrobenzene	0.1	60		Benzene, sulfuric acid, nitric acid	
Sulfonation	Surfactants	0.1	40–50		Alkylbenzenes/sulfur trioxide	
Chlorination	Chlorobenzene	0.1	30–40	Aluminum chloride/Iron chloride	Benzene, chlorine	
Condensation	Biphenol A	0.1	50–90	HCl	Phenol, acetone	

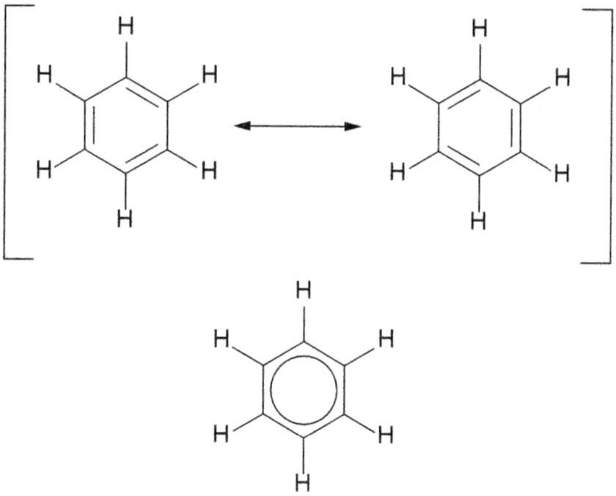

Benzene showing the electronic configurations that stabilize the ring system.

In the early days of the organic chemicals industry, benzene was obtained from the carbonization of coal, which produces combustible gas, coke, combustible gas, as well as a number of by-products, including benzene. Carbonization of coal to produce illuminating gas dates back in England to the very early years of the 19th century and the process is still employed in some countries, but more use is being made of natural gas. The carbonizing process is also used (with some slight modifications) to produce metallurgical coke, indispensable for the manufacture of iron and hence steel.

Toluene differs from benzene in that one of the hydrogen atoms is replaced by a special combination of carbon and hydrogen called a methyl group (CH_3).

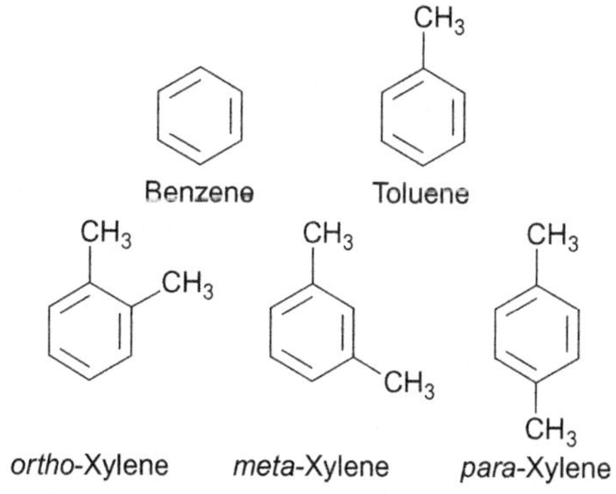

Benzene, toluene and the isomeric xylenes.

Toluene is also used as a solvent—the substance dissolved is usually also an organic compound. The xylene isomers have two methyl groups in different positions in the benzene ring, and thus all aromatics are to some extent interchangeable. In fact, one of the uses for toluene is to produce benzene by removing the methyl group. All of these hydrocarbons are useful as gasoline additives because of their antiknock properties (high octane numbers).

Benzene itself is perhaps the industrial chemical with the most varied uses of all. There are several routes to phenol, itself an important industrial chemical. In transforming benzene to the products obtained from it, other raw materials are required; for example, ethylene for the production of styrene, and sulfuric acid for the production of benzene sulfonic acid.

Cumene (isopropyl benzene) is produced by the Friedel-Crafts alkylation of benzene with propylene using an acid catalyst.

$$C_6H_6 + CH_3CH{=}CH_2 \rightarrow C_6H_5CH(CH_3)_2$$

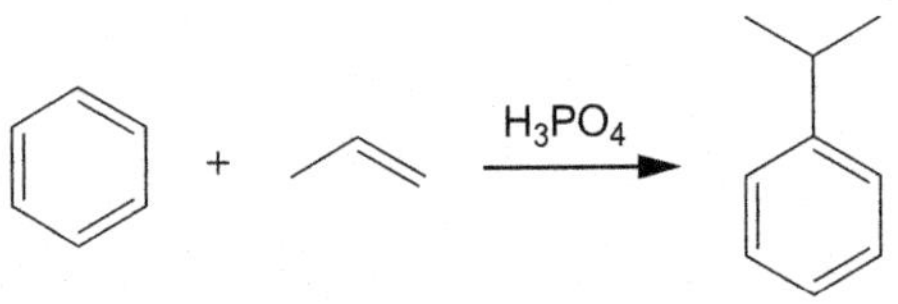

Although cumene is a high-octane automotive fuel and the high octane number makes it desirable in gasoline, most cumene is easily oxidized to the hydroperoxide, which is readily cleaved in dilute acid to phenol and acetone.

$$C_6H_5CH(CH_3)_2 \rightarrow C_6H_5CO_2H(CH_3)_2 \rightarrow C_6H_5OH + (CH_3)_2C{=}O$$

Thus, almost all of the cumene produced is used to make phenol (C_6H_5OH) and acetone [$(CH_3)_2CO$] which have a number of important commercial uses, but they also have an important use together. Phenol and acetone can be condensed to form bisphenol A, which is used in the production of polycarbonate and epoxy resins.

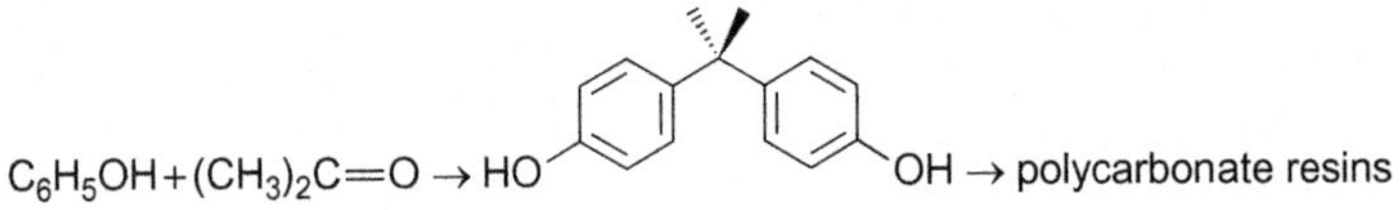

$$C_6H_5OH + (CH_3)_2C{=}O \rightarrow HO\text{———}OH \rightarrow \text{polycarbonate resins}$$

Acetone [$(CH_3)_2C{=}O$], a highly volatile and flammable organic chemical, is irritating to the eyes, nose, and throat. Symptoms of exposure to large quantities of acetone may include headache, unsteadiness, confusion, lassitude, drowsiness, vomiting, and respiratory depression. Reactions of acetone in the lower atmosphere contribute to the formation of ground-level ozone. Ozone (a major component of urban smog) can affect the respiratory system, especially in sensitive individuals such as asthmatics or allergy sufferers. If released into water, acetone will be degraded by microorganisms or will evaporate into the atmosphere, although degradation by microorganisms will be the primary removal mechanism. Once acetone reaches the troposphere (the layer known as the *lower atmosphere*), it will react with other gases, contributing to the formation of ground-level ozone (O_3) and other air pollutants.

Benzene is also an important starting material for the manufacture of cyclohexane (C_6H_{12}). The process involves the hydrogenation of benzene (over a nickel or platinum catalyst). However, most of the cyclohexane is converted to adipic acid [$(CH_2)_4(COOH)_2$, hexanedioic acid, hexane-1,6-dicarboxylic acid, hexane-1,6-dioic acid]:

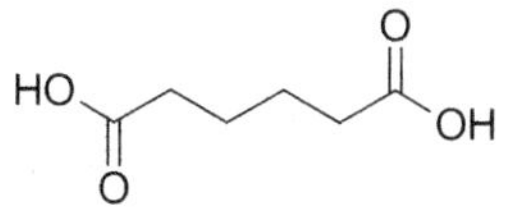

Adipic acid

The process is an oxidation process in which the intermediate chemicals cyclohexanol and cyclohexanone are oxidized to the adipic acid:

$$HOC_6H_{11} + [O] \rightarrow C_6H_{10}O_4 + HNO_2 + H_2O$$
Cyclohexanol   Adipic acid

$$O{=}C_6H_{10} + [O] \rightarrow C_6H_{10} + H_2O$$
Cyclohexanone   Adipic acid

In addition, adipic acid can be reacted with 1,6-hexamethylenediamine ($H_2NCH_2CH_2CH_2CH_2CH_2CH_2NH_2$) to produce nylon-6,6, a very strong synthetic fiber.

$$n\text{HOOC}(CH_2)_4\text{COOH} + n\text{H}_2\text{N}(CH_2)_6\text{NH}_2$$
Adipic acid   Hexamethylenediamine

$$\rightarrow \left[-OC(CH_2)_4CONH(CH_2)_6NH- \right]_n + 2n\text{H}_2\text{O}$$
Nylon−6,6

Most carpets are made of nylon, as are many silk-like garments, some kinds of rope, and many injection-molded articles. Caprolactam ($C_6H_{11}NO$) is also used to make nylon. Nylon-6,6 is made by direct polymerization of caprolactam, often obtained by reaction of cyclohexanone with hydroxylamine, followed by rearrangement of the oxime. Although nylon-6,6 is the dominant nylon produced in the United States, nylon-6 is the leading nylon product in Europe.

Aniline ($C_6H_5NH_2$) is made by nitration of benzene to nitrobenzene followed by hydrogenation of the nitrobenzene using a copper-based catalyst in which the copper is suspended in silica (Cu/SiO_2) to produce aniline:

$$C_6H_6 \rightarrow C_6H_5NO_2$$
Nitrobenzene

$$C_6H_5NO_2 \rightarrow C_6H_5NH_2$$
Aniline

The major use of aniline is in making diisocyanates (RN=C=O), which are used in producing polyurethane materials (such as home insulation).

Alkylbenzene sulfonates ($RC_6H_5SO_3Na$) are important surfactant compounds used in laundry detergents. In the manufacturing process, alkylbenzenes (made by the Friedel-Crafts alkylation of benzene using linear olefin molecules that have up to 12 carbon atoms) are sulfonated, and the sulfonic acids are then neutralized with sodium hydroxide (NaOH).

Toluene diisocyanate (TDI) is polymerized with diols to produce polyurethanes, which are used to make flexible foam for furniture cushions, mattresses, and carpet pads. Trinitrotoluene (TNT) is made via a stepwise nitration of toluene in the 2, 4, and 6 positions—TNT is a high explosive and missile propellant. Phthalic anhydride is made by air oxidation of *ortho*-xylene. Approximately 50% w/w of the phthalic anhydride produced is used to make plasticizers, especially the compound dioctyl phthalate [also called bis(2-ethylhexyl) phthalate, di-2-ethylhexyl phthalate, diethylhexyl phthalate, DEHP] for softening PVC plastic. Phthalic anhydride is also used to make unsaturated polyester resins and alkyd paints.

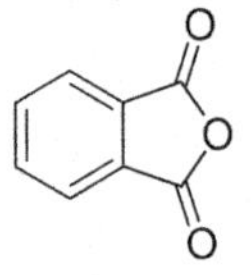

Phthalic anhydride [bis(2-ethylhexyl) phthalate, di-2-ethylhexyl phthalate, diethylhexyl phthalate, DEHP]

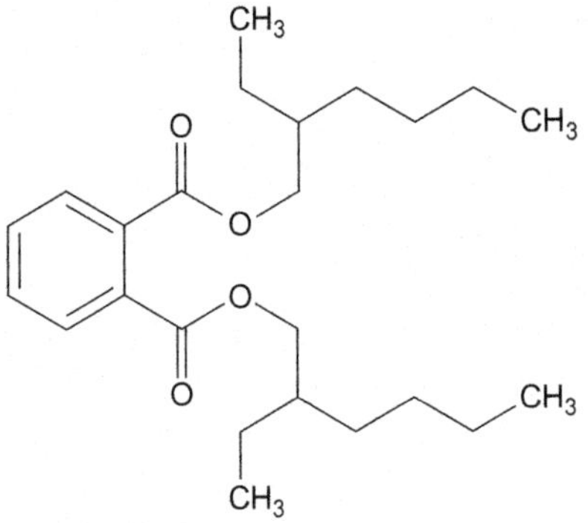

Dioctyl phthalate

3.6 Other Chemical Reactions

In addition to the C-1, C-2, C-3, and C-4 chemistries, there is a wide variety of chemical reactions that are used to produce organic chemicals, some of which are specific to one or two products, whilst others (e.g., oxidation, halogenation, hydrogenation) are used widely in many processes. For this reason, the majority of emissions from the industrial production of organic chemicals originate from a relatively few, but commonly used, unit processes. These reactions are presented here (alphabetically rather than by order of inferred importance) for information purposes.

3.6.1 Alkylation

Alkylation is the introduction of an alkyl group into an organic compound by substitution or addition. There are six types of alkylation reaction: (1) substitution for hydrogen bound to carbon, such as ethylbenzene ($C_6H_5C_2H_5$) from benzene (C_6H_6) and ethylene (CH_2=CH_2), (2) substitution for hydrogen attached to nitrogen, (3) substitution for hydrogen in a hydroxyl group of an alcohol or phenol, and (4) addition to a tertiary amine to form a quaternary ammonium compound. Acylation reactions that fall outside of the context of this text—such as the addition to a metal to form a carbon-metal bond as well as additions to sulfur or silicon—are not included in this list. The greatest use of the alkylation process is in refineries for the production of alkylates that are used as a blending stock to produce gasoline (Speight, 2014a,b). Other major alkylation products include ethylbenzene ($C_6H_5CH_2CH_3$), cumene [$C_6H_5CH(CH_3)_2$], and linear alkylbenzene derivatives [$C_6H_5(CH_2)_nCH_3$, where n is typically 3 or greater].

Alkylation is commonly carried out in liquid phase at temperatures higher than 200°C (390°F) at above atmospheric pressures. Sometimes vapor phase alkylation is more effective. Alkylation agents are usually olefins, alcohols, alkyl sulfates, or alkyl halides and the typical catalysts are hydrofluoric acid (HF), sulfuric acid (H_2SO_4), or phosphoric acid (H_3PO_4, which is also known as orthophosphoric acid or phosphoric(V) acid). Higher process temperatures cause the expected lowering of product specificity and increased by-product formation. Some more recent alkylation processes (such as used for the production of ethylbenzene and cumene) use zeolite catalysts as they can be more efficient and may have lower emissions. Lewis acids, such as aluminum chloride ($AlCl_3$) or boron trifluoride (BF_3), may also be used as catalysts.

There are environmental issues, related to the emission of volatile organic compounds (VOCs), which arise during alkylation processes and which include although based on data for the production of ethylbenzene, cumene, and linear alkylbenzene, the emission of VOCs from alkylation reactions tend to be low compared to the emission of VOCs from other refinery processes. However, the by-products and waste disposal of alkyl halides and sulfate derivatives can be problematic.

3.6.2 Ammonolysis

Ammonolysis is the process of forming amines using, as amminating agents, ammonia or primary and secondary amines. These reactions may also include hydroammonolysis in which amines are formed directly from carbonyl compounds using an ammonia-hydrogen mixture and a hydrogenation catalyst.

The four main ammonolysis reaction types are: (1) double decomposition in which the ammonia (NH_3) is split into the amino function (NH_2) which becomes part of the amine, and a hydrogen atom which reacts with a radical that is being substituted, (2) dehydration, in which the ammonia reacts to produce water and amines, (3) simple addition, in which both fragments of the ammonia molecule ($-NH_2$ and $-H$) become part of the new amine, and (4) multiple activity, in which ammonia reacts with the amine products to form secondary amine derivatives and tertiary amine derivatives. Nevertheless, typically, the major products of ammonolysis are carbamic acid (H_2NCOOH), ethanolamine derivatives ($HOCH_2CH_2NHR$), and alkylamine (RNH_2) derivatives.

In the case of the production of ethanolamine, the emission of VOCs arising from the reactors is minimal although there tends to be waste gases associated with any distillation. Any off-gas containing ammonia or amine derivatives is washed or incinerated in order to avoid odor problems and any hydrogen cyanide that is produced during the production of acrylonitrile (CH_2=$CHCN$) can may be recovered.

In the case of water contamination, unreacted ammonia can be recovered from alkaline effluents by steam stripping and recycled back to the process. Ammonia remaining in the effluent can be neutralized with sulfuric acid to produce a precipitate of ammonium sulfate [$(NH_4)_2SO_4$] that can be separated for use as fertilizer or biologically treated. In addition, wastewater containing impurities such as methanol and amine derivatives can be disposed of either by incineration or by biological treatment. Solid waste materials from base of the stripping unit are incinerated.

3.6.3 Ammoxidation

In the organic chemicals industry, ammoxidation is a process for the production of nitrile derivatives involving the use of ammonia and oxygen—the usual feedstock is an alkene. The process is an important application in the production of acrylonitrile (CH_2=$CHCN$) and process involves the gas phase oxidation of olefins, such as propylene with ammonia in the presence of oxygen and vanadium or molybdenum based catalysts:

$$2CH_3CH=CH_2 + 2NH_3 + 3O_2 \rightarrow 2CH_2=CHCN + 6H_2O$$

Acetonitrile ($CH_3C\equiv N$) is a by-product of this process. This colorless liquid is the simplest organic nitrile and is used as a polar aprotic solvent in organic synthesis and in the purification of butadiene. Acetonitrile is also a common two-carbon building block in the synthesis of organic chemicals—the most important applications are in the production of acrylic fibers, thermoplastics, and adiponitrile ($N\equiv CCH_2CH_2CH_2CH_2C\equiv N$, a viscous, colorless liquid, an important precursor to the polymer Nylon-6,6), as well as specialty polymers.

3.6.4 Carbonylation

Carbonylation (carboxylation) is the combination of an organic compound with carbon monoxide and carbonylation refers to reactions that introduce carbon monoxide into organic and inorganic compounds. The carbonylation reaction is used to make aldehydes and alcohols containing one additional carbon atom—the major product involves the introduction of the carbonyl group ($>C=O$) into the starting materials to produce aldehydes ($-CHO$), carboxylic acids ($-CO_2H$), and esters ($-CO_2R$). In the laboratory and in industry, carbonylation reactions are the basis of two main types of reactions, *hydroformylation* and *Reppe Chemistry*.

The hydroformylation reaction involves the addition of both carbon monoxide and hydrogen to unsaturated organic compounds, usually alkene derivatives to produce aldehyde derivatives:

$$RCH=CH_2 + H_2 + CO \rightarrow RCH_2CH_2CHO$$

The reaction requires the presence of metal catalysts to bond the carbon monoxide and the hydrogen to the olefin. The hydroformylation ("oxo" process) is a variant where olefins are reacted with carbon monoxide and hydrogen ("synthesis gas") in the presence of a cobalt or rhodium catalyst (such as in the production of *n*-butyraldehyde from propylene):

$$CH_3CH=CH_2 + H_2 + CO \rightarrow CH_3CH_2CH_2CHO$$

On the other hand, the Reppe reaction involves the addition of carbon monoxide and an acidic hydrogen donor with the organic substrate. Large-scale applications of this type of carbonylation are the conversion of methanol to acetic acid:

$$CH_3OH + CO \rightarrow CH_3CO_2H$$

In a related hydrocarboxylation, alkene derivatives are converted to carboxylic acid derivative in the presence of metal catalysts:

$$RCH=CH_2 + H_2O + CO \rightarrow RCH_2CH_2CO_2H$$

For example, as part of the industrial synthesis of ibuprofen, a benzyl alcohol derivative is converted to the corresponding carboxylic acids by way of palladium-catalyzed carbonylation reaction:

$$ArCH(CH_3)OH + CO \rightarrow ArCH(CH_3)CO_2H$$

Also, propanoic acid is mainly produced by the hydrocarboxylation of ethylene using nickel carbonyl [nickel tetracarbonyl, $Ni(CO)_4$] as the catalyst:

$$CH_2=CH_2 + H_2O + CO \rightarrow CH_3CH_2CO_2H$$

Hydroesterification is similar to hydrocarboxylation, but uses alcohol derivatives instead of water. Other related industrially oriented reactions include the Koch reaction, which involves the addition of carbon monoxide to unsaturated compounds in the presence of a catalyst, such as in the production of the important intermediate, glycolic acid:

$$\underset{\text{Formaldehyde}}{HCHO} + CO + H_2O \rightarrow \underset{\text{Glycolic acid}}{HOCH_2CO_2H}$$

Alkyl, benzyl, vinyl, aryl, and allyl halides can also be carbonylated in the presence carbon monoxide and suitable catalysts such as manganese (Mn), iron (Ni), or nickel (Ni).

In terms of environmental issues of the carbonylation processes, the processes typically generate vent streams containing VOCs in addition to carbon dioxide as well as other nonVOCs. Residual gas is recovered and used as fuel or flared. In addition, heavy metals (from the catalyst) must be removed from wastewater prior to biological treatment. The solid waste is typically the spent catalysts but caution is advised when planning the disposal process because of the potential for adsorbed material, which might be leached from the solid during periods of heavy rain, melting snow, or acid rain.

Waste: Spent catalysts.

3.6.5 Condensation

A condensation reaction, is a chemical reaction in which two reactants (each reactant contains a functional group) combine to form a higher molecular weight product together with the loss of a lower molecular weight species, such as water, hydrogen chloride, methanal, or acetic acid—typically, the most common lower molecular weight species is water.

A common type of industrial condensation reaction is a condensation polymerization reaction in which a series of condensation steps takes place whereby monomers or monomer chains add to each other to form longer chains.

$$\underset{\text{Di-acid}}{HOOCR^1COOH} + \underset{\text{Diamine}}{H_2NR^2NH_2} \rightarrow \underset{\text{Polyamide}}{-\left(COR^1CONHR^2NH\right)} -_n + nH_2O$$

In this reaction, R^1 and R^2 are alkyl or aryl moieties and the —CONH— moiety is the amid bond. The reaction is also known as *step-growth polymerization* and is used in processes such as the synthesis of polyesters or nylons. This reaction may be either (1) homopolymerization such as in the use of a single monomer with two different end groups that condense or (2) copolymerization of two comonomers with the necessary functional groups on each of the monomers (as shown in the equation above). Small molecules are usually liberated in these condensation steps, unlike polyaddition reactions such as in the polymerization of ethylene which is an addition reaction and there is no elimination of lower molecular weight by-products:

$$nCH_2{=}CH_2 \rightarrow -(CH_2CH_2)-_n$$

Environmental issues that arise during the use of condensation processes are usually limited to reactor emissions that are generally small and can be mitigated in a combustion unit. In addition, waste water volumes are generally low and the effluents mainly consist of reaction water if recycling after phase separation is not possible. The effluent is composed of high-boiling components (condensation products/by-products) that often show moderate or poor biodegradability, and low-boiling components that typically are more susceptible to biodegradation.

3.6.6 Dealkylation

Dealkylation is a chemical process through which alkyl groups are removed from a given compound such as the dealkylation of toluene to produce benzene or the conversion of 1,2,4-trimethylbenzene to xylene:

$$C_6H_5CH_3 + H_2 \rightarrow C_6H_6 + CH_4$$
$$1,2,4 - C_6H_3(CH_3)_3 + H_2 \rightarrow C_6H_4(CH_3)_2 + CH_4$$

When hydrogen is used (above equation) the process is known as hydrodealkylation. The hydrodealkylation process typically requires a high temperature and a high pressure as well as the presence of catalyst—the catalyst is predominantly a transition metal-containing catalyst using metal derivatives of chromium or molybdenum.

The dealkylation process is a common process in the organic chemicals industry, especially in industries such as crude oil refining and the pharmaceuticals industry. In fact, in the pharmaceuticals industry dealkylation can lead to activation of certain compounds and can also promote better absorption and efficacy of the pharmaceutical. The dealkylation process is also frequently employed by manufacturers of fertilizers and pesticides.

Many dealkylation processes use oxidative dealkylation (also known as *O*-dealkylation) in which an oxide is used to assist in removal of the alkyl group from the organic substrate through a reduction-oxidation reaction (redox reaction).

3.6.7 Dehydration

The dehydration of organic chemicals is a decomposition reaction in which a new compound is formed by the expulsion of water. The dehydration process is a subprocess of the condensation process that requires a catalyst. Examples of common dehydrating reactions include:

Reactions	Equations
Conversion of alcohols to ethers	$2ROH \rightarrow ROR + H_2O$
Conversion of alcohols to alkene	$RCH_2CHOH\text{-}R \rightarrow RCH{=}CHR + H_2O$
Conversion of carboxylic acids to acid anhydrides	$2RCOOH \rightarrow (RCO)_2O + H_2O$
Conversion of amides to nitriles	$RCONH_2 \rightarrow RCN + H_2O$

The reverse of a dehydration reaction is a hydration reaction in which water is added to a substrate such as an olefin.

Dehydration:

$$C_2H_5OH \rightarrow CH_2{=}CH_2 + H_2O$$

Hydration:

$$CH_2{=}CH_2 + H_2O \rightarrow C_2H_5OH$$

Typical dehydrating agents used in organic synthesis include concentrated sulfuric acid (H_2SO_4), concentrated phosphoric acid (H_3PO_4), and aluminum oxide (Al_2O_3). In the related condensation reaction, water is released from two different reactants.

3.6.8 Dehydrogenation

Dehydrogenation is the process by which hydrogen is removed from an organic compound to form a new chemical (e.g., to convert saturated into unsaturated compounds). It is used to produce aldehydes and ketones by the dehydrogenation of alcohols. Important products include acetone, cyclohexanone, methyl ethyl ketone, and styrene.

Dehydrogenation is most important in the refinery cracking process, where saturated hydrocarbons are converted into olefins (Speight, 2014a,b, 2016). The process is applied to appropriate hydrocarbon feedstocks (e.g., naphtha) in order to produce the very large volumes of ethylene, propylene, butene derivatives, and butadiene derivatives that are required as feedstocks for the organic chemicals industry.

Cracking (thermal decomposition) of organic compounds may be achieved by thermal (noncatalytic) processes or by catalytic processes and provides a process to convert higher boiling fractions into saturated, nonlinear paraffinic compounds, naphthenes, and aromatics. The concentration of olefin derivatives in the product stream is very low, so this method is more useful for the preparation of blending stocks (such as naphtha and kerosene) for the production of fuels (gasoline and diesel fuel, respectively). Olefin derivatives are more widely produced by the steam cracking of petroleum fractions—in the process a hydrocarbon stream is heated, mixed with steam and, depending on the feedstock, further heated to a cracking-temperature on the order of 600–650°C (1110–1200°F). The conversion of saturated hydrocarbon streams to unsaturated compounds is highly endothermic, and so the process requires a high energy input. High-temperature cracking is also used to produce pyrolysis gasoline from naphtha, gas oil, or high-boiling refinery streams.

Environmental issues that arise from the of dehydrogenation processes include the potential for hydrogen-rich vent streams that are produced as a result of the process but which can be employed as a hydrogen feedstock for other processes or as a refinery fuel. Any volatile hydrocarbons that occur in purge and vent gases will require collection and treatment and can be combined with beneficial energy production. On the other hand, quench water, dilution steam, decoking water, and flare water discharges will require treatment and wastewater streams with a high content of pollutants will require pretreatment prior to acceptance in a biological degradation plant.

3.6.9 Esterification

Esterification typically involves the formation of esters from an organic acid and an alcohol. Esters often have a characteristic pleasant, fruity odor which makes them appropriates for extensive use in the fragrance and flavor industry.

The most common method of esterification is the reaction of a concentrated alcohol and a concentrated carboxylic acid with the elimination of water:

$$\underset{\text{Acid}}{R_1CO_2H} + \underset{\text{Alcohol}}{R_2OH} \leftrightarrow \underset{\text{Ester}}{R_1CO_2R_2} + H_2O$$

In this equation, R_1 and R_2 are alkyl moieties. Only strong carboxylic acids react sufficiently quickly without a catalyst, so a strong mineral acid (such as sulfuric acid or hydrogen chloride) must usually be added to aid the reaction. Acid anhydrides are also used, e.g., in dialkyl phthalate production. The sulfonic acid group can be bound chemically to a polymeric material and so cation exchangers, such as sulfonated polystyrene, enable esterification under mild conditions.

The equilibrium of the reaction can be shifted to the ester by increasing the concentration of one of the reactants, usually the alcohol. In production scale esterification the reaction mixture is refluxed until all the condensation water is formed, and the water or the ester product is continuously removed from the equilibrium by distillation. The main products from esterification reactions are dimethyl terephthalate, ethyl acrylate, methyl acrylate, and ethyl acetate which have considerable economic importance in many applications (such as for fibers, films, adhesives and plastics). Some volatile esters are used as aromatic materials in perfumes, cosmetics, and foods.

Also, the reaction of alcohol derivatives with carboxylic acid derivatives is not the only process for producing ester derivatives. Alcohols react with acyl chlorides (acid chlorides) and acid anhydrides to produce esters:

$$R_1COCl + R_2OH \rightarrow R_1CO_2R_2 + HCl$$
$$(R_1CO)_2O + R_2OH \rightarrow R_1CO_2R_2 + R_1CO_2H$$

Again, in these equations R_1 and R_2 are alkyl moieties. The reactions are irreversible, simplifying and driving the process to completion. Since acyl chlorides and acid anhydrides also react with water, anhydrous conditions are preferred. The analogous acylation of amine derivatives to yield amide derivatives are less sensitive because amines are stronger nucleophiles and react more rapidly than does water.

Finally, ethylene, acetic acid, and oxygen react (in the presence of palladium-based catalysts) to yield vinyl acetate:

$$2CH_2{=}CH_2 + 2CH_3CO_2H + O_2 \rightarrow 2C_2H_3O_2CCH_3 + 2H_2O$$

Direct routes (such as the alcohol-acid reaction) to this same ester are not possible because vinyl alcohol ($CH_2=CHOH$) is unstable and has the propensity under normal conditions (ambient temperature and ambient pressure) to convert to acetaldehyde (CH_3CHO).

Environmental issues related to the esterification process relate to solvent vapor which can be collected and treated (such as by incineration, adsorption). The generation of aqueous effluents is not extensive since water is the only by-product of the esterification reaction. In addition, the choice of solid polymer based ion exchange resins for wastewater treatment avoids the need for extensive treatment facilities. Most esters possess low toxicity because they are easily hydrolyzed on contact with water or moist air, and so the properties of the acid and alcohol components are more important. Furthermore, waste streams can be reduced by recovering (and reusing) any organic solvents, water, and alcohol components. Any wastes from waste water treatment can be incinerated (high-boiling wastes) or recovered by distillation for reuse (low-boiling wastes).

3.6.10 Halogenation

Generally, halogenation is the reaction of a halogen with an alkane in which the introduction of halogen atoms occurs into the organic molecule by an addition reaction or by a substitution reaction. In organic synthesis this may involve the addition of molecular halogens: chlorine, bromine, iodine, or fluorine (Cl_2, Br_2, I_2, or F_2) or hydrohalogenation using: hydrogen chloride, hydrogen bromide, hydrogen iodide, or hydrogen fluoride (HCl, HBr, HI, or HF) to carbon-carbon double bonds. Substitution reactions involve replacing hydrogen atoms in olefin derivatives, paraffin derivatives, or aromatic derivatives with halogen atoms. Chlorination is the most important industrial halogenation reaction and chlorinated organic products include chlorinated aromatic derivatives, chlorinated methane derivatives, and chlorinated ethane derivatives but caution is advised since toxicity issues will demand additional control measures. Fluorination is used almost exclusively in the manufacture of fluorocarbons.

Several pathways exist for the halogenation of organic compounds, including free radical halogenation, electrophilic halogenation, and the halogen addition reaction. For example, saturated hydrocarbon derivatives (alkanes) typically do not add halogens but undergo free radical halogenation which involves the substitution of a hydrogen atom (or hydrogen atoms) by a halogen atom (or halogen atoms). The chemistry of the halogenation of alkane derivatives is usually determined by the relative weakness of the available carbon-hydrogen (C—H) bonds. Free radical halogenation is used for the industrial production of chlorinated methane derivatives:

$$CH_4 + Cl_2 \rightarrow CH_3Cl + HCl$$

Rearrangement often accompanies such free radical reactions. On the other hand, unsaturated compounds, especially alkene derivatives and alkyne derivatives add halogens across the unsaturated bond:

$$R^1CH=CHR^2 + X_2 \rightarrow R^1CHXCHXR^2$$

However, aromatic compounds are subject to electrophilic halogenation but addition to the ring system can occur under extreme conations:

$$RC_6H_5 + X_2 \rightarrow RC_6H_4X + HX$$

The ease of the reaction is influenced by the halogen—fluorine and chlorine are more electrophilic and, as a result, are more aggressive halogenating agents. On the other hand, bromine is a weaker halogenating agent than both fluorine and chlorine, while iodine is least reactive halogenating agent of the halogens. Accordingly, the ease of hydrogenolysis (removal of the halogen with hydrogen as HX) follows the reverse trend: iodine is most easily removed from organic compounds and organo fluorine compounds are most stable organo-halogen compounds.

Environmental issues of halogenation processes involve the treatment of waste gases which requires a distinction between acidic streams, reaction gases, and neutral waste streams. Air streams from tanks, distillation columns, and process vents can be collected and treated using such techniques as low temperature condensation or incineration. However, the treatment of acid gas streams is more complex because any equipment in contact with acid gases and water must be constructed from acid-resistant materials or internally coated to prevent corrosion (Speight, 2014c). The halogen content of the waste gas may represent a valuable raw material and pollution control techniques offer an opportunity for its recovery and reuse (either as hydrogen-halogen or aqueous solutions). The techniques may include: (1) product recovery (by vapor stripping of liquid streams followed by recycling to the process), (2) scrubbing the acid gas with an easily halogenated compound preferably a raw material used in the process, (3) absorbing the acid gas in water to give aqueous acid which is often followed by caustic scrubbing for environmental protection, (4) washing out organic constituents with organic solvents, and (5) condensing out organic by-products for use as feedstock in another process (Mokhatab et al., 2006; Speight, 2007, 2014a).

Environmental issues also arise with wastewater streams because the biological degradability (biodegradability) of halogenated hydrocarbons (especially aromatic derivatives) decreases as the halogen content increases. Only chlorinated hydrocarbon derivatives with a low degree of chlorination are degradable in biological waste water treatment plants but only if the concentration of the chlorinated hydrocarbon derivatives does not exceed certain levels. Prior to biological treatment, wastewater containing chlorinated organic compounds usually requires preliminary purification by stripping, extraction, and adsorption (using activated carbon or suitable polymeric resins). Wastewater contamination can be substantially reduced by avoiding the water quenching of reaction gases to separate hydrogen chloride (for example in the production of chlorinated ethane derivatives and chlorinated ethylene derivatives).

Finally, solid waste materials as a result of the halogenation process may arise from sources such as reactor residues or spent catalyst. Incineration is a common method for destruction of the organic components of the solid wastes but considerable attention must be paid to incineration conditions in order to avoid the formation of dioxins. If incineration is used, there is the need for an efficient flue gas scrubbing operation.

3.6.11 Hydrogenation

Catalytic hydrogenation refers to the addition of hydrogen to an organic molecule in the presence of a catalyst. The process can involve direct addition of hydrogen to the double bond of an unsaturated molecule; amine formation by the replacement of oxygen in nitrogen-containing compounds; and alcohol production by addition to aldehydes and ketones. These reactions are used to readily reduce many functional groups; often under mild conditions and with high selectivity.

Reactant	Product
Alkene: $R_2C{=}CR'_2$	Alkane: $R_2CHCHR'_2$
Alkyne: RCCR	Alkene: cis-$RHC{=}CHR'$
Aldehyde: RCHO	Primary alcohol: RCH_2OH
Ketone: R_2CO	Secondary alcohol: R_2CHOH
Ester: RCO_2R'	Mixed alcohols: $RCH_2OH + R'OH$
Imine: $RR'CNR''$	Amine: $RR'CHNHR''$
Amide: $RC(O)NR'_2$	Amine: $RCH_2NR'_2$
Nitrile: RCN	Primary amine: RCH_2NH_2
Nitro: RNO_2	Amine: RNH_2

Hydrogenation is an exothermic reaction and the equilibrium usually lies far towards the hydrogenated product under most operating temperatures. It is used to produce a wide variety of chemicals such as cyclohexane, aniline, n-butyl alcohol, hexamethylene diamine, as well as ethyl hexanol, and important isocyanate derivatives such as TDI and methylene diphenyl isocyanate both of which are used to produce urethane derivatives and thence urethane polymers.

Hydrogenation catalysts may be heterogeneous or homogeneous—heterogeneous catalysts are solids and form a distinct phase in the gases or liquids. Many metals and metal oxides have general hydrogenation activity—nickel, copper, cobalt, chromium, zinc, iron, and the platinum group are among the elements most frequently used as commercial hydrogenation catalysts.

Generally, the emission of VOCs for hydrogenation processes is relatively low since hydrogen-rich vent streams are typically sent to combustion units. The main issues with hydrogen are likely to arise from sulfur impurities in the process feedstocks or from the dust and ash by-products of the hydrogen production itself. Small quantities of sulfur compounds (such as sulfur dioxide,

SO_2, and hydrogen sulfide, H_2S) can be absorbed in dilute caustic solutions or adsorbed on activated charcoal as part of a gas cleaning operation while larger quantities would probably have to be converted to liquid sulfur or to solid sulfur (Mokhatab et al., 2006; Speight, 2007).

The hydrogenation of oxygenated compounds may generate water, which ends up as waste water, but volume of wastewater produced from hydrogenation reactions is not excessive. Moreover, the products often show good biodegradability and low toxicity whereas aniline compounds (from a hydrodenitrogenation process) will need disposal measures that are additional to the bio treatment technologies. In addition, the spent catalysts may be sent to disposal or may be treated for reclamation of any precious metals.

3.6.12 Hydrolysis

Hydrolysis involves the reaction of an organic chemical with water to form two or more new substances and usually means the cleavage of chemical bonds by the addition of water. In fact, Hydrolysis can be the reverse of a condensation reaction in which two molecules join together into a larger one and eject a water molecule. Thus hydrolysis adds water to break down, whereas condensation builds up by removing water. Hydration is the process variant where water reacts with a compound without causing its decomposition. These routes are used in the manufacture of alcohols (e.g., ethanol), glycols (e.g., ethylene glycol, propylene glycol), and propylene oxide.

Acid-base-catalyzed hydrolysis reactions and processes are very common—one example is the hydrolysis of ester derivatives or amide derivatives. The hydrolysis reaction occurs when the nucleophilic reactant (a nucleus-seeking agent, e.g., water or hydroxyl ion) attacks the carbon of the carbonyl group of the ester or the amide using an aqueous base medium since hydroxyl ions are better nucleophiles than polar molecules such as water. In acidic medium, the carbonyl group becomes protonated, and this leads to a much easier nucleophilic attack. The products for both hydrolyses are carboxylic acid derivatives. The oldest commercially practiced example of ester hydrolysis is the saponification reaction which results in the formation of soap and involves the hydrolysis of a triglyceride (fat) with an aqueous base such as sodium hydroxide (NaOH). During the process, glycerol ($CH_2OHCHOHCH_2OH$) is formed:

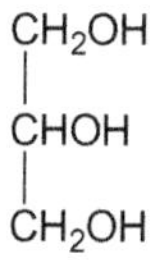

$$
\begin{array}{l}
CH_2OH \\
| \\
CHOH \\
| \\
CH_2OH
\end{array}
$$

glycerol
(glycerine)

The carboxylic acids react with the base, converting them to salts.

In the hydrolysis process, there are generally low-to-no emission of VOCs emanating from the reactor and, in most cases, the products of the hydrolysis process and the hydration process are biodegradable.

3.6.13 Nitration

The nitration process is a general class of chemical process for the introduction of a nitro group ($—NO_2$) into an organic chemical compound. The term is also applied (somewhat incorrectly) to the different process of forming nitrate esters between alcohol derivatives and nitric acid, as occurs in the synthesis of nitroglycerin. The difference between the resulting structure of nitro compounds and nitrates is that the nitrogen atom in nitro compounds is directly bonded to a non-oxygen atom, typically carbon or another nitrogen atom, whereas in nitrate esters, also called organic nitrates, the nitrogen is bonded to an oxygen atom that in turn usually is bonded to a carbon atom. There are many major industrial applications of nitration in the strict sense; the most important by volume are for the production of nitroaromatic compounds such as nitrobenzene. Nitration reactions are notably used for the production of explosives, for example the conversion of toluene to TNT (2,4,6-trinitrotoluene).

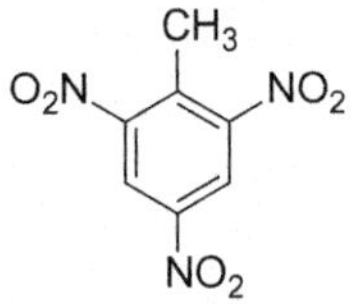

Trinitrotoluene

However, explosives aside, the nitro compounds are of wide importance as chemical intermediates and precursors.

Thus, nitration involves the replacement of a hydrogen atom (in an organic compound) with one or more nitro groups ($—NO_2$). By-products may be unavoidable due to the high reaction temperatures and the highly oxidizing environment, although many nitration reactions are carried out at low temperature for safety reasons. The nitration reaction can be carried out with aliphatic compounds (to produce nitroparaffin derivatives) but the nitration of aromatics is more commercially important (to produce explosives and propellants such as nitrobenzene and nitrotoluene derivatives). This is effected with nitric acid (HNO_3) or, in the case of aromatic nitration reactions, a mixture of nitric and sulfuric acids. Nitration is used in the first step of TDI production.

Environmental issues of nitration processes (excluding the more obvious potential explosive properties) relate to the occurrence of acid vapors (largely nitric or sulfuric acid) from the reaction and quenching as well as any unreacted nitrating agent arising from the use of an excess of the agent to carry the nitration reaction to completion. There is also the potential for the emission of VOCs as well as other gas streams that contain the various oxides of nitrogen. In terms of water pollutants, the nitration of aromatic feedstocks may produce large quantities of waste mixed acid that requires neutralization and disposal, or recovery (e.g., by distillation) and reuse. The products and by-products of the nitration process often are slow to biodegrade (if they are at all biodegradable) and toxic, so additional treatment of the waste products (such as extraction or incineration of aqueous wastes) may be required.

3.6.14 Oxidation

The term oxidation includes many different processes, but in general it describes the addition of one or more oxygen atoms to a compound. Catalytic oxidation process are processes that utilize catalysts to enhance the oxidation reaction—typical oxidation catalysts are metal oxides and metal carboxylates. The catalysis of the oxidation process occurs by the use of both heterogeneous catalysis and homogeneous catalysis (Table 3.8). In the heterogeneous processes, gaseous substrate and oxygen (or air) are passed over solid catalysts. Typical catalysts are platinum, redox-active oxides of iron, vanadium, and molybdenum. In many cases, catalysts are modified with the suitable choice (from an extensive list) of additives or promoters that enhance the reaction rate or the product selectivity.

The important homogeneous catalysts for the oxidation of organic compounds are the carboxylic acid salts of cobalt, iron, and manganese. To confer good solubility in the organic solvent, these catalysts are often derived from naphthenic acids and ethylhexanoic acid which are highly lipophilic. These catalysts initiate radical chain reactions that produce organic radicals that combine with oxygen to give hydroperoxide intermediates. Generally, the selectivity of oxidation is determined by the bond energy—for example, benzylic carbon hydrogen (C_6H_5CH—H) bonds are replaced by oxygen faster than aromatic carbon hydrogen (C_6H_5—H) bonds.

Common applications of the process involve oxidation of organic compounds by the oxygen in air. Such processes are conducted on a large scale for the remediation of pollutants, production of valuable chemicals, and the production of energy. An illustrative catalytic oxidation is the conversion of methanol to the more valuable compound formaldehyde using aerial oxygen:

$$2CH_3OH + O_2 \rightarrow 2HCHO + 2H_2O$$

This conversion is very slow in the absence of catalysts.

Atmospheric oxygen is by far the most important, and the cheapest, oxidizing agent although the inert nitrogen component will dilute products and generate waste gas streams. Other oxidizing agents include nitric acid, sulfuric acid, oleum, hydrogen peroxide, organic peroxides, and pure oxygen. In general terms, organic materials can be oxidized either by heterolytic or homolytic reactions, or by catalytic reactions (where the oxidizing agent is reduced and then reoxidized). Heterogeneous catalysts based on noble metals play a dominant role in industrial scale oxidations and an important example is the silver

TABLE 3.8 Examples of Industrial Oxidation Processes

Substrate	Process	Catalyst	Product	Application
Butane	Maleic anhydride process	Vanadium phosphates (heterogeneous)	Maleic anhydride	Plastics, alkyd resins
Cyclohexane	K-A process	Co and Mn salts (homogeneous)	Cyclohexanol, cyclohexanone	Nylon precursor
Ethylene	Epoxidation	Mixed Ag oxides (heterogeneous)	Ethylene oxide	Basic chemicals, surfactants
Ethylene	OMEGA process		Ethylene glycol	
Ethylene	Wacker process	Pd and Cu salts (homogeneous)	Acetaldehyde	Basic chemicals
Methanol	Formox process	Fe-Mo-oxides (heterogeneous)	Formaldehyde	Basic chemicals, alkyd resins
Propylene	Allylic oxidation	Mo-oxides (heterogeneous)	Acrylic acid	Plastic precursor
Propylene, ammonia	SOHIO process	Bi-Mo-oxides (heterogeneous)	Acrylonitrile	Plastic precursor
p-Xylene	Terephthalic acid synthesis	Mn and Co salts (homogeneous)	Terephthalic acid	Plastic precursor

catalyzed gas phase reaction between ethylene and oxygen to form ethylene oxide. Ethylene is still the only olefin that can be directly oxidized to the corresponding epoxide with high selectivity. Other important industrial oxidation processes are the production of acetic acid, formaldehyde, phenol, acrylic acid, acetone, and adipic acid. Oxidation reactions are exothermic and heat can be reused in the process to generate steam or to preheat other component streams. Fire and explosion risks exist with heterogeneously catalyzed direct oxidation processes (e.g., ethylene oxide process) and reactions involving concentrated hydrogen peroxide or organic peroxides.

In terms of the environmental aspects of the oxidation process, the oxidation of organic compounds produces a number of by-products (including water) and wastes from partial and complete oxidation. In the organic chemical industry, such compounds as aldehydes, ketones, acids, and alcohols are often the final products of partial oxidation of hydrocarbons. Careful control of partial oxidation reactions is usually required to prevent the material from oxidizing to a greater degree than desired as this produces carbon dioxide and many undesirable gaseous, liquid, or semisolid toxic by-products.

In addition, the emissions of volatile organics can arise from losses of unreacted feed, by-products, and products such as aldehydes and acids. Carbon dioxide is an ever-present by-product in the oxidation of organic compounds since it is difficult (if not impossible in some cases) to prevent the complete oxidation of some carbon. Aldehyde derivatives (especially formaldehyde, HCHO) require strict handling to minimize exposure and this limits atmospheric emissions. Acid gases usually require removal from waste streams. Also, to enable biological degradation in a wastewater treatment plant it will be necessary to neutralize any acidic components and to remove/destroy any chlorinated species that may inhibit biological activity.

3.6.15   Oxyacetylation

Acetylation is a reaction that introduces an acetyl functional group (acetoxy group, $CH_3C{=}O$) into an organic chemical compound—namely the substitution of the acetyl group for a hydrogen atom—while deacetylation is the removal of an acetyl group from an organic chemical compound. Thus, oxyacetylation involves the addition of oxygen and an acetyl group to an olefin to produce an unsaturated acetate ester. It is used to produce vinyl acetate from ethylene, acetic acid, and oxygen.

3.6.16   Reforming

Reforming is the decomposition (cracking) of hydrocarbon gases or low octane petroleum fractions by heat and pressure. Catalytic reforming is used to convert naphtha (having low octane ratings) into a high-octane liquid product (*reformate*) which is a premium blending stock for the production of high-octane gasoline. The process converts low-octane linear hydrocarbons (paraffins) into branched alkanes (isoparaffins) and cycloalkane derivatives, which are then partially dehydrogenated to produce high-octane aromatic hydrocarbon derivatives. The dehydrogenation also produces significant amounts of by-product hydrogen, which is used in other refinery processes such as hydrocracking. A side reaction in the reforming process is hydrogenolysis, which produces low-boiling hydrocarbons, such as methane, ethane, propane, and butanes. The nature of the final product is influenced by the source (and composition) of the feedstock. The four major catalytic reforming reactions are:

(1) The dehydrogenation of cycloalkane derivative (naphthenes) to aromatics:

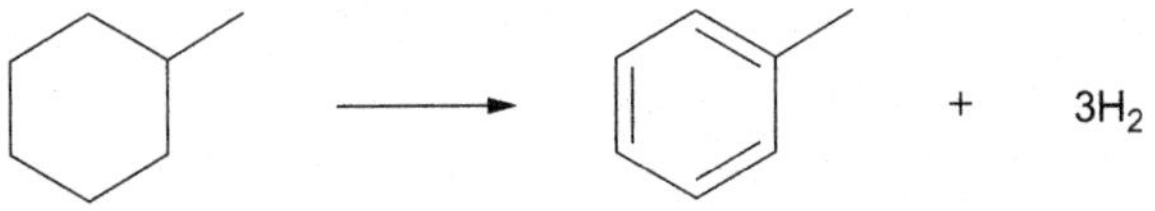

Methylcyclohexane   Toluene   + $3H_2$

(2) The isomerization of normal paraffin derivatives to isoparaffin derivatives:

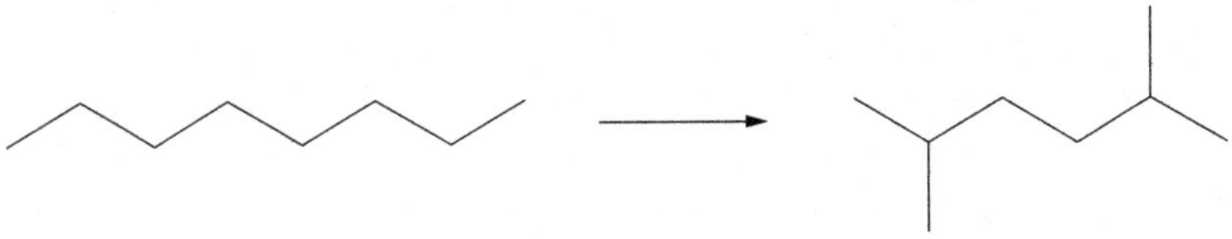

n-Octane   2,5-Dimethylhexane

(3) The dehydrogenation and aromatization of paraffin derivatives to aromatic products (known as dehydrocyclization): as exemplified in the conversion of normal heptane to toluene, as shown below:

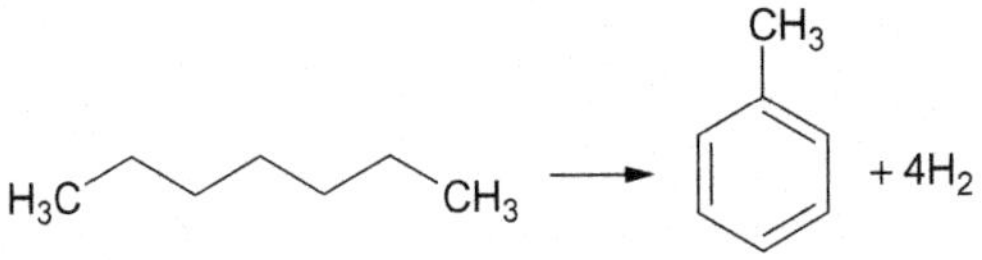

n-Heptane   Toluene

(4) The hydrocarbon of paraffin derivatives into lower molecular weight products

$$CH_3CH_2CH_2CH_2CH_2CH_2CH_3 \rightarrow (CH_3)_2CHCH_2CH_3 + CH_3CH_3$$
$$\underset{n-\text{Heptane}}{} \qquad \underset{\text{Isopentane}}{} \qquad \underset{\text{Ethane}}{}$$

The hydrocracking of paraffin derivatives is only one of the above four major reforming reactions that consumes hydrogen. The isomerization of normal paraffins does not consume or produce hydrogen but, moreover, both the dehydrogenation of naphthene derivatives and the dehydrocyclization of paraffin derivatives produce hydrogen. In many petroleum refineries, the net hydrogen produced in catalytic reforming supplies a significant part of the hydrogen used elsewhere in the refinery (for example, in hydrodesulfurization processes).

In a variation of the reforming process, the steam reforming is a process for producing hydrogen, carbon monoxide, or other useful products from hydrocarbon feedstocks such as natural gas, which is predominantly methane, hence the alternate name for the process: steam-methane reforming. The conversion is achieved in a reactor (reformer) in which the methane reacts at high temperature with the steam. At high temperatures (700–1100°C, 1290–2010°F) and in the presence of a metal-based catalyst (nickel), steam reacts with methane to yield carbon monoxide and hydrogen:

$$CH_4 + H_2O \rightleftharpoons CO + 3H_2$$

Thus:

$$C_nH_m + nH_2O \rightleftharpoons (n + {}^m/_2)H_2 + nCO$$

Additional hydrogen can be recovered by a lower-temperature gas-shift reaction with the carbon dioxide produced, in the presence of a copper-based or iron-based catalyst:

Water-gas shift reaction:

$$CO + H_2O \rightleftharpoons CO_2 + H_2$$

The first reaction is strongly endothermic (consumes heat) while the second reaction is mildly exothermic (produces heat).

3.6.17 Sulfonation

Sulfonation is the process by which a sulfonic acid group (or corresponding salt or sulfonyl halide) is attached to a carbon atom and the process is used to produce detergents (by sulfonating mixed linear alkyl benzenes with sulfur trioxide or oleum). In the process, a hydrogen atom on an aromatic ring is replaced by a sulfonic acid functional group by an electrophilic aromatic substitution reaction:

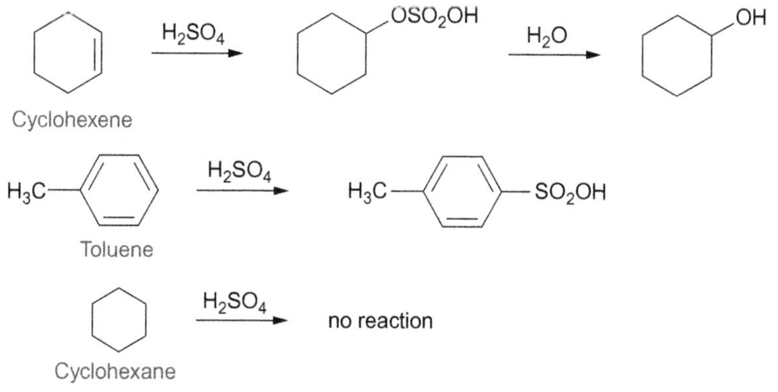

Thus, the general equation for sulfonation of the aromatic ring is:

$$ArH + H_2SO_4 \rightarrow ArSO_3H + H_2O$$
Oleum

The most widely used sulfonating agent for linear alkylbenzenes is oleum (fuming sulfuric acid—a solution of sulfur trioxide in sulfuric acid, $H_2SO_4 \cdot SO_3$). Sulfuric acid alone is effective in sulfonating the benzene ring but the acid content of the sulfuric acid (solution) must be in excess of 75% v/v. The excess sulfur trioxide in oleum removes the water of reaction thereby preventing the sulfuric acid from descending below the minimum acid content level for an efficient reaction and promotes higher yields of the desired product. Separating the product sulfonates from the reaction mixture is often difficult—the mother liquor (the solution remaining after the reaction) after product separation raises an environmental issue.

The acid vapor from the reaction and quenching as well as unreacted sulfonating agent arising from the excess use to drive the reaction, pose serious disposal problems. Also, acidic wastewater from the reactor and dilute acidic wash waters (from washing the product on the filter) require neutralization. In addition, the filtrate from the separation stage is contaminated with unreacted raw material and acid. Finally, oleum is an extremely strong oxidizing agent and produces tar by-products that also require disposal.

REFERENCES

Aiello-Mazzarri, C., Coward-Kelly, G., Agbogbo, F.K., Holtzapple, M.T., 2005. Conversion of municipal solid wastes into carboxylic acids by anaerobic countercurrent fermentation—effect of using intermediate lime treatment. Appl. Biochem. Biotechnol. 127 (2), 79–93.

Aiello-Mazzarri, C., Agbogbo, F.K., Holtzapple, M.T., 2006. Conversion of municipal solid wastes into carboxylic acids using a mixed culture of mesophilic microorganisms. Bioresour. Technol. 97 (1), 47–56.

Chadeesingh, R., 2011. The Fischer-Tropsch process. In: Speight, J.G. (Ed.), The Biofuels Handbook. The Royal Society of Chemistry, London, pp. 476–517 (Part 3, Chapter 5).

Chan, W.N., Holtzapple, M.T., 2003. Conversion of municipal solid wastes to carboxylic acids by thermophilic fermentation. Appl. Biochem. Biotechnol. 111 (2), 93–112.

Chang, V.S., Holtzapple, M.T., 2000. Fundamental factors affecting biomass enzymatic reactivity. Appl. Biochem. Biotechnol. 84 (6), 5–37.

Chang, V.S., Kaar, W.E., Burr, B., Holtzapple, M.T., 2001a. Simultaneous saccharification and fermentation of lime-treated biomass. Biotechnol. Lett. 23 (16), 1327–1333.

Chang, V.S., Nagwani, M., Kim, C.H., Holtzapple, M.T., 2001b. Oxidative lime-pretreatment of high-lignin biomass—poplar wood and newspaper. Appl. Biochem. Biotechnol. 94 (1), 1–28.

Chenier, P.J., 1992. Survey of industrial chemistry, second revised ed. Wiley-VCH, New York.

Gibson, J., Gregory, D.H., 1971. Carbonization of Coal. Mills and Boon, London.

Hammerschlag, R., 2006. Ethanol's energy return on investment: a survey of the literature 1990–present. Environ. Sci. Tech. 40, 1744–1750.

Hirsch, R.L., Bezdek, R., Wendling, R., 2006. Peak of world oil production and its mitigation. AICHE J. 52 (1), 2–8.

Kroschwitz, J.I., Howe-Grant, M., Kirk, R.E., Othmer, D.F. (Eds.), 1991. Encyclopedia of Chemical Technology, fourth ed. Wiley, Hoboken, NJ.

Lynd, L.R., 1996. Overview and evaluation of fuel ethanol from cellulosic biomass: technology, economics, the environment and policy. Ann. Rev. Energy Environ. 21, 403–465.

McNeil, D., 1966. Coal Carbonization Products. Pergamon Press, London.

Mokhatab, S., Poe, W.A., Speight, J.G., 2006. Handbook of Natural Gas Transmission and Processing. Elsevier, Amsterdam.

Priyadarsan, S., Annamalai, K., Sweeten, J.M., Muhktar, S., Holtzapple, M.T., 2004. Fixed bed gasification of feedlot manure and poultry litter biomass. Trans. ASAE 47 (5), 1689–1696.

Priyadarsan, S., Annamalai, K., Sweeten, J.M., Holtzapple, M.T., Muhktar, S., 2005. Cogasification of blended coal and feedlot and chicken litter biomass. Proc. Combust. Inst. 30, 2973–2980.

Ragauskas, A.J., Williams, C.K., Davison, B.H., Britovsek, G., Caimey, J., Eckert, C.A., Frederick Jr., W.J., Hallet, J.P., Leak, D.J., Liotta, C.L., Mielenz, J.R., Murphy, R., Templer, R., Tschalpliski, T., 2006. The path forward for biofuels and biomaterials. Science 311, 484–489.

Speight, J.G., 2002. Chemical Process and Design Handbook. McGraw-Hill, New York.

Speight, J.G., 2007. Natural Gas: A Basic Handbook. GPC Books, Houston, TX.

Speight, J.G., 2011a. An Introduction to Petroleum Technology, Economics, and Politics. Scrivener, Salem, MA.

Speight, J.G., 2011b. The Refinery of the Future. Elsevier, Oxford.

Speight, J.G. (Ed.), 2011c. The Biofuels Handbook. The Royal Society of Chemistry, London.

Speight, J.G., 2013. The Chemistry and Technology of Coal, third ed. CRC Press, Boca Raton, FL.

Speight, J.G., 2014a. The Chemistry and Technology of Petroleum, fifth ed. CRC Press, Boca Raton, FL.

Speight, J.G., 2014b. Gasification of Unconventional Feedstocks. Elsevier, Oxford.

Speight, J.G., 2014c. Oil and Gas Corrosion Prevention. Elsevier, Oxford.

Speight, J.G., 2016. Handbook of Petroleum Refining. CRC Press, Boca Raton, FL.

Speight, J.G., Islam, M.R., 2016. Peak Energy—Myth or Reality. Scrivener, Salem, MA.

Sricharoenchaikul, V., Frederick Jr., W.J., Agrawal, P., 2002. Black liquor gasification characteristics. 2. Measurement of condensable organic matter (tar) at rapid heating conditions. Ind. Eng. Chem. Res. 41, 5650–5658.

Thanakoses, P., Black, A.S., Holtzapple, M.T., 2003a. Fermentation of corn stover to carboxylic acids. Biotechnol. Bioeng. 83 (2), 191–200.

Thanakoses, P., Mostafa, N.A.A., Holtzapple, M.T., 2003b. Conversion of sugarcane bagasse to carboxylic acids using a mixed culture of mesophilic microorganisms. Appl. Biochem. Biotechnol. 105, 523–546.

Wilson Jr., P.J., Wells, J.H., 1950. Coal, Coke, and Coal Chemicals. McGraw-Hill, New York.

Wittkoff, H.A., Reuben, B.G., Plotkin, J.S., 2012. Industrial Organic Chemicals, third ed. Wiley, Hoboken, NJ.

Chapter 4

Sources and Types of Organic Pollutants

1 INTRODUCTION

It is doubtful if anyone (even though there may be claims to the contrary) who can state with any degree of accuracy (although there is always someone who can make a statement with a high degree of uncertainty) when the Earth was last pristine and unpolluted. Yet, to attempt to return the environment to such a mythical time might have a severe effect on the current indigenous life, perhaps a form of pollution in reverse! However, there is the possibility that through the judicious use of resources and the application of the principles of environmental science, environmental engineering, and environmental analysis (disciplines involved in the study of the environment as well as determining the *purity* of the environment) (Woodside, 1999; Speight and Lee, 2000; Manahan, 2010), a state can be reached where pollution is minimal and does not pose a threat to the future. Such a program will not only involve well-appointed suites of analytical tests but also subsequent studies that cover the effects of changes in the environmental conditions on the flora and fauna of a region. These studies can include aspects of chemistry, chemical engineering, microbiology, and hydrology as they can be applied to solve environmental problems (Pickering and Owen, 1994; Speight and Lee, 2000; Schwarzenbach et al., 2003; Tinsley, 2004).

The potential for pollution of organic chemicals starts during the production stage. The typical organic chemical synthesis process involves combining one or more feedstocks in a series of unit process operations. Commodity chemicals tend to be synthesized in a continuous reactor, while specialty chemicals usually are produced in batches. Most reactions take place at high temperatures, involve metal catalysts, and include one or two additional reaction components. The yield of the organic chemical will partially determine the kind and quantity of by-products and releases. In fact, many specialty organic chemicals require a series of two or three reaction steps, each involving a different reactor system and each capable of producing by-products. Once the reaction is complete, the desired product must be separated from the by-products by a second unit operation. A number of separation techniques such as settling, distillation, or refrigeration may be used. The final product may be further processed, such as by spray

Environmental Organic Chemistry for Engineers. http://dx.doi.org/10.1016/B978-0-12-804492-6.00004-6

drying or pelletizing, to produce the saleable item. Frequently, by-products are also sold, and their value can influence the economics of the process.

In spite of numerous safety protocols that are in place and the care taken to avoid environmental incidents that are harmful to the environment, every industry suffers accidents that lead to contamination by chemicals. It is therefore often helpful to be aware of the nature (the chemical and physical properties) of the chemical contaminants and the products arising therefrom (when ecosystem parameters interact with the chemicals) in order to understand not only the nature of the chemical contamination but also chemical changes to the contaminants following from which cleanup methods can be chosen.

In the past, the existence and source of such information was unknown and, if known, was not always consulted. When the existence and sources of the relevant information are known, decisions must be made in order for environmental scientists and engineers to make an informed, and often quick, decision on the next steps, even if it is decided at a later time not to use the information for a particular application. However, on the basis that it is better to know than to not know, knowing about the relevant data gives investigators and analysts the ability to assess whether or not a chemical discharge into the environment should be addressed or whether the environment can take care of itself through biodegradation of the chemical. This is especially true for scientists and engineers involved in site cleanup operations, assessment of ecological risk, and assessment of ecological damage. Modern data bases relating to the properties of chemicals, especially organic chemicals (which are the reason for the current text), there can be no reasons (or excuses) for not knowing or understanding the fundamental aspects of the behavior of organic chemicals that pollute the environment.

By way of clarification, an organic pollutant is an organic chemical that is released into an ecosystem and which causes pollution (however temporary or permanent) insofar as the chemical is harmful to or is destructive to the flora and/or the fauns of the ecosystem. Typically, and by virtue of the name, a pollutant is a chemical that is not indigenous to the ecosystem. However, if the discharged chemical is indigenous to the ecosystem (i.e., the organic chemical is a naturally occurring compound), it can be (should be) classed as a pollutant when it is released into the system in amounts that are in excess of the natural concentration of the organic chemical in the ecosystem, and by this increased concentration the chemical can cause harm to (or is destructive to) the flora and/or the fauna of the ecosystem.

Given time, some organic chemicals are removed from the ecosystem by natural events, such as attack by indigenous bacteria (biodegradation) or by increasing the concentration of natural-occurring bacteria to remove the chemical from the ecosystem (bioremediation) (Speight and Arjoon, 2012). However, there are organic chemicals that are known as *persistent organic pollutants* (POPs) which are compounds that are resistant to environmental degradation through the various chemical and biological processes (Jacob, 2013). POPs, as the name implies, are not easily degraded in the environment due to their stability and low decomposition rates and, thus, have a long life in various ecosystems and often require other forms of removal such as physical or chemical methods of cleanup as well as the addition of nonindigenous microbes for cleanup (Speight and Lee, 2000; Speight and Arjoon, 2012). POPs also have the ability for long-range transport, and environmental contamination by POPs is extensive, even in areas where these chemicals have never been used, and will remain in these environments for a considerable time (even years) and after restrictions implemented due to their resistance to degradation.

POPs, like any organic chemical pollutant, can enter an ecosystem through the gas phase, the liquid phase, or solid phase and which can resist degradation and are mobile over considerable distances (especially in the gas phase or through transportation in river systems) before being redeposited in a location that is remote to the location of their introduction into the ecosystem. Furthermore, POPs can be present as vapors in the atmosphere or bound to (adsorbed on) the surface of soil or mineral particles and also have variable solubility in water.

Many POPs are currently (or were in the past) arose from the extensive use of agrochemicals (agricultural chemicals) such as pesticides, herbicides, and biocides, solvents, pharmaceuticals, and various industrial chemicals (Chapter 3). Although some POPs arise naturally, for example, from various biosynthetic pathways, most are products of human industry and tend to have higher concentrations and are eliminated more slowly. If not removed and because of their properties, POPs will bioaccumulate and have significant impacts on and the flora and fauna of the environment. The most frequently used measure of the potential for bioaccumulations and persistence of an organic compound in the environment are the result of the physicochemical properties (such partition coefficients and reaction rate constants) (Mackay et al., 2001).

Furthermore, the capacity of the environment to absorb the effluents and other impacts of process technologies is not unlimited, as some would have us believe. The environment should be considered to be an extremely limited resource, and discharge of chemicals into it should be subject to severe constraints. Indeed, the declining quality of raw materials dictates that more material must be processed to provide the needed fuels. And the growing magnitude of the products and effluents from industrial processes has moved above the line where the environment has the capability to absorb such process effluents without disruption.

As a result of the increasing concern about pollution (especially pollution by organic chemicals), in May 1995, the United Nations Environment Program Governing Council investigated POPs and placed a global ban on those organic chemicals that were particularly harmful and toxic to the environment among which were many pesticides, herbicides, and fungicides which are historically or commercially important (Appendix: Table A6) (Hites, 2007) and required the participating governments to take measures to eliminate or reduce the release of POPs in the environment. On May 22, 2001, the Stockholm Convention was adopted and put into practice by the United Nations Environment Program. The purpose of the statement of the agreement is to protect the environment from POPs and the members at the recognized the potential for environmental toxicity of POPs which had the potential for long range transport resulting in bioaccumulation and biomagnification (Speight and Lee, 2000; Manahan, 2010; Speight and Arjoon, 2012).

As a commencement to this process of data examination and ingestion, this chapter introduces the terminology of environmental technology as it pertains to the sources and types of organic pollutants. Briefly, a *contaminant*, which is not usually classified as a pollutant unless it has some detrimental effect, can cause deviation from the normal composition of an environment. A *receptor* is an object (animal, vegetable, or mineral) or a locale that is affected by the pollutant. A *chemical waste* is any solid, liquid, or gaseous waste material that, if improperly managed or disposed of, may pose substantial hazards to human health and the environment (Table 4.1). At any stage of the management process, a chemical waste may be designated by law as a *hazardous waste*.

TABLE 4.1 Examples of the Types of Organic Chemical Waste

Source	Waste Type
Chemical manufacturing	Spent solvents
	Reactive materials
Cleaning agents	Spent solvents
Construction industry	Spent solvents
Cosmetics manufacturing	Ignitable materials
	Flammable solvents
Crude oil recovery and refining	Drilling mud spills
	Spilled solvents
	Process sludge
Furniture manufacturing and refinishing	Ignitable materials
	Spent solvents
Leather products	Waste solvents
Power generation	Gases and coal dust
	Combustion waste (ash and slag)
Printing industry	Spent solvents
Vehicle maintenance	Ignitable materials
	Spent solvents

TABLE 4.2 Effects of Organic Solvents

Solvent	Affected Parts of Human Body
Aliphatic hydrocarbons	
Pentanes, hexanes, heptanes, octanes	Central nervous system and liver
Halogenated aliphatic hydrocarbons	
Methylene chloride	Central nervous system, respiratory system
Chloroform	Liver
Carbon tetrachloride	Liver and kidneys
Aromatic hydrocarbons	
Benzene	Blood, immune system
Toluene	Central nervous system
Xylene	Central nervous system
Alcohols	
Methyl alcohol (methanol and toxic metabolites)	Optic nerve
Isopropyl alcohol	Central nervous system
Glycols	
Ethylene glycol (and toxic metabolites)	Central nervous system

Improper disposal of these waste streams, such as organic solvents (Table 4.2), in the past has created hazards to human health and the need for very expensive cleanup operations (Tedder and Pohland, 1993). Correct handling of these chemicals (NRC, 1981), as well as dispensing with many of the myths related to chemical processing (Kletz, 1990), can mitigate some of the problems that will occur, especially problems related to the flammability of organic liquids (Table 4.3), that will occur when incorrect handling is practiced. Chemical waste is also defined and classified into various subgroups (Table 4.1).

2 AEROSOLS

An aerosol is a suspension of liquid or solid particles in a gas, with particle diameters in the range of 10^{-9} to 10^{-4} m. In atmospheric science, however, the term aerosol traditionally refers to suspended particles that contain a large proportion of condensed matter other than water, whereas clouds are considered as separate phenomena (Pöschl, 2005). Aerosols give rise to a class of

TABLE 4.3 Flammability of Selected Organic Liquids

Liquid	Flash Point(°C)[a]	Volume Percent in Air	
		LFL[b]	UFL[b]
Diethyl ether	−43	1.9	36
Pentane	−40	1.5	7.8
Acetone	−20	2.6	13
Toluene	4	1.3	7.1
Methanol	12	6.0	37
Gasoline (2,2,4-trimethylpentane)	–	1.4	7.6
Naphthalene	157	0.9	5.9

[a]Closed-cup flash point test.
[b]LFL, lower flammability limit; UFL, upper flammability limit at 25°C (77°F).

compounds known as volatile organic compounds (VOCs) which can arise from various sources.

In addition to the emissions of VOCs from vegetation, large quantities of organic compounds are emitted into the atmosphere from anthropogenic (man-made) sources, largely from combustion of petroleum products such as gasoline and diesel fuels and from other sources such as solvent use and the use of consumer products. Although the theory is that the combustion of such chemicals (alkanes in particular) should proceed completely to produce quantitative yields of carbon dioxide and water:

cyclohexane
$$C_6H_{12} \; + \; 9\,O_2 \; \longrightarrow \; 6\,CO_2 \; + \; 6\,H_2O \; + \; energy$$

cyclohexene
$$C_6H_{10} \; + \; 8.5\,O_2 \; \longrightarrow \; 6\,CO_2 \; + \; 5\,H_2O \; + \; energy$$

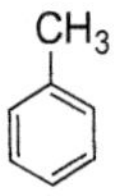

toluene
$$C_7H_8 \; + \; 9\,O_2 \; \longrightarrow \; 7\,CO_2 \; + \; 4\,H_2O \; + \; energy$$

This is not always the case, and the reference of heteroatoms (nitrogen, oxygen, and/or sulfur) complicates the process by converting the heteroatoms to the gases oxide (such as nitrogen oxides and sulfur oxides) that are gaseous pollutants. In addition, incomplete combustion (i.e., combustion in a dearth of oxygen) will produce polynuclear aromatic producers that appear in the atmosphere (as particulate matter) or in the soil.

In the atmosphere organic compounds are partitioned between the gaseous and particulate phases, with the chemicals being at least partially in the gas phase for liquid-phase vapor pressures of at least 10^{-6} Torr at atmospheric temperature (1 Torr is a unit of pressure based on an absolute scale and is 1/760 of a standard atmosphere; thus 1 Torr = 1 mmHg pressure). In the atmosphere, these gaseous organic compounds are transformed by photolysis and/or reaction with hydroxyl (OH) radicals, nitrate (NO_3) radicals, and ozone (O_3). Emissions of organic compounds and their subsequent in situ atmospheric transformations lead to a number of adverse effects, including: (1) the formation—in the presence of oxides of nitrogen, NOx—of ozone, a criteria air pollutant; (2) the formation of secondary organic minute particles—aerosols—resulting in loss of visibility and risks to human health; and (3) the in situ atmospheric formation of toxic air contaminants, including, for example, formaldehyde HCHO, peroxy-acetyl nitrate, and nitrated aromatic species.

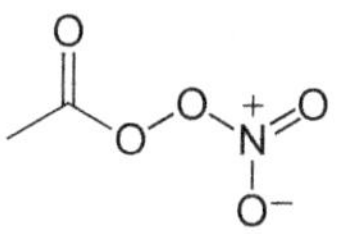

Peroxy-acetyl nitrate

Atmospheric aerosol particles originate from a wide variety of natural and anthropogenic sources, including biomass. Primary particles are directly emitted as liquids or solids from sources such as biomass burning, incomplete combustion of fossil fuels, volcanic eruptions, and wind-driven or traffic-related suspension of road, soil, and mineral dust, sea salt, and biological materials (such as plant fragments, microorganisms, and pollen) (Oliveira et al., 2011). This also include the commercial conversion of biomass to a variety of chemicals, some of which may not be beneficial when released to the environment (Werpy and Peterson, 2004; Kim and Holtzapple, 2005, 2006a,b; Speight, 2011). Another example of a primary aerosol is the carbonaceous soot formed during incomplete combustion processes—diesel soot is a classic example of this form of primary carbonaceous aerosol. In spite of claims to the contrary, diesel fuel is not a clean fuel—it might be clean insofar as sulfur content is concerned—try following a diesel fuel vehicle under full load and/or up an incline. The emission of black fumes becomes very evident and very uncomfortable.

Another example of primary aerosols (although not organic in nature) is the fly ash from coal combustion systems or the incineration of wastes. This material is typically generated by high-temperature processes that result in the production of condensed inorganic materials formed as small spherical beads, typically high in silicates or iron oxides. Anthropogenic sources of primary aerosol and particulate material include mechanical abrasion that produces construction and industrial dusts, as well as abrasion of tires and pavement materials on roads.

Secondary aerosol particles, on the other hand, are formed by gas-to-particle conversion in the atmosphere (new particle formation by nucleation and condensation of gaseous precursors). The most important aerosol-generating inorganic gases that are released into the atmosphere by human combustion of fossil fuels are the nitrogen oxides (nitric oxide and nitrogen dioxide) and sulfur dioxide. Nitric oxide (NO) is oxidized to nitrogen dioxide (NO_2) and subsequently to nitric acid (HNO_3) by the reaction of hydroxyl radical with nitrogen dioxide. The nitric acid can in turn react with ammonia to form ammonium nitrate, a white solid. The chemical equations are often subject to debate but can be represented simply as:

$$NO + O_2 \rightarrow NO_2$$
$$NO + O_3 \rightarrow NO_2 + O_2$$
$$NO_2 + OH \rightarrow HNO_3$$
$$HNO_3 + NH_3 \rightarrow NH_4NO_3$$

Sulfur dioxide—a common product from the combustion of sulfur-containing organic fuels—in the atmosphere reacts with hydroxyl radical in the gas phase and with hydrogen peroxide and ozone in the aqueous phase to form sulfuric acid (H_2SO_4), which is water soluble and also has a very low vapor pressure, so it rapidly forms aerosol once it is formed. Sulfuric acid can also react with ammonia to form ammonium bisulfate (NH_4HSO_4) and ammonium sulfate [$(NH_4)_2SO_4$]. These species are all fairly water soluble, and at high relative humidity values they will grow by adding water vapor to their surfaces.

Thus, airborne particles undergo various physical and chemical interactions and transformations (often referred to as *atmospheric aging*) which involves changes of particle composition, particle size, and particle structure through chemical reaction, gas uptake, and restructuring. Particularly efficient particle aging occurs in clouds, which are formed by condensation of water vapor on preexisting aerosol particles. Most clouds reevaporate, and modified aerosol particles are again released from the evaporating cloud droplets or ice crystals (*cloud processing*). If, however, the cloud particles cause precipitation which reaches the surface of the Earth, not only the condensation nuclei but also other aerosol particles that are scavenged on the way to the surface are removed from the atmosphere (*wet deposition*). Particle deposition without precipitation airborne water particles—that is *dry deposition*—is less important on a global scale but is highly relevant with respect to air quality. Depending on the properties of the aerosol and meteorological conditions, the characteristic residence times (lifetimes) of aerosol particles in the atmosphere can range from hours to weeks (Pöschl, 2005).

Depending on the origin of organic aerosols, the components can be classified as primary or secondary. Primary organic aerosol components are directly emitted in the condensed phase (as liquid particles or as solid particles) or as semivolatile vapors which are condensable under atmospheric conditions. The main sources of primary organic aerosol particles and components are natural and anthropogenic biomass burning (forest fires, slashing and burning, domestic heating), fossil-fuel combustion (domestic heating, industrial operations, traffic density), and wind-driven or the traffic-related suspension of soil and road dust, biological materials (such as plant debris, animal debris, pollen, and spores), sea spray, and spray from other surface waters that contain dissolved organic chemicals.

Secondary organic aerosol components are formed by chemical reaction and gas-to-particle conversion of VOCs in the atmosphere, which may proceed through different chemical and physical pathways, such as: (1) new particle formation; (2) formation of semivolatile organic compounds—SVOCs—by gas-phase reactions and participation of the SVOCs in the nucleation and growth of new aerosol particles; (3) gas-particle partitioning, which results in the formation of SVOCs by gas-phase reactions and uptake through adsorption or by absorption by preexisting aerosol or cloud particles; and (4) heterogeneous or multiphase reactions: formation of low-volatile organic compounds or nonvolatile organic compounds by chemical reaction of VOCs or SVOCs at the surface or in the bulk of aerosol or cloud particles.

Thus, in summary, aerosols caused by either the entry of organic chemicals into the environment or the reactivity of organic chemicals in the environment are of major importance for atmospheric chemistry and physics, the biosphere, and climate. The airborne solid and liquid particles in the nanometer (1×10^{-9} m) to micrometer (1×10^{-6} m) size range influence the energy balance of the Earth, the hydrological cycle, atmospheric circulation, and the abundance of greenhouse gases and reactive trace gases. Moreover, aerosols play an important role in the reproduction of biological organisms and the primary parameters that determine the environmental effects of aerosol particles are (1) the concentration of the particles, (2) particle size, (3) particle structure, and (4) the chemical composition of the particles. These parameters, however, are spatially and temporally highly variable.

3 AGROCHEMICALS

Agrochemicals (agricultural chemicals, agrichemicals) are the various chemical products that are used in agriculture. In most cases, the term *agrochemical* refers to the broad range of pesticide chemicals, including insecticide chemicals, herbicide chemicals, fungicide chemicals, and nematicides chemicals (chemicals used to kill round worms). The term may also include synthetic fertilizers, hormones, and other chemical growth agents, as well as concentrated stores of raw animal manure.

Typically, agrochemicals are toxic and when stored in bulk storage systems may pose significant environmental risks, particularly in the event of accidental spills. As a result, in many countries, the use of agrochemicals has become highly regulated and government-issued permits for purchase and use of approved agrichemicals may be required. Significant penalties can result from misuse, including improper storage resulting in chemical leaks, chemical leaching, and chemical spills. Wherever these chemicals are used, proper storage facilities and labeling; emergency cleanup equipment; emergency cleanup procedures; safety equipment; as well as safety procedures for handling, application, and disposal are often subject to mandatory standards and regulations.

While agrochemicals increase plant and animal crop production, they can also damage the environment. Excessive use of fertilizers has led to the contamination of groundwater with nitrate, a chemical compound that in large concentrations is poisonous to humans and animals. In addition, the runoff (or leaching from the soil) of fertilizers into streams, lakes, and other surface waters (the aquasphere) can increase the growth of algae, which can have an adverse effect on the life-cycle of fish and other aquatic animals.

Pesticides that are sprayed on entire fields using equipment mounted on tractors, airplanes, or helicopters often drift away (due to wind or air convection patterns) from the targeted field, settling on nearby plants and animals. Some older pesticides, such as the powerful insecticide DDT (dichlorodiphenyltrichloroethane), remain active in the environment for many years (Table 4.4), contaminating virtually all wildlife, well water, food, and even humans with whom it comes in contact. Although many of these pesticides have been banned (Chapter 1), some newer pesticides still cause severe environmental damage.

TABLE 4.4 Harmful Chemicals Identified by the United Nations Environment Program Governing Council[a]

Aldrin: an insecticide used in soils to kill insects such as termites, grasshoppers, and western corn rootworm

Chlordane: an insecticide used to control termites and on a range of agricultural crops; a chemical that remains in the soil with a reported half-life of 1 year

Chlordecone: a synthetic chlorinated organic compound that is primarily used as an agricultural pesticide

Dichlorodiphenyltrichloroethane (DDT): used as insecticide during WWII to protect against malaria and typhus; after the war, used as an agricultural insecticide; can persists in the soil for 10-15 years after application

Dieldrin: a pesticide used to control termites, textile pests, insect-borne diseases, and insects living in agricultural soils; half-life is approximately 5 years

TABLE 4.4 Harmful Chemicals Identified by the United Nations Environment Program Governing Council—cont'd

Dioxins: by-products of high-temperature processes, such as incomplete combustion and pesticide production also emitted from the burning of hospital waste, municipal waste, and hazardous waste as well as automobile emissions, combustion of peat, coal, and wood

Endosulfans: insecticides used to control pests on crops such coffee, cotton, rice, and sorghum and soybeans, tsetse flies, ectoparasites of cattle; also used as a wood preservative

Endrin: an insecticide sprayed on the leaves of crops, and used to control rodents; half-life is up to 12 years

Heptachlor: a pesticide primarily used to kill soil insects and termites, along with cotton insects, grasshoppers, other crop pests, and malaria-carrying mosquitoes

Hexabromocyclododecane (HBCD): a brominated flame retardant used as a thermal insulator in the building industry; persistent, toxic, and ecotoxic with bioaccumulative properties and long-range transport properties

Hexabromodiphenyl ether (hexaBDE) and heptabromodiphenyl ether: the main components of commercial octabromodiphenyl ether (octaBDE); highly persistent in the environment

Hexachlorobenzene: a fungicide used as a seed treatment, especially on wheat to control the fungal disease bunt; also a by-product produced during the manufacture of chlorinated solvents and other chlorinated compounds

α-Hexachlorocyclohexane (α-HCH) and β-hexachlorocyclohexane (β-HCH): insecticides as well as by-products in the production of lindane; highly persistent in the water of colder regions

Lindane, also known as gamma-hexachlorocyclohexane, (γ-HCH), gammaxene, Gammallin, and sometimes incorrectly called benzene hexachloride (BHC): a chemical variant of hexachlorocyclohexane that has been used as an agricultural insecticide

Mirex: an insecticide used against ants and termites or as a flame retardant in plastics, rubber, and electrical goods; half-life is up to 10 years

Pentachlorobenzene (PeCB): a pesticide and also used in polychlorobiphenyl products, dyestuff carriers, as a fungicide, a flame retardant, and a chemical intermediate

Perfluorooctane sulfonic acid (PFOS) salts of the acid: used in the production of fluoropolymers; extremely persistent in the environment through bioaccumulation and biomagnification

Polychlorinated biphenyls (PCBs): used as heat exchange fluids in electrical transformers and capacitors; also used as additives in paint, carbonless copy paper, and plastics; a half-life up to 10 years

Polychlorinated dibenzofurans: by-products of high-temperature processes, such as incomplete combustion after waste incineration, pesticide production, and polychlorinated biphenyl production

Continued

TABLE 4.4 Harmful Chemicals Identified by the United Nations Environment Program Governing Council—cont'd

Tetrabromodiphenyl ether (tetraBDE) and pentabromodiphenyl ether (pentaBDE): industrial chemicals and the main components of commercial pentabromodiphenyl ether (pentaBDE)

Toxaphene: an insecticide used on cotton, cereal, grain, fruits, nuts, and vegetables, as well as for tick and mite control in livestock; a half-life up to 12 years in soil

[a]*Listed alphabetically and not by effects; all chemicals listed are harmful to flora and fauna, including humans.*

There is now an awareness of the health hazards of pesticides and related chemicals due to the pioneering work that commenced in the latter half of the 20th century and has continued into the 21st century (Carson, 1962; Carson and Mumford, 1988, 1995, 2002). These materials are carefully regulated, and the safety requirements for every pesticide product are spelled out in detail. Most fertilizers have been in an opposite category, considered useful, safe, and inert. These and other environmental effects have prompted the search for nonchemical methods of enhancing soil fertility and dealing with crop pests. These alternatives, however, are still emerging and are not yet in widespread use.

4 CHEMICAL WASTE

Chemical waste is a general term and covers many types of materials but is generally recognized as a waste that is composed of harmful chemicals. Thus, by this definition, organic chemical waste is composed of harmful chemicals. However, all chemical wastes are not hazardous wastes, and organic chemical waste may or may not be classed as hazardous waste. An organic chemical hazardous waste is a gaseous, liquid, or material that displays either a *hazardous characteristic* or is specifically listed by name as a hazardous waste (Appendix: Tables A2–A6). There are four characteristics chemical wastes that may have to be considered as hazardous are (1) ignitability, (2) corrosivity, (3) reactivity, and (4) toxicity. This type of hazardous waste must be categorized as to its identity, constituents, and hazards so that it may be safely handled and managed.

The United States Environmental Protection Agency (US EPA) designates more than 450 chemicals or chemical wastes that are specific substances or classes of substances known to be hazardous. Each such chemical or waste is assigned a hazardous waste number in the format of a letter followed by three numerals, where a different letter is assigned to substances from each of the following list: (1) F-type: chemicals or chemical wastes from nonspecific sources (Appendix: Table A2), (2) K-type: chemicals or chemical wastes from specific sources (Appendix: Table A3), (3) P-type: chemicals or chemical wastes that are hazardous and that are mostly specific chemical species such as fluorine (Appendix: Table A4), and (4) U-type: generally hazardous chemicals or chemical wastes that are predominantly specific compounds (Appendix: Table A5).

The Comprehensive Environmental Response, Compensation, and Liability Act (CERCLA) gives a broader definition of hazardous substances that includes the following: (1) any element, compound, mixture, solution, or substance, the release of which may substantially endanger public health, public welfare, or the environment; (2) any element, compound, mixture, solution, or substance in reportable quantities designated by CERCLA Section 102; (3) certain substances or toxic pollutants designated by the Water Pollution Control Act; (4) any hazardous air pollutant listed under Section 112 of the Clean Air Act; (5) any imminently hazardous chemical substance or mixture that has been the subject of government action under Section 7 of the Toxic Substances Control Act (TSCA); and (6) any hazardous chemical or chemical waste listed or having characteristics identified by the Resource Conservation Recovery Act, with the exception of those suspended by Congress under the Solid Waste Disposal Act.

In terms of quantity by weight, more wastes than all others combined are those from categories designated by hazardous waste numbers preceded by F and K. The F categories are those wastes from nonspecific sources. The K-type hazardous wastes are those from specific sources produced by industries such as, in the context of organic chemicals, the manufacture of organic chemicals, pesticides, explosives, as well as from processes such as wood preservation petroleum refining or wood preservation.

Some refinery wastes that might exhibit a degree of hazard are exempt from the Resource Conservation Recovery Act regulation by legislation and include the following: (1) ash and scrubber sludge from thermal generation or power generation by utilities, (2) oil field and gas field drilling mud, (3) by-product brine from petroleum production, and (4) catalyst dust (Speight, 2005a). Eventual reclassification of these kinds of low-hazard wastes could increase the quantities of regulated wastes several-fold. Thus, as stated earlier, an organic chemical waste is considered hazardous if it exhibits one or more of the following characteristics: *ignitability*, *corrosivity*, *reactivity*, and *toxicity*. Under the authority of the Resource Conservation and Recovery Act (RCRA) and the United States Environmental Protection Agency (EPA), a hazardous substance has one or more of the earlier characteristics.

Briefly, *ignitability* is that characteristic of chemicals that are volatile liquids and the vapors are prone to ignition in the presence of an ignition sources. Nonliquids that may catch fire from friction or contact with water and which burn vigorously or are persistently ignitable compressed gases and oxidizers also fall under the mantle of ignitable chemicals. Examples include solvents, friction-sensitive substances, and pyrophoric solids that may include catalysts and metals isolated from various refining processes. Organic solvents are indigenous to the petroleum industry and release to the atmosphere as vapor and can pose a significant inhalation hazard. Improper storage, use, and disposal can result in the contamination of land systems as well as groundwater and drinking water (Barcelona et al., 1990; Speight, 2005a).

Often, the term *ignitable chemical* (ignitable organic chemical, such as naphtha or gasoline) is used in the same sense as the term *flammable organic chemical* insofar as it is a chemical that will burn readily but a *combustible organic chemical* (any higher boiling hydrocarbon product of refining but which can include naphtha or gasoline) often requires relatively more persuasion to burn, that is, the chemical is less flammable. Most petroleum products that are likely to burn accidentally are low-boiling liquids that form vapors that are usually denser than air and thus tend to settle in low spots. The tendency of a liquid to ignite is measured by a test in which the liquid is heated and periodically exposed to a flame until the mixture of vapor and air ignites at the liquid's surface. The temperature at which this occurs is called the *flash point* (Speight, 2015).

There are several standard tests for determining the flammability of materials. For example, the upper and lower concentration limits for the *flammability* of chemicals and waste can be determined by standard test methods (ASTM D4982, 2016; ASTM E681, 2016) as can the *combustibility* and the *flash point* (ASTM D1310, 2016; ASTM E176, 2016; ASTM E502, 2016). With these definitions in mind it is possible to divide ignitable materials into four subclasses. Thus:

1. A *flammable solid* is a solid that can ignite from friction or from heat remaining from its manufacture, or which may cause a serious hazard if ignited. Explosive materials are not included in this classification.
2. A *flammable liquid* is a liquid having a flash point below 37.8°C (100°F) (ASTM D92, 2016; ASTM D1310, 2016). A *combustible liquid* has a flash point in excess of 37.8°C (100°F), but below 93.3°C (200°F). Gases are substances that exist entirely in the gaseous phase at 0°C (32°F) and 1 atm pressure (14.7 psi) pressure. A *flammable compressed gas* (such as liquefied petroleum gas, LPG, or any liquefied hydrocarbon gas or petroleum product) meets specified criteria for lower flammability limit, flammability range, and flame projection.

In considering the ignition of vapors, two important concepts are *flammability limit and flammability range*. Values of the vapor/air ratio below which ignition cannot occur because of insufficient fuel define the lower flammability limit. Similarly values of the vapor/air ratio above which ignition cannot occur because of insufficient air define the upper flammability limit. The difference between upper and lower flammability limits at a specified temperature is the flammability range. In addition, explosions that are not due to the flammability of an organic chemical can also occur. Dust explosions (ASTM E789, 2016) can occur during catalytic reactor shutdown and cleaning are due to production of finely divvied solids through attrition. Many catalyst dusts can burn explosively in air. Thus, control of dust generated by catalyst attrition is essential (Mody and Jakhete, 1988). Organic chemicals that catch fire spontaneously in air without an ignition source are called *pyrophoric organic chemicals*, all of which may occur on a refinery site. Moisture in air is often a factor in *spontaneous ignition*.

Corrosivity is that characteristic of chemicals that exhibit extremes of acidity or basicity or a tendency to corrode steel. Such chemicals, as used in various refining (treating) processes, are acidic and are/or capable of corroding metal such as tanks, containers, drums, and barrels. On the other hand, *reactivity* is a violent chemical change (an explosive substance is an obvious example) that can result to pollution and/or harm to indigenous flora and fauna. Such wastes are unstable under ambient conditions insofar as they can create explosions, toxic fumes, gases, or vapors when mixed with water. Finally, *toxicity* (defined in terms of a standard extraction procedure followed by chemical analysis for specific substances) is a characteristic of all chemicals be the petroleum or nonpetroleum in origin. Toxic wastes are harmful or fatal when ingested or absorbed and, when such wastes are disposed of on land, the chemicals may drain (leach) from the waste and pollute groundwater. Leaching of such chemicals from contaminated soil may be particularly evident when the area is exposed to acid rain. The acidic nature of the water may impart mobility to the waste by changing the chemical character of the waste or the character of the minerals to which the waste species are adsorbed.

As with flammability, there are many tests that can be used to determine corrosivity (ASTM D1838, 2016; ASTM D2251, 2016). Most corrosive substances belong to at least one of the four following nonorganic chemical classes: (1) strong acids, (2) strong bases, (3) oxidants, or (4) dehydrating agents, which are all are used in the refining industry (Speight, 2005a). For example, sulfuric acid is a prime example of a corrosive substance (ASTM C694, 2016). As well as being a strong acid, concentrated sulfuric acid is also a dehydrating agent and oxidant. The heat generated when water and concentrated sulfuric acid are mixed illustrates the high affinity of sulfuric acid for water. If this is done incorrectly by adding water to the acid, localized boiling and spattering can occur and result in personal injury. The major destructive effect of sulfuric acid on skin tissue is removal of water with accompanying release of heat. Contact of sulfuric acid with tissue results in tissue destruction at the point of contact. Inhalation of sulfuric acid fumes or mists damages tissues in the upper respiratory tract and eyes. Long-term exposure to sulfuric acid fumes or mists has caused erosion of teeth, as well as destruction of other parts of the body!

Reactive chemicals are those that tend to undergo rapid or violent reactions under certain conditions. Such substances include those that react violently or form potentially explosive mixtures with water, such as some of the common oxidizing agents. Explosives (Sudweeks et al., 1983; Austin, 1984) constitute another class of reactive chemicals. For regulatory purposes, those substances are also classified as reactive that react with water, acid, or base to produce toxic fumes, particularly hydrogen sulfide or hydrogen cyanide.

Heat and temperature are usually very important factors in reactivity since many reactions require energy of activation to get them started. The rates of most reactions tend to increase sharply with increasing temperature, and most chemical reactions give off heat. Therefore, once a reaction is started in a reactive mixture lacking an effective means of heat dissipation, the rate will increase exponentially with time (doubling with every 10° rise in temperature), leading to an uncontrollable event. Other factors that may affect the reaction rate include the physical form of reactants, the rate and degree of mixing of reactants, the degree of dilution with a nonreactive medium (e.g., an inert solvent), the presence of a catalyst, and pressure.

Toxicity is of the utmost concern in dealing with chemicals and their disposal. This includes both long-term chronic effects from continual or periodic exposures to low levels of toxic chemicals and acute effects from a single large exposure (Zakrzewski, 1991). Not all toxins are immediately apparent. For example, living organisms require certain metals for physiological processes. These metals when present at concentrations above the level of homeostatic regulation can be toxic (ASTM E1302, 2016). In addition, there are metals that are chemically similar to, but higher in molecular weight than, the essential metals (heavy metals). Metals can exert toxic effects by direct irritant activity, blocking functional groups in enzymes, altering the conformation of biomolecules, or displacing essential metals in a metalloenzyme.

5 COAL AND COAL PRODUCTS

Coal (the term is used generically throughout the book to include all types of coal) is a black or brownish-black *organic sedimentary rock* of biochemical origin which is combustible and occurs in rock strata (coal beds, coal seams) and is composed primarily of carbon with variable proportions of hydrogen, nitrogen, oxygen, and sulfur. Coal occurs in seams or strata. In terms of *coal grade*, the grade of a coal establishes its economic value for a specific end use. Grade of coal refers to the amount of mineral matter that is present in the coal and is a measure of coal quality. Sulfur content, ash fusion temperature (i.e., the temperature at which measurement the ash melts and fuses), and quantity of trace elements in coal are also used to grade coal. Although formal classification systems have not been developed around grade of coal, grade is important to the coal user.

Coal is a naturally occurring combustible material with varying composition, and it not surprising that the properties of coal vary considerably from coal type to coal type and even from sample to sample within a specific coal types. This can only be ascertained by application of a series of standard test methods (Zimmerman, 1979; Speight, 2005b).

The constituents of coal can be divided into two groups: (1) the organic fraction, which can be further subdivided into soluble and insoluble fractions as well as microscopically identifiable macerals and (ii) the inorganic fraction, which is commonly identified as ash subsequent to combustion. Because of this complex heterogeneity, it might be expected that the properties of coal can vary considerable, even within a specific rank of coal (Speight, 2005b, 2013).

5.1 Coal

Coal is one of the many vital commodities that contributes on a large scale to energy supply and, unfortunately to environmental pollution, including acid rain, the greenhouse effect, and *allegedly* global warming (global climate change) (Bell, 2011; Speight, 2013). Whatever the effects, the risks attached to the coal-fuel cycle could be minimized by the introduction of new clean coal technologies (Speight, 2013), remembering that there is no single substitute for coal fuel in the generation of energy.

Coal itself is harmless and presents no risk when it is in situ where it was deposited millions of years ago. When involved in coal-related activities, however, its environmental impacts are deleterious if the coal is utilized in the wrong place at the wrong time and in the wrong amounts. At one time, oil-fuel and then nuclear power were considered to be the answer to the world energy demands. These assumptions were to be proved inadequate: (1) because of the unrest and armed hostile conflicts in the Middle East affecting oil supplies and (2) second, the catastrophic nuclear accidents in various parts of the world, which have (justifiably or unjustifiably—it is not the purpose of this text to decide on the viability of energy from nuclear sources) posed serious questions on the viability and safety of the nuclear industry.

By comparison, coal offers substantial opportunities for diversification of energy supply. Coal reserves are abundant, and it is well dispersed geographically. This makes it an invaluable source of energy and fundamental raw material for the generation of electrical power. However, the use of coal does pose serious environment questions some of which have been answered with satisfaction and others which have not been answered to the satisfaction of everyone.

In addition to the adverse effects that can occur during mining, cleaning, and transportation, the major issues of coal use (in the current context) arise from combustion, coke production (coal carbonization), and gasification (Chakrabartty and Selucky, 1985; Speight, 2013). Large amounts of coal are consumed in generating electricity, and the emissions from power stations and similar industrial sources represent a potential, and considerable, environmental hazard. These power plants and the accompanying flue-gas desulfurization (FGD) processes emit effluents, which often are pollutants, and which by mere contact with the external environment or by (generally) simple atmospheric chemical transformations, may form secondary pollutants that are more harmful than the initial effluent/pollutant.

5.2 Coal Products

Coal-based processes involved in conversion facilities release gaseous and liquid effluents as well as solid effluents deleterious to the environment and human health. The preference can be made from the following alternatives (1) removing the pollutant from the process effluent, for example, passing polluted air through a series of dust collectors which filter the fine particulates; (2) removing the pollutant from the process input, desulfurization of coal; (3) controlling the process, lowering combustion temperature to minimize the generation of nitrogen oxides and their emission; (4) replacing the process with one that does not generate or will minimize the pollutant, for example, pressurized fluidized bed combustion instead of pulverized coal burning; and (5) selecting a type of coal-fuel that eliminates the pollutant, for example, use of low-sulfur coal.

Coal-based processes involved in combustion and conversion facilities release gaseous and liquid effluents as well as solid effluents deleterious to the environment and human health. The preference can be made from the following alternatives: (1) removing the pollutant from the process effluent, e.g., passing polluted air through a series of dust collectors which filter the fine particulate matter; (2) removing the pollutant from the process input, desulfurization of coal; (3) controlling the process, lowering combustion temperature to minimize the generation of nitrogen oxides and their emission; (4) replacing the process with one that does not generate or will minimize the pollutant, for example, pressurized fluidized bed combustion instead of pulverized coal burning; and (5) selecting a type of coal-fuel that eliminates the pollutant, for example, use of low-sulfur coal.

Coal combustion products (CCPs), also called coal combustion wastes (CCW) or coal combustion residuals, are categorized in four groups, each based on physical and chemical forms derived from coal combustion methods and emission controls: (1) fly ash, (2) FGD products, and (3) bottom ash.

Fly ash, which can carry with it carbonaceous organic products, is captured after coal combustion by filters, bag houses, electrostatic precipitators, and other air pollution control devices. It comprises 60% (w/w) of all CCW (labeled here as CCPs). It is most commonly used as a high-performance substitute for Portland cement or as clinker for Portland cement production. Cements blended with fly ash are becoming more common. Building material applications range from grouts and masonry products to cellular concrete and roofing tiles. Many asphaltic concrete pavements contain fly ash. Geotechnical applications include soil stabilization, road base, structural fill, embankments, and mine reclamation. Fly ash also serves as filler in wood and plastic products, paints, and metal castings.

FGD materials are produced by chemical scrubber emission control systems that remove sulfur and oxides from power plant flue gas streams. FGD comprises 24% of all CCW. Residues vary, but the most common are FGD gypsum (or "synthetic" gypsum) and spray dryer absorbents. FGD gypsum is used in almost 30% of the gypsum panel products manufactured in the United States. It is also used in agricultural applications to treat undesirable soil conditions and to improve crop performance. Other FGD materials are used in mining and land reclamation activities.

Bottom ash and boiler slag can be used as a raw feed for manufacturing Portland cement clinker, as well as for skid control on icy roads. The two materials comprise 12% and 4% (w/w) of CCW, respectively. These materials are also suitable for geotechnical applications such as structural fills and land reclamation. The physical characteristics of bottom ash and boiler slag lend themselves as replacements for aggregate in concrete masonry products. Boiler slag is also used for roofing granules and as blasting grit.

The majority of CCPs are landfilled, placed in mine shafts or stored on site at coal-fired power plants. Approximately 43% (w/w) of CCPs were recycled for beneficial uses. The chief benefit of recycling is to stabilize the environmental harmful components of the CCPs such as arsenic, beryllium, boron, cadmium, chromium, cobalt, lead, manganese, mercury, molybdenum, selenium, strontium, thallium, and vanadium, along with dioxins and polynuclear aromatic compounds.

Combustion of coal produces environmentally harmful emissions which are typically carbon dioxide, the oxides of sulfur and nitrogen that contribute to acid rain if not removed from the gaseous effluents.

$$CO_2 + H_2O \rightarrow H_2CO_3 \,(\text{carbonic acid})$$
$$SO_2 + H_2O \rightarrow H_2SO_3 \,(\text{sulfurous acid})$$
$$2SO_2 + O_2 \rightarrow 2SO_3$$
$$SO_3 + H_2O \rightarrow H_2SO_4 \,(\text{sulfuric acid})$$
$$NO + H_2O \rightarrow HNO_2 \,(\text{nitrous acid})$$
$$2NO + O_2 \rightarrow NO_2$$
$$NO_2 + H_2O \rightarrow HNO_3 \,(\text{nitric acid})$$

However, some gases produced from burning coal are organic compounds (such as methane) which are known as greenhouse gases because they trap the earth's heat like the roof of a greenhouse and may contribute to possible global warming. Other emissions from coal combustion can lead to air and water pollution.

More pertinent to the present text, coal processing is a source of polynuclear aromatic hydrocarbons (PNAs, also called polycyclic aromatic hydrocarbons, PAHs), which are a large group of organic compounds with two or more fused aromatic rings. They have a relatively low solubility in water but are highly lipophilic. Most of the polynuclear aromatic hydrocarbons with low vapor pressure in the air are adsorbed on particles. When dissolved in water or adsorbed on particulate matter, polynuclear aromatic hydrocarbons can undergo photodecomposition when exposed to ultraviolet light from solar radiation. In the atmosphere, polynuclear aromatic hydrocarbons can react with pollutants such as ozone, nitrogen oxides and sulfur dioxide, yielding diones, nitro-polynuclear aromatic hydrocarbons and dinitro-polynuclear aromatic hydrocarbons, and sulfonic acids, respectively. Polynuclear aromatic hydrocarbons may also be degraded by some microorganisms in the soil.

Polynuclear aromatic hydrocarbons also occur as part of the carbonaceous deposits found on spent catalysts from crude oil refineries and are also deposited on the ash from combustion as well as from incomplete combustion of biomass and fossil fuel in the absence of oxygen and are often referred to as black carbon (Shrestha et al., 2010). Thus, black carbon is the collective term for a range of carbonaceous substances encompassing partly charred plant residues to highly graphitized soot. Depending on its form, condition of origin, storage, and surrounding environmental conditions, black carbon can influence the environment at local, regional, and global scales in different ways.

The toxicity of polynuclear aromatic hydrocarbons is perhaps one of the most serious long-term problems associated with the use of crude oil. They comprise a large class of crude oil compounds containing two or more benzene rings. Polynuclear aromatic hydrocarbons are formed in nature by long-term, low-temperature chemical reactions in sedimentary deposits of organic materials and in high-temperature events such as volcanoes and forest fires. The major source of this pollution is, however, human activity. Polynuclear aromatic hydrocarbons accumulate in soil, sediment, and biota. At high concentrations, they can be acutely toxic by disrupting membrane function. Many cause sunlight-induced toxicity in humans and fish and other aquatic organisms. In addition, long-term chronic toxicity has been demonstrated in a wide variety of organisms. Through metabolic activation, some polynuclear aromatic hydrocarbons *form* reactive intermediates that bind to deoxyribonucleic acid. For this reason, many of these hydrocarbons are *mutagenic* (tending to cause mutations), *teratogenic* (tending to cause developmental malformations), or *carcinogenic* (tending to cause cancer).

6 CRUDE OIL

Petroleum (also known as *crude oil*) is perhaps the most important substance consumed in modern society. It provides not only raw materials for the ubiquitous plastics and other products but also fuel for energy, industry, heating, and transportation. From a chemical standpoint, petroleum is an extremely complex mixture of hydrocarbon compounds, usually with minor amounts of nitrogen-, oxygen-, and sulfur-containing compounds as well as trace amounts of metal-containing compounds (Speight, 2014, 2015).

Petroleum is a carbon-based resource. Therefore, the geochemical carbon cycle is also of interest to fossil fuel usage in terms of petroleum formation, use, and the buildup of atmospheric carbon dioxide. Thus, the more efficient use of petroleum and is of paramount importance. Petroleum technology, in one form or another, is with us until suitable alternative forms of energy are readily available (Ramage, 1997). For example, the fuels that are derived from petroleum supply more than half of the world's total supply of energy. Gasoline, kerosene, and diesel oil provide fuel for automobiles, tractors, trucks, aircraft, and ships. Fuel oil and natural gas are used to heat homes and commercial buildings, as well as to generate electricity. Petroleum products are the basic materials used for the manufacture of synthetic fibers for clothing and in plastics, paints, fertilizers, insecticides, soaps, and synthetic rubber. The uses of petroleum as a source of raw material in manufacturing are central to the functioning of modern industry.

6.1   Crude Oil

Crude oil and the equivalent term *petroleum* cover a wide assortment of materials consisting of mixtures of hydrocarbons and other compounds containing variable amounts of sulfur, nitrogen, and oxygen, which may vary widely in volatility, specific gravity, and viscosity. Metal-containing constituents, notably those compounds that contain vanadium and nickel, usually occur in the more viscous crude oils in amounts up to several thousand parts per million and can have serious consequences during processing of these feedstocks (Speight, 2014). Because crude oil is a mixture of widely varying constituents and proportions, its physical properties also vary widely and the color from colorless to black.

Indeed, crude oil reservoirs have been found in vastly different parts of the world, and their chemical composition varies greatly. Consequently, no single composition of crude oil can be defined. Thus crude oil-derived inputs to the environment vary considerably in composition and the complexity of crude oil composition is matched by the range of properties of the components and the physical, chemical, and biochemical processes that contribute to the distributive pathways and determine the fate of the inputs. Put simply, crude oil is a naturally occurring mixture of hydrocarbons, generally in a liquid state, which may also include compounds of sulfur nitrogen oxygen metals and other elements (Speight, 2014).

In terms of the elemental composition of crude oil, the carbon content is relatively constant, while the hydrogen and heteroatom contents are responsible for the major differences between crude oil from different sources. The nitrogen, oxygen, and sulfur can be present in only trace amounts in some crude oil, which as a result consists primarily of hydrocarbons. On the other hand, a crude oil containing 9.5% (w/w) heteroatoms may contain essentially no true hydrocarbon constituents insofar as the constituents contain *at least one or more* nitrogen, oxygen, and/or sulfur atoms within the molecular structures.

Crude oil use is a necessary part of the modern world, hence the need for stringent controls over the amounts and types of emissions from the use of crude oil and its products. So it is predictable that crude oil will be a primary source of energy for the next several decades and, therefore, the message is clear. The challenge is for the development of technological concepts that will provide the maximum recovery of energy from crude oil not only cheaply but also efficiently and with minimal detriment to the environment.

The use of crude oil has significant social and environmental impacts, from accidents and routine activities such as seismic exploration, drilling, and generation of polluting wastes. Crude oil from subterranean and submarine reservoirs extraction can be environmentally damaging. Crude oil and refined fuel spills from tanker ship accidents have damaged fragile ecosystems. Burning oil releases carbon dioxide and hydrocarbons into the atmosphere. The general prognosis for emission cleanup is not pessimistic and can be looked upon as being quite optimistic. Indeed, it is considered likely that most of their environmental impact of crude oil refining can be substantially abated. A considerable investment in retrofitting or replacing existing facilities and equipment might be needed. However, it is possible and a conscious goal must be to improve the efficiency with which crude oil is transformed and consumed.

Considering the composition of crude oil and crude oil products (Speight, 2014), it is not surprising that crude oil and crude oil-derived chemicals are environmental pollutants (Speight, 2005a,b). The world's economy is highly dependent on crude oil for energy production and widespread use has led to enormous releases to the environment of crude oil, crude oil products, exhaust from internal combustion engines, emissions from oil-fired power plants, and industrial emissions where fuel oil is employed (Speight, 2014).

6.2 Crude Oil Products

Crude oil is rarely used in the form produced at the well but is converted in refineries into a wide range of products, such as gasoline, kerosene, diesel fuel, jet fuel, domestic fuel oil, and industrial fuel oil, together with petrochemical feedstocks such as ethylene, propylene, the butene, butadiene, and isoprene. Crude oil is refined, that is, separated into useful products (Table 4.5) from which saleable products are produced by additional refining. Refining consists of, initially unless properties dictate otherwise (Speight, 2014, 2016a), dividing the crude oil into fractions of different boiling ranges by distillation. Other forms of treatment are utilized during the refining process to remove undesirable components

TABLE 4.5 Boiling Ranges of Crude Oil Products

Boiling Range		Product
°C	°F	
1–205	32–400	Naphtha
		Straight-run gasoline
205–345	400–655	Middle distillates: kerosene, jet fuel, heating oil, diesel fuel
345–565	655–1050	Gasoil, including lubricating (lube) oil and wax
565+	1050+	Residuum

of the crude oil. The fractions themselves are often further distilled to produce the desired commercial product. A variety of additives may be incorporated into some of the refined products to adjust the octane ratings or improve engine performance characteristics.

The lowest boiling (lightest) constituents of crude oil are gases at room temperature and are collected and used as heating gas mixtures and in the petrochemical industry or as a refinery fuel. The next lightest hydrocarbons occur in molecule that contain four-to-nine carbon atoms and have a boiling range (also known as the light and heavy naphtha fraction) and are used in gasoline formulation. Constituents boiling in the middle ranges or *middle distillates* are used for production of kerosene diesel fuel, jet fuel, and fuel oil. These fuels contain paraffins (alkanes), cycloparaffins (cycloalkanes), aromatics, and olefins from approximately 9- to 20-carbons molecular range.

A *residuum* (*pl. residua*, also shortened to *resid, pl. resids*) is the residue obtained from crude oil after nondestructive distillation has removed all the volatile materials. The temperature of the distillation is usually maintained below 350°C (660°F) since the rate of thermal decomposition of crude oil constituents is minimal below this temperature, but the rate of thermal decomposition of crude oil constituents is substantial above 350°C (660°F) (Speight, 2014). *Residua* are black, viscous materials and are obtained by distillation of a crude oil under atmospheric pressure (atmospheric residuum) or under reduced pressure (vacuum residuum). They may be liquid at room temperature (generally atmospheric residua) or almost solid (generally vacuum residua) depending upon the cut-point of the distillation or depending on the nature of the crude oil (Speight, 2014, 2016a).

The highest boiling molecular weight compounds that do not distill under refinery conditions vaporize at all are residua or paraffin derivatives, depending on the source of the crude oil. The highest boiling fractions are high-molecular-weight hydrocarbons suitable for lubricants and heating oil. Lubricants may contain hydrocarbons ranging from 18 to 25 carbon atoms per molecule. Paraffin wax and crude oil jelly typically contain 28 to 38 carbon atoms per molecule. Other crude oil products include a wide variety of solvents and other refined oils (such as lubricating oils) which may also include a number of additives such as gelling inhibitors that are added to diesel fuels during cold weather (Speight and Exall, 2014). Certain additives may be of special concern in an injury assessment, either because they are toxic themselves or because they significantly change the behavior of the oil products.

Crude oil products have a vast array of uses. In approximate order of importance, the uses are: fuels for vehicles and industry, heating oils, lubricants, raw materials in manufacturing petrochemicals and pharmaceuticals, and solvents. By a wide margin, most of the products derived from crude oil find use as fossil fuels to run vehicles, produce electricity, and to heat homes and business. About 65% (v/v) of the crude oil used as fuel is consumed as gasoline in automobiles. Thus, crude oil products are ubiquitous in the modern environment that leads to contamination problems both for the environment and in sampling activities. In particular, the toxicity of polynuclear aromatic hydrocarbons is perhaps one of the most serious long-term problems associated with the use of crude oil. They comprise a large class of crude oil compounds containing three or more condensed benzene rings and are concentrated mainly in crude oil residua (Speight, 2014).

In terms of the composition of crude oil, it contains compounds that are composed of carbon and hydrogen only which do not contain any heteroatoms (nitrogen, oxygen, and sulfur as well as compounds containing metallic constituents, particularly vanadium, nickel, iron, and copper). The hydrocarbons found in crude oil are classified into the following types: (1) *paraffin derivatives*, which are saturated hydrocarbons with straight or branched chains, but without any ring structure; (2) *cycloparaffin derivatives* also called *naphthene derivatives* but more correctly known as *alicyclic hydrocarbons*, which are saturated hydrocarbons containing one or more rings, each of which may have one or more paraffinic side-chains; and (3) *aromatic derivatives*, which are hydrocarbons containing one or more aromatic nuclei such as the benzene ring system, the naphthalene ring system, and the phenanthrene ring system that may be linked up with (substituted) naphthalene rings and/or paraffinic side-chains.

Another way to describe or characterize crude oil products is by generalized spill cleanup categories and the following categories are in use by the National Oceanic and Atmospheric Administration (NOAA) to identify cleanup options: (1) gasoline-type products, (2) diesel-type products, (3) intermediate products, (4) fuel oil products, and (5) residua.

Gasoline-type products are highly volatile products that evaporate quickly (often completely) within 1–2 days. They are narrow cut fraction with no residue, low viscosity, that spreads rapidly to a thin sheen on water or on to the land. They are highly toxicity to biota, will penetrate the substrate, and are nonadhesive. *Diesel-like products* (jet fuel, diesel, No. 2 fuel oil, kerosene) are moderately volatile products that can evaporate with no residue. They have a low-to-moderate viscosity and spread rapidly into thin films as well as form stable emulsions. These products also have a moderate-to-high (usually high) toxicity to biota and the specific toxicity is often related to type and concentration of aromatic compounds. They have the ability to penetrate substrate, but fresh (unoxidized) spills are nonadhesive. *Intermediate products* (No. 4 fuel oil, lube oil) are products that are less volatile than the two previous categories—up to one-third will evaporate within 24 h. They have a moderate-to-high viscosity and a variable toxicity that depends on amount of the lower boiling components. These products may penetrate the substrate and, therefore, cleanup most effective if conducted quickly.

Fuel oil (heavy industrial fuel oil) is a medium viscosity product that is highly variable and often blended with lower boiling products. The blends may be unstable and the oil may separate when spilled on to the ground or on to a waterway. The oil may be buoyant or sink in water depending on water density. The sunken oil has little potential for evaporation and may accumulate on bottom (of the waterway) under calm conditions. However, the sunken oil may be resuspended during storm events providing shoreline oiling (contamination). These products weather (oxidize) slowly. *Residual products* (No. 6 fuel oil, Bunker C oil): these products have little (usually/no) ability to evaporate. When spilled, persistent surface and intertidal area contamination is likely with long-term contamination of the sediment. The products are very viscous to semisolid and often become less viscous when warmed and such products weather oxidize (often referred to as *weathering*) slowly and may form tar balls that can sink in waterways (depending on product density and water density). They are highly adhesive to soil. Heavy oil, a highly viscous crude oil, and bitumen (isolated from tar sand deposits) also come into this category of contaminant.

Also, the petrochemical industry uses fossil fuels (e.g., natural gas) or petroleum refinery products (e.g., naphtha) as feedstocks. Petrochemicals are chemical products derived from crude oil. The two most common petrochemical classes are olefins (including ethylene, $CH_2=CH_2$, and propylene, $CH_3CH=CH_2$) and aromatic derivatives (including benzene, toluene, and xylene isomers). Primary petrochemicals are divided into three groups depending on their chemical structure: (1) olefins, which ethylene, propylene, and butadiene, $CH_2=CHCH=CH_2$; (2) aromatics, which includes benzene, C_6H_6, toluene $C_6H_5CH_3$, and xylene isomers, $CH_3C_6H_4CH_3$; and (3) synthesis gas, which is a mixture of carbon monoxide and hydrogen that is to make ammonia (NH_3) and methanol (methyl alcohol, CH_3OH).

6.3 Refinery Waste

There are several hundred individual hydrocarbon chemicals defined as petroleum-based. Furthermore, each petroleum product has its own mix of constituents because (Chapter 2) petroleum varies in composition from one reservoir to another, and this variation may be reflected in the finished product(s).

The chemicals in petroleum vary from (chemically speaking) simple hydrocarbons of low-to-medium molecular weight to organic compounds containing sulfur, oxygen, and nitrogen, as well as compounds containing metallic constituents, particularly vanadium nickel, iron, and copper. Many of these latter compounds are of indeterminate molecular weight. Residua, that are produced by distillation that is a concentration process, contain significantly less hydrocarbon constituents than the original crude oil. The constituents of residua, depending on the crude oil, may be molecular entities of which the majority contains at least one heteroatom.

Typical refinery products include (1) natural gas and LPG, (2) solvent naphtha, (3) kerosene, (4) diesel fuel, (5) jet fuel, (6) lubricating oil, (7) various fuel oils, (8) wax, (9) residua, and (10) asphalt (Chapter 3). A single refinery does not necessarily produce all of these products. Some refineries are dedicated to particular products, for example, the production of gasoline or the production of lubricating oil or the production of asphalt. However, the issue is that refineries also produce a variety of waste products (Table 4.1) that must be disposed in an environmentally acceptable manner.

Waste treatment processes also account for a significant area of the refinery, particularly sulfur compounds in gaseous emissions together with various solid and liquid extracts and wastes generated during the refining process. The refinery is therefore composed of a complex system of stills, cracking units, processing and blending units and vessels in which the various reactions take place, as well as packaging units for products for immediate distribution to the retailer, for example, lubricating oils. Bulk storage tanks usually grouped together in tank farms are used for storage of both crude and refined products. Other tanks are used in the processes outlined, for example, treating, blending, and mixing, while others are used for spill and fire control systems. A boiler and electrical generating system usually operate for the refinery as a whole.

Petroleum hydrocarbons are environmental contaminants, but they are not usually classified as hazardous wastes. Soil and groundwater petroleum hydrocarbon contamination has long been of concern and has spurred various analytical and site remediation developments, for example, risk-based corrective actions. In some instances, it may appear that such cleanup operations were initiated with an incomplete knowledge of the charter and behavior of the contaminants. The most appropriate first assumption is that the spilled constituents are toxic to the ecosystem. The second issue is an investigation of the products of the spilled material to determine an appropriate cleanup method. The third issue is whether or not the chemical nature of the constituents has changed during the time since the material was released into the environment. If it has, a determination must be made of the effect of any such changes on the potential cleanup method.

Despite the large number of hydrocarbons found in petroleum products and the widespread nature of petroleum use and contamination, many of the lower boiling constituents are well characterized in terms of physical properties, but only a relatively small number of the compounds are well characterized for toxicity. The health effects of some fractions can be well characterized, based on their components or representative compounds (e.g., light aromatic fraction benzene-toluene-ethylbenzene-xylenes). However, higher-molecular-weight (higher boiling) fractions have far fewer well-characterized compounds.

This section deals with the toxicity of petroleum and petroleum products, and toxicity, the effects of petroleum constituents on the environment, and the individual process wastes, and the means by which petroleum, petroleum products, and process wastes are introduced into the environment. The processes are restricted to those processes by which the common products are produced (Chapter 3).

6.3.1 Process Wastes

Petroleum refineries are complex, but integrated, unit process operations that produce a variety of products from various feedstock blends (Speight, 2005a,b, 2014, 2016a). During petroleum refining, refineries use and generate an enormous amount of chemicals, some of which are present in air emissions, wastewater, or solid wastes. Emissions are also created through the combustion of fuels, and as by-products of chemical reactions occurring when petroleum fractions are upgraded. A large source of air emissions is, generally, the process heaters and boilers that produce carbon monoxide, sulfur oxides, and nitrogen oxides, leading to pollution and the formation of acid rain.

$$CO_2 + H_2O \rightarrow H_2CO_3 \,(\text{carbonic acid})$$
$$SO_2 + H_2O \rightarrow H_2SO_3 \,(\text{sulfurous acid})$$
$$2SO_2 + O_2 \rightarrow 2SO_3$$
$$SO_3 + H_2O \rightarrow H_2SO_4 \,(\text{sulfuric acid})$$
$$NO + H_2O \rightarrow HNO_2 \,(\text{nitrous acid})$$
$$2NO + O_2 \rightarrow NO_2$$
$$NO_2 + H_2O \rightarrow HNO_3 \,(\text{nitric acid})$$

Hence, there is the need for gas-cleaning operations on a refinery site so that such gases are cleaned from the gas stream prior to entry into the atmosphere.

In addition, some processes create considerable amounts of particulate matter and other emissions from catalyst regeneration or decoking processes. Volatile chemicals and hydrocarbons are also released from equipment leaks, storage tanks, and wastewaters. Other cleaning units such as the installation of filters, electrostatic precipitators, and cyclones can mitigate part of the problem.

Process wastewater is also a significant effluent from a number of refinery processes. Atmospheric and vacuum distillation create the largest volumes of process wastewater, about 26 gallons per barrel of oil processed. Fluid catalytic cracking and catalytic reforming also generate considerable amounts of wastewater (15 and 6 gallons per barrel of feedstock, respectively). A large portion of wastewater from these three processes is contaminated with oil and other impurities and must be subjected to primary, secondary, and sometimes tertiary water treatment processes, some of which also create hazardous waste.

Wastes, residua, and by-products are produced by a number of processes. Residuals produced during refining are not necessarily wastes. They can be recycled or regenerated, and in many cases do not become part of the waste stream but are useful products. For example, processes utilizing caustics for neutralization of acidic gases or solvent (e.g., alkylation, sweetening/chemical treating, lubricating oil manufacture) create the largest source of residuals in the form of spent caustic solutions. However, nearly all of these caustics are recycled.

The treatment of oily wastewater from distillation, catalytic reforming, and other processes generates the next largest source of residuals in the form of biomass sludge from biological treatment and pond sediments. Water treatment of oily wastewater also produces a number of sludge materials associated with oil-water separation processes. Such sludge is often recycled in the refining process and is not considered wastes.

Catalytic processes (fluid catalytic cracking, catalytic hydrocracking, hydrotreating, isomerization, ethers manufacture) also create some residuals in the form of spent catalysts and catalyst fines or particulates. The latter are sometimes separated from exiting gases by electrostatic precipitators or filters. These are collected and disposed in landfills or may be recovered by off-site facilities.

6.3.2 Spills

It is almost impossible to transport, store, and refine crude oil without spills and losses. It is difficult to prevent spills resulting from failure or damage on pipelines. It is also impossible to install control devices for controlling the ecological properties of water and the soil along the length of all pipelines. The soil suffers the most ecological damage in the damage areas of pipelines. Crude oil spills from pipelines lead to irreversible changes of the soil properties. The most affected soil properties by crude oil losses from pipelines are filtration, physical and mechanical properties. These properties of the soil are important for maintaining the ecological equilibrium in the damaged area.

Principal sources of releases to air from refineries include: (1) combustion plants, emitting sulfur dioxide, oxides of nitrogen, and particulate matter; (2) refining operations, emitting sulfur dioxide, oxides of nitrogen, carbon monoxide, particulate matter, VOCs, hydrogen sulfide, mercaptans, and other sulfurous compounds; (3) bulk storage operations and handling of VOCs (various hydrocarbons). In light of this, it is necessary to consider (1) regulatory requirements—air emission permits stipulating limits for specific pollutants, and possibly health and hygiene permit requirements; (2) requirement for monitoring program; and (3) requirements to upgrade pollution abatement equipment.

6.3.3 Storage and Handling of Petroleum Products

Large quantities of environmentally sensitive petroleum products are stored in (1) tank farms (multiple tanks), (2) single above-ground storage tanks (ASTs), (3) semiunderground or underground storage tanks (USTs). Smaller quantities of materials may be stored in drums and containers of assorted compounds (such as lubricating oil, engine oil, other products for domestic supply). In light of this, it is also necessary to consider (1) secondary containment of tanks and other storage areas and integrity of hard standing (without cracks, impervious surface) to prevent spills reaching the wider environment: also secondary containment of pipelines where appropriate; (2) age, construction details, and testing program of tanks; (3) labeling and environmentally secure storage of drums (including waste storage); (4) accident/fire precautions, emergency procedures; and (5) disposal/recycling of waste or "out of spec" oils and other materials.

There is a potential for significant soil and groundwater contamination to have arisen at petroleum refineries. Such contamination consists of (1) petroleum hydrocarbons including lower boiling, very mobile fractions (paraffins, cycloparaffins, and volatile aromatics such as benzene, toluene, ethylbenzene, and xylenes) typically associated with gasoline and similar boiling range distillates; (2) middle distillate fractions (paraffins, cycloparaffins, and some polynuclear aromatics) associated with diesel, kerosene, and lower boiling fuel oil, which are also of significant mobility; (3) higher boiling distillates (long-chain paraffins, cycloparaffins, and polynuclear aromatics) are associated with lubricating oil and heavy fuel oil; (4) various organic compounds associated with petroleum hydrocarbons or produced during the refining process, for example, phenols, amines, amides, alcohols, organic acids, nitrogen, and sulfur containing compounds; (5) other organic additives, for example, antifreeze (glycols), alcohols, detergents, and various proprietary compounds; (6) organic lead, associated with leaded gasoline and other heavy metals.

Key sources of such contamination at petroleum refineries are at (1) transfer and distribution points in tankage and process areas, also general loading and unloading areas; (2) land farm areas; (3) tank farms; (4) individual ASTs and particularly individual USTs; (5) additive compounds; and (6) pipelines, drainage areas as well as on-site waste treatment facilities, impounding basins, lagoons, especially if unlined.

While contamination may be associated with specific facilities the contaminants are relatively highly mobile in nature and have the potential to migrate significant distances from the source in soil and groundwater. Petroleum hydrocarbon contamination can take several forms: free-phase product, dissolved phase, emulsified phase, or vapor phase. Each form will require different methods of remediation so that cleanup may be complex and expensive. In addition, petroleum hydrocarbons include a number of compounds of significant toxicity, for example, benzene and some polyaromatics are known carcinogens. Vapor-phase contamination can be of significance in terms of odor issues. Due to the obvious risk of fire, refineries are equipped with sprinkler or spray systems that may draw upon the main supply of water, or water held in lagoons, or from reservoirs or neighboring water courses. Such water will be polluting and require containment.

Refining facilities require significant volumes of water for on-site processes (e.g., coolants, blow-downs, etc.) as well as for sanitary and potable use. Wastewater will derive from these sources (process water) and from storm water runoff. The latter could contain significant concentrations of petroleum product. Petroleum hydrocarbons, dissolved, emulsified or occurring as free-phase, will be the key constituents although wastewater may also contain significant concentrations of phenols, amines, amides, alcohols, ammonia, sulfide, heavy metals, and suspended solids.

Wastewaters may be collected in separate drainage systems (for process, sanitary, and storm water) although industrial water systems and storm water systems may in some cases be combined. In addition, ballast water from bulk crude tankers may be pumped to receiving facilities at the refinery site prior to removal of floating oil in an interceptor and treatment as for other wastewater streams. On-site treatment facilities may exist for wastewater or treatment may take place at a public wastewater treatment plant. Storm water/process water is generally passed to a separator or interceptor prior to leaving the site which takes out free-phase oil (i.e., floating product) from the water prior to discharge, or prior to further treatment, e.g., in settling lagoons). Discharge from wastewater treatment plants is usually passed to a nearby watercourse. Other wastes that are typical of a refinery include (1) waste oils, process chemicals, still resides; (2) nonspecification chemicals and/or products; (3) waste alkali (sodium hydroxide); (4) waste oil sludge (from interceptors, tanks, and lagoons); and (5) solid wastes (cartons, rags, catalysts, and coke).

The pollution of ecosystems, either inadvertently or deliberately, has been a fact of life for millennia (Pickering and Owen, 1994). In recent times, the evolution of industrial operations has led to issues related to the production and disposal of a wide variety of organic chemicals (Chapter 3) (Easterbrook, 1995). Chemical wastes that were once exotic have become commonplace and hazardous (Tedder and Pohland, 1993). Recognition of this makes it all the more necessary that steps be taken to terminate the pollution, preferably at the source or before it is discharged into the environment. It is also essential that the necessary tests be designed to detect the pollution and its effect on living forms.

Any chemical substance, if improperly managed or disposed of, may pose a danger to living organisms, materials, structures, or the environment, by explosion or fire hazards, corrosion, toxicity to organisms, or other detrimental effects. In addition, many chemical substances, when released to the environment, can be classified as hazardous or nonhazardous. Consideration must be given to the distribution of chemical wastes on land systems, in water systems, and in the atmosphere.

In general terms, the origin of chemical wastes refers to their points of entry into the environment. Point-source leaks and spills (i.e., sources that release emissions through a confined vent (stack) or opening and nonpoint-source emissions) have resulted in environmental contamination from crude oil and crude oil products. Spills of crude oil and fuels have caused wide-ranging damage in the marine and freshwater environments. Oil slicks and tars in shore areas and beaches can ruin the esthetic value of entire regions. Other sources of environmental leakage as it affects the crude oil industry may consist of (1) deliberate addition to soil, water, or air by humans, for example, the disposal of used engine oil; (2) evaporation or wind erosion from emissions into the atmosphere; (3) leaching from waste dumps into groundwater, streams, and bodies of water;

(4) leakage, such as from USTs or pipelines; (5) accidents, such as fire or explosion; and (6) emissions waste treatment or storage facilities.

In terms of waste definition, there are three basic approaches (as it pertains to crude oil, crude oil products, and noncrude oil chemicals) to defining crude oil or a crude oil product as hazardous: (1) a qualitative description of the waste by origin, type, and constituents; (2) classification by characteristics based upon testing procedures; and (3) classification as a result of the concentration of specific chemical substances.

However, various countries use different definitions of chemical waste and, many times, there are several inconsistencies in the definitions. Usually the definition involves qualification of whether or not the material is hazardous. For example, in some countries, a hazardous waste is any material that is especially hazardous to human health, air, or water, or which are explosive, flammable, or may cause diseases. Poisonous waste is material that is poisonous, noxious, or polluting and whose presence on the land is liable to give rise to an environmental hazard. But, in more general terms (in any country), hazardous waste is waste material that is unsuitable for treatment or disposal in municipal treatment systems, incinerators, or landfills and that therefore requires special treatment.

Moreover, and somewhat paradoxically, measures taken to reduce air and water pollution may actually increase production of chemical wastes. As examples, disposal of crude oil wastes by water treatment processes can yield a chemical sludge or concentrated liquor that require stabilization and disposal (Cheremisinoff, 1995). Scrubbing to remove hydrogen sulfide, sulfur oxides, and low-boiling organic sulfides as well as carbon dioxide (gas cleaning) are not immune to process waste, even though the chemistry of the cleaning processes is, in theory, reversible (Speight, 2007). Sludge is often produced, and the disposal of this material became a major environmental issue that cannot be ignored. In addition, electrostatic precipitators, used to remove metals from flue gases (Speight, 2007) also yield significant quantities of solid by-products; some of which are hazardous.

7 FLAME RETARDANTS

Flame retardant chemicals are used in commercial and consumer products (such as furniture and building insulation) to meet flammability standards. Not all flame retardants present concerns, but the following types often do: (1) halogenated flame retardants, also known as organo-halogen flame retardants that contain chlorine or bromine bonded to carbon and (2) organo-phosphorous flame retardants that contain phosphorous bonded to carbon.

Flame retardants inhibit or delay the spread of fire by suppressing the chemical reactions in the flame or by the formation of a protective layer on the surface of a material. They may be mixed with the base material (additive flame retardants) or chemically bonded to it (reactive flame retardants). Mineral flame retardants are typically additive, while organohalogen and organophosphorus compounds can be either reactive or additive.

Many flame retardants, while having measurable or considerable toxicity, degrade into compounds that are also toxic, and in some cases, the degradation products may be the primary toxic agent. For example, halogenated compounds with aromatic rings can degrade into dioxin derivatives, particularly when heated, such as during production, a fire, recycling, or exposure to sun. In addition, polybrominated diphenyl ethers with higher numbers of bromine atoms, such as decabromodiphenyl ether (decaBDE), are less toxic than pentabromodiphenyl ether derivatives with lower numbers of bromine atoms (Table 4.4). However, as the higher-order pentabromodiphenyl ether derivatives degrade biotically or abiotically, bromine atoms are removed, resulting in more toxic pentabromodiphenyl ether derivatives.

In addition, when some of the halogenated flame retardants such as pentabromodiphenyl ether derivatives are metabolized, they form hydroxylated metabolites that can be more toxic than the parent compound. These hydroxylated metabolites, for example, may compete more strongly to bind with transthyretin or other components of the thyroid system, can be more potent estrogen mimics than the parent compound, and can more strongly affect neurotransmitter receptor activity.

When products with flame retardants reach the end of their usable life, they are typically recycled, incinerated, or landfilled. Recycling can contaminate workers and communities near recycling plants, as well as new materials, with halogenated flame retardants and their breakdown products. Electronic waste, vehicles, and other products are often melted to recycle their metal components, and such heating can generate toxic dioxins and furans. Brominated flame retardants may also change the physical properties of plastics, resulting in inferior performance in recycled products. Poor-quality incineration similarly generates and releases high quantities of toxic degradation products. Controlled incineration of materials with halogenated flame retardants, while costly, substantially reduces release of toxic by-products.

Many products containing halogenated flame retardants are sent to landfills. Additive, as opposed to reactive, flame retardants are not chemically bonded to the base material and leach out more easily. Brominated flame retardants, including pentabromodiphenyl ether derivatives, have been observed leaching out of landfills in some countries. Landfill designs must allow for leachate capture, which would need to be treated, but these designs can degrade with time.

8 INDUSTRIAL CHEMICALS

Organic chemistry chemicals are some of the important starting materials for a great number of major chemical industries (Chapter 3). The production of organic chemicals as raw materials or reagents for other applications is a major sector of manufacturing polymers, pharmaceuticals, pesticides, paints, artificial fibers, food additives, etc. Organic synthesis on a large scale, compared to the laboratory scale, involves the use of energy, basic chemical ingredients from the petrochemical sector, catalysts and after the end of the reaction, separation, purification, storage, packaging, distribution, etc. During these processes there are many problems of health and safety for workers in addition to the environmental problems caused by their use and disposition as waste.

The industrial organic chemical sector produces organic chemicals used as either chemical intermediates or end-products (Chapter 3) (Sheldon, 2010). This categorization corresponds to Standard Industrial Classification (SIC) code 286 established by the Bureau of Census to track the flow of goods and services within the economy. The 286 category includes gum and wood chemicals (SIC 2861), cyclic organic crudes and intermediates, organic dyes and pigments (SIC 2865), and industrial organic chemicals not elsewhere classified (SIC 2869). By this definition, the industry does not include plastics, drugs, soaps and detergents, agricultural chemicals or paints, and allied products which are typical end-products manufactured from industrial organic chemicals. In 1993, there were 987 establishments in SIC 286 of which the largest 53 firms (by employment) accounted for more than 50% of the industry's value of shipments. The SIC 286 may include a small number of integrated firms that are also engaged in petroleum refining and manufacturing of other types of chemicals at the same site although firms primarily engaged in manufacturing coal tar crudes or petroleum refining are classified elsewhere.

The industrial organic chemical industry uses feedstocks derived from petroleum and natural gas (about 90%) and from recovered coal tar condensates generated by coke production (about 10%) (Chapter 3) (Speight, 2013, 2014). The chemical industry produces raw materials and intermediates, as well as a wide variety of finished products for industry, business, and individual consumers. The important classes of products within SIC code 2861 are hardwood and softwood distillation products, wood and gum naval stores, charcoal, natural dyestuffs, and natural tanning materials.

The chemicals industry is very diverse, comprising basic or commodity chemicals; specialty chemicals derived from basic chemicals (adhesives and sealants, catalysts, coatings, electronic chemicals, plastic additives, etc.); products derived from life sciences (pharmaceuticals, pesticides, and products of modern biotechnology); and consumer care products (soap, detergents, bleaches, hair and skin care products, fragrances, etc.). The global chemicals industry today produces tens of thousands of substances (some in volumes of millions of tons, but most of them in quantities of less than 1000 tons per year). The substances can be mixed by the chemicals industry and sold and used in this form, or they can be mixed by downstream customers of the chemicals industry (e.g., retail stores which sell paint). It is important to note that most of the output from chemical companies is used by other chemical companies or other industries (e.g., metal, glass, electronics), and chemicals produced by the chemicals industry are present in countless products used by consumers (e.g., automobiles, toys, paper, clothing).

The chemical industry involves the use of chemical processes such as chemical reactions and refining methods to produce a wide variety of solid, liquid, and gaseous materials. Most of these products serve to manufacture other items, although a smaller number goes directly to consumers. Solvents, pesticides, lye, washing soda, and Portland cement provide a few examples of product used by consumers. The industry includes manufacturers of and organic-industrial chemicals, petrochemicals, agrochemicals, polymers and rubber (elastomers), oleo-chemicals (oils, fats, and waxes), explosives, fragrances, and flavors (Table 4.6) (Chapter 3).

Chemical processes such as chemical reactions operate in chemical plants to form new substances in various types of reaction vessels. In many cases the reactions take place in special corrosion-resistant equipment at elevated temperatures and pressures with the use of catalysts. The products of these reactions are separated using a variety of techniques including distillation, especially fractional distillation, precipitation, crystallization, adsorption, filtration, sublimation, and drying (Speight, 2002).

The processes and product or products are usually tested during and after manufacture by dedicated instruments and on-site quality control laboratories to ensure safe operation and to assure that the product will meet required specifications. More organizations within the industry are implementing chemical compliance software to maintain quality products and manufacturing standards. The products are packaged and delivered by many methods, including pipelines, tank-cars and tank-trucks (for both solids and liquids), cylinders, drums, bottles, and boxes. Chemical companies often have a research-and-development laboratory for developing and testing products and processes.

TABLE 4.6 Examples of Industrial Organic Chemicals

Product Type	Examples
Organic industrial	Acrylonitrile, phenol, ethylene oxide, urea
Petrochemicals	Ethylene, propylene, benzene, styrene
Agrochemicals	Fertilizers, insecticides, herbicides
Polymers	Polyethylene, Bakelite, polyester
Elastomers	Polyisoprene, neoprene, polyurethane
Oleo-chemicals	Lard, soybean oil, stearic acid
Explosives	Nitroglycerin, ammonium nitrate, nitrocellulose
Fragrances and flavors	Benzyl benzoate, coumarin, vanillin
Industrial gases	Acetylene, olefin derivatives alkane derivatives

These facilities may include pilot plants, and such research facilities may be located at a site separate from the production plant(s).

Industrial organic chemical manufacturers use and generate both large numbers and quantities of chemicals (Chapter 3). The types of pollutants a single facility will release depend on the feedstocks, processes, equipment in use, and maintenance practices. These can vary from hour to hour and can also vary with the part of the process that is underway. For example, for batch reactions in a closed vessel, the chemicals are more likely to be emitted at the beginning and end of a reaction step (associated with vessel loading and product transfer operations) than during the reaction.

Industrial organic synthesis, followed a largely *stoichiometric* line of evolution that can be traced back to the synthesis of mauveine by Perkin, the subsequent development of the dyestuffs industry based on coal tar, and the fine chemicals and pharmaceuticals industries, which can be regarded as spin-offs from the dyestuffs industry. Consequently, fine chemicals and pharmaceuticals manufacture, which is largely the domain of synthetic organic chemists, is rampant with classical stoichiometric processes (Chapter 3).

The desperate need for more catalytic methodologies in industrial organic synthesis is nowhere more apparent than in oxidation chemistry. For example, as any organic chemistry textbook will note that the reagent of choice for the oxidation of secondary alcohols to the corresponding ketones, a pivotal reaction in organic synthesis, is the Jones reagent. The latter consists of chromium trioxide and sulfuric acid and is reminiscent of the phloroglucinol process referred to earlier. The introduction of the storage-stable pyridinium chlorochromate and pyridinium dichromate in the 1970s represented a practical improvement, but the stoichiometric amounts of carcinogenic chromium(VI) remain a serious problem. Obviously there is a definite need in the fine chemical and pharmaceutical industry for catalytic systems that are green and scalable and have broad utility.

9 NATURAL GAS

As with other fuels, natural gas also affects the environment when it is produced, stored, and transported (Renesme et al., 1992; Speight, 2007, 2014). Because natural gas is made up mostly of methane (another greenhouse gas), small amounts of methane can sometimes leak into the atmosphere from wells, storage tanks, and pipelines. The natural gas industry is working to prevent any methane from escaping. Exploring and drilling for natural gas will always have some impact on land and marine habitats. But new technologies have greatly reduced the number and size of areas disturbed by drilling, sometimes called "footprints." Satellites, global positioning systems, remote sensing devices, and 3D and 4D seismic technologies make it possible to discover natural gas reserves while drilling fewer wells. Plus, use of horizontal drilling and directional drilling make it possible for a single well to produce gas from much bigger areas.

While the primary constituent of natural gas is methane (CH_4), it may contain smaller amounts of other hydrocarbons, such as ethane (C_2H_6) and various isomers of propane (C_3H_8), butane (C_4H_{10}), and pentane (C_5H_{12}), as well as trace amounts of higher boiling hydrocarbons up to octane (C_8H_{18}). Nonhydrocarbon gases, such as carbon dioxide (CO_2), helium (He), hydrogen sulfide (H_2S), nitrogen (N_2), and water vapor (H_2O), may also be present. At the pressure and temperature conditions of the source reservoir, natural gas may occur as free gas (bubbles) or be dissolved in either crude oil or brine.

The major constituent of natural gas, methane, also directly contributes to the greenhouse effect. Its ability to trap heat in the atmosphere is estimated to be 21 times greater than that of carbon dioxide, so although methane emissions amount to only 0.5% (v/v) of the emissions of carbon dioxide in the United States, they account for approximately 10% (v/v) of the greenhouse effect of these emissions.

There is a great deal of uncertainty about the precise methods and the amounts by which hazardous pollution is being emitted into the air during the development and processing of natural gas. Methane is a potent greenhouse gas, far more warming than carbon dioxide. Methane also adds to ozone levels. Some methane—it is unclear exactly how much—leaks out of natural gas pipelines and fracking equipment. This is unintentional and can happen at many points along the system.

The fracking process (Speight, 2016b) can release VOCs, such as benzene, toluene, and methane, into the air, where they contribute to ozone formation. Ozone is formed when the sun reacts with VOCs and nitrogen oxides (NO_x) in the atmosphere—the presence of these gases play a role in the formation of ozone, which is a powerful oxidant that can irritate the airways, causing a burning sensation, coughing, wheezing, and shortness of breath.

Natural gas is the cleanest of all the fossil fuels. Composed primarily of methane, the main products of the combustion of natural gas are carbon dioxide and water vapor, the same compounds we exhale when we breathe. Coal and oil are composed of much more complex molecules, with a higher carbon ratio and higher nitrogen and sulfur contents. This means that when combusted, coal and oil release higher levels of harmful emissions, including a higher ratio of carbon emissions, nitrogen oxides (NO_x), and sulfur dioxide (SO_2). Coal and fuel oil also release ash particles into the environment, substances that do not burn but instead are carried into the atmosphere and contribute to pollution. The combustion of natural gas, on the other hand, releases very small amounts of sulfur dioxide and nitrogen oxides, virtually no ash or particulate matter, and lower levels of carbon dioxide, carbon monoxide, and other reactive hydrocarbons.

Natural gas, as the cleanest of the fossil fuels, can be used in many ways to help reduce the emissions of pollutants into the atmosphere. Burning natural gas in the place of other fossil fuels emits fewer harmful pollutants, and an increased reliance on natural gas can potentially reduce the emission of many of these most harmful pollutants. Pollutants from the combustion of fossil fuels have led to the development of many pressing environmental problems. Natural gas, emitting fewer harmful chemicals into the atmosphere than other fossil fuels, can help to mitigate some of these environmental issues. These issues include: (1) greenhouse gas emissions and (2) smog, air quality and acid rain.

9.1   Greenhouse Gas Emissions

Global warming, or the "greenhouse effect," is an environmental issue that deals with the potential for global climate change due to increased levels of atmospheric "greenhouse gases." There are certain gases in our atmosphere that serve to regulate the amount of heat that is kept close to the earth's surface. Scientists theorize that an increase in these greenhouse gases will translate into increased temperatures around the globe, which would result in many disastrous environmental effects.

The principle greenhouse gases include water vapor, carbon dioxide, methane, nitrogen oxides, and some engineered chemicals such as chlorofluorocarbon derivatives. While most of these gases occur in the atmosphere naturally, levels have been increasing due to the widespread burning of fossil fuels by growing human populations. The reduction of greenhouse gas emissions has become a primary focus of environmental programs in countries around the world.

One of the principle greenhouse gases is carbon dioxide. Although carbon dioxide does not trap heat as effectively as other greenhouse gases (making it a less potent greenhouse gas), the sheer volume of carbon dioxide emissions into the atmosphere is very high, particularly from the burning of fossil fuels. In fact, according to the Energy Information Administration in its Dec. 2009 report *Emissions of Greenhouse Gases* in the United States, 81.3% of greenhouse gas emissions in the United States in 2008 came from energy-related carbon dioxide.

One issue that has arisen with respect to natural gas and the greenhouse effect is the fact that methane, the principle component of natural gas, is itself a potent greenhouse gas. Methane has an ability to trap heat almost 21 times more effectively than carbon dioxide. Sources of methane emissions in the United States include the waste management and operations industry, the agricultural industry, as well as leaks and emissions from the oil and gas industry itself.

9.2   Smog, Air Quality, and Acid Rain

Smog and poor air quality is a pressing environmental problem, particularly for large metropolitan cities. Smog, the primary constituent of which is ground level ozone, is formed by a chemical reaction of carbon monoxide, nitrogen oxides, VOCs, and heat from sunlight. As well as creating that familiar smoggy haze commonly found surrounding large cities, particularly in the summer time, smog and ground level ozone can contribute to respiratory problems ranging from temporary discomfort to long-lasting, permanent lung damage. Pollutants contributing to smog come from a variety of sources, including vehicle emissions, smokestack emissions, paints, and solvents. Because the reaction to create smog requires heat, smog problems are the worst in the summertime.

The use of natural gas does not contribute significantly to smog formation, as it emits low levels of nitrogen oxides, and virtually no particulate matter. For this reason, it can be used to help combat smog formation in those areas where ground level air quality is poor. The main sources of nitrogen oxides are electric utilities, motor vehicles, and industrial plants. Increased natural gas use in the electric generation sector, a shift to cleaner natural gas vehicles, or increased industrial natural gas use could all serve to combat smog production, especially in urban centers where it is needed the most. Particularly in the summertime, when natural gas demand is the lowest and smog problems are the greatest, industrial plants and electric generators could use natural gas to fuel their operations instead of other, more polluting fossil fuels. This would effectively reduce the emissions of smog-causing chemicals, and result in clearer, healthier air around urban centers.

Particulate emissions also cause the degradation of air quality in the United States. These particulates can include soot, ash, metals, and other airborne particles. Natural gas emits virtually no particulates into the atmosphere: in fact, emissions of particulates from natural gas combustion are 90% lower than from the combustion of oil, and 99% lower than burning coal. Thus increased natural gas use in place of other dirtier hydrocarbons can help reduce particulate emissions in the United States.

Acid rain is another environmental problem that affects much of the Eastern United States, damaging crops, forests, wildlife populations, and causing respiratory and other illnesses in humans. Acid rain is formed when sulfur dioxide and nitrogen oxides react with water vapor and other chemicals in the presence of sunlight to form various acidic compounds in the air. The principle source of acid rain-causing pollutants, sulfur dioxide and nitrogen oxides, are coal fired power plants. Since natural gas emits virtually no sulfur dioxide, and up to 80% less nitrogen oxides than the combustion of coal, increased use of natural gas could provide for fewer acid rain-causing emissions.

Other aspects of the development and use of natural gas need to be considered as well in looking at the environmental consequences related to natural gas. For example, the major constituent of natural gas, methane, also directly contributes to the greenhouse effect through venting or leaking of natural gas into the atmosphere. This is because methane is 21 times as effective in trapping heat as is carbon dioxide. Although methane emissions amount to only 0.5% of US emissions of carbon dioxide, they account for about 10% of the greenhouse effect of US emissions. A major transportation-related environmental advantage of natural gas is that it is not a source of toxic spills. But, because there are about 300,000 miles of high-pressure transmission pipelines in the United States and its offshore areas, there are corollary impacts. For instance, the construction right-of-way on land commonly requires a width of 75 to 100 ft along the length of the pipeline; this is the area disturbed by trenching, soil storage, pipe storage, vehicle movement, etc. This area represents between 9.1 and 12.1 acres per mile of pipe which is, or has been, subject to intrusion.

10   VOLATILE ORGANIC COMPOUNDS

Organic compounds that evaporate easily are collectively referred to as VOCs. Microbial volatile organic compounds are a variety of compounds formed in the metabolism of fungi and bacteria (Korpi et al., 2009). Typically, a VOC is any organic compound that will evaporate at the temperature of use and which, by a photochemical reaction, will cause oxygen in the air to be converted into smog-promoting ozone under favorable climatic conditions.

Thus, VOCs are gases that are emitted into the air from products or processes (Table 4.7). Some are harmful by themselves, including some that cause cancer. In addition, they can react with other gases and form other air pollutants after they are in the air. Many VOCs have been classified as toxic and

TABLE 4.7 Sources of Volatile Organic Compounds

Indoor Sources

- Tobacco smoke
- Paint, paint remover
- Cleaning products, varnishes, wax
- Pesticides
- Air fresheners
- Personal care products such as cosmetics
- Hobby products such as glue
- Office equipment including printers and copiers
- Wood burning stoves
- Fuel oil, gasoline
- Furniture or building products such as flooring, carpet, pressed wood products
- Car exhaust in an attached garage

Outdoor Sources

- Gasoline emissions
- Diesel emissions
- Wood burning
- Crude oil production and refining
- Natural gas production and refining
- Industrial emissions

carcinogenic. They are found in a wide variety of commercial, industrial, and residential products including fuel oils, gasoline, solvents, cleaners and degreasers, paints, inks, dyes, refrigerants, and pesticides. When VOCs are spilled or improperly disposed, they can soak into the soil and eventually end up in groundwater. VOCs are found in almost all natural and synthetic materials and are commonly used in fuels, fuel additives, solvents, perfumes, flavor additives, and deodorants. Potential health hazards and environmental degradation resulting from the widespread use of VOCs have prompted increasing concern among scientists, industry, and the general public (Table 4.8).

Typically, VOCs are organic chemicals that have a high vapor pressure at ordinary room temperature. The high vapor pressure results from a low boiling point, which causes large numbers of molecules to evaporate from the liquid form or sublime from the solid form of the compound and enter the surrounding air. VOCs are numerous, varied, and ubiquitous. They include both human-made and naturally occurring chemical compounds. Many VOCs are dangerous to human health or cause harm to the environment. Anthropogenic VOCs are regulated by law, especially indoors, where concentrations are the highest. Harmful VOCs typically are not acutely toxic but have compounding long-term health effects.

Nonmethane volatile organic compounds (NMVOCs) are a collection of organic compounds that differ widely in their chemical composition but display similar behavior in the atmosphere. NMVOCs are emitted into the atmosphere

TABLE 4.8 Harmful Effects of Selected Constituents of Volatile Organic Compounds

Constituent	Effects
Benzene	Exposure to benzene can cause skin and respiratory irritation, and long-term exposure can lead to cancer and blood, developmental, and reproductive disorders
Toluene	Long-term exposure to toluene can cause skin and respiratory irritation, headaches, dizziness, birth defects, and damage the nervous system
Ethylbenzene	Can cause irritation of the throat and eyes, and dizziness and long-term exposure can cause blood disorders
Xylene	High levels of xylene exposure have numerous short-term impacts, including nausea, gastric irritation, and neurological effects, and long-term exposure can negatively impact the nervous system
n-Hexane	Exposure can cause dizziness, nausea, and headaches, while long-term exposure can lead to numbness, muscular atrophy, blurred vision, and fatigue

from a large number of sources including combustion activities, solvent use, and production processes. NMVOCs contribute to the formation of ground level (tropospheric) ozone. In addition, certain NMVOC species or species groups such as benzene and 1,3 butadiene are hazardous to human health. Quantifying the emissions of total NMVOCs provides an indicator of the emission trends of the most hazardous NMVOCs.

11 WOOD SMOKE

Wood smoke forms when wood is made up of a complex mixture of gases and fine particles (particulate matter, PM). In addition to particle pollution, wood smoke contains several toxic harmful air pollutants including: benzene, formaldehyde, acrolein, and PAHs.

Wood smoke is by far the most compelling argument against wood heating. In cold climates, a frigid, stagnant air mass traps the smoke close to the ground. In the process of wood combustion, wood vaporizes when heated into gases and tar particles. If the temperature is high enough, the tar particles vaporize into chemicals and carbon particles. If the temperature is higher still and there is oxygen present, the gases and particles burn in bright flames.

Chemically, wood is about half carbon, and the rest is mostly made up of oxygen and hydrogen. When a piece of wood is heated, it starts to smoke and turn black at the same time. This is because the other products vaporize under intense heat faster than the carbon burns, so smoking leaves much of the carbon behind until only charcoal, which is just about pure carbon, is left. The smoke that vaporizes out of the wood is a cloud of nasty, gooey little droplets of a tar-like liquid. Chemically, these droplets are actually big, gooey, complicated hydrocarbon molecules that take a number of different forms, mostly bad.

When wood is burned correctly in a bright, hot, turbulent fire, what is observed is the tar droplets rising off the wood into a zone of extreme heat where they revaporize, cracking into their basic, mostly gaseous, constituents, and oxidize. That is to say they burn. This leaves carbon dioxide, some carbon monoxide, and a number of other gases, water vapor and some not quite completely oxidized hydrocarbon products, which are the particulate emissions that US Environmental Protection Agency regulates.

The sentiment that wood smoke, being a natural substance, must be benign to humans is still sometimes heard. It is now well established, however, that wood-burning stoves and fireplaces as well as wildland and agricultural fires emit significant quantities of known health-damaging pollutants, including several carcinogenic compounds (Naeher et al., 2007). Two of the principal gaseous pollutants in wood smoke, carbon monoxide (CO) and the oxides of nitrogen (NOx), add to the atmospheric levels of these regulated gases emitted by other combustion sources. Health impacts of exposures to these gases and some of the other woodsmoke constituents (e.g., benzene) are well characterized in thousands of publications. As these gases are indistinguishable no matter where they come from, there is no urgent need to examine their particular health implications in woodsmoke. With this as the backdrop, this review approaches the issue of why woodsmoke may be a special case requiring separate health evaluation through two questions. The first issue is whether or not woodsmoke should be regulated and/or managed separately, even though some of the individual constituents are already regulated in many jurisdictions. The second issue is whether or not woodsmoke particles pose different levels of risk than other ambient particles of similar size.

Wood smoke interferes with normal lung development in infants and children. It also increases children's risk of lower respiratory infections such as bronchitis and pneumonia. Exposure to wood smoke can depress the immune system and damage the layer of cells in the lungs that protect and cleanse the airways. According to the Environmental Protection Agency (US EPA), toxic air pollutants are components of wood smoke. Wood smoke can cause coughs, headaches, and eye and throat irritation in otherwise healthy people. For vulnerable populations, such as people with asthma and chronic respiratory disease, and those with cardiovascular disease, wood smoke is particularly harmful—even short exposures can prove dangerous.

The particle matter wood smoke is extremely small and therefore is not filtered out by the nose or the upper respiratory system. Instead, these small particles end up deep in the lungs where they remain for months, causing structural damage and chemical changes. Wood smoke's carcinogenic chemicals adhere to these tiny particles, which enter deep into the lungs.

12 EFFECTS ON THE ENVIRONMENT

To start with an extremely relevant definition, *environmental technology* is the application of scientific and engineering principles to the study of the environment, with the goal of the improvement of the environment. Furthermore, issues related to the pollution of the environment are relative. Any organism is exposed to an *environment*, even if the environment is predominantly many members of the same organism. An example is a bacterium in a culture that is exposed to many members of the same species. Thus the environment is all external influences, abiotic (physical factors) and biotic (actions of other organisms), to which an organism is exposed. The environment affects basic life functions, growth, and reproductive success of organisms and determines their local and geographic distribution patterns. A fundamental idea in *ecology* is that the environment changes in time and space, and living organisms respond to these changes.

Since ecology is that branch of science related to the study of the relationship of organisms to their environment, an *ecosystem* is an ecological community (or living unit) considered together with the nonliving factors of its environment as a unit. By way of brief definition, *abiotic factors* include such influences as light radiation (from the sun), ionizing radiation (cosmic rays from outer space), temperature (local and regional variations), water (seasonal and regional distributions), atmospheric gases, wind, soil (texture and composition), and catastrophic disturbances. These latter phenomena are usually unpredictable and infrequent, such as fire, hurricanes, volcanic activity, landslides, major floods, and any disturbance that drastically alters the environment and thus changes the species composition and activity patterns of the inhabitants. On the other hand, *biotic factors* include natural interactions (e.g., predation and parasitism) and anthropogenic stress (e.g., the effect of human activity on other organisms). Because of the abiotic and biotic factors, the environment to which an organism is subjected can affect the life functions, growth, and reproductive success of the organism and can determine the local and geographic distribution patterns of an organism.

Living organisms respond to changes in the environment by either adapting or becoming extinct. The basic principles of the concept that living organisms respond to changes in the environment were put forth by Darwin and Lamarck. The former noted the slower adaptation (evolutionary trends) of living organisms, while the latter noted the more immediate adaptation of living organisms to the environment. Both essentially espoused the concept of the survival of the fittest, alluding to the ability of an organism to live in harmony with its environment. This was assumed to indicate that the organism that competed successfully with environmental forces would survive. However, there is the alternate thought that the organism that can live in a harmonious symbiotic relationship with its environment has an equally favorable chance of survival. The influence of the environment on organisms can be viewed on a large scale (i.e., the relationship between regional climate and geographic distribution of organisms) or on a smaller scale (i.e., some highly localized conditions determine the precise location and activity of individual organisms).

Organisms may respond differently to the frequency and duration of a given environmental change. For example, if some individual organisms in a population have adaptations that allow them to survive and to reproduce under new environmental conditions, the population will continue but the genetic composition will have changed (Darwinism). On the other hand, some organisms have the ability to adapt to the environment (i.e., to adjust their physiology or morphology in response to the immediate environment) so that the new environmental conditions are less (certainly no more) stressful than the previous conditions. Such changes may not be genetic (Lamarckism).

In terms of *anthropogenic stress* (the effect of human activity on other organisms), there is the need for the identification and evaluation of the potential impacts of proposed projects, plans, programs, policies, or legislative actions upon the physical-chemical, biological, cultural, and socioeconomic components of the environment. This activity is also known as *environmental impact assessment* and refers to the interpretation of the significance of anticipated changes related to a proposed project. The activity encourages consideration of the environment and arriving at actions that are environmentally compatible.

Identifying and evaluating the potential impact of human activities on the environment require the identification of mitigation measures. *Mitigation* is the sequential consideration of the following measures: (1) avoiding the impact by not taking a certain action or partial action; (2) minimizing the impact by limiting the degree or magnitude of the action and its implementation; (3) rectifying the impact by repairing, rehabilitating, or restoring the affected environment; (4) reducing or eliminating the impact over time by preservation and maintenance operations during the life of the action; and (5) compensating for the impact by replacing or providing substitute resources or environments.

Nowhere is the effect of anthropogenic stress felt more than in the development of natural resources of the earth. Natural resources are varied in nature and often require definition. For example, in relation to mineral resources, for which there is also descriptive nomenclature, the terms related to the available quantities of the resource must be defined. In this instance, the term *resource* refers to the total amount of the mineral that has been estimated to be ultimately available. *Reserves* are well-identified resources that can be profitably extracted and utilized by means of existing technology. In many countries, fossil fuel resources are often classified as a subgroup of the total mineral resources.

In some cases, environmental pollution is a clear—cut phenomenon, whereas in others it remains a question of degree. The ejection of various materials into the environment is often cited as pollution, but there is the ejection of the so-called beneficial chemicals that can assist the air, water, and land to perform their functions. However, it must be emphasized that the ejection of chemicals into the environment, even though they are indigenous to the environment, in quantities above the naturally occurring limits can be extremely harmful. In fact, the timing and the place of a chemical release are influential in determining whether a chemical is beneficial, benign, or harmful! Thus, what may be regarded as a pollutant in one instance can be a beneficial chemical in another instance. The phosphates in fertilizers are examples of useful (beneficial) chemicals, while phosphates generated as by-products in the metallurgical and mining industries may, depending upon the specific industry, be considered pollutants (Chenier, 1992). In this case, the means by which such pollution can be prevented must be recognized (Boyle Breen and Dellarco, 1992). Thus, increased use of the earth's resources as well as the use of a variety of chemicals that are nonindigenous to the earth have put a burden on the ability of the environment to tolerate such materials.

Finally, some recognition must be made of the term carcinogen since many of the environmental effects referenced in this text can lead to cancer. *Carcinogens* are cancer—causing substances and there is a growing awareness of the presence of carcinogenic materials in the environment. A classification scheme is provided for such materials (Table 4.9) (Zakrzewski, 1991; Milman and

TABLE 4.9 Weight-of-Evidence Carcinogenicity Classification Scheme as Determined by the United States Environmental Protection Agency

Group	Description
A	Human carcinogen
B1	Probable human carcinogen; limited human data are available
B2	Probable human carcinogen
C	Possible human carcinogen
D	Not classifiable as a human carcinogen
E	No carcinogenic activity in humans

Weisburger, 1994). The number of substances with which a person comes in contact is in the tens of thousands, and there is not a full understanding of the long-term effects of these substances in their possible propensity to cause genetic errors that ultimately lead to carcinogenesis. *Teratogens* are those substances that tend to cause developmental malformations.

Pollution is the introduction of indigenous (beyond the natural abundance) and nonindigenous (artificial) gaseous, liquid, and solid contaminants into an ecosystem. The atmosphere and water and land systems have the ability to cleanse themselves of many pollutants within hours or days especially when the effects of the pollutant are minimized by the natural constituents of the ecosystem. For example, the atmosphere might be considered to be self-cleaning as a result of rain. However, removal of some pollutants from the atmosphere (e.g., sulfates and nitrates) by rainfall results in the formation of *acid rain* that can cause serious environmental damage to ecosystems within the water and land systems (Pickering and Owen, 1994).

Briefly, lakes in some areas of the world are now registering a low pH (acidic) reading because of excess acidity in rain. This was first noticed in Scandinavia and is now prevalent in eastern Canada and the northeastern United States. Normal rainfall has a pH of 5.6 and the slight acidity (neutral water has a pH equal to 7.0) because of carbon dioxide (CO_2) in the air that, with water, forms carbonic acid (H_2CO_3):

$$CO_2 + H_2O \rightarrow H_2CO_3$$

The increased use of hydrocarbon fuels in the last five decades is slowly increasing the concentration of carbon dioxide in the atmosphere, which produces more carbonic acid leading to an imbalance in the natural carbon dioxide content of the atmosphere that, in turn, leads to more acidity in the rain. In addition, there is a so-called *greenhouse effect*, and the average temperature of the earth may be increasing.

In addition, excessive use of fuels with high sulfur and nitrogen content cause sulfuric and nitric acids in the atmosphere from the sulfur dioxide and nitrogen oxide products of combustion that can be represented simply as:

$$SO_2 + H_2O \rightarrow H_2SO_3 \text{(sulfurous acid)}$$
$$2SO_2 + O_2 \rightarrow 2SO_3$$
$$SO_3 + H_2O \rightarrow H_2SO_4 \text{(sulfuric acid)}$$
$$NO + H_2O \rightarrow HNO_2 \text{(nitrous acid)}$$
$$2NO + O_2 \rightarrow NO_2$$
$$NO_2 + H_2O \rightarrow HNO_3 \text{(nitric acid)}$$

A *pollutant* is a substance (for simplicity most are referred to as chemicals) present in a particular location when it is not indigenous to the location or is in a greater-than-natural concentration. The substance is often the product of human activity. The pollutant, by virtue of its name, has a detrimental effect on the environment, in part or *in toto*. Pollutants can also be subdivided into two classes: (1) primary and (2) secondary.

$$\text{Source} \rightarrow \text{Primary pollutant} \rightarrow \text{Secondary pollutant}$$

Primary pollutants are those pollutants emitted directly from the source. In terms of atmospheric pollutants by crude oil constituents, examples are hydrogen sulfide, carbon oxides, sulfur dioxide, and nitrogen oxides from refining operations (see earlier text).

The question of classifying nitrogen dioxide and sulfur trioxide as primary pollutants often arises, as does the origin of the nitrogen. In the former case, these higher oxides can be formed in the upper levels of the combustors. The nitrogen, from which the nitrogen oxides are formed, does not originate solely from the fuel but may also originate from the air used for the combustion.

Secondary pollutants are produced by interaction of primary pollutants with another chemical or by dissociation of a primary pollutant, or other effects within a particular ecosystem. Again, using the atmosphere as an example, the formation of the constituents of acid rain is an example of the formation of secondary pollutants (see earlier text).

REFERENCES

ASTM C694, 2016. Standard test method for weight loss (mass loss) of sheet steel during immersion in sulfuric acid solution. In: Annual Book of Standards. ASTM International, West Conshohocken, PA.

ASTM D1310, 2016. Standard test method for flash point and fire point of liquids by tag open-cup apparatus. In: Annual Book of Standards. ASTM International, West Conshohocken, PA.

ASTM D1838, 2016. Standard test method for copper strip corrosion by liquefied petroleum (LP) gases. In: Annual Book of Standards. ASTM International, West Conshohocken, PA.

ASTM D2251, 2016. Standard test method for metal corrosion by halogenated organic solvents and their admixtures. In: Annual Book of Standards. ASTM International, West Conshohocken, PA.

ASTM D4982, 2016. Standard test methods for flammability potential screening analysis of waste. In: Annual Book of Standards. ASTM International, West Conshohocken, PA.

ASTM D92, 2016. Standard test method for flash and fire points by Cleveland open cup tester. In: Annual Book of Standards. ASTM International, West Conshohocken, PA.

ASTM E1302, 2016. Standard guide for acute animal toxicity testing of water-miscible metalworking fluids. In: Annual Book of Standards. ASTM International, West Conshohocken, PA.

ASTM E176, 2016. Standard terminology of fire standards. In: Annual Book of Standards. ASTM International, West Conshohocken, PA.

ASTM E502, 2016. Standard test method for selection and use of ASTM standards for the determination of flash point of chemicals by closed cup methods. In: Annual Book of Standards. ASTM International, West Conshohocken, PA.

ASTM E681, 2016. Standard test method for concentration limits of flammability of chemicals (vapors and gases). In: Annual Book of Standards. ASTM International, West Conshohocken, PA.

ASTM E789, 2016. Standard test method for explosibility of dust clouds. In: Annual Book of Standards. ASTM International, West Conshohocken, PA.

Austin, G.T., 1984. In: Shreve's Chemical Process Industries, fifth ed. McGraw-Hill, New York (Chapter 22).

Barcelona, M., Wehrmann, A., Keeley, J.F., Pettyjohn, J., 1990. Contamination of Ground Water. Noyes Data Corp, Park Ridge, NJ.

Bell, L., 2011. Climate of Corruption: Politics and Power Behind the Global Warming. Greenleaf book Group Press, Austin, TX.

Boyle Breen, J.J., Dellarco, M.J. (Eds.), 1992. Pollution prevention in industrial processes. In: Symposium Series No. 508. American Chemical Society, Washington, DC.

Carson, R., 1962. Silent Spring. Houghton Mifflin Company/Houghton Mifflin Harcourt International, Geneva, IL.

Carson, P., Mumford, C., 1988. The Safe Handling of Chemicals in Industry. vols. 1 and 2 John Wiley & Sons Inc, New York.

Carson, P., Mumford, C., 1995. The Safe Handling of Chemicals in Industry. vol. 3. John Wiley & Sons Inc, New York.

Carson, P., Mumford, R., 2002. Hazardous Chemicals Handbook, second ed. Butterworth-Heinemann, Oxford.

Chakrabartty, S.K., Selucky, M.L., 1985. Characteristics of in situ coal gasification liquids from a cretaceous coal. Nature 316, 244–246.

Chenier, P.J., 1992. Survey of Industrial Chemistry, second ed. VCH Publishers Inc., New York.

Cheremisinoff, P., 1995. Handbook of Water and Wastewater Treatment Technology. Marcel Dekker Inc., New York.

Easterbrook, G., 1995. A Moment on the Earth: The Coming Age of Environmental Optimism. Viking Press, New York.

Hites, R.A., 2007. Elements of Environmental Chemistry. In: John Wiley & Sons Inc, Hoboken, NJ, pp. 155–179 (Chapter 6).

Jacob, J., 2013. A review of the accumulation and distribution of persistent organic pollutants in the environment. Int. J. Biosci. Biochem. Bioinforma. 3 (6), 657–661.

Kim, S., Holtzapple, M.T., 2005. Lime pretreatment and enzymatic hydrolysis of corn stover. Bioresour. Technol. 96 (18), 1994–2006.

Kim, S., Holtzapple, M.T., 2006a. Effect of structural features on enzyme digestibility of corn stover. Bioresour. Technol. 97 (4), 583–591.

Kim, S., Holtzapple, M.T., 2006b. Delignification kinetics of corn stover in lime pretreatment. Bioresour. Technol. 97 (5), 778–785.

Kletz, T.A., 1990. Improving Chemical Engineering Practices, second ed. Hemisphere Publishers/ Taylor& Francis, Washington, DC.

Korpi, A., Järnberg, J., Pasanen, A.L., 2009. Microbial volatile organic compounds. Crit. Rev. Toxicol. 39 (2), 139–193.

Mackay, D., McCarty, L.S., MacLeod, M., 2001. On the validity of classifying chemicals for persistence, bioaccumulation, toxicity, and potential for long-range transport. Environ. Toxicol. Chem. 20 (7), 1491–1498.

Manahan, S.E., 2010. Environmental Chemistry, ninth ed. CRC Press/Taylor & Francis Group, Boca Raton, FL.

Milman, H.A., Weisburger, E.K., 1994. Handbook of Carcinogen Testing, second ed. Noyes Data Corp., Park Ridge, NJ.

Mody, V., Jakhete, R., 1988. Dust Control Handbook. Noyes Data Corp., Park Ridge, NJ.

Naeher, L.P., Brauer, M., Lipsett, M., Zelikoff, J.T., Simpson, C.D., Koenig, J.Q., Smith, K.R., 2007. Woodsmoke health effects: a review. Inhal. Toxicol. 19, 67–106.

NRC, 1981. Prudent Practices for Handling Hazardous Chemicals in Laboratories. National Academy Press/National Research Council, Washington, DC.

Oliveira, B.F., Ignotti, E., Hacon, S.S., 2011. A systematic review of the physical and chemical characteristics of pollutants from biomass burning and combustion of fossil fuels and health effects in brazil. Cad. Saude Publica 27 (9), 1678–1698.

Pickering, K.T., Owen, L.A., 1994. Global Environmental Issues. Routledge Publishers, New York.

Pöschl, U., 2005. Atmospheric aerosols: composition, transformation, climate and health effects. Angew. Chem. Int. Ed. 44, 7520–7540.

Ramage, J., 1997. Energy: A Guidebook. Oxford University Press, Oxford.

Renesme, G., Saintjust, J., Muller, Y., 1992. Transportation fuels and chemicals directly from natural-gas—how expensive? Catal. Today 13 (2–3), 371–380.

Schwarzenbach, R.P., Gschwend, P.M., Imboden, D.M., 2003. Environmental Organic Chemistry, second ed. John Wiley & Sons Inc., Hoboken, NJ.

Sheldon, R., 2010. Introduction to green chemistry, organic synthesis and pharmaceuticals. In: Dunn, P.J., Wells, A.S., Williams, M.T. (Eds.), Green Chemistry in the Pharmaceutical Industry. Wiley-VCH Verlag GmbH & Co. KGaA, Weinheim.

Shrestha, G., Traina, S.J., Swanston, C.W., 2010. Black Carbon's properties and role in the environment: a comprehensive review. Sustainability 2, 294–320.

Speight, J.G., 2002. Chemical Process and Design Handbook. McGraw-Hill Publishers, New York.

Speight, J.G., 2005a. Environmental Analysis and Technology for the Refining Industry. John Wiley & Sons Inc., Hoboken, NJ.

Speight, J.G., 2005b. Handbook of Coal Analysis. John Wiley & Sons Inc., Hoboken, NJ.

Speight, J.G., 2007. Natural Gas: A Basic Handbook. GPC Books/Gulf Publishing Company, Houston, TX.

Speight, J.G. (Ed.), 2011. The Biofuels Handbook. Royal Society of Chemistry, London.

Speight, J.G., 2013. The Chemistry and Technology of Coal, third ed. CRC Press/Taylor & Francis Group, Boca Raton, FL.

Speight, J.G., 2014. The Chemistry and Technology of Petroleum, fifth ed. CRC Press/Taylor & Francis Group, Boca Raton, FL.

Speight, J.G., 2015. Handbook of Petroleum Product Analysis, second ed. John Wiley & Sons Inc., Hoboken, NJ.

Speight, J.G., 2016a. Handbook of Petroleum Refining. CRC Press/Taylor & Francis Group, Boca Raton, FL.

Speight, J.G., 2016b. Handbook of Hydraulic Fracturing. John Wiley & Sons Inc., Hoboken, NJ.

Speight, J.G., Arjoon, K.K., 2012. Bioremediation of Petroleum and Petroleum Products. Scrivener Publishing, Salem, MA.

Speight, J.G., Exall, D.I., 2014. Refining Used Lubricating Oils. CRC Press/Taylor and Francis Group, Boca Raton, FL.

Speight, J.G., Lee, S., 2000. Handbook of Environmental Technology, second ed. Taylor & Francis, Washington, DC.

Sudweeks, W.B., Larsen, R.D., Balli, F.K., 1983. In: Kent, J.A. (Ed.), Riegel's Handbook of Industrial Chemistry. Van Nostrand Reinhold, New York, p. 700.

Tedder, D.W., Pohland, F.G. (Eds.), 1993. Emerging technologies in hazardous waste management III. In: Symposium Series No. 518. American Chemical Society, Washington, D.C.

Tinsley, I.J., 2004. Chemical Concepts in Pollutant Behavior, second ed. John Wiley & Sons Inc., Hoboken, NJ.

Werpy, T., Peterson, G (Eds.), 2004. Top value added chemicals from biomass, volume I: results of screening for potential candidates from sugars and synthesis gases. In: Pacific Northwest National Laboratory, National Renewable Energy Laboratory, DOE Office of Biomass Program, Energy Efficiency Renewable Energy. August.

Woodside, G., 1999. Hazardous Materials and Hazardous Waste Management, second ed. John Wiley & Sons Inc., Hoboken, NJ.

Zakrzewski, S.F., 1991. Principles of Environmental Toxicology. ACS Professional Reference Book/American Chemical Society, Washington, DC.

Zimmerman, R.E., 1979. Evaluating and Testing the Coking Properties of Coal. Miller Freeman & Co., San Francisco.

Chapter 5

Properties of Organic Compounds

1 INTRODUCTION

Organic chemistry is the chemistry of carbon compounds and all organic compounds contain carbon but not all compounds that contain carbon are organic compounds (Bailey and Bailey, 2000; Atkins and Carey, 2002; Brown and Foote, 2002). There are some compounds of carbon that are not classified as organic such as carbonate minerals (e.g., sodium carbonate, Na_2CO_3, and calcium carbonate, $CaCO_3$) and cyanide compounds (such as potassium cyanide, KCN, or any of the metallic cyanide derivative) which are designated as inorganic compounds. Therefore, a more useful and less confusing description of organic chemistry might be that chemistry of compounds of carbon that usually contain hydrogen as well as carbonaceous compounds that, in addition to carbon and hydrogen, may also contain other elements such as oxygen, nitrogen, sulfur, phosphorus, or any of the halogens (fluorine, F, chlorine, Cl, bromine, Br, or iodine, I) as well as a host of other carbonaceous derivatives (Chapter 2). On a more general basis, the majority of the carbon-containing compounds are organic compounds and organic chemistry is the chemistry of these compounds.

The field of organic chemistry includes more than 20 million compounds for which properties have been determined and recorded in the various literary sources and many more compounds are added every year. Some of the newly added properties pertain to the properties of organic compounds that have been isolated from plants or animals while others are produced by modifying naturally occurring chemicals while another group of the newly added organic compounds are synthesized in various academic and industrial laboratories.

Organic compounds exist because of the uniqueness of carbon, which forms very strong bonds with other carbon atoms and with hydrogen atoms (Chapter 2). The bonds are so strong that carbon can form long chains, some containing thousands of carbon atoms, while no other elements have the same capability. A carbon atom forms four bonds, therefore carbon not only can form long chains, but also forms chains that have branches. It is a major reason why carbon compounds exhibit so much isomerism. Isomers are molecules with the same molecular formula but different structures. There is only one structure for methane, ethane, or propane; but butane, C_4H_{10}, can have either of two different structures:

Environmental Organic Chemistry for Engineers. http://dx.doi.org/10.1016/B978-0-12-804492-6.00005-8

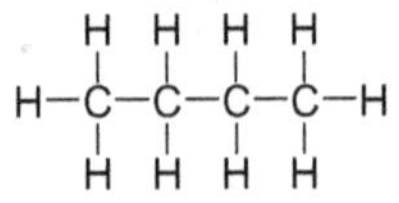

$$H-\overset{\overset{\displaystyle H}{|}}{\underset{\underset{\displaystyle H}{|}}{C}}-\overset{\overset{\displaystyle H}{|}}{\underset{\underset{\displaystyle H}{|}}{C}}-\overset{\overset{\displaystyle H}{|}}{\underset{\underset{\displaystyle H}{|}}{C}}-\overset{\overset{\displaystyle H}{|}}{\underset{\underset{\displaystyle H}{|}}{C}}-H$$

n-Butane: $CH_3CH_2CH_2CH_3$

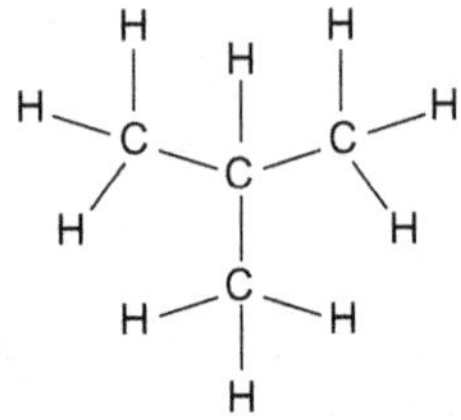

Iso-butane: $(CH_3)_2CHCH_3$

The linear molecule (1) is called butane, or *normal* butane (*n*-butane), whereas the branched molecule (2) is methylpropane, rather than 2-methylpropane, since the methyl group has to be in a 2-position and is also called iso-butane. If the methyl group of iso-butane was attached to a terminal carbon, the resultant molecule would be the same as *n*-butane. In addition, there are differences between the properties of the various isomers, which are often substantial when boiling points are compared. Thus, using butane as the example:

Property	Butane	Iso-butane
Formula	C_4H_{10}	C_4H_{10}
Boiling point, °C	−0.4	−11.75
Relative density		
Water=1	0.58	0.60
Air=1	2.00	2.07
Vapor pressure		
kPa, 21°C	215.1	310.0
psi	4492	6474

If the only constituents of organic compounds were the hydrocarbons, the complexity is further illustrated by the number of potential isomers, i.e., molecules having the same atomic formula, that can exist for a given number of paraffinic carbon atoms and that increases rapidly as molecular weight increases. For example, pentane (C_5H_{12}) has three isomers: pentane (or *n*-pentane), methylbutane (or *iso*-pentane), and dimethyl propane (or *neo*-pentane). Hexane (C_6H_{14}) has five isomers: hexane, 2-methylpentane, 3-methylpentane, 2,2-dimethylbutane, and 2,3-dimethylbutane. Heptane (C_7H_{16}) has 9 different isomers, octane (C_8H_{18}) has 18 isomers, nonane (C_9H_{20}) has 35 isomers, and decane ($C_{10}H_{22}$) has 75 different isomers. There are more than 4000 isomers (4347 isomers to be precise) of $C_{15}H_{32}$ and more than 366,000 (366,319) isomers of $C_{20}H_{42}$. The formula $C_{30}H_{62}$ has more than 4 billion isomers (4,111,846,763) most of which have never been isolated as pure compounds.

This same increase in number of isomers with molecular weight also applies to the other molecular types present. Since the molecular weights of organic compounds can vary from that of methane (CH_4; molecular weight $= 16$) to several thousand, it is clear that the higher molecular weight species can contain (theoretically) unlimited numbers of molecules. However, in reality the number of molecules in any specified fraction is limited by the nature of the precursors of organic compounds, their chemical structures, and the physical conditions that are prevalent during the maturation (conversion of the precursors) processes. Furthermore, for molecules other than alkane hydrocarbons, still other kinds of isomers are possible. For example, the simple formula C_2H_6O can represent ethyl alcohol (C_2H_5OH) or dimethyl ether (CH_3OCH_3) and C_3H_8O could include organic compounds such as alcohol [$CH_3CH_2CH_2OH$, $(CH3)_2CHOH$] or an ether ($CH_3CH_2OCH_3$):

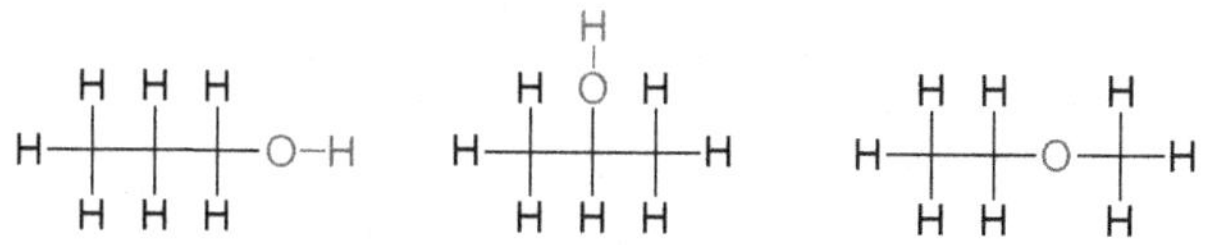

I	II	III
n-Propyl alcohol	Isopropyl alcohol	Ethyl methyl ether

as well as other possible organic chemicals. Organic molecules with higher molecular weights have a greater number of atoms and, therefore, a greater variety of structures which includes a greater number of functional groups leading to a higher number of potential isomers.

In addition, carbon atoms can be bonded by double bonds ($>C=C<$) or by triple bonds ($-C\equiv C-$) as well as single bonds which is also a characteristic of carbon that is much more prevalent with carbon than with any other element. Also, carbon atoms can form rings of various sizes. The rings may be saturated or unsaturated. The unsaturated 6-membered ring known as the benzene ring is the basis for an entire subfield of *aromatic organic chemistry*. Furthermore, carbon atoms form strong bonds not only with other carbon atoms but also with atoms of other elements. In addition to hydrogen, many carbon compounds also contain oxygen. Nitrogen, sulfur, phosphorus, and the halogens also frequently occur in carbon compounds. As a result, there are various kinds of functional groups that can exist within the range of organic and many different kinds of isomers are possible.

The physical state (Table 5.1) as well as the physical and chemical properties of an organic contaminant are dependent upon the molecular structure of the compound (within the chemical series, such as, for example, the alkane series) and are critical to understanding and modeling the fate and transport of the contaminant in the environment (Table 5.2). For example, the molecular formula can only provide a limited amount of information about an organic chemical

TABLE 5.1 The Natural States of Selected Alkane Hydrocarbons

Formula		Name	State
CH_4	CH_4	Methane	Gas
C_2H_6	CH_3CH_3	Ethane	Gas
C_3H_8	$CH_3CH_2CH_3$	Propane	Gas
C_4H_{10}	$CH_3CH_2CH_2CH_3$	Butane	Gas
C_5H_{12}	$CH_3(CH_2)_3CH_3$	Pentane	Liquid
C_6H_{14}	$CH_3(CH_2)_4CH_3$	Hexane	Liquid
C_7H_{16}	$CH_3(CH_2)_5CH_3$	Heptane	Liquid
C_8H_{18}	$CH_3(CH_2)_6CH_3$	Octane	Liquid
C_9H_{20}	$CH_3(CH_2)_7CH_3$	Nonane	Liquid
$C_{10}H_{22}$	$CH_3(CH_2)_8CH_3$	Decane	Liquid
$C_{11}H_{24}$	$CH_3(CH_2)_9CH_3$	Undecane	Liquid
$C_{12}H_{26}$	$CH_3(CH_2)_{10}CH_3$	Dodecane	Liquid
$C_{13}H_{28}$	$CH_3(CH_2)_{11}CH_3$	Tridecane	Liquid
$C_{14}H_{30}$	$CH_3(CH_2)_{12}CH_3$	Tetradecane	Liquid
$C_{15}H_{32}$	$CH_3(CH_2)_{13}CH_3$	Pentadecane	Liquid
$C_{16}H_{34}$	$CH_3(CH_2)_{14}CH_3$	Hexadecane	Liquid
$C_{17}H_{36}$	$CH_3(CH_2)_{15}CH_3$	Heptadecane	Liquid
$C_{18}H_{38}$	$CH_3(CH_2)_{16}CH_3$	Octadecane	Solid
$C_{19}H_{40}$	$CH_3(CH_2)_{17}CH_3$	Nonadecane	Solid
$C_{20}H_{42}$	$CH_3(CH_2)_{18}CH_3$	Eicosane	Solid

after which supposition and assumption takes over, which are not always accurate—some observers would say *never accurate*. On the other hand, molecular size (molecular weight) is a determinant in terms of solubility and leachability of organic chemicals from the same series from mineral deposits. Density gives an indication of whether or not the organic chemical will float or sink in water while surface tension is an indicator of the relative energy between an organic chemical and water. The final example, viscosity is an indicator of the freedom of the organic compound, if it is a liquid, to flow under conditions of added stress. From these properties, an assessment can be made of the relative behavior of different compounds after release into the environment.

TABLE 5.2 Selected Properties of Industrial Organic Chemicals[a]

Name	Molecular Formula	Molecular Weight	Density	Surface Tension	Viscosity cP^b	Viscosity cS^b
Acetic acid (ethanoic acid)	$C_2H_4O_2$	60.05	1.043	27.0	1.06	1.0
Acetone (propanone)	C_3H_6O	58.08	0.786	23.0	0.31	0.4
Acetonitrile (ethane nitrile)	C_2H_3N	41.05	0.779	28.7	0.37	0.5
Acetophenone (1-phenylethanone)	C_8H_8O	120.15	1.024	39.0	1.68	1.6
Acrylonitrile (ethenylnitrile)	C_3H_3N	53.06	0.801	26.7	0.34	0.4
2-Aminoethanol (ethanolamine)	C_2H_7NO	61.08	1.014	48.3	21.10	20.8
Aniline	C_6H_7N	93.13	1.018	42.4	3.85	3.8
Benzaldehyde	C_7H_6O	106.12	1.040	38.3	1.40	1.4
Benzene	C_6H_6	78.11	0.873	28.2	0.60	0.7
Benzonitrile	C_7H_5N	103.12	1.001	38.8	1.27	1.3
Benzyl alcohol	C_7H_8O	108.14	1.041	36.8	5.47	5.3
Bis(2-ethylhexyl) phthalate (BEHP, dioctyl phthalate)	$C_{24}H_{38}O_4$	390.56	0.980	31.1	80.00	82.0
1,3-Butanediol	$C_4H_{10}O_2$	90.12	1.002	47.1	98.30	98.1
1-Butanol (n-butyl alcohol)	$C_4H_{10}O$	74.12	0.806	25.0	2.54	3.2
2-Butanol (sec-butanol)	$C_4H_{10}O$	74.12	0.805	22.6	3.10	3.9

Continued

TABLE 5.2 Selected Properties of Industrial Organic Chemicals—cont'd

Name	Molecular Formula	Molecular Weight	Density	Surface Tension	Viscosity	
					cP	cS
2-Butanone (methyl ethyl ketone)	C_4H_8O	72.11	0.799	24.0	0.41	0.5
2-Butoxyethanol (ethylene glycol monobutyl ether)	$C_6H_{14}O_2$	118.17	0.896	26.6	6.40	7.1
Butyl acetate	$C_6H_{12}O_2$	116.16	0.876	24.8	0.69	0.8
Cyclohexane	C_6H_{12}	84.16	0.773	24.7	0.89	1.2
Cyclohexanol	$C_6H_{12}O$	100.16	0.960	33.4	57.50	59.9
Cyclohexanone	$C_6H_{10}O$	98.14	0.942	34.4	2.02	2.1
Dibutyl phthalate	$C_{16}H_{22}O_4$	278.34	1.043	37.4	16.60	15.9
1,2-Dichlorobenzene	$C_6H_4Cl_2$	147.00	1.301	35.7	1.32	1.0
1,2-Dichloroethane (ethylene dichloride)	$C_2H_4Cl_2$	98.96	1.246	32.6	0.78	0.6
Dichloromethane (methylene chloride, DCM)	CH_2Cl_2	84.93	1.318	27.8	0.41	0.3
Diethylene glycol	$C_4H_{10}O_3$	106.12	1.114	55.1	30.20	27.1
Diethyl ether (ethoxyethane)	$C_4H_{10}O$	74.12	0.708	16.7	0.22	0.3
Diiodomethane (methylene iodide)	CH_2I_2	267.84	3.306	50.8	2.60	0.8
1,2-Dimethoxyethane (DME) (ethylene glycol dimethyl ether)	$C_4H_{10}O_2$	90.12	0.865	20.0	1.10	1.3

Dimethoxymethane (methylal)	$C_3H_8O_2$	76.10	0.854	18.8	0.33	0.4
Dimethylethanolamine (2-(dimethylamino)ethanol)	$C_4H_{11}NO$	89.14	0.882	51.6	4.08	4.6
N,N-Dimethylformamide (DMF)	C_3H_7NO	73.09	0.945	34.4	0.79	0.8
Dimethyl sulfoxide (DMSO)	C_2H_6OS	78.13	1.095	42.9	1.99	1.8
1,4-Dioxane	$C_4H_8O_2$	88.11	1.029	32.9	1.18	1.2
Epichlorohydrin (chloromethyloxirane)	C_3H_5ClO	92.52	1.174	36.3	1.07	0.9
Ethanol (ethyl alcohol)	C_2H_6O	46.07	0.787	22.0	1.07	1.4
2-Ethoxyethanol (ethylene glycol monoethyl ether)	$C_4H_{10}O_2$	90.12	0.925	28.8	2.10	2.3
Ethyl acetate (ethyl ethanoate)	$C_4H_8O_2$	88.11	0.894	23.2	0.42	0.5
Ethylbenzene	C_8H_{10}	106.17	0.865	28.6	0.63	0.7
Ethylene glycol	$C_2H_6O_2$	62.07	1.111	48.4	16.10	14.5
2-Ethyl-1-hexanol	$C_8H_{18}O$	130.23	0.830	27.7	6.27	7.6
Formamide (methanamide)	CH_3NO	45.04	1.129	57.0	3.34	3.0
2-Furanmethanol (furfuryl alcohol)	$C_5H_6O_2$	98.10	1.127	53.3	4.62	4.1
Glycerol	$C_3H_8O_3$	92.09	1.257	76.2	934.00	743.0
n-Heptane	C_7H_{16}	100.20	0.682	19.8	0.39	0.6
2-Heptanone (methyl n-amyl ketone)	$C_7H_{14}O$	114.19	0.811	26.1	0.71	0.9
Hexadecane	$C_{16}H_{34}$	226.44	0.770	27.1	–	–

Continued

TABLE 5.2 Selected Properties of Industrial Organic Chemicals—cont'd

Name	Molecular Formula	Molecular Weight	Density	Surface Tension	Viscosity cP	Viscosity cS
Isophorone	$C_9H_{14}O$	138.21	0.920	35.5	2.33	2.5
Isopropyl acetate (isopropyl ethanoate)	$C_5H_{10}O_2$	102.13	0.871	22.3	0.52	0.6
Methanoic acid (formic acid)	CH_2O_2	46.03	1.214	37.7	1.61	1.3
Methanol (methyl alcohol)	CH_4O	32.04	0.787	22.1	0.54	0.7
2-Methoxyethanol (ethylene glycol monomethyl ether)	$C_3H_8O_2$	76.10	0.960	42.8	1.70	1.8
Methyl acetate	$C_3H_6O_2$	74.08	0.927	24.5	0.36	0.4
Methyl methacrylate	$C_5H_8O_2$	100.12	0.937	24.2	0.57	0.6
4-Methyl-2-pentanone (methyl isobutyl ketone)	$C_6H_{12}O$	100.16	0.796	23.5	0.55	0.7
3-Methylphenol (m-Cresol)	C_7H_8O	108.14	1.030	35.8	12.90	12.5
2-Methylpropyl alcohol (isobutanol)	$C_4H_{10}O$	74.12	0.797	22.6	3.95	5.0
2-Methylpropyl ethanoate (isobutyl acetate)	$C_6H_{12}O_2$	116.16	0.869	23.0	0.68	0.8
N-Methyl-2-pyrrolidine	C_5H_9NO	99.13	1.025	44.6	1.67	1.6
Morpholine	C_4H_9NO	87.12	0.997	38.8	2.02	2.0
Nitromethane	CH_3NO_2	61.04	1.129	36.3	0.63	0.6
Octanoic acid (caprylic acid)	$C_8H_{16}O_2$	144.21	0.903	27.9	5.02	5.6

1,2-Propanediol (propylene glycol)	$C_3H_8O_2$	76.10	1.033	45.6	40.40	39.1
1-Propanol (n-propanol, n-propyl alcohol)	C_3H_8O	60.10	0.802	20.9	1.95	2.4
2-Propanol (isopropyl alcohol)	C_3H_8O	60.10	0.783	23.3	2.04	2.6
Propyl acetate (propyl ethanoate)	$C_5H_{10}O_2$	102.13	0.883	23.9	0.54	0.6
Propylene carbonate	$C_4H_6O_3$	102.09	1.200	40.9	2.50	2.1
Pyridine	C_5H_5N	79.10	0.979	36.7	0.88	0.9
Styrene (phenylethene, phenylethylene)	C_8H_8	104.15	0.900	32.0	0.70	0.8
Tetrachloromethane (carbon tetrachloride)	CCl_4	153.82	1.583	26.3	0.91	0.6
Tetrahydrofuran	C_4H_8O	72.11	0.880	26.7	0.46	0.5
Toluene	C_7H_8	92.14	0.865	27.9	0.56	0.7
Trichloroethylene (TCE, trichloroethane, trichloroethylene)	C_2HCl_3	131.39	1.458	28.7	0.55	0.4
1,1,1-Trichloroethane (methyl chloroform)	$C_2H_3Cl_3$	133.40	1.330	25.0	0.79	0.6
Trichloromethane (chloroform)	$CHCl_3$	119.38	1.480	26.7	0.54	0.4
Triethanolamine	$C_6H_{15}NO_3$	149.19	1.120	51.5	609.00	543.0
o-Xylene	C_8H_{10}	106.17	0.876	29.6	0.76	0.9
p-Xylene	C_8H_{10}	106.17	0.861	27.9	0.60	0.7

[a]For the most part, the IUPAC (International Union of Pure and Applied Chemistry) form of nomenclature has been used except where the IUPAC designation is not widely used in industry in which case the more familiar name is used.
[b]Centipoises and centistokes.

Furthermore, isomers do not necessarily share similar properties, unless they also have the same functional groups (Chapter 2). Therefore, chemical and physical property data should be selected on a chemical-by-chemical basis—properties of a mixture should not be averaged based on the properties of the individual constituents. For, example, the flash point of a hydrocarbon mixture is the flash point of the most reactive constituent. In addition, many observers make the mistake of stating that natural gas is lighter than air and therefore completely volatile under ambient (temperature and pressure) conditions.

In another case, density might be the property of choice for decision making. Briefly, density is a physical property of matter and each organic chemical has a unique density associated with it. The density is defined in a qualitative manner as the measure of the relative heaviness of objects with a constant volume. For gases the density may vary with the number of gas molecules in a constant volume. However, what may be more important in the case of organic mixtures, such as natural gas or process gas, is the relative density data (i.e., density of the constituents of natural gas relative to the density of air) which shows that of all the constituents of natural gas only methane is lighter than air:

Constituent	Density Relative to Air at NTP[a]
Air	1.00
Methane	0.55
Ethane	1.00 (0.99)
Propane	1.56
n-Butane	2.07
n-Pentane	2.49
n-Hexane	2.97
n-Heptane	3.45
n-Octane	3.86
Benzene	2.89
Toluene	3.41

[a]NTP: normal temperature and pressure: 20°C and 1 atm pressure.

Thus, if natural gas is allowed to escape over undulating ground, volatility in air will not be complete (even though methane is the major constituent) and the higher boiling constituents (especially any condensate constituents, such as alkanes like pentane and any higher molecular weight alkanes) could remain in wind-sheltered areas leading to the potential for ignition and a flammable explosion. This means that first responders and cleanup crews should read very carefully after such a spillage or release incident.

For some properties, such as molecular weight, there will be very little variation in measurements reported in the scientific literature. However other properties, such as the octanol-water partition coefficient, may be subject to considerable variation due to differences in the experimental method or conditions used to measure its value or in the calculation used to derive the estimate from other thermodynamic properties (Alberty and Reif, 1988; Chao et al., 1990; Domalski and Hearing, 1993). This variability may have a significant effect on the predicted contaminant behavior and ultimately on the estimated human exposure.

Just as the chemical and physical properties of organic compounds offer challenges in selecting and designing optimal production processes, these properties also affect disposal schemes because of the potential for the longevity of such compounds in the environment and the effects of these compounds on the environment. In particular, predicting the fate of the polynuclear aromatic systems, the heteroatom systems (principally compounds containing nitrogen and sulfur), and the metal-containing systems (principally compounds of vanadium, nickel, and iron) in the feedstocks is the subject of many studies and migration models. These constituents generally cause processing problems and knowledge of the behavior of these elements is essential for process improvements, process flexibility, and environmental compliance.

Because of the variation in the amounts of chemical types and bulk fractions, it should not be surprising that organic compounds exhibit a wide range of physical properties and several relationships can be made between various physical properties. Whereas the properties such as viscosity, density, boiling point, and color of organic compounds may vary widely, the ultimate or elemental analysis varies, as already noted, over a narrow range for a large number of organic compounds samples. The carbon content is relatively constant, while the hydrogen and heteroatom contents are responsible for the major differences between organic compounds. The nitrogen, oxygen, and sulfur can be present in only trace amounts in some organic compounds, which as a result consists primarily of hydrocarbons. On the other hand, organic compounds containing 9.5% heteroatoms may contain essentially no true hydrocarbon constituents insofar as the constituents contain *at least one or more* nitrogen, oxygen, and/or sulfur atoms within the molecular structures. Coupled with the changes brought about to the feedstock constituents by refinery operations, it is not surprising that organic compounds characterization is a monumental task.

Thus, initial inspection of the nature of the organic compounds will provide deductions about the most logical means of clean up and any subsequent environmental effects. Indeed, careful evaluation of organic compounds from physical property data is a major part of the initial study of any organic compounds that has been released to the environment. Proper interpretation of the data resulting from the inspection of organic compounds requires an understanding of their significance.

Consequently, various standards organizations, such as the American Society for Testing and Materials (ASTM, 2016) in North America, have devoted considerable time and effort to the correlation and standardization of methods for the inspection and evaluation of cured oil and crude oil products. Just as these test methods can be applied to products from crude oil and other fossil fuels, they can also be applied with justification to organic compounds and mixtures of organic compounds. As a result, the data derived from any one, or more of the evaluation techniques present an indication of the nature of organic compounds and its products. The data can be employed to give the environmental scientist or engineer an indication of the means by which the spilled material can be, or should be, recovered. Other properties (Speight, 2014) may also be required for further evaluation, or, more likely, for comparison on before and after scenarios even though they may not play any role in dictating which cleanup operations are necessary.

This chapter presents some of the methods that are generally applied to study the properties of organic chemicals in terms of chemical structures as well as chemical reactivity. There are, of course, many analytical methods that can be applied to the study of the behavior of organic chemicals and chemical mixtures but they vary with composition of the mixture as well as with environmental conditions. More specifically, this chapter deals with the more common methods used to define the chemical and physical properties of organic chemicals. In addition, any of the analytical method cited in this chapter (although developed predominantly for the analysis of products derived from crude oil, coal, or natural gas) might also be applied to the analysis of sample for environmental purposes (Speight, 2005; Speight and Arjoon, 2012).

Thus, this chapter will deal with the various aspects of the relevant properties of organic compounds and mixture of organic compounds as well as some of the test methods that can be applied in order to help predefine predictability of organic compounds and mixtures of organic compounds and the behavior of these chemicals when released into the environment. To take this one step further, it may then be possible to develop preferred cleanup methods from one (but preferably more) of the physical properties as determined by the evaluation test methods.

2 COMPOSITION

An important step in assessing the effects of organic compounds products that have been released into the environment is to evaluate the nature of the particular mixture and eventually select an optimum remediation technology for that mixture. As a general rule, organic compounds (unrefined organic compounds) is either a single complex compound or a complex mixture (such as exists when petroleum products naphtha, kerosene, gas oil are considered) composed of different isomers and functional group types but it must be recognized that the quantities of the individual constituents differ in organic compound mixtures from different locations. This rule of thumb implies that the quantities of some compounds can be zero to the majority constituent in a mixture of compounds from a specific location or production facility.

In very general terms, organic compounds are a mixture of: (a) hydrocarbon types, (b) nitrogen compounds, (c) oxygen compounds, (d) sulfur compounds, and (e) metallic constituents. Organic compounds products are less well defined in terms of heteroatom compounds and are better defined in terms of the hydrocarbon types present. However, this general definition is not adequate to describe the true composition as it relates to the behavior of the organic compounds, and its products, in the environment. For example, the occurrence of amphoteric species (i.e., compounds having a mixed acid/base nature) is not always addressed nor is the phenomenon of molecular size or the occurrence of specific functional types that can play a major role in.

In the present context, the composition of mixtures of organic compounds is defined in terms of (1) the elemental composition of the mixture, (2) the chemical composition of the mixture, and (3) the fractional composition of the mixture. All three are interrelated although the closeness of the elemental composition makes it difficult to relate precisely to the chemical composition and the fractional composition. The chemical composition and the fractional composition are somewhat easier to relate because of the quantities of heteroatoms (nitrogen, oxygen, sulfur, and metals) that form the basis of the functional group types and the functional group content in some mixtures.

2.1   Elemental Composition

Chemically, a mixture is a material system made up of two or more different substances which are mixed but are not combined chemically. Thus, in the current context, a mixture refers to the physical combination of two or more organic chemicals substances on which the identities are retained and are mixed in the form of (1) a solution, (2) a suspension, or (3) a colloid. A mixture of organic chemicals is a product of a mechanical blending operation such as gasoline and diesel fuel prepared for final without chemical bonding or any chemical change, so that each organic chemical constituent within the mixture retains its own individual chemical properties and makeup. However, despite the lack of chemical changes to the constituents, the physical properties of a mixture, such as its melting point may differ from the melting points of the individual constituents due to intermolecular interactions between the constituents or intramolecular interaction that are no longer operative. As indicated above by using natural gas as the example, averaging the properties of a mixture based on the properties of the individual constituents is not always a wise decision and the outcome can be erroneous.

Mixtures can either be homogenous or heterogeneous and typically can have any amounts of ingredients (constituents). A homogeneous mixture is a type of mixture in which the chemical composition is uniform (such as a mixture of the alkane hydrocarbons) while a heterogeneous mixture is a type of mixture in which the constituents are chemically different (such as alkane hydrocarbons mixed with organic compounds that contain a variety of functional groups and there may even be two or more phases present in such mixtures. A homogeneous mixture in which there is both a solute and solvent present is also a solution).

The elemental composition of organic compounds is determined by the elemental analysis, which is a process where a sample of some material is analyzed to determine the amount (typically as the % w/w) of the various elements in the compound. Elemental analysis can be qualitative (determining what elements are present), and it can be quantitative (determining how much of each are present). Elemental analysis is a first (optional) step in determining the composition of a mixture, but only in terms of the elements present but not in terms of any structural features or physical or chemical properties of the constituents of the mixture. Thus, in the discipline of organic chemistry, elemental analysis refers to the composition of an analyte in terms of the proportions of the elements (as % w/w) carbon, hydrogen, nitrogen, halogens, and other heteroatoms (such as oxygen, sulfur, and metals; CHNX analysis where X typically represents the elements (such as halogens) other than carbon, hydrogen, nitrogen, and the metals). While this information is important to help determine the structure of an unknown individual organic compound, as well as to help ascertain the structure and purity of a synthesized compound, it is not the same for a mixture but may be used as a guide for the next set of analyses.

The elemental composition of organic compounds varies greatly from organic compounds to organic compounds. Many organic compounds in mixtures (especially if the mixture is a petroleum-derived mixture) are typically types of hydrocarbons (usually more than 75%, v/v) accompanied by other constituents that also contain sulfur (0–8%, w/w), nitrogen (0–1%, w/w), and oxygen (0.0–0.5%, w/w). If the liquids are produced from coal, the proportion of oxygen can be expected to be higher or if the liquids are produced by the Fischer-Tropsch process the proportion of hydrocarbon can be as high as 1005 v/v (Chadeesingh, 2011; Speight, 2013, 2014).

The elemental analysis (often referred to as the *ultimate analysis*) of organic compounds for the percentages of carbon, hydrogen, nitrogen, oxygen, and sulfur is perhaps the first method used to examine the general nature, and perform an evaluation, of a feedstock. The atomic ratios of the various elements to carbon (i.e., H/C, N/C, O/C, and S/C) are frequently used for indications of the overall character of the feedstock. It is also of value to determine the amounts of trace elements, such as vanadium and nickel, in a feedstock since these materials can have serious deleterious effects on catalyst performance during refining by catalytic processes.

For example, *carbon content* can be determined by the method designated for coal and coke or by the method designated for municipal solid waste (ASTM E777). There are also methods designated for *hydrogen content* (ASTM D1018; ASTM D3343; ASTM D3701; ASTM E777; ASD\TM E191), *nitrogen content* (ASTM D3228; ASTM E258; ASTM E778), *oxygen content* (ASTM E385), and *sulfur content* (ASTM D1266; ASTM D1552; ASTM D1757; ASTM D4045; ASTM D4294). Of the data produced by elemental analysis, more focus is often on the heteroatom (N, O, S) content which is an initial indicator for the potential presence of one or more functional groups within an organic molecule and a functional group can play an important role in the disposition of the organic compound or the organic mixture in the environment. However, the analytical data do not point to the type of functional group that may be present in the molecule—to determine the functional group type, further analysis is needed.

Heteroatoms (*nitrogen, oxygen, sulfur,* and *metals*) are found in every organic compound and the concentrations have to be reduced to convert the oil to transportation fuel. The reason is that if nitrogen and sulfur are present in the final fuel during combustion, nitrogen oxides (NO_x) and sulfur oxides (SO_x) form, respectively. In addition, metals affect many upgrading processes adversely, poisoning catalysts in refining and causing deposits in combustion. In addition, heteroatoms also play a major role in environmental issues and can cause the organic compounds to adhere to the soil ensuring long-term contamination.

A variety of test methods (ASTM D1318; ASTM D1548; ASTM D3341; and ASTM D3605) have been designated for the determination of metals in organic compounds and organic compounds products. At the time of writing, the specific test for the determination of metals in whole feeds has not been designated. However, this task can be accomplished by combustion of the sample so that only inorganic ash remains. The ash can then be digested with an acid and the solution examined for metal species by atomic absorption (AA) spectroscopy or by inductively coupled argon plasma spectrometry.

2.2 Chemical Composition

Chemical composition refers to the arrangement, type, and ration of atoms in organic chemical substances. If the substance is a single chemical compound, the composition will remain constant (unless oxidation occurs when the compound is exposed to the air). On the other hand, the chemical composition of a mixture can be varied when organic chemicals are added or subtracted from the mixture, when the ratio of substances changes, or when other chemical changes occur in organic chemicals.

Chemical composition is a concept in chemistry that has different, but similar, meanings if referred to a single pure substance or a mixture. The chemical composition of a pure substance corresponds to the relative amounts of the elements that constitute the substance itself (elemental composition). The chemical composition of a mixture depends on the distribution of the individual organic constituents that make up and the properties of the mixture is often surmised (correctly or incorrectly) to be due to the concentration of each organic constituents within the mixture. This latter statement is subject to the lack of any intermolecular interactions between the constituents of the mixture or the lack of any intramolecular interactions within the constituents of the mixture. Because there are different ways to define the concentration of a component, as a consequence there are also different ways to define the composition of a mixture. For example, it can be expressed as (1) the mole fraction, also known as the molar fraction, which is the amount of a constituent expressed in moles divided by the total amount of all constituents in the mixture, (2) the volume fraction, which is the volume of a constituent divided by the volume of all constituents of the mixture prior to mixing and usually expressed as % v/v, (3) the mass fraction, which is the weight of a constituent divided by the weight of all constituents of the mixture prior to mixing and usually expressed as % w/w, (4) the molality, also called the molal concentration, which is a measure of the concentration of a solute in a solution in terms of the amount of the organic chemical in a specified amount of mass of the solvent which contrasts with the definition of molarity which is based on a specified volume of the solution, (5) the molarity, which is a measure of the concentration of a solute in a solution, or of any chemical species, in terms of the amount of the substance in a given volume, or (6) the normality, which is the molar concentration divided by an equivalence factor that is dependent upon the constitution of the mixture.

Thus in order to estimate, assess, or predict the behavior of mixtures of organic compounds in the environment, it is first necessary to understand that the chemical composition of mixtures of organic compounds varies over a wide range, as is observed in the fossil fuel industry. A broad functional definition of organic hydrocarbon compounds is that hydrocarbon mixtures are primarily composed of many organic compounds of natural origin and low water solubility. Remembering that hydrocarbons, are (by definition) compounds containing carbon and hydrogen *only*, the hydrocarbon content may be as high as 97%, w/w by weight in a lower boiling petroleum-derived products like lighter paraffinic organic compounds or low as about 50%, w/w in higher boiling petroleum products, such as asphalt.

However, within the hydrocarbon series of constituents, there is variation of chemical type (Chapter 2) which leads to differences in behavior when released to the environment. Organic compounds hydrocarbons may be paraffinic (alkane derivatives and cycloalkane derivatives), alkene derivatives, or aromatic derivatives which occur in varying concentrations within the different petroleum products and fuels (as well as similar products derived from coal). Thus, the constituents of organic compounds that occur in varying amounts depend on the source and character of the mixture are (Table 5.3) (Speight, 2001, 2015):

(1) *Alkanes* (also called *normal paraffins* or *n-paraffins*)

These constituents are characterized by branched or unbranched chains of carbon atoms with attached hydrogen atoms, and contain no carbon-carbon double bonds (they are saturated). Examples of alkanes are pentane (C_5H_{12}) and heptane (C_7H_{16}).

(2) *Cycloalkanes* or *cycloparaffins* (also called *naphthenes*)

These constituents are characterized by the presence of simple closed rings of carbon atoms, such as the cyclopentane ring or the cyclohexane ring.

TABLE 5.3 Compound Types in Various Classes of Organic Compounds

Class	Compound Types
Saturated hydrocarbons	n-Paraffins
	iso-Paraffins and other branched paraffins
	Cycloparaffins (naphthenes)
	Condensed cycloparaffins (including steranes, hopanes)
	Alkyl side chains on ring systems
Unsaturated hydrocarbons	Olefins not indigenous to organic compounds; present in products of thermal reactions
Aromatic hydrocarbons	Benzene systems
	Condensed aromatic systems
	Condensed aromatic-cycloalkyl systems
	Alkyl side chains on ring systems
Saturated heteroatomic systems	Alkyl sulfides
	Cycloalkyl sulfides
	Alkyl side chains on ring systems
Aromatic heteroatomic systems	Furans (single-ring and multiring systems)
	Thiophenes (single-ring and multiring systems)
	Pyrroles (single-ring and multiring systems)
	Pyridines (single-ring and multiring systems)
	Mixed heteroatomic systems
	Amphoteric (acid-base) systems
	Alkyl side chains on ring systems

(3) *Alkenes* (also called *olefins*)

These constituents are characterized by the presence of branched or unbranched chains of carbon atoms. Alkenes are not generally found in organic compounds but are common in refined products, such as naphtha (a precursor to gasoline). Common gaseous alkenes include ethylene ($CH_2\!=\!CH_2$) and propylene (also called propene, $CH_3CH\!=\!CH_2$).

(4) *Single-ring aromatics*

Aromatic constituents are characterized by the presence of rings with six carbon atoms and are considered to be the most acutely toxic component of organic compounds constituents because of their association with chronic and carcinogenic effects. Many low molecular weight aromatics may have a measurable solubility in water thereby increasing the potential for exposure to aquatic flora and fauna.

(5) *Multiring aromatics*

Aromatic derivatives with two or more condensed rings are referred to as polynuclear aromatic hydrocarbons (PNAs) and naphthalene (two-ring) and phenanthrene (three-ring). The most abundant aromatic hydrocarbon families have two and three fused rings with one to four carbon atom alkyl group substitutions. Condensed aromatic constituents with more than two condensed rings are also present in the higher boiling fractions of organic mixtures.

Any of the higher boiling fractions of crude oil (or coal-derived liquids) while comprising typical organic compounds may also contain a small proportion of PNAs. These polynuclear aromatic compounds can vary from two-ring (naphthalene) derivatives to the more carcinogenic PNAs such as benzo(a)pyrene derivatives:

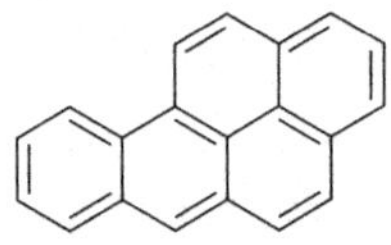

Benzo(a)pyrene

In addition, mixtures of organic hydrocarbons may also contain a fraction of volatile hydrocarbon compounds (VOCs), some of which may pose a threat to human health such as benzene, toluene, xylenes, and other dingle-ring aromatic derivatives. However, the relative mass fraction of volatile hydrocarbon derivatives in petroleum-based compounds is significantly less than that found in organic compounds distillate products such as gasoline. The volume percentage of benzene in gasoline may range up to 3% (30,000 ppm), while the benzene content of organic compounds is ~0.2% (2000 ppm). As a result of the lower percentage of volatile aromatics, vapor emissions from organic compounds-contaminated soils are expected to be much less than potential emissions from gasoline-contaminated soils.

2.3 Composition by Volatility

In the realm of organic chemistry, volatility (the tendency of an organic chemical to vaporize which is directly related to the vapor pressure of the organic chemical) is often used as a mean of separation or determination of the composition of mixtures. Thus, at a given temperature, an organic chemical with higher vapor pressure vaporizes more readily than an organic chemical with a relatively lower vapor pressure. The term *volatility* is primarily applied to liquids but it may also be used to describe the process of sublimation which is associated with organic compounds that are solid in the natural state in which the compounds can change directly from the solid state to the vapor state, thus by-passing the liquid state.

Briefly, the vapor pressure of a substance is the pressure at which the gas phase is in equilibrium with the condensed phase (liquid or solid) and is a measure of the tendency of molecules and atoms to escape from a liquid or a solid. The atmospheric pressure boiling point of an organic compound corresponds to the temperature at which the vapor pressure is equal to the surrounding atmospheric pressure (*normal boiling point* or *STP boiling point*). The higher the vapor pressure of a liquid at a given temperature, the higher the volatility and the lower the normal boiling point of the liquid.

On the other hand, the *relative volatility* is a measure comparing the vapor pressures of the components in a liquid mixture of organic chemicals and is widely used in designing large industrial distillation processes, such as the distillation units in a crude oil refinery (Speight, 2014, 2016). In effect, the relative volatility is also an indication of the efficiency (or inefficiency) of using a distillation unit to separate the more volatile components of a mixture from the less volatile components of the mixture. In addition to being used for design of all types of distillation processes, relative volatility data are also used in the design of other separation or absorption processes that involve the contacting of vapor and liquid phases in a series of equilibrium stages. However, relative volatility data are not used in separation or absorption processes that involve components reacting with each other (i.e., where there are intermolecular interactions between the constituents of the mixture).

Distillation is used to separate mixtures of liquids by exploiting the differences in the volatility (boiling points) of the different constituents of organic mixtures (fractional distillation). Fractional distillation is the most common form of separation technology used in petroleum refineries, petrochemical plants, chemical plants, natural gas processing plants (natural gas refining), and cryogenic air separation plants. In most cases, the distillation is operated at a continuous steady state insofar as fresh feedstock is continuously added to the distillation column and, at the same time, products are continuously being removed. Unless the process is disturbed due to changes in feed, heat, ambient temperature, or condensing, the amount of feed being added and the amount of product being removed are normally equal.

Industrial distillation is typically performed in a large, vertical cylindrical column (distillation tower, fractionating towers, distillation column) with diameters ranging from inches to yards and height ranging from ~20 to 85 ft or even higher. The distillation tower has liquid outlets at intervals up the column which allow for the withdrawal of different products or different fractions having different boiling points or boiling ranges, respectively. By increasing the temperature of the ascending product stream inside the distillation column, the different products or streams of hydrocarbons are separated. The lightest products (those products or streams with the lowest boiling point or boiling range) exit from the top of the column and the heaviest products (those products or streams with the highest boiling point or boiling range) exit from the bottom of the column.

The technique of distillation has been practiced for many centuries, and the stills that have been employed have taken many forms and the technique is widely used in industry, for example, in the commercial manufacture and purification of not only organic compounds but also in commercial operations such as of nitrogen, oxygen and the rare gases. However, one of its best known uses is the refining of crude oil into its main fractions, including naphtha, kerosene, and gas oil which is the first stage of converting oil into compounds which are then used to manufacture a wide variety of commercial products (Speight, 2014, 2015). Distillation is the first and the most fundamental step in the separation process (after the mixture of organic compounds has been cleaned and any remnants of water or nonorganic chemicals removed) and the mixture is separated by maintaining the enrapture under control so that there is no thermal decomposition of the constituents (Table 5.4). Distillation involves the separation of the different hydrocarbon compounds that occur naturally in organic compound mixtures, into a number of different fractions. In the atmospheric distillation process (Speight, 2014, 2016), the heated mixture is separated in a distillation column (distillation tower, fractionating tower, atmospheric pipe still) into a variety of boiling range fractions (cuts, streams) that are then purified, transformed, adapted, and treated in a number of subsequent refining processes, into products for the refinery market. The lower boiling products separate out higher up the column, whereas the heavier, less volatile, products settle out toward the bottom of the distillation column. The fractions produced in this manner (using a petroleum refinery distillation unit as the example) are known as *straight run fractions* ranging (in the atmospheric tower) from gas, naphtha, and low-boiling kerosene, atmospheric gas oil, and atmospheric residuum (the nonvolatile constituents, also called the reduced crude).

The atmospheric residuum is then fed to the vacuum distillation unit at the pressure of 10 mmHg where light vacuum gas oil, heavy vacuum gas oil, and vacuum residue are the products. The fractions obtained by vacuum distillation of the reduced crude (atmospheric residuum) from an atmospheric distillation

TABLE 5.4 Separation by Distillation

Process Name	Action	Method	Purpose
Atmospheric distillation	Separation	Thermal	Separate fractions without cracking
Vacuum distillation	Separation	Thermal	Separate fraction without cracking

unit depend on whether or not the unit is designed to produce lubricating or vacuum gas oils. In the former case, the fractions include (1) heavy gas oil, which is an overhead product and is used as catalytic cracking stock or, after suitable treatment, a light lubricating oil, (2) lubricating oil, usually three fractions: light lubricating oil, intermediate lubricating oil, and heavy lubricating oil, which are obtained as side-stream products, and (3) asphalt (or residuum), which is the bottom product and may be used directly as, or to produce, asphalt, and which may also be blended with gas oils to produce a heavy fuel oil. Operating conditions for vacuum distillation are usually 50–100 mmHg (atmospheric pressure is equivalent to 760 mmHg). The product fractions from the atmospheric tower and the vacuum tower are often used as feedstocks to the second stage refinery processes that (thermally or catalytically) decompose the constituents of the fractions and bring about a series of basic chemical changes in the nature of a particular hydrocarbon (or nonhydrocarbon) constituents to produce specific products.

While distillation in the crude oil refinery has been used as the prime example of the separation of organic mixtures on the basis of relative volatility, there are many example of distillation being used as the separation process in various industries. In fact, commercially and in addition to the fossil fuel industry, distillation has many applications. For example, in the field of industrial organic chemistry, large ranges of crude liquid products from the synthesis of organic chemicals are distilled to separate them, either from other products, or from impurities, or from unreacted starting materials. Also, the distillation of fermented products produces distilled beverages (such as whiskey) with a high alcohol content, or separates out other fermentation products of commercial value.

Thus, distillation is a common method for the fractionation of organic compounds that is used in the laboratory as well as in refineries. Distillation is the most important industrial method of phase separation. Distillation involves the partial evaporation of a liquid phase followed by condensation of the vapor. This separates the starting mixture (the feed) into two fractions with different compositions; namely a condensed vapor (the condensate or distillate) that is enriched in the more volatile components and a remaining liquid phase (the residue, also call the residuum) that is depleted of volatile constituents. The actual distillation process can be divided into subcategories: (1) batch distillation or continuous distillation, according to the operating mode of the distillation unit, (2) distillation at atmospheric or under vacuum or pressurized distillation, according to the operating pressure, (3) single stage or multistage distillation, depending upon the number of distillation stages, (4) steam-assisted distillation, which involves the introduction of steam or other inert gases into the unit to aid separation, and (5) azeotropic distillation or extractive distillation, which involves the use of additional compounds added to the mixture to aid separation. In all cases, heat is required at the bottom of the distillation column for evaporating the feedstock and condensation energy is needed at the top of the column. The condensation energy is often transferred into cooling water or air, and this may provide an opportunity for energy recovery.

By way of explanation, in batch distillation unit, the composition of the organic mixture, the vapors of the volatile constituents, and the distillate change during the progress of the distillation process. In batch distillation process, a still is charged (supplied) with a specific amount (depending upon the capacity of the still) of the organic mixture, which is then separated into its component fractions and which are collected sequentially from most volatile to less volatile, with the nonvolatile fraction (bottoms) remaining in the still pot and removed after the distillation process is terminated. The still can then be recharged and the process repeated. On the other hand, in the continuous distillation process, the organic mixture, the organic vapors, and the organic distillate are maintained at a constant composition by carefully replenishing the source organic mixture and removing the fractions from both the vapor streams and the liquid streams as they emerge from the still pot and move up the distillation column. This results in a better-controlled and more efficient separation process.

Only a limited number of separation problems may be solved by simple distillation and it is unsuitable for the separation of organic mixtures that contain components with similar boiling temperatures. Higher efficiency can be achieved by increasing the contact surface area or by contacting the liquid and vapor phases. Rectification columns constructed in the plate-column design or packed-column design may involve more than 100 distillation stages in order to provide efficient mass transfer between the multiple stages (column plates or the column packing) by the repeated countercurrent contact of the vapor with liquid streams. The internal structure provides a large mass transfer contact surface which is maximized by ensuring that the column packing is (and remains) fully wetted.

In terms of environmental issues of the distillation process, a distillation column may contribute to emissions in three ways: (1) by allowing impurities to remain in the product, (2) through polymer formation in the still due to excessive temperature, and (3) by inadequate condensation of the volatile constituents. Off-gases from distillation may contain volatile organic material in the form of vapor or entrained droplets/mist, although this can be reduced by the use of additional condensing areas. Noncondensable substances (such as oxygen, nitrogen, carbon dioxide, low-boiling organic constituents) are not usually cooled to their condensation temperature and will exit the condenser. Emission points from distillation are typically: the condenser, accumulator, hot wells, steam jet ejectors, vacuum pump, and pressure relief valve. The total volume of gases emitted from a distillation operation depends upon: (1) leakage of air into the column, which increases with reduced pressure in the column and increased size of the column, (2) the volume of the inert carrier gas, (3) gases dissolved in the feedstock, (4) efficiency of the condenser or other recovery equipment, and (5) the physical properties of the organic constituents of the mixture.

Depending on the boiling point of the components, effluents may result from aqueous bottom residues or from the top, after condensation. Discharges depend on the efficiency of the distillation process and of additional steps for phase separation (preferably fractionated-condensation of top effluent, stripping of bottom residues). Highly concentrated still bottoms (resids, residua) are often incinerated if recovery of the organic constituents for these still bottoms is not possible.

2.4   Composition by Fractionation

Fractionation of organic compounds by volatility, informative as it might be, does not give any indication of the chemical types or chemical and physical properties of the constituents of an organic mixture. Recognition of the chemical types or chemical and physical properties of the constituents of an organic mixture is more often achieved by subdivision of the organic compounds into bulk fraction that are separated by a variety of solvent and adsorption methods.

An understanding of the chemical types (or composition) in organic compounds can lead to an understanding of the chemical aspects of organic compounds behavior. Indeed, this is not only a matter of knowing the elemental composition of a feedstock; it is also a matter of understanding the bulk properties as they relate to the chemical or physical composition of the material. For example, it is difficult to understand, *a priori*, the behavior of organic compounds and organic compounds products from the elemental composition alone and more information is necessary to understand environmental behavior.

Thus, in the simplest sense, organic compounds and organic compounds products can be considered to be composites of four major fractions (saturates, aromatics, resins, and asphaltenes) (SARA analysis) in varying amounts depending upon the type of organic compounds and the type of organic compounds product (Speight, 2001, 2014, 2015). In fact, the lower boiling product may only be compositions of one or two fractions (saturates and aromatics). Whatever the case, it must never be forgotten that the nomenclature of these fractions lies within the historical development of organic compounds science and that the fraction names are operational and related more to the general characteristics of the fraction rather than to the identification of specific compound types.

Insoluble constituents and sediment can comprise a large fraction of organic compounds, making the mixtures very dense and viscous. Composition is dependent upon source (these structures have the highest individual molecular weight of all organic compound components and are basically colloidal aggregates). Asphaltenes are substances in organic compounds that are insoluble in solvents of low molecular weight such as pentane or heptane. These compounds are composed of very polynuclear aromatic and heterocyclic molecules and are solids at normal temperatures. Consequently, oils that have high asphaltene contents are very viscous, have a high pour point, and are generally nonvolatile in nature. The porphyrins, asphaltene, and resin compounds are considered the residual oil, or residuum. During the weathering process, this fraction is the last to degrade, and its persistence over years has been noted.

There are also two other operational definitions that should be noted at this point and these are the terms *carbenes* and *carboids*. Both such fractions are, by definition, insoluble in benzene (or toluene) but the carbenes are soluble in carbon disulfide whereas the carboids are insoluble in carbon disulfide (Speight, 2014). Only traces of these materials occur in organic compounds and none occur in the products, unless the product is a high-boiling product of thermal treatment (such as visbroken feedstocks). On the other hand, oxidized organic compounds and oxidized high-boiling product that have been susceptible to oxidation though a spill may contain such fractions. But again, it must be remembered that the fraction separated by the various techniques are based on solubility or adsorption properties and not on specific chemical types. The terminology used for the identification of the various methods might differ. However, it must be recognized that the fractions produced by the use of different adsorbents will differ in content and will also be different from fractions produced by solvent separation techniques.

The variety of fractions isolated by these methods and the potential for the differences in composition of the fractions makes it even more essential that the method is described accurately and that it is reproducible not only in any one laboratory but also between various laboratories (Speight, 2001, 2015).

The evaluation of mixtures of organic compounds must of necessity involve a study of composition because of the interrelationship of the physical properties and composition as part of the overall evaluation of different feedstocks. There are several ASTM procedures for the evaluation of organic mixtures. These are:

(1) Separation of aromatic and nonaromatic fractions from high-boiling oils (ASTM D2549).
(2) Determination of hydrocarbon groups in rubber extender oils by clay-gel adsorption (ASTM D2007).
(3) Determination of hydrocarbon types in liquid organic compounds products by a fluorescent indicator adsorption (FIA) test (ASTM D1319).

Gel permeation chromatography is an attractive technique for the determination of the number of average molecular weight (Mn) distribution of organic compounds fractions, especially the heavier constituents, and organic compounds products.

Ion exchange chromatography is also widely used in the characterization of organic compounds constituents and products. For example, cation exchange chromatography can be used primarily to isolate the nitrogen constituents in organic compounds thereby giving an indication of how the organic compounds (singly or in a mixture) feedstock might behave in the environment and also an indication of any potential deleterious effects on the soil.

Liquid chromatography (also called *adsorption chromatography*) has helped to characterize the group composition of organic compounds since the beginning of 20th century. The type and relative amount of certain hydrocarbon classes in the matrix can have a profound effect on the quality and performance of the hydrocarbon product. The FIA method (ASTM D1319) has been used to measure the paraffinic, olefinic, and aromatic content of gasoline, jet fuel, and liquid products in general.

High-performance liquid chromatography (*HPLC*) has found great utility in separating different hydrocarbon group types and identifying specific constituent types. Of particular interest, is the application of the HPLC technique to the identification of the molecular types in the heavier feedstocks, especially the molecular types in the asphaltene fraction. This technique is especially useful for studying such materials on a *before-processing* and *after-processing* basis.

Several recent high-performance liquid chromatographic separation schemes are applicable since they also incorporate detectors not usually associated with conventional hydrocarbon group-type analyses. The general advantages of HPLC method are: (1) each sample may be analyzed *as received* even though the boiling range may vary over a considerable range, (2) the total time per analysis is usually on the order of minutes; and, perhaps most important, and (3) the method can be adapted for any recoverable organic compounds sample or product.

In recent years, *supercritical fluid chromatography* has found use in the characterization and identification of organic compounds constituents and products. A supercritical fluid is defined as a substance above its critical temperature. A primary advantage of chromatography using supercritical mobile phases results from the mass transfer characteristics of the solute. The increased diffusion coefficients of supercritical fluids compared with liquids can lead to greater speed in separations or greater resolution in complex mixture analyses. Another advantage of supercritical fluids compared with gases is that they can solubilize thermally labile and nonvolatile solutes and, upon expansion (decompression) of this solution, introduce the solute into the vapor phase for detection.

Currently, supercritical fluid chromatography is leaving the stages of infancy. The indications are that it will find wide applicability to the problems of characterization and identification of the higher molecular weight species in organic compounds thereby adding an extra dimension to our understanding of refining chemistry. It will still retain the option as a means of product characterization although the use may be somewhat limited because of the ready availability of other characterization techniques.

3 PROPERTIES

The chemical composition of mixtures of organic chemicals spilled into the environment is, for the most part complex—petroleum-derived products (with the exception of single petrochemicals—are complex mixtures of organic compounds which, when, discharged into the environment, require careful planning for cleanup according to the nature of the constituents in the mixture. This makes it essential that the most appropriate analytical methods are selected from a comprehensive list of methods and techniques that are used for to determine the properties not only of individual organic chemicals but also mixtures of organic chemicals (Dean, 1998; Miller, 2000; Budde, 2001; Speight, 2001, 2005; Speight and Arjoon, 2012; Speight, 2014, 2015). Furthermore, samples may be disturbed during sampling, storage, and pretreatment and, although most laboratory experiments and storage facilities impose specific environmental conditions, many chemicals exhibit dynamic behavior while in an indoor climate (Speight and Arjoon, 2012). However, the manner in which the corrections are applied must be quality data (ASTM D6299; ASTM D6792) and must be beyond reproach or claims of falsification of the data will be the most likely result.

In the early days of environmental science and cleanup, there did not appear to be the need to understand the character and behavior of organic chemicals and mixtures of organic chemicals in the detail that is currently required. Cleanup involved simple methods and as long as the chemicals was removed form the air, water, and land the deed was done. However, with the startling demands made by various levels of environmental legislation and the emergence of (even the demand for) a clean environment, the organic chemicals industry (including the various fossil fuel processing industries) needed to guard and protect the environment from the various chemicals.

Thus, the organic chemicals industry had to take on the role of technological watchdog as new and better processes were invented to produce new and better organic chemicals as advances in the use of materials for reactors were developed. In addition, there became a necessity to find out more about organic chemicals so that manufactures and first responders could enjoy the luxury of predictability of chemical behavior in the environment and plan to mitigate the potential for spills of chemicals and cleanup programs that were based on the properties and environmental behavior of the chemicals. A difficult task at first, but as planning evolved the mitigation of chemical spills and the ensuing cleanup were no longer insurmountable hurdles. In fact, when the character of the organic chemical or the mixture of organic chemicals was known and understood, the concept of a clean environment came closer to reality that had, at first, been believed.

The cleanup of spills of organic chemicals and mixtures of organic chemicals requires not only knowledge of the chemical and physical properties of the chemicals but also knowledge of its chemical and physical reactivity in the environment (Chapter 7). The chemical and physical properties of organic chemicals are presented in this chapter because of the need for an understanding of these properties which dictates the behavior of the chemicals after discharge of any chemicals into the environment and also the need to understand the relationship of reactivity to structure and functionality as well as the effects of the environmental conditions of reactivity (Chapters 6 and 7) is dealt with elsewhere in this book. Because organic chemicals vary markedly in properties and composition according to the source (Chapter 3), there is also variance of chemicals in terms of chemical and physical reactivity. Thus, knowledge of reactivity is required for optimization of existing cleanup procedures as well as for the development and design of new cleanup processes.

For example, when considering the effects of chemical mixtures (such as liquid petroleum products) valuable information can be obtained from the true boiling point (TBP) curve which is a function of percent weight distilled and temperature, that is, a boiling point distribution (Speight, 2001, 2014, 2015). However, the boiling range does not convey much detail about chemical reactivity of a chemical or a mixture of chemicals. In addition to the boiling point distribution, it is possible to measure bulk physical properties, such as specific gravity and viscosity that have assisted in the establishment of certain empirical relationships for transportation in an ecosystem processing from the *TBP* curve. Many of these relationships include assumptions that are based on experience with a range of organic chemicals (and petroleum products). However, the chemical aspects of single organic chemicals and mixtures of organic chemicals that contain different proportions of chemicals emphasizes the need for more definitive data that would enable more realistic predictions to be made of behavior (such as reactivity and chemical transformation) when such chemicals are discharged into any environmental ecosystem.

Thus, the physical and chemical properties of a contaminant are critical to understanding and modeling its fate and transport in the environment. Chemical data should be selected on a chemical-by-chemical basis and even for mixtures there is no such reliable conclusion about behavior in the environment that can be derived from average properties. For some properties, such as molecular weight, there will be very little variation in measurements reported in the scientific and engineering literature. However other properties, such as the octanol-water partition coefficient, may be subject to considerable variation due to differences in the experimental method or conditions used to measure its value or in the calculation used to derive the estimate from other thermodynamic properties. This variability in parameter may have a significant effect on the predicted contaminant behavior and ultimately on the estimated human exposure. Therefore, when deriving an assessment of the behavior of organic compounds in the environment, it is important to adopt authoritative values for property data from reviews of the scientific and technical literature. All choices should be fully justified and any description should include the sources examined, the range of values found, and where relevant, the temperature, and pressure and other relevant conditions under which experimental data were defined.

Organic compounds are typically described in terms of their physical properties (such as density and pour point) and chemical composition (such as percent composition of various organic compounds like hydrocarbons, asphaltenes, and sulfur). Although very complex in makeup, crude can be broken down into four basic classes of organic compounds hydrocarbons. Each class is distinguished on the basis of molecular composition. In addition, properties important for characterizing the behavior of organic compounds and organic compounds products when spilled into the waterways or on to land and/or released into the air, include flash point, density (or specific gravity), viscosity, emulsion formation in waterways, and adhesion to soil.

Generally, the properties of organic compounds constituents vary over the boiling range 0 to >565°C (32–1050°F). Variations in density also occur over the range 0.6–1.3 and pour points can vary from <0 to >100°C. Although these properties may seem to be of lower consequence in the grand scheme of environmental cleanup, they are important insofar as these properties influence (1) evaporation rate, (2) ability of the organic compounds constituents or organic compounds product to float on water, and (3) fluidity or mobility of the organic compounds or organic compounds product at various temperatures.

3.1 Density and Specific Gravity

The density (the volumetric mass density) of an organic compound is the mass per unit volume:

$$\rho\,(\text{density}) = \text{mass/volume}\ (m/V)$$

In some cases, for instance, in the crude oil and gas industries of the United States, density is generally represented as the weight per unit volume (specific weight).

The specific gravity of an organic compound is the ratio of the density of the compound to the density of a reference substance; equivalently, it is the ratio of the mass of the compound to the mass of a reference substance (typically water) for the same given volume. The *apparent* specific gravity is the ratio of the weight of a volume of the substance to the weight of an equal volume of the reference substance. The reference substance is nearly always water at its densest (4°C) for liquids; for gases it is air at room temperature (21°C). Nevertheless, the temperature and pressure must be specified for both the sample and the reference. Pressure is nearly always 1 atm (760 mmHg). Specific gravity is commonly used as a simple means of obtaining information about the concentration of solutions of various materials such as brine solutions, hydrocarbon solutions, sugar solutions, and acids.

In general, density can be changed by changing either the pressure or the temperature. Increasing the pressure always increases the density of a material. Increasing the temperature generally decreases the density, but there are, as might be expected, exceptions to this generalization.

The *density* and *specific gravity* of organic compounds (ASTM D287; ASTM D1217; ASTM D1298; ASTM D1555) are two properties that have found wide use for preliminary assessment of the character of the organic compounds. *Density* is the mass of a unit volume of material at a specified temperature and has the dimensions of grams per cubic centimeter (a close approximation to grams per milliliter). *Specific gravity* is the ratio of the mass of a volume of the substance to the mass of the same volume of water and is dependent on two temperatures, those at which the masses of the sample and the water are measured. When the water temperature is 4°C (39°F), the specific gravity is equal to the density in the centimeter-gram-second (cgs) system, since the volume of 1 g of water at that temperature is, by definition, 1 mL. Thus the density of water, for example, varies with temperature, and its specific gravity at equal temperatures is always unity. The standard temperatures for a specific gravity in the organic compounds industry in North America are 60/60°F (15.6/15.6°C).

Although density and specific gravity are used extensively, the API (American Petroleum Institute) gravity is the preferred property.

$$\text{Degrees API} = 141.5/\text{sg at } 60/60°F - 131.5$$

The specific gravity of organic compounds usually ranges from about 0.8 (45.3°API) for the light organic compounds 0.9 (26°API) for higher molecular weight compounds to over 1.0 (10°API) for tar sand bitumen.

Density or specific gravity or API gravity may be measured by means of a hydrometer (ASTM D287 and D1298) or by means of a pycnometer (ASTM D1217). The variation of density with temperature, effectively the coefficient of expansion, is a property of great technical importance, since most organic compounds products are sold by volume and specific gravity is usually determined at the prevailing temperature (21°C, 70°F) rather than at the standard temperature (60°F, 15.6°C). The tables of gravity corrections (ASTM D1555) are based on an assumption that the coefficient of expansion of all organic compounds products is a function (at fixed temperatures) of density only. Recent work has focused on the calculation and predictability of density using new mathematical relationships.

3.2   Functional Groups

In organic chemistry, *functional groups* are specific groups (moieties) of atoms within molecules that are responsible for the characteristic chemical reactions and physical behavior of those molecules (Table 5.5) (Chapter 2). The same functional group will undergo the same or similar chemical reaction(s) regardless of the size of the molecule in which the functional group occurs. An example of a functional group is the hydroxyl group (—OH) which is present in alcohols (ROH, where R is a nonaromatic group) and phenols (C_6H_5OH, where C_6H_5 is the phenyl group or any aromatic group).

Organic reactions are facilitated and controlled by the functional groups of the reactants. In general, alkyls are unreactive and difficult to get to react selectively at the desired positions, with few exceptions. In contrast, unsaturated carbon functional groups, and carbon-oxygen and carbon-nitrogen functional

TABLE 5.5 Representation of Various Functional Groups

Functional Group	Type	Compound	Example	IUPAC Name	Common Name
C=C	Double bond	Alkene	$H_2C=CH_2$	Ethene	Ethylene
C≡C	Triple bond	Alkyne	$HC≡CH$	Ethyne	Acetylene
—OH	Hydroxyl	Alcohol	CH_3OH	Methanol	Methyl alcohol
—O—	Oxy	Ether	H_3COCH_3	Methoxymethane	Methyl ether
>C=O	Carbonyl	Aldehyde	$H_2C=O$	Methanal	Formaldehyde
>C=O	Carbonyl	Ketone	CH_3COCH_3	Propanone	Acetone
$-CO_2^-$	Carboxyl	Carboxylic acid	$HCOOH$	Methanoic acid	Formic acid
$-CO_2^-$	Carboxyl	Ester	$HCOOCH_2CH_3$	Ethyl methanoate	Ethyl formate
$-NH_2$	Amino	Amine	CH_3NH_2	Aminomethane	Methylamine
—CN	Cyano	Nitrile	CH_3CN	Ethanenitrile	Acetonitrile
—X	Halogen	Haloalkane	CH_3Cl	Chloromethane	Methyl chloride

IUPAC, International Union of Pure and Applied Chemistry.

groups have a more diverse array of reactions that are also selective. It may be necessary to create a functional group (functionalization) in the molecule to make it react.

Functionalization is the addition of functional groups onto the surface of a material by chemical synthesis methods. The functional group added can be subjected to ordinary synthesis methods to attach virtually any kind of organic compound onto the surface. Functionalization is employed for surface modification of industrial materials in order to achieve desired surface properties such as water repellent coatings for automobile windshields and nonbiofouling, hydrophilic coatings for contact lenses. In addition, functional groups are used to covalently link functional molecules to the surface of chemical and biochemical devices such as microarrays and microelectromechanical systems. Thus, the properties conferred upon organic compounds by the presence of functional groups can have a significant influence on the behavior of such compounds in the environment.

As an example of functional groups behavior in the environment, phenol derivatives are widely used in nonionic surfactants in many different industrial processes, such as laundering, textile processing (wetting and scouring), pulp and paper processing, paint and resin formulation, oil and gas recovery, steel manufacturing, pest control, and power generation for over years (Staples et al., 1998; Bettinetti et al., 2002; Kannan et al., 2003; Michałowicz and Duda, 2007). The wide use of phenol derivatives made these compounds frequently detected in the environment where the derivative enters the aquatic environment via various pathways such as municipal and industrial wastewater discharges and sewage treatment plant effluents (Fytianos et al., 1997; Ying et al., 2002; Vazquez-Duhalt et al., 2005; Michałowicz and Duda, 2007; Soares et al., 2008; Sun et al., 2014).

There is a large variety of reactions that can be used to convert one organic compound to another. Some reactions involve addition of one molecule to another; some involve decomposition of molecules; some involve substitution of one atom or group by another; and some even involve the rearrangement of molecules, with some atoms moving into new positions. Some reactions require energy in the form of heat or radiation; and some require a special kind of catalyst but not all organic reactions are highly successful. One reaction might be a very simple one giving essentially 100% of the desired product; but another might be a complex multistep process yielding less than 5% overall of the wanted product.

3.3   Liquefaction and Solidification

Organic compounds and the majority of organic compounds products are liquids at ambient temperature, and problems that may arise from solidification during normal use are not common. Nevertheless, the *melting point* is a test (ASTM D127) that often serves to determine the state of the organic compounds or the product under various weather conditions or under applied conditions, such as steam stripping of the material from the soil. The reverse process, *solid-ification*, has received attention, again to determine the behavior of the material in nature. However, solidification of organic compounds and mixtures of organic compounds have been differentiated into four categories, namely, freezing point, congealing point, cloud point, and pour point.

Organic compounds become more or less a plastic solid when cooled to suf-ficiently low temperatures. This is due to the congealing of the various hydro-carbons that constitute the oil. The cloud point of organic compounds (or a product) is the temperature at which paraffin wax or other solidifiable com-pounds present in the oil appear as a haze when the oil is chilled under definitely prescribed conditions (ASTM D2500). As cooling is continued, organic com-pounds become more viscous and the pour point is the lowest temperature at which the oil pours or flows under definitely prescribed conditions when it is chilled without disturbance at a standard rate (ASTM D97).

The solidification characteristics of organic compounds and its products depend on its grade or kind. For pure or essentially pure hydrocarbons, the solid-ification temperature is the freezing point, the temperature at which a hydrocar-bon passes from a liquid to a solid state (ASTM D1015; ASTM D1016). For grease and residua, the temperature of interest is that at which fluidity occurs, commonly known as the dropping point. The dropping point of grease is the temperature at which the grease passes from a plastic solid to a liquid state and begins to flow under the conditions of the test (ASTM D566; ASTM D2265). For another type of plastic solid, including petrolatum and microcrys-talline wax, both melting point and congealing point are of interest.

3.4 Spectroscopic Properties

The identification and characterization of the organic chemicals of unknown structure and origin is an important part of environmental organic chemistry. Although it is often possible to establish the structure of a compound on the basis of physical and chemical data (elemental analysis, physical state and prop-erties such as melting point, boiling point, solubility, odor, and color), the prop-erty data often need to be supplemented with other information about the organic chemical (or the mixture of organic chemicals) compound, physical, and chemical properties that data often need to be supplemented with data from spectroscopic investigations (such as infrared (IR) spectroscopy and nuclear magnetic resonance (NMR) spectroscopy) which will provide confirmatory data for the present and types of functional groups.

Every organic molecule has unique properties and reactivity and it is diffi-cult to predict the properties and reactivity for each new or unknown molecule. One method to simplify the process of identification and prediction of reactivity is to identify the functional groups (common arrangements of atoms) within a

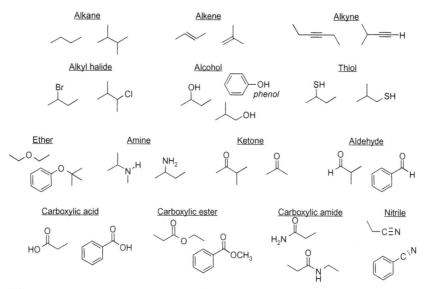

FIG. 5.1 Representation of the various organic functional groups.

molecule (Fig. 5.1) (Chapter 2). Typically, a specific functional group will exhibit similar properties and reactivity which will provide the environmental scientist or engineer to predicting other molecular attributes. In addition, in the alkane series of hydrocarbons (Chapter 2), each carbon in an alkane can be classified as primary (1°), secondary (2°), tertiary (3°), or quaternary (4°):

	Primary (1°)	Secondary (2°)	Tertiary (3°)	Quaternary (4°)
Alkanes				
Alkyl halides				
Alcohols				
Amines				

This offers another form of reactivity insofar as some of the carbon atoms that are shielded from external reactants will either (1) be slow to react (or) (2) will not react because of steric interference by the remainder of the molecule. This steric interference can also influence the reactivity of a functional group in a sterically shielded location.

In terms of functional group identification, spectroscopic studies have played an important role in the evaluation of organic compounds and of organic compounds products for the last three decades and many of the methods are now used as standard methods of analysis of organic compounds and its products before and after a spill. Application of these methods to organic compounds and its and products is a natural consequence for the environmental scientist and engineer.

3.4.1 IR Spectroscopy

An IR spectrum is a graph of the absorbed energy versus the wavenumber (v), which is the reciprocal of the wavelength (i.e., $1/\lambda$) and is measured in units of cm^{-1} and is proportional to the frequency or energy of the radiation—thus, the higher the wavenumber, the higher the energy. Most stretching vibrations occur in the region 3600–1000 cm^{-1} and bending vibrations are restricted to the region below 1600 cm^{-1}—the normal range for an IR spectrum is 4000–600 cm^{-1}. Thus, conventional IR spectroscopy yields information about the functional features of various organic compounds constituents. In order to generate an IR spectrum, different frequencies of IR light are passed through a sample, and the transmittance of light at each frequency is measured.

IR spectroscopy is useful in organic chemistry because it enables identification of different functional groups—each type of functional group contains certain bonds, and these bonds always show up in the same places in the IR spectrum (Table 5.6). Thus, different functional groups produce bond absorptions at different locations and intensities on the IR spectrum and recognition of the various absorption bands that are generated by the common functional groups occur that will enable interpretation of the IR spectrum and the type of functional group (or function groups) present in the molecule (Fessenden and Györgyi, 1991).

For example, IR spectroscopy will aid in the identification of amino-functions or imino-functions (N—H) and hydroxyl (O—H) functions, the nature of polymethylene chains, the types of carbon-hydrogen bonds, as well as the nature of any polynuclear aromatic systems (Speight, 2014). In addition, *IR spectroscopy* is used for the determination of benzene in motor and/or aviation gasoline (ASTM D4053) whilst ultraviolet spectroscopy is employed for the evaluation of mineral oils (ASTM D2269) and for determining the naphthalene content of aviation turbine fuels (ASTM D1840).

With the recent progress of *Fourier transform infrared (FTIR) spectroscopy*, quantitative estimates of the various functional groups can also be made. This aspect of the technique is particularly important for application to the higher molecular weight solid constituents of organic compounds (i.e., the asphaltene fraction) that form strong bonding arrangements (adsorption) with minerals such as clay.

TABLE 5.6 Absorption Bands of Various Functional Groups in the Infrared Spectrum

Functional Group	Absorption Location (cm^{-1})	Absorption Intensity
Alkane (C—H)	2850–2975	Medium to strong
Alcohol (O—H)	3400–3700	Strong, broad
Alkene (C=C) (C=C—H)	1640–1680 3020–3100	Weak to medium Medium
Alkyne (C≡C) (C≡C—H)	2100–2250 3300	Medium Strong
Nitrile (C≡N)	2200–2250	Medium
Aromatics	1650–2000	Weak
Amines (N—H)	3300–3350	Medium
Carbonyls (C=O)		Strong
Aldehyde (CHO)	1720–1740	
Ketone (RCOR)	1715	
Ester (RCOOR)	1735–1750	
Acid (RCOOH)	1700–1725	

3.4.2 Nuclear Magnetic Resonance

NMR spectroscopy is an analytical chemistry technique used in quality control for determining the content and purity of an organic compound as well as the molecular structure of the compound. The technique involves the detection of nuclei. In proton magnetic resonance (PMR), an external magnetic field is applied to force protons into two possible orientations which are not of equal energy. A spectrum can be obtained by measuring the energy absorbed or the energy emitted.

Thus, NMR spectroscopy can quantitatively analyze mixtures containing a variety of known organic chemicals. For unknown organic chemicals, NMR spectroscopy can either be used to match against spectral libraries or to infer the basic structure directly. Once the basic structure is known, NMR spectroscopy can be used to determine molecular conformation in solution as well as studying physical properties at the molecular level such as conformational exchange, phase change, solubility, and diffusion.

On an operational note, NMR spectroscopy is a technique that exploits the magnetic properties of certain atomic nuclei as a means of identification. The technique is a means of determining the physical and chemical properties of atoms or the molecules in which the atoms are contained and can provide detailed information about the structure, dynamics, reaction state, and chemical environment of molecules. The intramolecular magnetic field around an atom in a molecule changes the resonance frequency, thus giving access to details of the electronic structure of a molecule.

Most frequently, NMR spectroscopy is used to investigate the properties of organic molecules, although it is applicable to any kind of sample that contains nuclei possessing spin. Suitable samples range from small compounds analyzed with one-dimensional proton (^{1}H) or carbon-13 (^{13}C) NMR spectroscopy to large high molecular weight polymers and proteins. The impact of NMR spectroscopy on environmental sciences and technology has been substantial because of the range of information and the diversity of samples, including solutions and solids.

The NMR spectra are unique to each kind of molecule and are analytically tractable as well as being predictable for small molecules. Thus, in environmental organic chemistry (as in general organic chemistry practice), NMR spectroscopy analysis is used to confirm the identity of an organic chemical, even organic mixtures. Different functional groups are distinguishable from one another and identical functional groups with differing neighboring substituents give distinguishable signals. NMR spectroscopy has largely replaced traditional wet chemistry analytical methods.

The precise resonant frequency of the energy transition is dependent on the effective magnetic field at the nucleus. This field is affected by electron shielding which is in turn dependent on the chemical environment. As a result, information about the nucleus' chemical environment can be derived from its resonant frequency. In general, the more electronegative the nucleus is, the higher the resonant frequency. Other factors such as ring currents (anisotropy) and bond strain affect the frequency of the chemical shift. It is customary to employ tetramethylsilane as the proton reference frequency because the precise resonant frequency shift of each nucleus depends on the magnetic field that is applied to the molecule (Fig. 5.2).

NMR has frequently been employed for general studies and for the structural studies of organic compounds constituents. In fact, PMR studies (along with IR spectroscopic studies) were, perhaps, the first studies of the modern era that allowed structural inferences to be made about the polynuclear aromatic systems that occur in the high molecular weight constituents of organic compounds. Furthermore, *NMR spectroscopy* has been developed as a standard method for the determination of hydrogen types in aviation turbine fuels (ASTM D3701). *X-ray fluorescence spectrometry* has been applied to the determination of lead in gasoline as well as to the determination of sulfur in various organic compounds products (ASTM D2622; ASTM D4294).

Carbon-13 (^{13}C) magnetic resonance (CMR) can play a useful role. Since carbon magnetic resonance deals with analyzing the carbon distribution types, the obvious structural parameter to be determined is the aromaticity, f_a. A direct

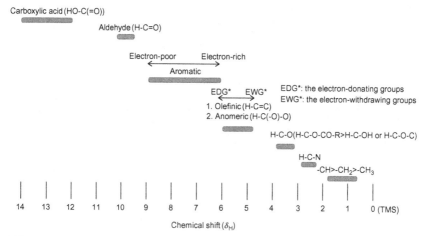

FIG. 5.2   Ranges of NMR chemical shifts for organic chemicals.

determination from the various carbon type environments is one of the better methods for the determination of aromaticity. Thus, through a combination of proton and carbon magnetic resonance techniques, refinements can be made on the structural parameters and, for the solid-state high-resolution carbon magnetic resonance technique, additional structural parameters can be obtained.

Of particular interest to the environmental science and environmental engineer, a variety of physical circumstances do not allow some organic molecules to be studied in solution, and at the same time these organic molecules are not amenable to study by other spectroscopic techniques to an atomic level, either. In solid-phase media, such as crystals, microcrystalline powders, gels, anisotropic solutions, or organic molecules adsorbed on minerals, it is in particular the dipolar coupling and chemical shift anisotropy that becomes dominant to the behavior of the nuclear spin systems (Fig. 5.2). In conventional solution-state NMR spectroscopy, these additional interactions would lead to a significant broadening of spectral lines. A variety of techniques allow establishing high-resolution conditions, that can, at least for ^{13}C spectra, be comparable to solution-state NMR spectra.

3.4.3   Mass Spectrometry

Mass spectrometry is an analytical technique that involves the ionization of chemical species and separates the ions based on their mass-to-charge ratio of each ion. The mass spectrum is obtained by ionizing a molecule to give a molecular ion which is then accelerated through a magnetic field and the ion is deviated according to its mass and charge. The result is a mass spectrum which is a graphical representation of the ion signal as a function of the mass-to-charge ratio. The mass spectra are used to determine (1) the elemental signature, (2) the isotopic signature of a sample, (3) the mass of each individual particle, (4) the molecular mass and of the constituent molecules if the sample is a mixture, and (5) elucidation of the chemical structure of the molecule or molecules, if the sample is a mixture.

In the analytical procedure, the sample (which may be gas, liquid, or solid) is introduced into the ionization chamber and bombarded with electrons. The resulting ions are then separated according to their mass-to-charge ratio, typically by accelerating the ions followed by application of an electric field or a magnetic field and, under these conditions, ions of the same mass-to-charge ratio will undergo the same amount of deflection while ions of different mass undergo a different amount of deflection. The ions are detected by a mechanism capable of detecting charged particles, such as an electron multiplier and the results are displayed as spectra of the relative abundance of detected ions as a function of the mass-to-charge ratio.

Mass spectrometry can play a key role in the identification of the constituents of feedstocks and products. The principal advantages of mass spectrometric methods are (1) high reproducibility of quantitative analyses; (2) the potential for obtaining detailed data on the individual components and/or carbon number homologues in complex mixtures; and (3) a minimal sample size is required for analysis. The ability of mass spectrometry to identify individual components in complex mixtures is unmatched by any modern analytical technique, perhaps the exception is gas chromatography.

The methods include the use of *mass spectrometry* to determine the (1) hydrocarbon types in middle distillates (ASTM D2425); (2) hydrocarbon types of gas oil saturate fractions (ASTM D2786); and (3) hydrocarbon types in low-olefin gasoline (ASTM D2789); (d) aromatic types of gas oil aromatic fractions (ASTM D3239). However, there are disadvantages arising from the use of mass spectrometry and these are: (1) the limitation of the method to organic materials that are volatile and stable at temperatures up to 300°C (570°F); and (2) the difficulty of separating isomers for absolute identification. The sample is usually destroyed, but this is seldom a disadvantage. Thermal decomposition increases the complexity of the mass spectrum but can be used to an advantage.

3.4.4 Other Techniques

The *crude oil assay* is a means of evaluating crude oil feedstocks (Speight, 2001, 2014, 2015). Each crude oil type has unique molecular and chemical characteristics—no two crude oils are identical which result in differences in crude oil quality. The results of crude oil assay testing provide extensive detailed analytical data for refiners and help determine if a crude oil feedstock is compatible for a particular petroleum refinery or if the crude oil could cause yield, quality, production, environmental, and other problems. The assay can be an inspection assay (involving the determination of 5 or 6 key properties) or comprehensive assay (involving the determination of 20 or more properties). Testing can include crude oil characterization of whole crude oils and the various boiling range fractions produced from physical or simulated distillation by various procedures. Information obtained from the petroleum assay is used for detailed refinery engineering and client marketing purposes. Feedstock assay data are an important tool in the refining process. When spills occur leading to environmental contamination, crude oil assay data are of extreme importance as the data are a *fingerprint* of the crude oil and can lead to determination of the source of the contamination.

Ultraviolet spectroscopy is one of the oldest forms of spectroscopy, but the use of this technique in identifying the structure of organic compounds has decreased with the introduction of more recent spectroscopic techniques. Nevertheless, ultraviolet spectroscopy can still be a worthwhile technique for use in organic chemistry, especially in the structural analysis of organic molecules which contain extended conjugated systems or the number of rings in a polynuclear aromatic system. This typically occurs when functional groups such as alkenes, ketones, aldehydes, carboxylic acids, esters, and aromatic rings are in conjugation with other unsaturated systems or the samples contain multiple condensed aromatic rings.

Other spectroscopic techniques also include the use of *flame emission spectroscopy* for determining trace metals in gas turbine fuels (ASTM D3605) and the use of *absorption spectrophotometry* for the determination of the alkyl nitrate content of diesel fuel (ASTM D4046). *Atomic absorption* has been employed as a means of measuring the lead content of gasoline (ASTM D3237) and also for the manganese content of gasoline (ASTM D3831) as well as for determining the barium, calcium, magnesium, and zinc contents of lubricating oils (ASTM D4628). *Flame photometry* has been employed as a means of measuring the lithium/sodium content of lubricating greases and the sodium content of residual fuel oil (ASTM D1318).

3.5 Surface and Interfacial Tension

Surface tension is a measure of the force acting at a boundary between two phases. If the boundary is between a liquid and a solid or between a liquid and a gas (air) the attractive forces are referred to as surface tension, but the attractive forces between two immiscible liquids are referred to as *interfacial tension*.

Thus, surface tension refers to the elastic tendency of a fluid surface which makes it acquire the least possible surface area. At liquid-air interfaces, surface tension results from the greater attraction of liquid molecules to each other (cohesion) than to the molecules in the air (adhesion). The overall effect is an inward force at the liquid surface that causes the liquid to behave as if the surface were covered with an elastic membrane. Because of the relatively high attraction of water molecules for each other, water has a higher surface tension (72.8 mN/m at 20°C, 68°F) compared to the surface tension of many other liquids. Surface tension is an important factor in the phenomenon of capillarity, which is an effects effect (capillary action) that occurs when liquids are discharged and come into contact with rocks or minerals that contain pore systems. Interfacial tension is somewhat similar to surface tension insofar as cohesive forces are also involved. However, the main forces involved in interfacial tension are adhesive forces (tension) between the liquid phase of one substance and either a gas, liquid, or solid phase of another substance. The interaction occurs at the surfaces of the substances involved, that is at the interface.

Temperature and molecular weight have a significant effect on surface tension (Table 5.7) (Speight, 2001, 2015). For example, in the normal hydrocarbon series, a rise in temperature leads to a decrease in the surface tension, but an increase in molecular weight increases the surface tension. A similar trend, that is, an increase in molecular weight causing an increase in surface tension, also occurs in the acrylic series and, to a lesser extent, in the alkylbenzene series.

TABLE 5.7 Surface Tension of Selected Hydrocarbons

		Surface Tension		
	C	*20*	*38*	*93*
Hydrocarbon	*F*	*68*	*100*	*200*
n-Pentane		16.0	14.0	8.0	dyn/cm
		16.0	14.0	8.0	mN/m
n-Hexane		18.4	16.5	10.9	dyn/cm
		18.4	16.5	10.9	mN/m
n-Heptane		20.3	18.6	13.1	dyn/cm
		20.3	18.6	13.1	mN/m
n-Octane		21.8	20.2	14.9	dyn/cm
		21.8	20.2	14.9	mN/m
Cyclopentane		22.4			dyn/cm
		22.4			mN/m
Cyclohexane		25.0			dyn/cm
		25.0			mN/m
Tetralin		35.2			dyn/cm
		35.2			mN/m
Decalin		29.9			dyn/cm
		29.9			mN/m

TABLE 5.7 Surface Tension of Selected Hydrocarbons—cont'd

		Surface Tension			
	C	20	38	93	
Hydrocarbon	F	68	100	200	
Benzene		28.8			dyn/cm
		28.8			mN/m
Toluene		28.5			dyn/cm
		28.5			mN/m
Ethylbenzene		29.0			dyn/cm
		29.0			mN/m
n-Butylbenzene		29.2			dyn/cm
		29.2			mN/m

The surface tension of organic compounds and organic compounds products has been studied for many years. The narrow range of values (24–38 dyn/cm) for such widely diverse materials as gasoline (26 dyn/cm), kerosene (30 dyn/cm), and the lubricating fractions (34 dyn/cm) has rendered the surface tension of little value for any attempted characterization. However, it is generally acknowledged that nonhydrocarbon materials dissolved in an oil reduce the surface tension: polar compounds, such as soaps and fatty acids, are particularly active. The effect is marked at low concentrations up to a critical value beyond which further additions cause little change; the critical value corresponds closely with that required for a monomolecular layer on the exposed surface, where it is adsorbed and accounts for the lowering. Recent work has focused on the predictability of surface tension using mathematical relationships:

$$\text{Dynamic surface tension} = 681.3/K\left(1 - T/13.488^{1.7654} \times \text{sg}^{2.1250}\right)^{1.2056}$$

in this equation, K is the Watson characterization factor, sg is the specific gravity, and T is the temperature in degrees Kelvin. Briefly, the Watson characterization factor is a method that has been used to define crude oil in terms of the relative amounts of paraffin constituent and aromatic constituents. A characterization factor on the order of 12.5 (or higher) indicates a crude oil that is predominantly paraffin in nature while a characterization factor on the order of 10 or lower indicates a crude oil that is predominantly aromatic in nature. The K factor is also referred to as the UOP K factor (Speight, 2014, 2015). The characterization factor may also be a part of the crude oil assay.

A high proportion of the complex phenomena shown by emulsions and foams, that are common when organic compounds enter the environment, can be traced to these induced surface tension effects. Dissolved gases, even hydrocarbon gases, lower the surface tension of oils, but the effects are less dramatic and the changes probably result from dilution. The matter is of some importance in environmental issues because the viscosity and surface tension of the organic compounds govern the amount of oil that migrates or can be recovered under certain conditions.

On the other hand, although organic compounds products show little variation in surface tension, within a narrow range the *interfacial tension* of organic compounds, especially of organic compounds products, against aqueous solutions provides valuable information (ASTM D971). Thus, the interfacial tension of organic compounds is subject to the same constraints as surface tension, that is, differences in composition, molecular weight, and so on. When oil-water systems are involved, the pH of the aqueous phase influences the tension at the interface; the change is small for highly refined oils, but increasing pH causes a rapid decrease for poorly refined, contaminated, or slightly oxidized oils.

A change in interfacial tension between oil and alkaline water has been proposed as an index for following the refining or deterioration of certain products, such as turbine and insulating oils. When surface or interfacial tensions are lowered by the presence of solutes, which tend to concentrate on the surface, time is required to obtain the final concentration and hence the final value of the tension. In such systems dynamic and static tension must be distinguished; the first concerns the freshly exposed surface having nearly the same composition as the body of the liquid; it usually has a value only slightly less than that of the pure solvent. The static tension is that existing after equilibrium concentration has been reached at the surface.

The interfacial tension between oil and distilled water provides an indication of compounds in the oil that have an affinity for water. The measurement of interfacial tension has received special attention because of its possible use in predicting when an oil in constant use will reach the limit of its serviceability. This interest is based on the fact that oxidation decreases the interfacial tension of the oil. Furthermore, the interfacial tension of turbine oil against water is lowered by the presence of oxidation products, impurities from the air or rust particles, and certain antirust compounds intentionally blended in the oil. Thus, a depletion of the antirust additive may cause an increase in interfacial tension, whereas the formation of oxidation products or contamination with dust and rust lowers the interfacial tension.

3.6 Viscosity

Viscosity is the force in dynes required to move a plane of 1 cm^2 area at a distance of 1 cm from another plane of 1 cm^2 area through a distance of 1 cm in 1 s. In the centimeter-gram-second (cgs) system the unit of viscosity is the poise (P) or centipoise (1 cP = 0.01 P). Two other terms in common use are *kinematic viscosity* and *fluidity*. The kinematic viscosity is the viscosity in centipoises divided by the specific gravity, and the unit is the stoke (cm^2/s), although centistokes (0.01 cSt) is in more common usage; fluidity is simply the reciprocal of viscosity.

The viscosity (ASTM D88; ASTM D341; ASTM D445; ASTM D2161; ASTM D2270) of organic compounds oils varies markedly over a very wide range (less than 10 cP at room temperature to many thousands of centipoises at the same temperature). For example, the viscosity of common organic liquids varies over a considerable range:

Liquid	Viscosity, cP[a]
Acetone	0.306
Benzene	0.604
Ethyl alcohol	1.074
Ethylene glycol	16.1
Glycerol (at 20°C)	1200
Methyl alcohol	0.544
Lubricating oil (SAE 10 at 20°C)	65
Lubricating oil (SAE 40 at 20°C)	319
Nitrobenzene	1.863
Propyl alcohol	1.945
Residuum	2.3×10^{11}
Water[b]	0.894

[a]*cP, centipoises.*
[b]*Included for comparison.*

Many types of instruments have been proposed for the determination of viscosity. The simplest and most widely used are capillary types (ASTM D445), and the viscosity is derived from the equation:

$$\mu = Br^4 P / 8nl$$

in this equation, r is the tube radius, l the tube length, P the pressure difference between the ends of a capillary, n the *coefficient of viscosity*, and the quantity discharged in unit time. Not only are such capillary instruments the simplest, but when designed in accordance with known principle and used with known necessary correction factors, they are probably the most accurate viscometers available. It is usually more convenient, however, to use relative measurements, and for this purpose the instrument is calibrated with an appropriate standard liquid of known viscosity. Batch flow times are generally used, that is: the time required for a fixed amount of sample to flow from a reservoir through a capillary is the datum actually observed. Some of the principal capillary viscometers in use are those of Cannon-Fenske, Ubbelohde, Fitzsimmons, and Zeitfuchs.

The Saybolt universal viscosity (Saybolt universal seconds, SUS) (ASTM D88) is the time in seconds required for the flow of 60 mL of organic compounds from a container, at constant temperature, through a calibrated orifice.

The Saybolt furol viscosity (Saybolt furol seconds, SFS) (ASTM D88) is determined in a similar manner except that a larger orifice is employed.

As a result of the various methods for viscosity determination, it is not surprising that much effort has been spent on interconversion of the several scales, especially converting Saybolt to kinematic viscosity (ASTM D2161):

$$\text{Kinematic viscosity} = a \times \text{Saybolt s} + b/\text{Saybolt s}$$

in this equation, a and b are constants.

The Saybolt universal viscosity equivalent to a given kinematic viscosity varies slightly with the temperature at which the determination is made because the temperature of the calibrated receiving flask used in the Saybolt method is not the same as that of the oil. Conversion factors are used to convert kinematic viscosity from 2 to 70 cSt at 38°C (100°F) and 99°C (210°F) to equivalent Saybolt universal viscosity in seconds. Appropriate multipliers are listed to convert kinematic viscosity over 70 cSt. For a kinematic viscosity determined at any other temperature the equivalent Saybolt universal value is calculated by use of the Saybolt equivalent at 38°C (100°F) and a multiplier that varies with the temperature:

$$\text{Saybolt s at } 100°\text{F} (38°\text{C}) = \text{cSt} \times 4.635$$

$$\text{Saybolt s at } 210°\text{F} (99°\text{C}) = \text{cSt} \times 4.667$$

Viscosity decreases as the temperature increases and the rate of change appears to depend primarily on the nature or composition of the organic compounds, but other factors, such as volatility, may also have an effect. The effect of temperature on viscosity is generally represented by the equation:

$$\log(n+c) = A + B \log T$$

in this equation, n is absolute viscosity, T is temperature, and A and B are constants. This equation has been sufficient for most purposes and has come into very general use. The constants A and B vary widely with different oils, but c remains fixed at 0.6 for all oils having a viscosity over 1.5 cSt; it increases only slightly at lower viscosity (0.75 at 0.5 cSt). The viscosity-temperature characteristics of any oil, so plotted, thus create a straight line, and the parameters A and B are equivalent to the intercept and slope of the line. To express the viscosity and viscosity-temperature characteristics of an oil, the slope and the viscosity at one temperature must be known; the usual practice is to select 38°C (100°F) and 99°C (210°F) as the observation temperatures.

Suitable conversion tables are available (ASTM D341), and each table or chart is constructed in such a way that for any given organic compounds or organic compounds product the viscosity-temperature points result in a straight line over the applicable temperature range. Thus, only two viscosity measurements need be made at temperatures far enough apart to determine a line on the appropriate chart from which the approximate viscosity at any other temperature can be read.

Since the viscosity-temperature coefficient of high-boiling fractions, such as lubricating oil, is an important expression of its suitability, a convenient number to express this property is very useful, and hence, a viscosity index (ASTM D2270) was derived. Thus:

$$\text{Viscosity index} = L - U/L - H \times 100$$

L and H are the viscosities of the zero and 100 index reference oils, both having the same viscosity at 99°C (210°F), and U is that of the unknown, all at 38°C (100°F). Originally the viscosity index was calculated from Saybolt viscosity data, but subsequently figures were provided for kinematic viscosity.

The viscosity of organic compounds fractions increases on the application of pressure, and this increase may be very large. The pressure coefficient of viscosity correlates with the temperature coefficient, even when oils of widely different types are compared. At higher pressures the viscosity decreases with increasing temperature, as at atmospheric pressure; in fact, viscosity changes of small magnitude are usually proportional to density changes, irrespective of whether these are caused by pressure or by temperature.

Because of the importance of viscosity in determining the transport properties of organic compounds, and this is particularly important in the migration of organic compounds and organic compounds products through soil, recent work has focused on the development of an empirical equation for predicting the dynamic viscosity of low molecular weight and high molecular weight hydrocarbon vapors at atmospheric pressure. The equation uses molar mass and specific temperature as the input parameters and offers a means of estimation of the viscosity of a wide range of organic compounds fractions. Other work has focused on the prediction of the viscosity of blends of lubricating oils as a means of accurately predicting the viscosity of the blend from the viscosity of the base oil components.

3.7   Volatility and Flammability

The *volatility* of an organic chemical (typically referring to a liquid or liquefied gas) may be defined as its tendency to vaporize, that is, to change from the liquid to the vapor or gaseous state. The volatility of an organic chemical is determined by the boiling point which is the *temperature at which the vapor pressure of a liquid is equal to the pressure of the atmosphere on the liquid*. Pure organic chemicals have a unique boiling point and this is often used to identify compounds in laboratory investigations. Mixtures of two or more compounds have a boiling point range.

The range of normal boiling points for organic chemicals is −162°C (−260°F) for methane to in excess of over 700°C (1290°F) for some polycyclic aromatic hydrocarbon derivatives, polychlorinated biphenyl derivatives and dioxin/furan derivatives. Many organic chemicals decompose at temperatures lower than the hypothetical boiling points, estimating this value for such thermally unstable compounds is an important aspect of environmental organic chemistry because of the relationship with other chemical properties. Thus, the boiling point of an organic chemical can (1) be used to define the uppermost temperature at which a chemical can exist as a liquid, (2) serve as an indicator of chemical volatility, with a higher boiling point associated with a lower volatility, and (3) be a key parameter in the temperature adjustment of other physical-chemical data including the air-water partition coefficient and vapor pressure.

The term *volatility* (as deduced from the boiling point) is primarily to be applied to liquids (such as petroleum-derived products) but may also be used to describe the process of sublimation which is associated with solid substances that can pass into the gas phase without involving any intermediate liquids phase. On the other hand, the *flammability* of an organic chemical is the ability of the chemical to burn or ignite, thereby causing fire or combustion.

The term *flammable* is used for organic chemicals which ignite more easily than other materials and thus are more dangerous and more highly regulated—inorganic chemicals are also flammable and the term *pyrophoric* is more often used in such cases. Less easily ignited organic chemicals or those chemicals which burn less vigorously are *combustible*. For example, in the United States flammable liquids by definition have a flash point below 38°C (100°F) while combustible liquids have a flash point above 38°C (100°F). Flammable solids are solids that are readily combustible, or may cause or contribute to fire through friction. Readily combustible chemicals are dangerous if they can be easily ignited by brief contact with an ignition source, such as a burning match, and if the flame spreads rapidly. Perhaps one of the best examples of volatility combined with flammability comes from the recent reports of train derailments where the train carried tank cars of volatile crude oil and the derailment led to explosion and fire (or fire and explosion).

The volatility of an organic chemical is directly related to the vapor pressure of the organic chemical. At a given temperature, an organic chemical with a higher vapor pressure will vaporize (volatilize) more readily than an organic chemical with a lower vapor pressure. The vapor pressure is also related to the intermolecular interactions and intramolecular interactions of the various functional groups, such as the occurrence of hydrogen binding between the various functional groups. For example, 2-hydroxypyridine, where the hydroxyl group on the carbon atom can form a hydrogen bond with the nitrogen atom in the same molecule (intramolecular hydrogen bonding), has a much lower boiling point than 4-hydroxypyridine in which the hydroxyl group can only form hydrogen bonds with nitrogen atoms in other molecules (intermolecular hydrogen bonding).

2-Hydroxypyridine, b.p. 280–281°C/760 mmHg

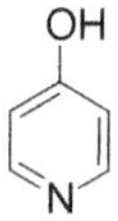

OH

4-Hydroxypyridine, b.p. 230–235°C/12 mmHg

Furthermore, 2-hydroxypyridine might be expected to form much stronger bonds with minerals because of the potential of the nitrogen-carbon-hydroxy system to form more thermodynamically stable organic chemical-mineral bonds.

However, for single organic compounds, the volatility will be determined by measurement of the boiling point while for mixtures of organic compounds (of which crude oil products are examples) the volatility is determined by the boiling range which is the range of boiling between the initial boiling point and the final boiling point. Refined crude oil products are often characterized by approximate boiling point range, which corresponds with the size (such as, number of carbon atoms) of the petroleum hydrocarbons in the refined oil (Speight, 2014, 2015).

The vaporizing tendencies of organic compounds and organic compounds products are the basis for the general characterization of liquid organic compounds fuels, such as liquefied organic compounds gas, natural gasoline, motor and aviation gasoline, naphtha, kerosene, gas oil, diesel fuel, and fuel oil (ASTM D2715). A test (ASTM D6) also exists for determining the loss of material when organic compounds and asphaltic compounds are heated. Another test (ASTM D20) is a method for the distillation of road tars that might also be applied to estimating the volatility of high molecular weight unknown residues.

For many environmental purposes it is necessary to have information on the initial stage of vaporization. To supply this need, flash and fire, vapor pressure, and evaporation methods are available. The data from the early stages of the several distillation methods are also useful. For other uses it is important to know the tendency of a product to partially vaporize or to completely vaporize, and in some cases to know if small quantities of high-boiling components are present. For such purposes, chief reliance is placed on the distillation methods.

The *flash point* of organic compounds or a mixture of organic compounds is the temperature to which the product must be heated under specified conditions to give of sufficient vapor to form a mixture with air that can be ignited momentarily by a specified flame (ASTM D56, ASTM D92, ASTM D93). The *fire point* is the temperature to which the product must be heated under the prescribed conditions of the method to burn continuously when the mixture of vapor and air is ignited by a specified flame (ASTM D92).

From the viewpoint of safety, information about the flash point is of most significance at or slightly above the maximum temperatures (30–60°C, 86–140°F) that may be encountered in storage, transportation, and use of liquid organic compounds products, in either closed or open containers. In this temperature range the relative fire and explosion hazard can be estimated from the flash point. For products with flash point below 40°C (104°F) special precautions are necessary for safe handling. Flash points above 60°C (140°F) gradually lose their safety significance until they become indirect measures of some other quality. The flash point of an organic compound (or a mixture of organic compounds) is also used to detect contamination. A substantially lower flash point than expected for a product is a reliable indicator that a product has become contaminated with a more volatile product, such as gasoline. The flash point is also an aid in establishing the identity of a particular organic compounds product.

A further aspect of volatility that receives considerable attention is the vapor pressure of organic compounds and its constituent fractions. The *vapor pressure* is the force exerted on the walls of a closed container by the vaporized portion of a liquid. Conversely, it is the force that must be exerted on the liquid to prevent it from vaporizing further (ASTM D323). The vapor pressure increases with temperature for any given gasoline, liquefied organic compounds gas, or other product. The temperature at which the vapor pressure of a liquid, either a pure compound or a mixture of many compounds, equals 1 atm pressure (14.7 psi, absolute) is designated as the boiling point of the liquid.

In each homologous series of hydrocarbons, the boiling points increase with molecular weight; structure also has a marked influence: it is a general rule that branched paraffin isomers have lower boiling points than the corresponding *n*-alkane. Furthermore, in any specific aromatic series, steric effects notwithstanding, there is an increase in boiling point with an increase in carbon number of the alkyl side chain. This particularly applies to alkyl aromatic compounds where alkyl-substituted aromatic compounds can have higher boiling points than polycondensed aromatic systems. And this fact is very meaningful when attempts are made to develop hypothetical structures for asphaltene constituents.

One of the main properties of organic compounds that serve to indicate the comparative ease with which the material evaporates after a spill is the volatility. Investigation of the volatility of organic compounds is usually carried out under standard conditions in which organic compounds or the product is subdivided by distillation into a variety of fractions of different *fractions* (Table 5.8) (Speight, 2001, 2015). In fact, distillation involves the general procedure of vaporizing the organic compounds liquid in a suitable flask either at *atmospheric pressure* (ASTM D86; ASTM D216; ASTM D2892) or at *reduced pressure* (ASTM D1160).

Simulated distillation (*simdis*) by gas chromatography is often applied to organic compounds to obtain TBP data for an organic compound or a mixture

TABLE 5.8 Typical Boiling Fractions From Crude Oil (Speight, 2014, 2015)

Product	Lower Carbon Limit	Upper Carbon Limit	Lower Boiling Point (°C)	Upper Boiling Point (°C)	Lower Boiling Point (°F)	Upper Boiling Point (°F)
Refinery gas	C1	C4	−161	−1	−259	31
Liquefied petroleum gas	C3	C4	−42	−1	−44	31
Naphtha	C5	C17	36	302	97	575
Gasoline	C4	C12	−1	216	31	421
Kerosene/diesel fuel	C8	C18	126	258	302	575
Aviation turbine fuel	C8	C16	126	287	302	548
Fuel oil	C12	>C20	216	421	>343	>649
Lubricating oil	>C20		>343		>649	
Wax	C17	>C20	302	>343	575	>649
Asphalt	>C20		>343		>649	
Coke	>C50[a]		>1000[a]		>1832[a]	

[a]Carbon number and boiling point is difficult to assess; inserted for illustrative purposes only.

of organic compounds and may also be useful in predicting the amount of material that can be (or needs to be) recovered after a spill. Two standardized methods (ASTM D2887; ASTM D3710) are available for the boiling point determination of organic compounds fractions and gasoline, respectively. The ASTM D2887 method utilizes nonpolar, packed gas chromatographic columns in conjunction with flame ionization detection. The upper limit of the boiling range covered by this method is ~540°C (1000°F) atmospheric equivalent boiling point. Recent efforts in which high temperature gas chromatography was used have focused on extending the scope of the ASTM D2887 method for higher boiling organic compounds materials to 800°C (1470°F) atmospheric equivalent boiling point.

Also related to volatility, the enthalpy or heat of vaporization (ΔHv) is one of the fundamental thermodynamic properties of a substance and is the quantity of energy required to convert a unit mass of a liquid into a vapor without a rise in temperature. The heat of vaporization is temperature-dependant, decreasing in value as temperature increases.

Finally, in terms of volatility and flammability, the US Government uses the Hazardous Materials Identification System standard for flammability ratings:

Rating	Degree of Flammability	Examples
0	Materials that will not burn	Water
1	Materials that must be preheated before they will ignite	Lubricating oil
2	Materials that must be moderately heated or exposed to relatively high ambient temperatures before they will ignite	Kerosene, diesel fuel
3	Liquids and solids that can ignite under almost all temperature conditions	Naphtha, gasoline, acetone
4	Materials which will rapidly vaporize at atmospheric pressure and normal temperatures, or are readily dispersed in air and which burn readily	Natural gas (methane), ethane propane, butane

4 USE OF THE DATA

The accurate evaluation of the properties and behavior of organic chemicals could be the difference between success and failure when dealing with mitigating the accidental spills or discharges of organic chemicals into the environment. An effective analysis of such properties can have a major impact on the strategic plan for any business—threats to the environment must not only be identified but also plans must be made for application of methods of clean up. Thus, environmental analysis (use of the data) is a process in which the various factors can have an impact on the environment. Some of these items that could have an impact on the business are political regulations as well as social and technical factors. Each aspect must be examined individually and determination made as to how it could affect the success of any environmental protection process.

Thus, in any situation involving organic chemicals (or any chemicals for that matter) it is essential to have in place a chemical database that relates to the properties of the chemicals, and, in the current context, the effect of the organic chemicals on the environment and how any such spills or discharges might be mitigated and/or cleaned up. Briefly, the term *database* refers to a set of related data and the way it is organized, which might be in the form of tables, reports, and other relevant information (e.g., Speight, 2012). The database must be organized to provide efficient retrieval of information—the collected information could be in any number of formats (electronic, printed, graphic, audio, statistical, combinations). Preferably, the information should be maintained on a designated server (that is protected and nonhackable) and should updated on a regular basis to incorporate recent transaction data from the operational systems of the business (or organization). No business or organization that manufactures, handles, or trades organic chemicals (in fact, any chemicals) should be immune from keeping such detailed and relevant records.

Formal access to these data is usually provided by a *database management system* (DBMS) consisting of an integrated set of computer software that allows users of the database to interact with one or more databases and provides access to all of the data contained in the database (although restrictions may exist that limit access to particular data). The DBMS provides various functions that allow entry, storage, and retrieval of large quantities of information and provides ways to manage how that information is organized. For security purposes (and in order to stand up to, and survive, legal scrutiny), the database should be (must be) a database that stores historical data on any organic chemicals that the business manufactures, uses, or might use as well as the behavior of the chemicals, inventory levels, and storage parameters. In addition, management access to insert new data, for example, should be limited to a number of designated named individuals who can enter or change data through a series of entry codes and there should also be a records log of any entries or changes made by these designated individuals. On the other hand, however, any person within the company or organization should be allowed to run queries and specific keyword reports through the database. The structure of the database should be such that the data can easily be retrieved (*data mining*) and, above all, user friendly.

Despite the nature of the environmental regulations and the precautions taken by the organic chemicals industry, the accidental release of organic chemicals (nonhazardous chemicals and hazardous chemicals) into the environment has occurred and, without being unduly pessimistic, will continue to occur (by all industries—not wishing to select any industry in particular as the industry that suffers accidental release of chemicals into the environment). It is a situation that, to paraphrase *chaos theory*, no matter how well one prepares, the unexpected is always inevitable. It is, at this point that the environmental scientist and/or the environmental engineer has to identity the nature of the chemicals and make predictions through an understanding of the properties of the chemical (or chemicals) of the potential effects on the environment.

Briefly, for environmental purposes, chemicals are subdivided into two classes: (1) *organic chemicals* and (2) *inorganic chemicals*. Furthermore, classification occurs insofar as organic chemicals are classified as *volatile organic compounds* or *semivolatile organic compounds* and nonvolatile compounds (on occasions, the word *chemicals* is substituted for the word *compounds* without affecting the definition).

The term *volatile organic compounds* (VOCs) is used to describe organic material in the vapor phase excluding methane. Volatile organic compounds are important in atmospheric chemistry for the formation of photochemical smog. There are many noncombustion sources that lead to the emission of volatile organic compounds of which the most important is the use of solvents, including those released from paints. Evaporative losses of gasoline during storage and distribution are also significant. Thus, the class known as *volatile organic compounds* (VOCs) is subdivided into *regulated compounds* and *unregulated compounds*. Regulated compounds have maximum contaminant levels, but unregulated compounds do not. Regulated compounds generally (but not always) have low-boiling points or low-boiling ranges, and some are gases. Many of these chemicals can be detected at extremely low levels by a variety of instrumentation, including the human nose! In the case of the volatile organic compounds, sources are typically crude oil-derived solvents and products (Speight, 2014, 2016) as well as industries that handle and produce volatile liquids (such as dry cleaning solvents, paint thinners, cleaning solvents).

Briefly, a volatile organic chemical is a chemical with a boiling point, or sublimation temperature, such that the chemical can exist to a significant extent in the gaseous phase under ambient conditions. Some commonly encountered volatile hydrocarbons are:

Aliphatic	Aromatic
Pentane derivatives	Benzene and benzene derivatives
Hexane derivatives	Toluene and toluene derivatives
Heptane derivatives	Ethylbenzene and ethylbenzene derivatives
Octane derivatives	Xylenes and xylene derivatives
Nonane derivatives	Naphthalene and naphthalene derivatives
Decane derivatives	Phenanthrene and phenanthrene derivatives
	Anthracene and anthracene derivatives
	Acenaphthylene and acenaphthylene derivatives

Release of gas condensate might be equated to the release of volatile constituents but are often named as such because of the specific constituents of the condensate, often with some reference to the gas condensate that is produced by certain petroleum and natural gas well. However, the condensate is often restricted to the benzene, toluene, ethyl benzene, and xylenes (BTEX) family of compounds of which the properties are:

	Benzene	Toluene	Ethylbenzene	p-Xylene	m-Xylene	o-Xylene
Molecular formula	C_6H_6	C_7H_8	C_8H_{10}	C_8H_{10}	C_8H_{10}	C_8H_{10}
Molecular mass	78.12	92.15	106.17	106.17	106.17	106.17
Boiling point (°C)	80.1	110.6	136.2	138.4	139.1	144.4
Melting point (°C)	5.5	−95.0	−95.0	13.3	−47.9	−25.2

In almost all cases of hydrocarbon contamination, some attention will have to be paid to the presence of semivolatile compounds and nonvolatile compounds.

The second class of organic compounds, the *semivolatile compounds*, typically have high-boiling points, or high-boiling ranges in the case of mixtures such as solvents, and are not always easily detected by the instrumentation that may be used to detect the volatile organic compounds (including the human nose). Some of the common sources of contamination are lubricating oils, pesticides, herbicides, fungicides, wood preservatives, and a variety of other chemicals that can be linked to the chemicals. Regulations (Chapter 6) are in place that set the maximum contamination concentration levels of organic chemicals that are designed to ensure public safety. There are primary and secondary standards for inorganic chemicals. Primary standards are for those chemicals that cause neurological damage, cancer, or blood disorders. Secondary standards are developed for other environmental reasons.

Another source of toxic compounds is combustion of organic materials such as coal (Speight, 2013). In fact, some of the greater dangers of fires are from toxic products and by-products of combustion. Generally, the most obvious of these is carbon monoxide (CO) which is not the subject of this text but needs to be mentioned because of the adverse effects of this gas such as serious illness or death. The gas forms carboxyhemoglobin with hemoglobin in the blood so that the blood is no longer able to carry oxygen efficiently to the body tissues. Toxic sulfur dioxide and hydrogen chloride are formed by the combustion of sulfur compounds and organic chlorine compounds, respectively. More pertinent to this text are the noxious organic compounds such as aldehydes (RCH=O) that are generated as by-products of combustion as are the PNAs (consisting of fused-ring structures) that are also produced during combustion. Some of these compounds, such as benzo(a)pyrene are precarcinogenic compounds, insofar as they are acted upon by enzymes in the body to yield cancer-producing metabolites.

Most investigations involving organic chemicals are regulated by various agencies that may require methodologies and action levels as well as cleanup. Indeed, the complex chemical composition of many organic chemical products can make it extremely difficult to select the most appropriate analytical test methods for evaluating environmental samples and to accurately interpret and use the data (Speight and Lee, 2000; Speight, 2005; Speight and Arjoon, 2012). Accordingly, general methods of environmental analysis (Smith, 1999), i.e., analysis are available for the identification of organic chemicals that have been released. The data determine whether or not a release of such chemicals will be detrimental to the environment and may lead to regulations governing the use and handling of such chemicals.

In summary, many of the specific organic chemicals handled in the various industries are hazardous because of their chemical reactivity, fire hazard, toxicity, and other properties. In fact, a simple definition of a hazardous chemical (or hazardous waste) is that it is an organic chemical (or organic chemical waste) that has been inadvertently released, discarded, abandoned, neglected, or designated as a waste material and has the potential to be detrimental to the environment. Alternatively, a hazardous organic chemical may be a chemical that may interact with other (chemical) substances to give a product that is hazardous to the environment. Whatever the case, the properties of the chemical (or chemical mixture) should be known in order for the scientist or engineer to understand the nature of the released organic chemical (or organic chemical waste) and from the data predict the potential hazard to the environment.

REFERENCES

Alberty, R.A., Reif, A.K., 1988. Standard chemical thermodynamic properties of polycyclic aromatic hydrocarbons and their isomer groups. 1. Benzene series. J. Phys. Chem. Ref. Data 17 (1), 214–253.

ASTM, 2016. Annual Book of Standards. ASTM International, West Conshohocken, PA.

Atkins, R.C., Carey, F.A., 2002. Organic Chemistry: A Brief Course, third ed. McGraw-Hill Education, New York, NY.

ASTM D6, 2016. Standard test method for loss on heating of oil and asphaltic compounds. Annual Book of Standards. ASTM International, West Conshohocken, PA.

ASTM D20, 2016. Standard test method for distillation of road tars. Annual Book of Standards. ASTM International, West Conshohocken, PA.

ASTM D56, 2016. Standard test method for flash point by tag closed cup tester. Annual Book of Standards. ASTM International, West Conshohocken, PA.

ASTM D86, 2016. Standard test method for distillation of petroleum products at atmospheric pressure. Annual Book of Standards. ASTM International, West Conshohocken, PA.

ASTM D88, 2016. Standard test method for Saybolt viscosity. Annual Book of Standards. ASTM International, West Conshohocken, PA.

ASTM D92, 2016. Standard test method for flash and fire points by Cleveland open cup tester. Annual Book of Standards. ASTM International, West Conshohocken, PA.

ASTM D93, 2016. Standard test methods for flash point by Pensky-Martens closed cup tester. Annual Book of Standards. ASTM International, West Conshohocken, PA.

ASTM D97, 2016. Standard test method for pour point of petroleum products. Annual Book of Standards. ASTM International, West Conshohocken, PA.

ASTM D127, 2016. Standard test method for drop melting point of petroleum wax, including petrolatum. Annual Book of Standards. ASTM International, West Conshohocken, PA.

ASTM D216, 2016. Standard practice for conversion of kinematic viscosity to Saybolt Universal Viscosity or to Saybolt Furol viscosity. Annual Book of Standards. ASTM International, West Conshohocken, PA.

ASTM D287, 2016. Standard test method for API gravity of crude petroleum and petroleum products (hydrometer method). Annual Book of Standards. ASTM International, West Conshohocken, PA.

ASTM D323, 2016. Standard test method for vapor pressure of petroleum products (Reid method). Annual Book of Standards. ASTM International, West Conshohocken, PA.

ASTM D341, 2016. Viscosity-temperature charts for liquid petroleum products. Annual Book of Standards. ASTM International, West Conshohocken, PA.

ASTM D445, 2016. Standard test method for kinematic viscosity of transparent and opaque liquids (and calculation of dynamic viscosity). Annual Book of Standards. ASTM International, West Conshohocken, PA.

ASTM D566, 2016. Standard test method for dropping point of lubricating grease. Annual Book of Standards. ASTM International, West Conshohocken, PA.

ASTM D971, 2016. Standard test method for interfacial tension of oil against water by the ring method. Annual Book of Standards. ASTM International, West Conshohocken, PA.

ASTM D1015, 2016. Standard test method for freezing points of high-purity hydrocarbons. Annual Book of Standards. ASTM International, West Conshohocken, PA.

ASTM D1016, 2016. Standard test method for purity of hydrocarbons from freezing points. Annual Book of Standards. ASTM International, West Conshohocken, PA.

ASTM D1018, 2016. Standard test method for hydrogen in petroleum fractions. Annual Book of Standards. ASTM International, West Conshohocken, PA.

ASTM D1160, 2016. Standard test method for distillation of petroleum products at reduced pressure. Annual Book of Standards. ASTM International, West Conshohocken, PA.

ASTM D1217, 2016. Standard test method for density and relative density (specific gravity) of liquids by Bingham Pycnometer. Annual Book of Standards. ASTM International, West Conshohocken, PA.

ASTM D1266, 2016. Standard test method for sulfur in petroleum products (lamp method). Annual Book of Standards. ASTM International, West Conshohocken, PA.

ASTM D1298, 2016. Standard test method for density, relative density, or API gravity of crude petroleum and liquid petroleum products by hydrometer method. Annual Book of Standards. ASTM International, West Conshohocken, PA.

ASTM D1318, 2016. Standard test method for sodium in residual fuel oil (flame photometric method). Annual Book of Standards. ASTM International, West Conshohocken, PA.

ASTM D1319, 2016. Standard test method for hydrocarbon types in liquid petroleum products by fluorescent indicator adsorption. Annual Book of Standards. ASTM International, West Conshohocken, PA.

ASTM D1548, 2016. Standard test methods for determination of nickel, vanadium, and iron in crude oils and residual fuels by inductively coupled plasma (ICP) atomic emission spectrometry. Annual Book of Standards. ASTM International, West Conshohocken, PA.

ASTM D1552, 2016. Standard test method for sulfur in petroleum products by high temperature combustion and IR detection. Annual Book of Standards. ASTM International, West Conshohocken, PA.

ASTM D1555, 2016. Standard test method for calculation of volume and weight of industrial aromatic hydrocarbons and cyclohexane [metric]. Annual Book of Standards. ASTM International, West Conshohocken, PA.

ASTM D1757, 2016. Standard test method for sulfur in ash from coal and coke. Annual Book of Standards. ASTM International, West Conshohocken, PA.

ASTM D1840, 2016. Standard test method for naphthalene hydrocarbons in aviation turbine fuels by ultraviolet spectrophotometry. Annual Book of Standards. ASTM International, West Conshohocken, PA.

ASTM D2007, 2016. Standard test method for characteristic groups in rubber extender and processing oils and other petroleum-derived oils by the clay-gel absorption chromatographic method. Annual Book of Standards. ASTM International, West Conshohocken, PA.

ASTM D2161, 2016. Standard practice for conversion of kinematic viscosity to Saybolt Universal viscosity or to Saybolt Furol viscosity. Annual Book of Standards. ASTM International, West Conshohocken, PA.

ASTM D2265, 2016. Standard test method for dropping point of lubricating grease over wide temperature range. Annual Book of Standards. ASTM International, West Conshohocken, PA.

ASTM D2269, 2016. Standard test method for evaluation of white mineral oils by ultraviolet absorption. Annual Book of Standards. ASTM International, West Conshohocken, PA.

ASTM D2270, 2016. Standard practice for calculating viscosity index from kinematic viscosity at 40°C and 100°C. Annual Book of Standards. ASTM International, West Conshohocken, PA.

ASTM D2425, 2016. Standard test method for hydrocarbon types in middle distillates by mass spectrometry. Annual Book of Standards. ASTM International, West Conshohocken, PA.

ASTM D2500, 2016. Standard test method for cloud point of petroleum products. Annual Book of Standards. ASTM International, West Conshohocken, PA.

ASTM D2549, 2016. Standard test method for separation of representative aromatics and nonaromatics fractions of high-boiling oils by elution chromatography. Annual Book of Standards. ASTM International, West Conshohocken, PA.

ASTM D2622, 2016. Standard test method for sulfur in petroleum products by wavelength dispersive X-ray fluorescence spectrometry. Annual Book of Standards. ASTM International, West Conshohocken, PA.

ASTM D2715, 2016. Standard test method for volatilization rates of lubricants in vacuum. Annual Book of Standards. ASTM International, West Conshohocken, PA.

ASTM D2786, 2016. Standard test method for hydrocarbon types analysis of gas-oil saturates fractions by high ionizing voltage mass spectrometry. Annual Book of Standards. ASTM International, West Conshohocken, PA.

ASTM D2789, 2016. Standard test method for hydrocarbon types in low olefinic gasoline by mass spectrometry. Annual Book of Standards. ASTM International, West Conshohocken, PA.

ASTM D2887, 2016. Standard test method for boiling range distribution of petroleum fractions by gas chromatography. Annual Book of Standards. ASTM International, West Conshohocken, PA.

ASTM D2892, 2016. Standard test method for distillation of crude petroleum (15-theoretical plate column). Annual Book of Standards. ASTM International, West Conshohocken, PA.

ASTM D3228, 2016. Standard test method for total nitrogen in lubricating oils and fuel oils by modified Kjeldahl method. Annual Book of Standards. ASTM International, West Conshohocken, PA.

ASTM D3237, 2016. Standard test method for lead in gasoline by atomic absorption spectroscopy. Annual Book of Standards. ASTM International, West Conshohocken, PA.

ASTM D3239, 2016. Standard test method for aromatic types analysis of gas-oil aromatic fractions by high ionizing voltage mass spectrometry. Annual Book of Standards. ASTM International, West Conshohocken, PA.

ASTM D3341, 2016. Standard test method for lead in gasoline—iodine monochloride method. Annual Book of Standards. ASTM International, West Conshohocken, PA.

ASTM D3343, 2016. Standard test method for estimation of hydrogen content of aviation fuels. Annual Book of Standards. ASTM International, West Conshohocken, PA.

ASTM D3605, 2016. Standard test method for trace metals in gas turbine fuels by atomic absorption and flame emission spectroscopy. Annual Book of Standards. ASTM International, West Conshohocken, PA.

ASTM D3701, 2016. Standard test method for hydrogen content of aviation turbine fuels by low resolution nuclear magnetic resonance spectrometry. Annual Book of Standards. ASTM International, West Conshohocken, PA.

ASTM D3710, 2016. Standard test method for boiling range distribution of gasoline and gasoline fractions by gas chromatography. Annual Book of Standards. ASTM International, West Conshohocken, PA.

ASTM D4045, 2016. Standard test method for sulfur in petroleum products by hydrogenolysis and rateometric colorimetry. Annual Book of Standards. ASTM International, West Conshohocken, PA.

ASTM D4046, 2016. Standard test method for alkyl nitrate in diesel fuels by spectrophotometry. Annual Book of Standards. ASTM International, West Conshohocken, PA.

ASTM D4053, 2016. Standard test method for benzene in motor and aviation gasoline by infrared spectroscopy. Annual Book of Standards. ASTM International, West Conshohocken, PA.

ASTM D4294, 2016. Standard test method for sulfur in petroleum and petroleum products by energy dispersive X-ray fluorescence spectrometry. Annual Book of Standards. ASTM International, West Conshohocken, PA.

ASTM D4628, 2016. Standard test method for analysis of barium, calcium, magnesium, and zinc in unused lubricating oils by atomic absorption spectrometry. Annual Book of Standards. ASTM International, West Conshohocken, PA.

ASTM D6299, 2016. Standard practice for applying statistical quality assurance and control charting techniques to evaluate analytical measurement system performance. Annual Book of Standards. ASTM International, West Conshohocken, PA.

ASTM D6792, 2016. Standard practice for quality system in petroleum products and lubricants testing laboratories. Annual Book of Standards. ASTM International, West Conshohocken, PA.

ASTM E191, 2016. Standard specification for apparatus for microdetermination of carbon and hydrogen in organic and organo-metallic compounds. Annual Book of Standards. ASTM International, West Conshohocken, PA.

ASTM E258, 2016. Standard test method for total nitrogen in organic materials by modified Kjeldahl method. Annual Book of Standards. ASTM International, West Conshohocken, PA.

ASTM E385, 2016. Standard test method for oxygen content using a 14-MeV neutron activation and direct-counting technique. Annual Book of Standards. ASTM International, West Conshohocken, PA.

ASTM E777, 2016. Standard test method for carbon and hydrogen in the analysis sample of refuse-derived fuel. Annual Book of Standards. ASTM International, West Conshohocken, PA.

ASTM E778, 2016. Standard test methods for nitrogen in refuse-derived fuel analysis samples. Annual Book of Standards. ASTM International, West Conshohocken, PA.

Bailey Jr., P.S., Bailey, C.A., 2000. Organic Chemistry: A Brief Survey of Concepts and Applications, sixth ed. Prentice Hall, Upper Saddle River, NJ.

Bettinetti, R., Cuccato, D., Galassi, S., Provini, A., 2002. Toxicity of 4-nonylphenol in spiked sediment to three populations of *Chironomus riparius*. Chemosphere 46, 201–207.

Brown, W.H., Foote, C.S., 2002. Organic Chemistry, third ed. Saunders Publishing, Elsevier, Amsterdam.

Budde, W.L., 2001. The Manual of Manuals. Office of Research and Development, Environmental Protection Agency, Washington, DC.

Chadeesingh, R., 2011. The Fischer-Tropsch process. In: Speight, J.G. (Ed.), The Biofuels Handbook. The Royal Society of Chemistry, London, pp. 476–517 (Part 3, Chapter 5).

Chao, J., Gadalla, N.A.M., Gammon, B.E., March, K.N., Rodgers, A.S., Somayajulu, G.R., Wilholt, R.C., 1990. Thermodynamic and thermophysical properties of organic nitrogen compounds. Part 1. Methanamine, ethanamine, 1- and 2-propananmine, benzenamine, 2,3,4-methylbenzenamine. J. Phys. Chem. Ref. Data 19 (6), 1547–1615.

Dean, J.R., 1998. Extraction Methods for Environmental Analysis. John Wiley & Sons, Inc., New York, NY.

Domalski, E.S., Hearing, E.D., 1993. Estimation of the thermodynamic properties of C-H-N-O-S-Halogen compounds at 298.15 K. J. Phys. Chem. Ref. Data 22 (4), 805–1159.

Fessenden, R.J., Györgyi, L., 1991. Identifying functional groups in IR spectra using an artificial neural network. J. Chem. Soc. Perkin Trans. 2 (11), 1755–1762.

Fytianos, K., Pegiadou, S., Raikos, N., Eleftheriadis, I., Tsoukali, H., 1997. Determination of nonionic surfactants (polyethoxylated-nonylphenols) by HPLC in waste waters. Chemosphere 35, 1423–1429.

Kannan, K., Keith, T.L., Naylor, C.G., Staples, C.A., Snyder, S.A., Giesy, J.P., 2003. Nonylphenol and nonylphenol ethoxylates in fish, sediment, and water from the Kalamazoo River, Michigan. Arch. Environ. Contam. Toxicol. 44, 77–82.

Michałowicz, J., Duda, W., 2007. Phenols—sources and toxicity. Pol. J. Environ. Stud. 16 (3), 347–362.

Miller, M. (Ed.), 2000. Encyclopedia of Analytical Chemistry. John Wiley & Sons Inc., Hoboken, NJ.

Smith, R.K., 1999. Handbook of Environmental Analysis, fourth ed. Genium Publishing, Schenectady, NY.

Soares, A., Guieysse, B., Jefferson, B., Cartmell, E., Lester, J., 2008. Nonylphenol in the environment: a critical review on occurrence, fate, toxicity and treatment in wastewaters. Environ. Int. 34, 1033.

Speight, J.G., Lee, S., 2000. Environmental Technology Handbook, second ed. Taylor & Francis, New York, NY.

Speight, J.G., 2001. Handbook of Petroleum Analysis. John Wiley & Sons Inc., Hoboken, NJ.

Speight, J.G., 2005. Environmental Analysis and Technology for the Refining Industry. John Wiley & Sons Inc., Hoboken, NJ.

Speight, J.G., 2012. Crude Oil Assay Database. Knovel, Elsevier, New York, NY. Available at: http://www.knovel.com/web/portal/browse/display?_EXT_KNOVEL_DISPLAY_bookid=5485&VerticalID=0.

Speight, J.G., Arjoon, K.K., 2012. Bioremediation of Petroleum and Petroleum Products. Scrivener Publishing, Salem, MA.

Speight, J.G., 2013. The Chemistry and Technology of Coal, third ed. CRC Press, Taylor & Francis Group, Boca Raton, FL.

Speight, J.G., 2014. The Chemistry and Technology of Petroleum, fifth ed. CRC Press, Taylor & Francis Group, Boca Raton, FL.

Speight, J.G., 2015. Handbook of Petroleum Product Analysis, second ed. John Wiley & Sons Inc., Hoboken, NJ.

Speight, J.G., 2016. Handbook of Petroleum Refining. CRC Press, Taylor & Francis Group, Boca Raton, FL.

Staples, C.A., Weeks, J., Hall, J.F., Naylor, C.G., 1998. Evaluation of aquatic toxicity and bioaccumulation of C8- and C9-alkylphenol ethoxylates. Environ. Toxicol. Chem. 17, 2470–2480.

Sun, H.W., Hu, H.W., Wang, L., Yang, Y., Huang, G.L., 2014. The bioconcentration and degradation of nonylphenol and nonylphenol polyethoxylates by *Chlorella vulgaris*. Int. J. Mol. Sci. 15, 1255–1270.

Vazquez-Duhalt, R., Marquez-Rocha, F., Ponce, E., Licea, A.F., Viana, M.T., 2005. Nonylphenol: an integrated vision of a pollutant. Appl. Ecol. Environ. Res. 4, 1–25.

Ying, G.G., Williams, B., Kookana, R., 2002. Environmental fate of alkylphenols and alkylphenol ethoxylates—a review. Environ. Int. 28, 215–226.

Chapter 6

Introduction Into the Environment

1 INTRODUCTION

When the use of organic chemical results in discharge into, and contamination of the environment, it becomes necessary to set standards for acceptable concentrations in water, air, soil, and biota (Chapter 8). Monitoring of these concentrations in the environment, and resultant biological effects, must be undertaken to ensure that the standards as set in any regulation are realistic and provide protection of the environment from all adverse effects. Furthermore, considerable attention continues to be focused on the regulation of the use of all organic chemicals and a primary aspect involves the prediction of the behavior and effects of a chemical, from its properties. With this concept the characteristics of the molecule govern the physicochemical properties of the compound which in turn influences transformation and distribution in the environment and the biological effects. This suggests that the transformation and distribution in the environment as well as any effects on the floral and faunal species can be predicted from the physicochemical properties of the chemical. However, the prediction of biological effects is the most complex of the set of predictions.

Industrial organic chemical manufacturers use and generate both large numbers and quantities of chemicals. In the past, the organic chemicals industry had introduced organic chemicals to all types of environmental ecosystems including air (through both fugitive emissions and direct emissions), water (direct discharge and runoff), and land (Table 6.1). The types of pollutants a single facility will release depend on (1) the type of process, (2) the process feedstocks, (3) the equipment in use, such as the reactor, and (4) the equipment and process maintenance practices, which can vary over short periods of time (such as from hour to hour) and can also vary with the part of the process that is underway. For example, for batch reactions in a closed vessel, the chemicals are more likely to be emitted at the beginning and end of a reaction step (that are associated with reactor or treatment vessel loading and product transfer operations) than during the reaction.

The organic chemical pollutants that are most likely to present ecological risks are those that are (1) highly bioaccumulative, building up to high levels

Environmental Organic Chemistry for Engineers. http://dx.doi.org/10.1016/B978-0-12-804492-6.00006-X

TABLE 6.1 Types of Releases From Industrial Processes

Release is an on-site discharge of a toxic chemical to the environment and include (1) emissions to the air, (2) discharges to bodies of water, (3) releases at the facility to land, as well as (4) the contained disposal into underground injection wells

    Releases to air (point and fugitive air emissions): these releases include all air emissions from industry activity; point emissions occur through confined air streams as found in stacks, ducts, or pipes while fugitive emissions include losses from equipment leaks or evaporative losses from impoundments, spills, or leaks.

    Releases to water (surface water discharges): these releases include any releases going directly to streams, rivers, lakes, oceans, or other bodies of water; any estimates for storm water runoff and nonpoint losses must also be included.

    Releases to land: these releases include disposal of toxic chemicals in waste to on-site landfills, land treated or incorporation into soil, surface impoundments, spills, leaks, or waste piles; these activities must occur within the facility's boundaries for inclusion in this category.

    Underground injection: this type of release is a contained release of a fluid into a subsurface well for the purpose of waste disposal.

Transfer is a transfer of toxic chemicals in wastes to a facility that is geographically or physically separate from the facility reporting under the toxic release inventory. The quantities reported represent a movement of the chemical away from the reporting facility and, except for off-site transfers for disposal, these quantities of chemicals do not necessarily represent entry of the chemicals into the environment

    Transfers to publicly owned treatment works: include waste waters transferred through pipes or sewers to a publicly owned treatment works (POTW); treatment and chemical removal depend on the nature of the chemical and the treatment methods employed; chemicals that are not treated or destroyed by the publicly owned treatments works are generally released to surface waters or land filled within the sludge.

    Transfers to recycling: these transfers include chemicals that are sent off-site for the purposes of regenerating or recovering still valuable materials; once these chemicals have been recycled, they may be returned to the originating facility or sold commercially.

    Transfers to energy recovery: these transfers include wastes combusted off-site in industrial furnaces for energy recovery; treatment of a chemical by incineration is not considered to be energy recovery.

    Transfers to treatment: these transfers are wastes moved off-site for either neutralization, incineration, biological destruction, or physical separation; in some cases, the chemicals are not destroyed but are prepared for further waste management.

    Transfers to disposal are wastes taken to another facility for disposal generally as a release to land or as an injection underground.

in floral and faunal tissues even when concentrations in the ecosystem remain relatively low, and (2) highly toxic, so that they cause harm at comparatively low doses. In addition, atmosphere-water interactions that control the input and outgassing of persistent organic pollutants (POPs) in aquatic systems are critically important in determining the life cycle and residence times of these compounds and the extent of contamination. Although the effects of various types of organic chemical pollutants are usually evaluated independently, many ecosystems are subject to multiple pollutants, and their fate and impacts are intertwined. For example, the effects of nutrient deposition in an ecosystem can alter the methods by which the organic contaminants are assimilated, bioaccumulated, and the means by which the organisms in the ecosystem are affected.

Of all the organic chemical pollutants released into the environment by anthropogenic activity, POPs are among the most dangerous and need extreme measures for removal (Chapter 1). POPs are chemical substances that persist in the environment, bioaccumulate through the food web, and pose a risk of causing adverse effects to human health and the environment. This group of priority pollutants consists of pesticides (such as DDT), industrial chemicals (such as polychlorinated biphenyls, PCBs), and unintentional by-products of industrial processes (such as dioxins and furans). POPs can be transported across international boundaries far from their sources, even to regions where they have never been used or produced. Consequently, POPs pose a threat to the environment and to human health, all on a global scale.

Releases into the environment of organic chemicals that persist in the ecosystem (rather than undergo some form of biodegradation) lead to an exposure level that is not only subject to the length of time the chemical remains in circulation (in the environment) but also on the number of times that the organic chemical is recirculated before it is ultimately removed from the ecosystem. Typically, POPs are pesticides, industrial chemicals, or unwanted by-products of industrial processes that have been used and disposed for decades (prior to the inception of the various regulations and often without due regard for the environment) but have more recently been found to share a number of significant characteristics that need consideration before disposal is planned, including: (1) persistence in the environment insofar as these organic chemicals resist degradation in air, water, and sediments, (2) bioaccumulation insofar as these organic chemicals accumulate in living tissues at concentrations higher than those in the surrounding environment, and (3) long-range transport insofar as these organic chemicals can travel great distances from the source of release through air, water, and the internal organs of migratory animals, any of which can result in the contamination of ecosystems thousands of significant distances (up to thousands of miles) away from the source of the chemicals.

Briefly and by way of explanation, bioaccumulation is a process by which persistent environmental pollution leads to the uptake and accumulation of one or more contaminants, by organisms in an ecosystem. The amount of a pollutant available for exposure depends on its persistence and the potential for its bioaccumulation. Any chemical (including an organic chemical) is considered to be capable of bioaccumulation if the chemical has a degradation half-life in excess of 30 days or if the chemical has a bioconcentration factor (BCF) >1000 if the log K_{ow} is >4.2:

$$BCF = (\text{concentration in biota})/(\text{concentration in ecosystem})$$

The BCF indicates the degree to which a chemical may accumulate in fish (and other aquatic animals, such as mussels, etc.) by transport across the gills or other membranes, excluding feeding. Bioconcentration is distinct from food-chain transport, bioaccumulation, or biomagnification. The BCF is a constant of proportionality between the chemical concentration in flora or fauna in an ecosystem. It is possible, for many organic chemicals, to estimate the BCF from the octanol-water partition coefficients (K_{ow}) (Bergen et al., 1993):

$$\text{Log (bioconcentration factor)} = m \log K_{ow} + b$$

In terms of actual numbers, for many lipophilic organic chemicals, the BCF can be calculated using the regression equation:

$$\log \text{BCF} = -2.3 + 0.76 \times (\log K_{ow})$$

Furthermore, empirical relationships between the octanol-water partition coefficients and the BCF can be developed on a chemical-by-chemical basis.

On this note, it is worth defining the source of chemical contaminant insofar as chemical contaminants can originate from a point source or a nonpoint source (NPS). Point source of pollution is a single identifiable source of pollution which may have negligible extent, distinguishing it from other pollution source geometries. On the other hand, NPS pollution generally results from land run-off, precipitation, atmospheric deposition, drainage, seepage, or hydrologic modification. NPS pollution, unlike pollution from industrial and sewage treatment plants, comes from many diffuse sources and is often caused by rainfall or snowmelt moving over and through the ground. As the runoff moves, it picks up and carries away natural and human-made pollutants, finally depositing them into lakes, rivers, wetlands, coastal waters, and groundwater.

Typically, POPs are highly toxic and long-lasting (hence the name *persistent*) and cause a wide range of adverse effects to environmental flora and fauna, including disease and birth defects in humans and animals—some of the severe human health impacts of POPs include (1) the onset of cancer, (2) damage to the central nervous system, (3) damage to the peripheral nervous system, (4) damage to the reproductive system, and (5) disruption of the immune system. Moreover, POPs do not respect international borders, the serious environmental and human health hazards created by these chemicals affect not only developing countries, where systems and technology for monitoring, tracking, and disposing them can be weak or nonexistent, but also affect developed countries. As long as the chemical can be transported by air, water, and land, no country is immune from the effects of these chemicals.

In the last four decades it has become increasingly clear that the chemical and allied industries, such as the pharmaceutical industry, has been faced with serious environmental problems. Many of the classical organic synthetic processes have broad scope but often generate extreme amounts of chemical waste.

As a result, the chemical industry, the pharmaceutical industry, the various fossil fuel industries, as well as any other allied chemical industries have been subjected to increasing environmental regulation pressure to minimize or, preferably, eliminate any form of chemical waste. An illustrative example is provided by the manufacture of phloroglucinol (1,3,5-trihydroxybenzene, $C_6H_6O_3$), a reprographic chemical and pharmaceutical intermediate. Until the latter part of the 20th century, phloroglucinol was produced mainly from 2,4,6-trinitrotoluene (TNT):

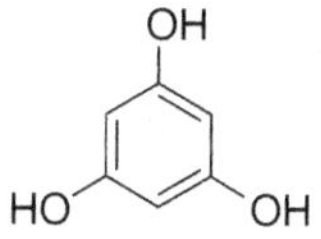

1,3,5-Trihydroxybenzene (phloroglucinol)

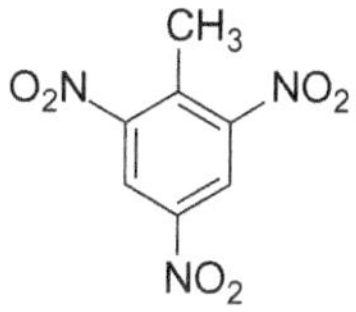

2,4,6-Trinitrotoluene (TNT)

Unfortunately, as a result of this process, for every kilogram of phloroglucinol produced, approximately 40 kg of solid waste were generated and also the waste contained such environmental nasties as chromium sulfate $[Cr_2(SO4)_3]$, ammonium chloride (NH_4Cl), ferrous chloride ($FeCl_2$), and potassium bisulfate ($KHSO_4$). This process was eventually discontinued as the costs associated with the disposal of this chromium-containing waste approached or exceeded the selling price of the product. This decision to terminate the process, seemingly based on process economics, boded well for the environment.

Indeed, an analysis of the amount of waste formed in processes for the manufacture of a range of fine chemicals and pharmaceutical intermediates has revealed that the generation of tens of kilograms of waste per kilogram of desired product was not exceptional in the organic chemical industry. This led to the introduction of the E (environmental) factor (kilograms of waste per kilogram of product) as a measure of the environmental footprint of manufacturing processes in various segments of the chemical industry. Thus

$$E = \text{kilograms of waste} / \text{kilogram of product}$$

The factor can be conveniently calculated from a knowledge of the number of tons of raw materials purchased and the number of tons of product sold, the calculation being for a particular product or a production site or even a whole company. A higher E factor means more waste and, consequently, a larger environmental footprint—thus, the ideal E factor for any process is 0.

However, in the context of environmental protection and to be all inclusive, the E factor is the total mass of raw materials plus ancillary process requirements minus the total mass of product, all divided by the total mass of product. Thus, the E factor should represent the *actual amount* of waste produced in the process, defined as everything but the desired product and takes the chemical yield into account and includes reagents, solvent losses, process aids, and (in principle) even the fuel necessary for the process. Water has been generally excluded from the E factor as the inclusion of all process water could lead to exceptionally high E factors in many cases and made meaningful comparisons of the technical factors (E factors excluding water use) for processes difficult. However, in the modern environmentally-conscious era, there is no reason for water requirement to be omitted since the disposal of process water is an environmental issue. Moreover, use of the E factor has been widely adopted by many of the chemical industries and the pharmaceutical industries in particular. Thus, a major aspect of process development recognized by process chemists and by process engineers is the need for determining an E factor—whether or not it is called by that name (i.e., E factor) but *chemicals in* vs. *chemical out* has become a major yardstick in many industries.

It is clear that the E factor increases substantially on going from bulk chemicals to fine chemicals and then to pharmaceuticals. This is partly a reflection of the increasing complexity of the products, necessitating not only processes which use multistep syntheses, but is also a result of the widespread use of stoichiometric amounts of the reagents (Chapters 2 and 3), i.e., the required amounts of the reagents (some observers would advocate the stoichiometric amounts of the reagents plus 10%) to accomplish conversion of the starting organic chemical to the product(s). A reduction in the number of steps of a process for the synthesis of organic chemicals will, in most cases (but not always), lead to a reduction in the amounts of reagents and solvents used and hence a reduction in the amount of waste generated. This has led to the introduction of the concepts of step economy and function oriented synthesis (FOS) of pharmaceuticals. The main issues behind the concept of FOS is that the structure of an active compound, which may be a natural product, can be reduced to simpler structures designed for ease of synthesis while retaining or enhancing the biological activity. This approach can provide practical access to new (designed) structures with novel activities while at the same time allowing for a relatively straightforward synthesis.

As noted above, a knowledge of the stoichiometric equation allows the process chemist or process engineer to predict the theoretical minimum amount of waste that can be expected. This led to the concept of *atom economy* or *atom utilization* to quickly assess the environmental acceptability of alternatives to a particular product before any experiment is performed. It is a theoretical number, that is, it assumes a chemical yield of 100% and exact stoichiometric amounts and disregards substances which do not appear in the stoichiometric equation. In short, the key to minimizing waste is precision or *selectivity* in organic synthesis which is a measure of how efficiently a synthesis is performed. The standard definition of selectivity is the yield of product divided by the amount of substrate converted, expressed as a percentage.

Organic chemists distinguish between different categories of selectivity: (1) chemoselectivity, which relates to competition between different functional groups and (2) regioselectivity, which is the selective formation of one regioisomer, for example, ortho vs. para substitution in aromatic ring systems. However, one category of selectivity was, traditionally, largely ignored by organic chemists: the *atom selectivity* or *atom utilization* or *atom economy*. The virtually complete disregard of this important parameter by chemists and engineers has been a major cause of the waste problem in the manufacture of organic chemicals. Quantification of the waste generated in chemicals manufacturing, by way of *E* factors, served to illustrate the omissions related to the production of chemical waste and focus the attention of fine chemical companies, the pharmaceutical companies, and the petrochemical companies on the need for a paradigm shift from a concept of process efficiency, which was exclusively based on chemical yield, to a need that more focused on the elimination of waste chemicals and maximization of raw materials utilization.

Many of the global environmental changes forced by human activities are mediated through the chemistry of the environment (Andrews et al., 1996; Schwarzenbach et al., 2003; Manahan, 2010; Spellman, 2016). Important changes include the global spread of air pollution, groundwater pollution, and pollution of the oceans which increase the concentration of tropospheric oxidants (including ozone), stratospheric ozone depletion, and global warming (the so-called greenhouse effect). Since the onset and establishment of the agricultural revolution and the industrial revolution, the delicate balance between physical, chemical, and biological processes within the Earth system has been perturbed as a result. Example of the causes of perturbation of the Earth systems include (1) the exponential growth in the world population, (2) the use of increasing amounts of fossil fuel, (3) fossil fuel-related emissions of carbon to the atmosphere, and (4) the intensification of agricultural practices including the more frequent use of fertilizers. The observed increase in the atmospheric abundance of carbon dioxide (CO_2) has been ascribed (correctly or incorrectly) mainly to fossil fuel burning (Speight and Islam, 2016), although biomass destruction is an important secondary source of carbon dioxide emissions. Atmospheric concentrations are additionally influenced by exchanges of carbon with the ocean and the continental biosphere (Firor, 1990; WMO, 1992; Calvert, 1994; Goody, 1995).

The progressive modification and fertilization of the terrestrial biosphere are believed to have caused the observed increase in atmospheric nitrous oxide (N_2O), a tropospheric greenhouse gas (GHG) and a source of reactive species in the stratosphere. Methane (CH_4), which also contributes to *greenhouse forcing* (the technical term for the influence of the greenhouse effect that causes a shift in the climate)—this can occur due to changes in the level of gasses that share two properties: they are transparent to visible light, but absorb the infrared, which we typically perceive as heat, and also plays an important role in the photochemistry of the troposphere and the stratosphere, is produced by biosphere-related processes (wetlands, livestock, landfills, biomass burning) as well as by leakage from gas distribution systems in various countries (Calvert, 1994). The global atmospheric concentration of methane has also grown in the past. Observed increases in the abundance of tropospheric ozone (O_3), which contribute to deteriorating air quality, result from complex photochemical processes involving industrial and biological emissions of nitrogen oxides, hydrocarbons, and certain other organic compounds. Ozone is a strong absorber of solar ultraviolet radiation and also contributes to greenhouse forcing. Anthropogenic emissions of sulfur resulting from combustion of sulfur-containing fuels and also from coal combustion without the necessary end-of-pipe gas cleaning protocols, coal burning in highly populated and industrialized regions of the Northern Hemisphere, and the related increase in the aerosol load of the troposphere, have contributed to regional pollution and have probably produced a cooling of the surface in these regions by backscattering a fraction of the incoming solar energy.

Finally, the bad news is that the rapid increase in the abundance of industrially manufactured chlorofluorocarbons in the atmosphere produced an observed depletion in stratospheric ozone and the formation each spring (since the late 1970s) of an *ozone hole* over Antarctica. By way of explanation, the ozone hole is not technically a *hole* where no ozone is present, but it is actually a region of exceptionally depleted ozone in the stratosphere over the Antarctic that happens at the beginning of Southern Hemisphere spring (August–October). The good news is that, as a result of environmental caution and a reduction in terms of the release of contaminants, the ozone hole has diminished in size (i.e., the depletion of ozone has stopped) and may even be on a turnaround to an increase in the amount of ozone in the area.

Although toxic and hazardous chemicals are produced by the various chemicals industries, frequent reference is made here to the chemical products produced by the crude oil refining industry, without any attempt to point to this industry as the major polluter. The chemicals produced in the various cured oil-derived products offer a wide range of properties and behavior that makes the crude oil-derived products suitable for this text.

In summary, many of the specific chemicals in petroleum are hazardous because of their chemical reactivity, fire hazard, toxicity, and other properties. In fact, a simple definition of a hazardous chemical (or hazardous waste) is that it is a chemical substance (or chemical waste) that has been inadvertently released, discarded, abandoned, neglected, or designated as a waste material and has the potential to be detrimental to the environment. Alternatively, a hazardous chemical may be a chemical that may interact with other (chemical) substances to give a product that is hazardous to the environment.

2 RELEASE INTO THE ENVIRONMENT

For the purposes of this text, it is assumed that any organic chemicals released into the environment are hazardous chemicals following from the lists of chemicals produced by the US Environmental Protection Agency as well as environmental agencies in many other countries (Appendix). Thus, it is not only safe, but necessary, to assume that any organic chemical (except chemicals that are indigenous to the ecosystem into which they exist but in quantities close to the indigenous amounts) have the potential to be hazardous to the environment.

Contamination by chemicals is a global issue and there is no single company that should shoulder all of the blame. Past laws and regulations (or the lack thereof) allowed unmitigated disposal of chemicals and discharge of chemicals into the environment. These companies were not breaking the law; it is a matter of there being insufficient laws (the fault of various level of government) that protected the environment. As a result, toxic chemicals are found practically in all ecosystems on Earth, thus affecting biodiversity, agricultural production, and water resources. At the end of the various chemical life cycles, chemicals are recycled or disposed as part of waste. The inappropriate management of such waste results in negative impacts on the environment.

As already stated, of all the pollutants released into the environment by human activity, POPs among the most dangerous to environmental flora and fauna are pesticides, various industrial chemicals, or unwanted by-products of industrial processes that have been used for decades but more recently been found to share a number of disturbing characteristics, including: (1) persistence, which means that the chemicals resist degradation in air, water, and sediments, (2) bioaccumulation, which means that the chemicals accumulate in living tissues at concentrations higher than those in the surrounding environment, and (3) long-range transport, which means that the chemicals can travel a considerable distance from the source of release through air, water, and migratory animals, often contaminating areas miles away from any known source. On the environmental side, POPs are highly toxic and long-lasting, and cause an array of adverse effects on flora and fauna.

Thus, many organic chemicals have toxic, carcinogenic, mutagenic, or teratogenic (causing developmental malformations) effects on environmental flora and fauna and are designated either as *Acutely Hazardous Waste* or *Toxic Waste* by the United States Environmental Protection Agency (https://www.epa.gov/hw). Substances found to be fatal to humans in low doses or, in the absence of data on human toxicity, have been shown to have an oral LD_{50} toxicity (lethal dose at 50% concentration) of <2 mg L^{-1}, or a dermal LD_{50} of <200 mg kg^{-1} or is otherwise capable of causing or significantly contributing to an increase in serious irreversible, or incapacitating reversible illness are designated as *Acute*

Hazardous Waste (https://www.epa.gov/hw). Materials containing any of the toxic constituents so listed are to be considered hazardous waste, unless, after considering the following factors it can reasonably be concluded (by the Department of Environmental Health and Safety) that (1) the waste is not capable of posing a substantial present or potential hazard to public health or (2) the waste is not capable of posing a substantial present or potential hazard to the environment when improperly treated, stored, transported or disposed of, or otherwise managed.

Briefly, within this text and for environmental purposes, chemicals are subdivided into two classes: (1) *organic chemicals* and (2) *inorganic chemicals.* Furthermore, classification occurs insofar as organic chemicals are classified as *volatile organic compounds (VOCs)* or *semivolatile organic compounds* (on occasion, the word *chemicals* is substituted for the word *compounds* without affecting the definition). The first class of organic compounds, the VOCs, is subdivided into *regulated compounds* and *unregulated compounds.* Regulated compounds have maximum contaminant levels, while unregulated compounds do not. Regulated compounds generally (but not always) have low boiling points, or low boiling ranges, and some are gases. Many of these chemicals can be detected at extremely low levels by a variety of instrumentation, including the human nose! In the case of organic chemicals, sources for VOCs typically are petroleum refineries, fuel stations, naphtha (i.e., dry cleaning solvents, paint thinners, cleaning solvents for auto parts) and, in some cases, refrigerants that are manufactured from petrochemicals.

The constituents of the second class of organic compounds, the *semivolatile compounds*, typically have high boiling points, or high boiling ranges, and are not always easily detected by the instrumentation that may be used to detect the VOCs (including the human nose). Some of the common sources of contamination are high boiling petroleum products (such as automotive lubricating oils and machinery lubricating oils), pesticides, herbicides, fungicides, wood preservatives, and a variety of other chemicals that can be linked to the organic chemicals industry.

Regulations are in place that set the maximum contamination concentration levels that are designed to ensure environmental and public safety. There are primary and secondary standards for inorganic chemicals. Primary standards are for those chemicals that cause neurological damage, cancer, or blood disorders. Secondary standards are developed for other environmental reasons. In some instances, the primary standards are referred to as the *Inorganic Chemical Group*. The secondary standards are referred to as the *General Mineral Group* and *General Physical Testing Group*.

However, despite the nature of the environmental regulations and the precautions taken by the organic chemicals industry, the accidental release of nonhazardous organic chemicals and hazardous organic chemicals into the environment has occurred and, without being unduly pessimistic, will continue to occur (by all industries—not wishing to select any particular industry as the only industry that suffers accidental release of organic chemicals into the environment). It is a situation that, to paraphrase *chaos theory*, no matter how well the preparation, the unexpected is always inevitable. It is, at this point that the environmental scientist and engineer has to identity (through careful analysis) the nature of the chemicals and their potential effects on the ecosystem(s) (Smith, 1999). Although petroleum itself and its various products are complex mixtures of many organic chemicals (Chapters 2 and 3), the predominance of one particular chemical or one particular class of chemicals may offer the environmental scientist or engineer an opportunity for predictability of behavior of the chemical(s).

Thus, when a spill of organic chemicals occurs the primary processes determining the fate of organic chemicals are (1) dispersion, (2) dissolution, (3) emulsification, (4) evaporation, (5) leaching, (6) sedimentation, (7) spreading, and (8) wind. These processes are influenced by the physical properties of the organic chemicals (especially if the chemicals are constituents of a mixture), spill characteristics, environmental conditions, and physicochemical properties of the spilled material after it has undergone any form of chemical transformation (Chapter 7).

2.1   Dispersion

For the purposes of this text, the term *dispersion* encompasses all phenomena which give rise to the proliferation of substances through the man-made and natural environment. Thus, the physical transport of oil droplets into the water column is referred to as dispersion. This is often a result of water surface turbulence, but also may result from the application of chemical agents (dispersants). These droplets may remain in the water column or coalesce with other droplets and gain enough buoyancy to resurface. Dispersed oil tends to biodegrade and dissolve more rapidly than floating slicks because of high surface area relative to volume. Most of this process occurs from about half an hour to half a day after the spill.

Emissions of many chemicals of concern occur into the air initially, from where they are dispersed into other media. In fact, air is one of the main carriers of chemical carcinogens to humans (Corvalán and Kjellström, 1996). Many chemicals emitted into the air, for instance from combustion processes, tend to become associated with particulate matter. Removal from the air occurs through a range of complex processes involving photo-degradation, and particle sedimentation and/or precipitation (known respectively as dry and wet deposition). Volatile organic chemicals and semivolatile organic chemicals may undergo several cycles of evaporation and precipitation, which can also make chemicals more accessible to photochemical or biodegradation.

The dispersion of chemicals released into the environment has been the focus of much attention because of the realization that the dispersion behavior of organic chemicals can be markedly different when different chemicals are considered. Accidents that involve organic chemicals give rise to a new class of problems in dispersion prediction for the following reasons: (1) the material is, in almost all cases, stored as a liquid, which may appear as a gas after a spill, (2) the modes of release can vary widely and geometry of the source can take many forms and the initial momentum of the spill may be significant, and (3) in some cases, a chemical transformation also takes place as a result of reaction with water vapor in the ambient atmosphere.

In addition, the physical properties of the organic chemical(s) usually result in interactions with the surrounding ecosystem, especially if the chemical is a reactive liquid and has the potential of interacting with the air, water, or soil. This reactivity will influence the dispersibility of the chemical. Moreover, if the release occurs over a short time-scale, compared to the steady-state releases characteristic of many chemical releases problems, this can give rise to the complication of predicting dispersion for time-varying releases and to uncertainty in individual predictions resulting from variability about the behavior of a mixture. Also, the dispersing chemical, which is typically denser that air, may form a low-level cloud that is sensitive to the effects of man-made and natural obstructions and of topography.

Wind (Aeolian) transport (relocation by wind) can also occur and is particularly relevant when dust from organic carbonaceous solids (such as coke dust) and catalyst dust are considered. Dust becomes airborne when winds traversing arid land with little vegetation cover pick up small particles such as catalyst dust, coke dust, and other refinery debris and send them skyward after which the movement of pollutants in the atmosphere is caused by transport, dispersion, and deposition. Dispersion results from local turbulence, that is, motions that last less than the time used to average the transport. Deposition processes, including precipitation, scavenging, and sedimentation, cause downward movement of pollutants in the atmosphere, which ultimately remove the pollutants to the ground surface.

2.2 Dissolution

The solubility (dissolution) characteristics of organic molecules in water are complex and very much dependent upon the structure and properties of the organic chemicals(s). Solubility is the property of organic chemicals which might be a gas, a liquid, or a solid (the solute) that dissolves in a solvent which might also be a gas, a liquid, or a solid organic chemical (the solvent). The solubility of a gas, a liquid, or a solid organic chemical depends on the physical and chemical properties of the solute and solvent as well as on temperature, pressure, and the pH of the solution. The extent of the solubility of a substance in a specific solvent is measured as the saturation concentration, where adding more solute does not increase the concentration of the solution and begins to precipitate the excess amount of solute. In terms of environmental organic chemistry, the solvent is typically a liquid, which can be a pure organic chemical (such as hexane or toluene) or a mixture of organic chemicals (such as naphtha or kerosene).

The extent of solubility of the solute ranges widely, from infinitely soluble (without limit) (i.e., the solute is fully miscible with the solvent, such as ethanol in water) to poorly soluble (such as some polynuclear aromatic systems in water)—the term *insoluble* is often applied to poorly or very poorly soluble compounds and (in the world of organic chemistry) a common threshold to describe an organic chemical as insoluble is a solubility <0.1 g per 100 mL of solvent. However, solubility of an organic compound in a solvent should not be confused with the ability to of a solvent to dissolve a solute, because the solution might also occur because of a chemical reaction (*reactive solubility*).

Solubility of a solute in a solvent applies not only to environmental organic chemistry but to areas of chemistry, such as (alphabetically): biochemistry, geo-chemistry, inorganic chemistry, physical chemistry, and organic chemistry. In all cases, solubility depends on the physical conditions (temperature, pressure, and concentration of the solute) and the enthalpy and entropy directly relating to the solvents and solutes concerned. By far the most common solvent in chemistry is water which is a solvent for most ionic compounds as well as a wide range of organic chemicals. This is a crucial factor in the chemical phenomena known as acidity and alkalinity as well as in much of the area known as environmental organic chemistry.

In addition, the term *dissolved organic carbon* (DOC) is used as a broad classification for organic molecules of varied origin and composition within aquatic systems (the aquasphere) and the dissolved fraction of organic carbon is an operational classification. The source of the DOC in marine systems and in freshwater systems depends on the body of water. In general, organic chemical compounds are a result of decomposition processes from dead organic matter such as plants or marine organisms. When water contacts highly organic soils, these components can drain into rivers and lakes as DOC. Whatever the source of the DOC, it is also extremely important in the transport of metals in aquatic systems—certain metals can form extremely strong organo-metallic complexes with DOC which enhances the solubility of the metal in aqueous systems while also reducing the bioavailability of the metal.

In terms of organic chemicals that are not naturally occurring, knowledge of the structure and properties can be used to examine relationships between the solubility properties of an organic chemical and its structure, and vice versa. In fact, structure dictates function which means that by knowing the structure of an organic chemical it may be (but not always) possible to predict the properties of the chemical such as its solubility, acidity or basicity, stability, and reactivity. In the context of environmental distribution of a chemical, predicting the solubility of an organic molecule is a useful component of knowledge.

At the molecular level, solubility is controlled by the energy balance of the intermolecular forces between solute-solute, solvent-solvent, and solute-solvent molecules. But without getting too fay into a study of such forces, the simple, very useful, and practical empirical rule that is quite reliable which is *like dissolves like* which is based on the polarity of the organic chemical systems insofar as polar organic chemicals typically dissolve in polar solvents (such as water, alcohols) and nonpolar organic chemicals typically dissolve in nonpolar solvents (such as nonpolar hydrocarbon solvents). The polarity of organic molecules is determined by the presence of polar bonds due to the presence of electronegative atoms (such as nitrogen and oxygen) in polar functional groups such as amine derivatives (RNH_2) and alcohol derivatives (ROH) (Chapter 2). Furthermore, since the polarity of an organic chemical is related to the presence of a functional group, the solubility characteristics of an unknown organic contaminant can provide evidence (as well as evidence form other analytical techniques) (Chapter 5) for the presence (or absence) of an organic functional group:

Solvent	Solubility or Complete Miscibility
Water	Alcohols, amines, acids, selected (but not all) esters, ketones, and aldehyde derivatives
5% aqueous $NaHCO_3$[a]	Carboxylic acid derivatives
5% aqueous $NaOH$[a]	Carboxylic acid derivatives and phenol derivatives
5% aqueous HCl[a]	Amine derivatives
Diethyl ether	Most organic chemicals

[a] *A 5% w/w solution of the chemical in water.*

Most organic molecules are relatively nonpolar and are usually soluble in organic solvents (such as diethyl ether, dichloromethane, chloroform, and hexane) but not in polar solvents (such as water). If the organic chemical is soluble in water, this denotes that the chemical is polar and therefore soluble in water which denotes a high ratio of polar group(s) to the nonpolar hydrocarbon chain, such as a low molecular weight organic chemical that contains a hydroxy function (—OH), or an amino function (—NH_2) or a carboxylic acid function (—CO_2H) group, or a higher molecular weight organic chemical that contains two or more functional (polar) groups. The presence of an acidic carboxylic acid or basic amino group in a water-soluble compound can be detected by measurement of the pH of the solution—a low pH (pH ≤ 7) indicates the presence of an acidic function while a high pH (pH ≥ 7) indicates the presence of a basic function. Thus, organic chemicals that are insoluble in water can become soluble in an aqueous environment if they form an ionic species when treated with an acid or a base—the ionic form (for example the sedum salt of a carbocyclic acid, $RCO_2^- \, Na^+$) is much more polar than the carboxylic acid.

The solubility of carboxylic acid derivatives and phenol derivatives in sodium aqueous hydroxide is due to the formation of the polar (ionic) carboxylate ions (—CO_2^-) or the polar (ionic) phenoxide ($C_6H_5O^-$) ions since they are much stronger acids than water, and therefore the acid-base equilibria lie far to the right of the equation:

$$RCO_2H + OH^- \leftrightarrow RCO_2^- + H_2O$$

$$ArOH + OH^- \leftrightarrow ArO^- + H_2O$$

Carboxylic acid derivatives, but not phenol derivatives, are also stronger acids than carbonic acid (H_2CO_3), and are therefore also soluble in aqueous sodium bicarbonate ($NaHCO_3$ solution):

$$RCO_2H + HCO_3^- \leftrightarrow RCO_2^- + H_2O + CO_2$$

The solubility of amine derivatives in dilute aqueous acid is in accordance with the amine derivatives being stronger bases than water, and the amine derivatives are converted to the polar ammonium ion by protonation (through reaction with a proton):

$$RNH_2 + H_3O^+ \leftrightarrow RNH_3^+ + H_2O$$

Amine derivatives are the only common class of organic compounds which are protonated in dilute aqueous acid.

Thus, dissolution is the solubilization of organic chemicals in water. Many of the acutely toxic components of oils such as benzene, toluene, and xylene will also dissolve into water but only to a minute extent. Nevertheless, when the volume of water (as in a lake or in running water such as a river) is large the transportation of the benzene, toluene, ethylbenzene, and the xylene isomers (BTEX) through dissolution can be substantial. This process also occurs quickly after a discharge of the chemical(s) but tends to be less important than evaporation. For example, in a typical discharge into the aquasphere, generally <5–10% v/v of the benzene is lost to dissolution while >90–95% v/v can be lost to evaporation. The polynuclear aromatic compounds (Chapter 2) offer a different scenario.

For alkylated polynuclear aromatic compounds, solubility is inversely proportional to the number of rings and extent of alkylation (i.e., the number of alkyl moieties attached to the polynuclear aromatic ring system). The dissolution process is thought to be much more important in rivers because natural containment within the river system (i.e., the water, the character of the river banks, and the river bed) may prevent spreading, thereby reducing the surface area of the chemical slick (the chemicals on the surface of the water) and thus retard evaporation. However, river turbulence must not be ignored since turbulence increases the potential for mixing and dissolution.

Groundwater that is contaminated with organic chemicals tends to be enriched in aromatic derivatives relative to other organic chemicals constituents. Relatively insoluble hydrocarbons may be entrained in water through adsorption on to clay (kaolinite) particles suspended in the water or as an agglomeration of oil droplets (microemulsion) in the water. In cases where groundwater contains only dissolved hydrocarbons, it may not be possible to identify the original organic chemicals because only a portion of the free product will be present in the dissolved phase. If a hydrocarbon mixture has been spilled into the aquasphere, initially the whole mixture floats on top of the groundwater but any constituents with a tendency to water-solubility will be extracted from the mixture. Groundwater containing entrained hydrocarbons will have a gas chromatographic fingerprint that is a combination of the total mixture plus enhanced amounts of the soluble aromatics minus losses to the water. Generally, dissolved aromatics may be found quite far from the origin of a spill but entrained hydrocarbons may be found in water close to the organic chemical source. Oxygenates, such as methyl-*t*-butyl ether, are even more water soluble than aromatics and are highly mobile in the environment.

2.3 Emulsification

An emulsion is a mixture of two or more liquids that are usually immiscible. Examples include crude oil and water which can form an oil-in-water emulsion, wherein the oil is the dispersed phase, and water is the dispersion medium. In addition, crude oil and water can also form a water-in-oil emulsion, wherein water is the dispersed phase and crude oil is the external phase. Whether an emulsion of oil and water exists as a water-in-oil emulsion or an oil-in-water emulsion depends on the volume fraction of both phases and the type of emulsifier (surfactant) present. Multiple emulsions are also possible, including a water-in-oil-in-water emulsion and an oil-in-water-in-oil emulsion. Emulsions, being liquids, do not exhibit a static internal structure—the droplets dispersed in the liquid matrix (the *dispersion medium*) are usually assumed to be statistically distributed.

The stability of an emulsion refers to the ability of the emulsion to resist change in its properties over time. There are three types of instability in emulsions: (1) flocculation, (2) creaming, and (3) coalescence. Flocculation occurs when there is an attractive force between the droplets, so they form flocs. Coalescence occurs when droplets bump into each other and combine to form a larger droplet, so the average droplet size increases over time. Creaming occurs when the droplets rise to the top of the emulsion under the influence of buoyancy. Use of a surface-active agent (surfactant) can increase the stability of an emulsion so that the size of the droplets does not change significantly with time and the emulsion is defined as *stable*. Most emulsions contain droplets with a mean diameter of more than around 1 μm (1 micron, 1×10^{-6} m) however mini-emulsions and nano-emulsions can be formed with droplet sizes in the 100–500 nm (nanometer, 1×10^{-9} m) range, and with proper formulation, highly stable microemulsion can be prepared having droplets as small as a few nanometers.

Crude oil and crude oil products as well as many other organic chemicals can form water-in-oil emulsions (where water is incorporated into oil) or in the form of a *mousse* as weathering occurs. This process is significant because, for example, the apparent volume of the organic chemical may increase dramatically, and the emulsification will slow the other weathering processes, especially evaporation of the volatile organic chemicals.

Emulsification in the aquasphere (especially in marine environment) depends primarily on the composition of the crude oil (or the crude oil product) and the turbulent regime of the water mass. The most stable emulsions such as water-in-oil contain from 30% to 80% v/v water and usually appear after strong storms in the zones of spills of heavy oil or the higher boiling crude oil product that have an increased content of nonvolatile fractions (especially asphaltene constituents) (Speight, 2014). These types of emulsions can exist in the marine environment for over several months (100 days or more) in the form of (what is colloquially called) a *chocolate mousse* (which is not recommended for human consumption!). The stability of these emulsions usually increases (demulsification) with decreasing temperature—the reverse emulsions, such as oil-in-water emulsion (where droplets of oil are suspended in water) are much less stable because surface-tension forces quickly decrease the dispersion of oil.

This demulsification process can be slowed with the help of emulsifiers which are surface-active substances with strong hydrophilic properties that are used to eliminate the prolonged effects of spills of crude oil and crude oil products. Emulsifiers help to stabilize oil emulsions and promote dispersing oil to form microscopic (invisible) droplets which accelerates the decomposition of oil products in the water.

2.4 Evaporation

Evaporation (the opposite of condensation) occurs when a liquid organic chemical becomes a gas (or vapor), sublimation is the phenomenon that occurs when a solid becomes a gas without the conversion of the solid to liquid and thence to a gas.

$$\text{Evaporation}: \text{Liquid} \rightarrow \text{gas}$$

$$\text{Sublimation}: \text{Solid} \rightarrow \text{gas}$$

Both phenomena are important aspects of the environmental organic chemical cycle since both processes involve disappearance of organic contaminant for the water or land *but* the contaminant does appear in the atmosphere. Of the two processes, the evaporation is the most common process since organic chemicals that go through the sublimation processes are not as obvious or as common as chemicals that evaporate.

Many factors affect the evaporation process. For example, if the air is already saturated with other chemicals (or the humidity is high) there is typically little chance of a liquid evaporating as quickly as when the air is not saturated (or humidity is low). In addition, the air pressure also affects the evaporation process since, under high air pressure, an organic chemical is much more difficult to evaporate. Temperature also affects the ability of an organic chemical to evaporate.

Evaporative processes are very important in the weathering of volatile organic chemical products, and may be the dominant weathering process for naphtha, gasoline, and other mixtures that contain low-boiling organic chemicals. Automotive gasoline, aviation gasoline, and some grades of jet fuel (e.g., JP-4) contain 20–99% highly volatile constituents (i.e., constituents with less than nine carbon atoms). The evaporative processes typically begin immediately after a volatile organic chemical is discharged into the environment. For spilled crude oil, the amount lost to evaporation can typically range from approximately 20% to 60% v/v. Some low-boiling petroleum-derived products such as one-ring and two-ring aromatic hydrocarbons and/or low molecular weight alkane derivatives (having <15 carbon atoms) may evaporate entirely. In fact, a substantial fraction of higher molecular weight organic chemicals may also evaporate.

The primary factors that control evaporation are the composition of the oil, slick thickness, temperature and solar radiation, wind speed, and wave height. While evaporation rates increase with temperature, this process is not restricted to warm climates. For the Exxon Valdez incident, which occurred in cold conditions (Mar. 1989), it has been estimated that appreciable evaporation occurred even before all the oil escaped from the ship, and that evaporation ultimately accounted for 20% v/v of the crude oil. Most of this process occurs within the first few days after the spill. However, it is not unusual for the evaporation process to be active simultaneously with other processes to remove any volatile organic chemicals.

2.5 Leaching

Leaching is a natural process by which water-soluble substances (such as water-soluble organic chemicals or hydrophilic organic chemicals) are washed out from soil or waste disposal areas (such as a landfill). These leached out chemicals (leachates) can cause pollution of surface waters (ponds, lakes, rivers, and the sea) and subsurface water (groundwater aquifers). Thus, leaching is the process by which (in the current context) organic contaminants are released from the solid phase into the water phase under the influence of dissolution, desorption, or complexation processes as affected by acidity or alkalinity (the pH value). The process itself is universal, as any material exposed to contact with water will leach components from its surface or its interior depending on the porosity of the material under consideration. Leaching often occurs naturally with soil contaminants such as organic chemicals with the result that the chemicals end up in potable waters.

Many organic chemicals occur in a mixture of different components in a solid (such as the carbonaceous deposits on petroleum coke). In order to separate the desired solute constituent or remove an undesirable solute component from the solid phase, the solid is brought into contact with a liquid during which time the solid and liquid are in contact and the solute or solutes can diffuse from the solid into the solvent, resulting in separation of the components originally in the solid (*leaching, liquid-solid leaching*). In addition, leaching may also be referred to as *extraction* because the solute is being extracted from the solid or the proceeds may also be referred to as *washing* because a contaminant is removed from a solid with water.

Leaching is affected by (1) soil texture, (2) structure, and (3) water content of the soil. For example, in terms of soil texture, the proportions of sand, silt, and clay affect the movement of water through the soil. Coarse-textured soil containing more sand particles have large pores and is highly permeable which allow the water to move rapidly through the soil and, in fact, organic chemicals (such as pesticides) carried by water through coarse-textured soil are more likely to reach and contaminate groundwater. On the other hand, clay-textured soils have low permeability and tend to retain more water and adsorb more organic chemicals from the water. This slows the downward movement of chemicals, helps increase the chance of degradation and adsorption to soil particles, and reduces the chance of groundwater contamination.

In terms of soil structure, loosely packed soil particles allow speedy movement of water through the soil while tightly compacted soil holds water back and does not allow the water to move freely through it. Plant roots penetrate soil, creating excellent water channels when they die and rot away. These openings and channels may permit relatively rapid water movement even through clay-containing soil. On the other hand, the amount of water already in the soil has a direct bearing on whether rain or irrigation results in the recharging of groundwater and possible leaching of organic chemicals into an aquifer. Soluble chemicals are more likely to reach groundwater when the water content of the soil approaches or is at saturation. Saturation is typical in the spring when rain and snowmelt occurs but when soil is dry the added water just fills the pores in the soil near the soil surface, making it unlikely that the water will reach the groundwater supply.

Leaching processes introduce hydrocarbon into the water phase by solubility and entrainment. Leaching processes of organic chemical products in soils can have a variety of potential scenarios. Part of the aromatic fraction of a spill of organic chemicals on to soil may partition into water that has been in contact with the contamination.

2.6 Sedimentation or Adsorption

Sedimentation occurs when particles in suspension settle out of the fluid in which they are entrained and come to rest against a barrier, which is typically the basement of a waterway. Settling is due to the motion of the particles through the fluid in response to the forces acting on them, which in an ecosystem, can be due to gravity. The term *sedimentation* is often used as the opposite of erosion, i.e., the terminal end of sediment transport. Settling occurs when suspended particles fall through the liquid, whereas sedimentation is the termination of the settling process.

Sedimentation can be generally classified into three different types that are all applicable to organic chemicals. Type 1 sedimentation is characterized by particles that settle discretely at a constant settling velocity and typically these particles settle as individual particles and do not flocculate or stick to others during settling. Type 2 sedimentation is characterized by particles that flocculate during sedimentation and because of this the particle size is constantly changing and therefore their settling velocity keeps changing. Type 3 sedimentation (also known as zone sedimentation) involves particles that are at a high concentration (e.g., >1000 mg L^{-1}) such that the particles tend to settle as a mass and a distinct clear zone and sludge zone are present. Zone settling occurs in active sludge sedimentation and sedimentation of sludge thickeners.

Sediments typically consist of a few weight percent of organic matter and the balance being shared by various mineral constituents. The organic carbon environment acts as an attractor or accumulator of hydrophobic compounds, such as polychlorobiphenyl derivatives (PCBs). However, organic carbon in sediment comes in different forms that may have very different sorption capacities for hydrophobic organic compounds. In addition to natural sources such as vegetative debris, decayed remains of plants and animals, and humic matter, sediment organic carbon is also derived from particles of coal, coke, charcoal, and soot that are known to have extremely high sorption capacities.

Many organic chemicals (especially crude oil and crude oil-derived products) are less dense than water or are buoyant in water. However, in areas with high levels of suspended sediment, organic chemicals may be transported to the river, lake, or ocean floor through the process of sedimentation. The organic chemicals may adsorb on to sediments and sink—most of this process occurs from about 2 to 7 days after the spill. Furthermore, hydrophobic organic compounds, such as polynuclear aromatic hydrocarbons (PNAs) and PCBs, bind strongly to sediments and can serve as a long-term source of contaminants in water bodies and biota long after the original source has been removed. The inherent heterogeneity of most sediments makes it difficult to describe in terms of bulk sediment physicochemical parameters such as total organic carbon content, surface area, and particle size distribution. Therefore, management of sediments and the control of sediment contaminants are among the most challenging and complex problems faced by the environmental scientists and the environmental engineers and will become increasingly more so as other organic contaminants enter the environment in greater quantities.

In terms of the sedimentation of organic chemicals (Crompton, 2012), some of the chemicals may be adsorbed on the suspended material (especially if the suspended material is clay or another highly adsorptive mineral) and deposited to the bottom. This mainly happens in shallow waters where particulate matter is abundant and water is subjected to intense mixing, usually through turbulence. Simultaneously, the process of biosedimentation can also occur when plankton and other organisms absorb the organic chemical. The suspended forms of the organic chemicals undergo intense chemical and biological (microbial in particular) decomposition in the water. However, this situation radically changes when the suspended organic chemical reaches the lake bed, river bed, or sea bed and the decomposition rate of the organic chemical(s) buried on the bottom abruptly drops. The oxidation processes slow down, especially under anaerobic conditions in the bottom environment and the organic chemical(s) accumulated inside the sediments can be preserved for many months and even years.

2.7 Spreading

The movement of organic chemical contaminants through the subsurface is complex and is difficult to predict since different types of contaminants react differently with soils, sediments, and other geologic materials and commonly travel along different flow paths and at different velocity. Contaminants released at the source areas (typically on the surface) infiltrate into the subsurface and migrate downward by gravity through the vadose zone (the zone that extends from the top of the ground surface to the water table). When low-permeability soil units are encountered, the contaminants can also spread laterally along the permeability contrast.

Most organic chemical contaminants are introduced into the subsurface by percolation through soil strata and the interactions between the soil and a chemical contaminant are important for assessing the fate and transport of the contaminant in the subsurface, especially in the groundwater system. Contaminants that are highly soluble, such as salts of carboxylic acid derivatives can move readily from surface soil to saturated materials below the water table and often occurs during and after rainfall events. Those contaminants that are not highly soluble may have considerably longer residence times in the surface strata (the soil zone). Some contaminants adsorb readily onto soil particles and slowly dissolve during precipitation events, resulting in migration into groundwater—this is typical of the mode of transport for chemicals such as trichloroethylene ($CCl_2{=}CHCl$). Liquids spilled onto surface soils can migrate downward or can evaporate, which limits their potential for reaching the water table. Once below the water table, organic chemical contaminants are also subject to dispersion (mechanical mixing with uncontaminated water) and diffusion (dilution by concentration gradients).

Many organic contaminants (such as crude oil and crude oil-derived products) begin to spread immediately after entry into the environment. The viscosity of the oil, its pour point, and the ambient temperature will determine how rapidly the oil will spread, but light oils typically spread more rapidly than heavy oils. The rate of spreading and ultimate thickness of the oil slick will affect the rates of the other weathering processes. For example, discharges that occur in geographically contained areas (such as a pond or slow-moving stream) will evaporate more slowly than if the oil is allowed to spread.

3 TYPES OF CHEMICALS

The ability to collect and preserve a sample that is representative of the site is a critically important step in identification of chemicals in the environment (Dean, 1998; Patnaik, 2004; Speight, 2005). Obtaining representative environmental samples is always a challenge due to the heterogeneity of different sample matrices. Additional difficulties are encountered with petroleum hydrocarbons due to the wide range in volatility, solubility, biodegradation, and adsorption potential of individual constituents and the procedures used for sample collection and preparation must be legally defensible.

As described above, organic chemicals can enter the air, water, and soil when they are produced, used, or disposed. The impact of these chemicals on the environment is determined by the amount of the chemical that is released, the type and concentration of the chemical, and where it is found. Some chemicals can be harmful if released to the environment even when there is not an immediate, visible impact. On the other hand, some chemicals are of concern as they can work their way into the food chain and accumulate and/or persist in the environment for many years.

On a beneficial note, various types of organic chemicals (including mixtures such as crude oil and various crude oil-derived products) are biodegraded faster than many individual and more complex chemicals (Appendix, Tables A2–A6). Nevertheless, different types of organic chemicals and crude oils (as well as the crude oil products even though given similar names) are biodegraded at different rates in the same ecosystem. A crude oil product is a complex mixture of organic chemicals and contains within it less persistent constituents and more persistent constituents. The range between these two extremes is often extensive since the different organic chemicals have different physical and chemical properties and estimating the behavior and fate of any organic chemical (or even a particular crude oil-derived product) is very difficult.

The relative proportion of hazardous constituents present in any collection of organic chemicals (crude oil-derived products included) is variable and rarely consistent because of site differences. Therefore, the extent of the contamination will vary from one site to another and, in addition, the farther a contaminant progresses from low molecular weight to high molecular weight the greater the occurrence of polynuclear aromatic hydrocarbons, complex ring systems (not necessarily aromatic ring systems), as well as an increase in the composition of the semivolatile organic chemicals or the nonvolatile organic chemicals. These latter organic chemical constituents (many of which are not so immediately toxic as the volatiles) can result in long-term/chronic impacts to the flora and fauna of the environment. Thus, any complex mixture of organic chemicals should be analyzed for the semivolatile compounds which may pose the greatest long-term risk to the environment.

In addition to large spills of organic chemicals, crude oil-based hydrocarbons are released into the environment from natural seeps as well as NPS urban runoffs. Acute impacts from massive one-time spills are obvious and substantial but the impact of organic chemicals from small spills and chronic releases are the subject of much speculation. Clearly, such inputs of chemicals have the potential for significant environmental impact, but the effects of chronic low-level discharges can be minimized by the net assimilative capacities of many ecosystems, resulting in little detectable environmental harm.

Short-term (acute) hazards of lighter, more volatile, and water-soluble aromatic compounds (such as benzenes, toluene, ethylbenzene, and the xylene isomers, BTEX) include potential acute toxicity to aquatic life (especially in relatively confined areas) as well as potential inhalation hazards to other faunal species (including humans). However, the compounds which pass through the water column often tend to do so in small concentrations and/or for short periods of time, and fish and other pelagic species or generally mobile species can often swim away to avoid impacts from spilled oil in open waters.

Briefly, pelagic species are species that frequent the pelagic zone which is any water in a sea or lake that is neither close to the bottom nor near to the shore. The pelagic zone can be described in terms of an imaginary water column that extends from the surface of the sea almost to the sea bed. Conditions differ deeper in the water column such that as pressure increases with depth, the temperature drops and less light penetrates. Depending on the depth, the water column, rather like the atmosphere of the Earth (Chapter 1) may be divided into different layers. Pelagic life decreases with increasing depth and is affected by light intensity, pressure, temperature, salinity, the supply of dissolved oxygen, the supply of nutrients, and the submarine topography (bathymetry). In deep water, the pelagic zone (sometimes referred to as the *open-ocean zone*) can be contrasted with water that is near to the coast or water above the continental shelf.

Long-term (chronic) potential hazards of lighter, more volatile, and water-soluble aromatic compounds include contamination of groundwater. Chronic effects of benzene, toluene, ethylbenzene, and the xylene isomers include changes in the liver and harmful effects on the kidneys, heart, lungs, and nervous system. At the initial stages of a release of these organic chemicals, when the benzene derived compounds are present at their highest concentrations, acute toxic effects are more common sooner rather than later. These noncarcinogenic effects include subtle changes in detoxifying enzymes and liver damage. Generally, the relative aquatic acute toxicity of petroleum will be the result of the fractional toxicities of the different hydrocarbons present in the aqueous phase. There are also indications that naphthalene-derived chemicals have a similar effect.

Organic chemicals are weathered according to the individual physical properties and chemical properties, but during this process living species within the local environment may be affected via one or more routes of exposure, including ingestion, inhalation, dermal contact, and, to a much lesser extent, bioconcentration through the food chain. Aromatic compounds of concern include alkylbenzene derivatives, toluene derivatives, naphthalene derivatives, and polynuclear aromatic hydrocarbons (PNAs). Moreover, the impact on both, atmosphere and hydrosphere must be assessed when considering the implications from a release of chemical containing significant quantities of these single-ring aromatic compounds.

By way of explanation, bioconcentration is the accumulation of a chemical in or on an organism and is also the process by which a chemical concentration in an organism exceeds that in the surrounding environment as a result of exposure of the organism to the chemical. Bioconcentration can be measured and assessed and these include: (1) octanol-water partition coefficient (K_{ow}), (2) the bioconcentration factor, BCF, (3) the bioaccumulation factor, BAF, and (4) the biota-sediment accumulation factor, BSAF. Each of these factors can be calculated using either empirical data or measurements as well as from mathematical models (Mackay, 1982). The BCF can also be expressed as the ratio of the concentration of a chemical in an organism to the concentration of the chemical in the surrounding environment and is a measure of the extent of chemical sharing between an organism and the surrounding environment. Thus

$$BCF = (\text{concentration in biota})/(\text{concentration in ecosystem})$$

The BCF can also be related to the octanol-water partition coefficient (K_{ow}) which is correlated with the potential for a chemical to bioaccumulate in floral and faunal organisms. The BCF can be predicted from the octanol-water partition coefficient (Bergen et al., 1993):

$$\text{Log}(\text{bioconcentration factor}) = m \log K_{ow} + b$$

$$K_{ow} = (\text{concentration in octanol})/(\text{concentration in water})$$
$$= C_o/C_w \text{ at equilibrium}$$

4 PHYSICAL PROPERTIES AND DISTRIBUTION IN THE ENVIRONMENT

To effectively monitor changes in the environmental behavior of organic chemicals that are of most concern, it is extremely important to understand how these chemicals typically behave in natural systems and, in particular, in specific ecosystems. Equally important is an understanding of how these chemicals might respond to specific best management practices. Some of the discharged organic chemicals only become problems at high concentrations that impair the beneficial uses of these ecosystems. An effective monitoring program explicitly considers how these chemicals may change as they move from a source into the groundwater, surface water, or into the soil. This includes an understanding of how a specific organic chemical may be introduced or mobilized within an ecosystem, how the chemical moves through an ecosystem, and the transformations that may occur during this process (Chapter 7).

However, the entry of chemicals into the environment and the distribution of these chemicals within the environment is often complex and there have been many occasions when a significant amount of a chemical (or a mixture of chemicals) has entered an ecosystem and the effects of contamination are well defined. Generally, the assumption is that the organic chemical (or a mixture thereof) does not rapidly diffuse away, but remains in the immediate vicinity at a noticeably high concentration or perhaps moves, but in such a way that concentration levels of the chemical remain high as it moves. Such cases would normally occur when large quantities of a substance were being stored, transported, or otherwise handled in concentrated form.

Thus, due to leakages, spills, improper disposal, and accidents during transport, organic compounds have become subsurface contaminants that threaten various resources, the most important of which are drinking water resources.

4.1 Chemical and Biochemical Properties

One strategy to remediate such polluted environments, especially the subsurface, is to make use of the degradative capacity of bacteria. It is often sufficient to supply the subsurface with nutrients containing elements such as nitrogen and phosphorus. However, anaerobic processes have advantages such as low biomass production and good electron acceptor availability, and they are sometimes the only possible solution for cleanup and protection of the environment.

Whereas hydrocarbons are oxidized and completely mineralized under anaerobic conditions in the presence of electron acceptors such as nitrate (NO_3), iron (Fe), sulfate (SO_4), and carbon dioxide (CO_2), chlorinated organic compounds and nitroaromatic compounds are reductively transformed to products that are not always benign. For the often persistent polychlorinated compounds, reductive dechlorination leads to harmless products or to compounds that are aerobically degradable. The nitroaromatic compounds $(ArNO_2)$ are first reductively transformed to the corresponding amines $(ArNH_2)$ and can subsequently be bound to the humic fraction of the soil in an aerobic process—humic constituents of soil are the major organic constituents of soil that are produced by biodegradation of dead organic matter.

Such new findings and developments give hope that in the near future contaminated aquifers can efficiently be remediated, a prerequisite for a sustainable use of the precious subsurface drinking water resources.

4.2 Partitioning

The increasing awareness of chemicals in the environment, their disposition, and their ultimate fate has created the need to find reliable mechanisms to assess the environmental behavior and effects of new chemicals. Whether or not a chemical will pose a hazard to the environment will depend on the concentration levels it will reach in various ecosystems and whether or not those concentrations are toxic to the floral and faunal species within the ecosystem. It is, therefore, important to determine expected environmental distribution patterns of chemicals in order to identify which areas will be of primary environmental concern (McCall et al., 1983).

In order to assess the behavior of an organic chemical in the environment, typical physical and chemical properties play a role but a property that is not often considered in the partitioning of the organic chemical in an ecosystem is a function of the chemical and physical structure of the chemical. This leads to a determination of the way in which the chemical or the mixture of chemicals are distributed among the different environmental phases (McCall et al., 1983). These phases may include air, water, organic matter, mineral solids, and even organisms.

More specifically, the focus of partitioning studies is on the equilibrium distribution of an organic chemical that is established between the phases the associated issue problem of calculating the distribution of a compound between the different phases (*partitioning equilibrium*). There are many situations in which it is correct to assume that phase transfer processes are fast compared to the other processes, such as chemical transformation of the organic contaminant to other (benign or hazardous) chemicals (Chapter 7) that play a role in determining the fate of contaminants. In such cases, it is appropriate to describe the distribution of the chemical as a change of phase by the equilibrium approach using the valid assumption that an equilibrium condition will be reached with the chemical passing to one phase or remaining distributed between different phases. By way of explanation, an example of phase equilibrium is the partitioning of a contaminant (such as carboxylic acid or a carboxylic acid derivative—that is, the sodium salt of the acid) between the pore water and other water in the bed of a sediment and solids in sediment beds. Thus, from the equilibrium partition coefficient it is possible to calculate the rate of transfer of an organic chemical contaminant across interfaces and the rate at which such transfer can be anticipated to occur.

However, before too much excitement is generated at the thought of the answers that a study of partitioning will provide, it must be recognized that there are many situations where an equilibrium of the chemical between the phases is not reached. However, some observers (justifiably) find the information useful insofar as the data are used (correctly or incorrectly) to characterize what the equilibrium distribution of the chemical would be if sufficient time (often difficult to define) is allowed. However, in such cases a quantitative description of the potential partitioning equilibrium can be employed to estimate the possible direction of the transport of the organic chemical contaminant from one environmental ecosystem to another. If this is the case, it may be possible to evaluate whether or not the chemical component(s) of a contaminant (such as a solvent or naphtha) will continue to dissolve in the groundwater and/or volatilize into the overlying soil. Both options are viable, but (hopefully) the data will provide the most likely option to occur, given the position of the chemical within the underground chemical and geological system. Thus, the goal of any partitioning study of the ability of an organic contaminant is to gain insight into the role played by the chemical (and physical) structure of a contaminant in determining the fate of the contaminant in the environment (McCall et al., 1983).

However, when dealing with phase partitioning parameters, it is necessary to develop an understanding of the intramolecular interactions and the intermolecular interactions (Chapter 5) between the organic chemical(s) and the specific molecular environment in which the chemical is spilled or disposed. Thus, there is the necessity to understand the means by which structural groups function within the contaminants and the means by which the functional groups are related to chemical and physical behavior (Chapters 2 and 5). Therefore, there must be an effort to understand (1) the interactions arising from contacts of functional groups within a molecule, i.e., the intramolecular interactions, (2) the influence of these intramolecular interactions on the existence of the molecule in the environment, (3) the interactions arising from contacts of functional groups with the functional groups of another molecule, i.e., the intermolecular interactions, and (4) the influence of these intermolecular interactions on the existence of the molecule in the environment. However, partition coefficients are not the only means of estimating the transfer of organic chemicals between phases. It is also valuable to correlate the partition coefficients on the basis of solubility of the chemical(s) which are convenient and readily understood measurable expressions of single-phase activity coefficients (Cole and Mackay, 2000).

Thus, this section presents an examination of the pertinent compound properties and environmental factors that are needed for quantifying such partitioning.

4.2.1 Acid-Base Partitioning

Acid-base partitioning (also called *pH partitioning*) is the tendency for acids to accumulate in basic fluid compartments and bases to accumulate in acidic regions. The reason for this phenomenon is that acids become negatively electric charged in basic fluids, since they donate a proton. On the other hand, bases become positively electric charged in acidic liquids because the base receives a proton (H^+). Since electric charge decrease the membrane permeability of substances, once an acid enters a basic fluid and becomes electrically charged, then it cannot escape that compartment with ease and therefore accumulates, and vice versa with bases. Thus, by manipulating the pH of the aqueous layer, the partitioning of a solute can be changed.

4.2.2 Air-Water Partitioning

The air-water partition coefficient (K_{aw}) is the constant of proportionality between the concentration of a chemical in air and its concentration in water at low partial pressures and below its saturation limits in either air or water.

The coefficient can be estimated from Henry's law constant and is sometimes referred to as the dimensionless.

It is essential to have reliable data for the air-water partition coefficient or Henry's law constant for the compounds under investigation for elucidating the environmental dynamics of many natural and anthropogenic compounds. When a compound (here referred to as the solute) is introduced into the environment, it tends to diffuse from phase to phase in the direction towards establishing equilibrium between all phases. Frequently, the physical-chemical properties of the solute dictate that it will partition predominantly into a different phase from the one into which it is normally emitted. For example, benzene emitted in waste water will tend to partition or transfer from that water into the atmosphere where it becomes subject to atmospheric photolytic degradation processes. A knowledge of the air-water partition characteristics of a solute is thus important for elucidating where the solute will tend to accumulate and also in calculating the rates of transfer between the phases. Conventionally these rates are expressed as the product of a kinetic constant such as a mass transfer coefficient (or diffusivity divided by a diffusion path length) and the degree of departure from equilibrium which exists between the two phases. Elucidating the direction and rate of transfer of such solutes thus requires accurate values for the Henry's law.

The transfer between the atmosphere and bodies of water is one of the key processes affecting the transport of many organic compounds in the environment. For neutral compounds, at dilute solution concentration in pure water, the air-water distribution ratio is referred to as the Henry's Law constant (K_H or K_{aw}). For real aqueous solutions (i.e., solutions that contain many other chemical species), the term *air-water distribution ratio* is often used which, for practical purposes, is approximated by the Henry's Law constant. The Henry's Law constant K_H can be approximated as the ratio of a compound's abundance in the gas phase to that in the aqueous phase at equilibrium.

$$X(\text{aq}) = X(\text{g})$$

$$K = P_i/C_w \text{ atm L mol}^{-1}$$

In this equation, P_i is the partial pressure of the gas phase of the chemical and C_w is the molar concentration of the chemical in water.

Knowledge of the air-water partitioning behavior of VOCs and GHGs is important in a number of environmental applications. For example, VOCs emitted from open process streams in a paper mill can promote ground level ozone and lead to respiratory problems in humans. Therefore, it is important to have reliable estimates of the amount of VOCs in the atmosphere in contact with open process streams. In global climate models, the partitioning of carbon dioxide and methane between the atmosphere and ocean water is of current interest since oceans represent a substantial storage reservoir for these gases.

4.2.3   Molecular Partitioning

Partition coefficients are the ratio of the concentration of an organic compound in two phases that are in equilibrium with each other. For example, in a two-layer system of water (bottom layer) and an organic solvent (top layer), an organic compound will be in one or the other of the layers. After stirring and allowing time for the layers to settle, the organic compound could well be in both phases, albeit to a different extent (concentration, C) in each phase. The partition coefficient (K) is

$$K = C_{\text{organic}}/C_{\text{water}}$$

From this equation, a high value of K suggests that the compound is not very water soluble but is more soluble in the organic solvent that is if the organic compound is lipophilic (or hydrophobic).

Molecular partitioning occurs when an organic compound dissolves in each of two immiscible solvent phases and is measured by the *partition coefficient* or *distribution coefficient* which is the ratio of concentrations of the compound in a mixture of the two immiscible phases at equilibrium. This ratio is therefore a measure of the difference in solubility of the compound in these two phases. Most commonly, one of the solvents is water while the second is hydrophobic such as 1-octanol (1-hydroxy octane, $CH_3CH_2CH_2CH_2CH_2CH_2CH_2CH_2OH$). Hence the partition coefficient measures how hydrophilic or hydrophobic a chemical substance is.

The hydrophobic nature (hydrophobicity) of a compound can give an indication of the relative ease that a compound might be taken up in groundwater to pollute waterways. The partition coefficient can also be used to predict the mobility of species in groundwater and, in the field of hydrogeology, the octanol-water partition coefficient (K_{ow}) is used to predict and model the migration of dissolved hydrophobic organic compounds in soil and groundwater.

4.2.4   Octanol-Water Partitioning

The distribution of nonpolar organic compounds between water and natural solids (e.g., soils, sediments, and suspended particles) or organisms, can be viewed in many cases as a partitioning process between the aqueous phase and the bulk organic matter present in natural solids or in biota. More recently, environmental chemists have found similar correlations with soil humus and other naturally occurring organic phases. These correlations exist because the same molecular forces controlling the distribution of compounds between water-immiscible organic solvents and water also determine environmental partitioning from water into natural organic phases.

The octanol-water partition coefficient is a key parameter in understanding and predicting the environmental fate and transport behavior of organic chemicals (Lyman et al., 1990; Boethling and Mackay, 2000). K_{ow} is often used as a surrogate for the lipophilicity of a chemical and its tendency to concentrate in organic phases such as within plant lipids or fish from the aqueous solution.

Chemicals with relatively low K_{ow} values are considered relatively hydrophilic and will tend to have high water-solubility and low BCFs (Lyman et al., 1990). It is also used in the prediction of other parameters including water solubility and the organic carbon-water partition coefficient.

Thus, in order to simulate lipids (fats) in organic media (biota), n-octanol ($CH_3CH_2CH_2CH_2CH_2CH_2CH_2CH_2OH$) was selected as model compound for partitioning experiments. Thus, the partition coefficient that best describes lipophilicity is the octanol-water partition coefficient (K_{ow}). The values of the octanol-water partition coefficient typically fall into the range 102–107 and it is often more convenient to use the common logarithm of K_{ow}. The octanol-water partition coefficient is defined simply by

$$K_{ow} = C_{octanol}/C_{water}$$

$C_{octanol}$ is the molar concentration of the organic compound in the octanol phase and C_{water} is the molar concentration of the organic compound in water when the system is at equilibrium.

For a series of neutral nonpolar compounds partitioning between octanol and water, the K_{ow} value is determined largely by the magnitude of the aqueous activity coefficient (a measure of the dissimilarity between the organic solute and the aqueous solvent). In other words, the major factor that determines the magnitude of the partition constant of a nonpolar or moderately polar organic compound between an organic solvent and water is the incompatibility of the compound with water. The nature of the organic solvent is generally of secondary importance.

4.2.5 Sorption Partitioning

The persistence of organic pollutants in topsoil, their migration to groundwater, and the evaluation of the degree of contamination expected in a groundwater system after an accidental spill or as consequence of the presence of a waste disposal site, are problems of particular environmental concern which require the knowledge of the sorption characteristics of the pollutants to be investigated as well as the knowledge of the type of soil and its characteristics. Sorption also affects volatility of organic pollutants, their bioavailability and bioactivity, phytotoxicity, and chemical or microbial transformations (Delle Site, 2001).

The sorption of an organic chemical on a natural solid is a very complicated process which involves many sorbent properties, besides the physicochemical properties of the chemical itself. These properties are especially reflective of the relative amount of the mineral and organic material in soil/sediment and their respective composition with associated physical characteristics. Also, different regions of a soil or sediment matrix may contain different types, amounts, and distributions of surfaces and of soil organic material, even at the particle scale.

The extent to which an organic chemical partitions itself between solid and solution phases is determined by several physical and chemical properties of both the chemical and the soil or sediment aqueous solutions. However, in most cases, the tendency of a chemical to be adsorbed or desorbed can be expressed in terms of the organic carbon partition coefficient (K_{OC}) which is largely independent of soil or sediment properties (Lyman et al., 1982). Thus, the organic carbon partition coefficient is a chemical specific adsorption parameter and may be determined as the ratio of the amount of chemical adsorbed per unit weight of organic carbon in the soil or sediment to the concentration of the chemical in solution at equilibrium:

$$K_{OC} = (\text{microgram adsorbed per gram of organic carbon}) / (\text{microgram per mL of solution})$$

Factors which affect measured values of the organic carbon partition coefficient include (1) temperature, (2) acidity or alkalinity, measures as the pH, (3) salinity, (4) the concentration of dissolved oxygen, (5) suspended particulates, and (6) the solids-to-solution ratio. The presence of other chemicals in a complex mixture could alter the activity coefficient of the chemical in the water, the pH of the water, or the solubility of the chemical in water, and, consequently, the sorption of the chemical to soils and sediments. The degree of adsorption will not only affect chemical mobility but will also affect volatilization, photolysis, hydrolysis, and biodegradation. Adsorption of organic chemicals will also occur on minerals free of organic matter. It may be significant under certain conditions such as: (1) clay minerals with a very large surface area, (2) situations where cation exchange occurs, such as for dissociated organic bases, (3) situations where clay-colloid-induced polymerization occurs, and/or (4) situations where chemisorption is a factor (Lyman et al., 1982).

The distribution of an organic solute between sorbent and solvent phases results from its relative affinity for each phase, which in turn relates to the nature of forces which exist between molecules of the solute and those of the solvent and sorbent phases. The type of interaction depends on the nature of the sorbent as well as the physicochemical features of the sorbate ~ hydrophobic or polar at various degrees.

The physical sorption processes involve interactions between dipole moments—permanent or induced—of the sorbate and sorbent molecules. The relatively weak bonding forces associated with physical sorption are often amplified in the case of hydrophobic molecules by substantial thermodynamic gradients for repulsion from the solution in which they are dissolved. Chemical interactions involve covalent bond and hydrogen bond. Finally, electrostatic interactions involve ion-ion and ion-dipole forces. In a more detailed way, the type of interactions and the approximate values of energy associated are (1) van der Waals interactions, (2) hydrophobic bonding, (3) hydrogen bonding, (4) charge transfer, (5) ligand-exchange and ion bonding, (6) direct and induced ion-dipole and dipole-dipole interactions, and (7) chemisorption by the formation of a covalent bond. Sorption of organic pollutants sometimes can be explained with the simultaneous contribution of two of more of these mechanisms, especially when the nonpolar or polar character of the compounds is not well defined.

4.3 Vapor Pressure

The vapor pressure (or equilibrium vapor pressure) of an organic compound is the pressure exerted by a vapor in thermodynamic equilibrium with the condensed phase (liquid or solid) at a given temperature in a closed system (Boethling and Mackay, 2000; Mackay et al., 2006). The equilibrium vapor pressure is an indication of the evaporation rate of the liquid and relates to the tendency of particles to escape from the liquid (or a solid). A substance with a high vapor pressure at normal temperatures is often referred to as *volatile*. As the temperature of a liquid increases, the kinetic energy of its molecules also increases and the number of molecules passing from the liquid phase to the vapor phase also increases, thereby increasing the vapor pressure.

Vapor pressure controls the volatility of a chemical from soil, and along with its water solubility, determines evaporation from water (Boethling and Mackay, 2000). Predicting the volatility of a chemical in soil systems is important for estimating chemical partitioning in the subsurface between sorbed phase, dissolved phase, and gas phase. Vapor pressure is an important parameter in estimating vapor transport of an organic chemical in air. It is often useful to determine and other properties of a chemical by assuming that it is a liquid or supercooled liquid at a temperature less than the melting point (Mackay et al., 2006). At very low environmental concentrations such as in liquid solutions or on aerosol particles, pure chemical behavior relates to the liquid rather than the solid state.

The vapor pressure of any substance increases nonlinearly with temperature according to the Clausius-Clapeyron relation, which equation allows an expression of the pressure, P, the enthalpy of vaporization, DH_{vap}, and the temperature, T, as:

$$P = A \exp\left(-DH_{vap}/RT\right)$$

In this equation, R ($=8.3145$ J mol^{-1} K^{-1}) and A are the gas constant and unknown constant, respectively. If P_1 and P_2 are the pressures at two temperatures T_1 and T_2, the equation has the form:

$$\ln P_1/P_2 = DH_{vap}/R(1/T_1 - 1/T_2)$$

The Clausius-Clapeyron equation allows an estimate of the vapor pressure at another temperature, if the vapor pressure is known at some temperature, and if the enthalpy of vaporization is known.

The atmospheric pressure boiling point (the normal boiling point) of a liquid is the temperature at which the vapor pressure equals the ambient atmospheric pressure. With any incremental increase in that temperature, the vapor pressure becomes sufficient to overcome atmospheric pressure and lift the liquid to form vapor bubbles inside the bulk of the substance. Bubble formation deeper in the liquid requires a higher pressure, and therefore higher temperature, because the fluid pressure increases above the atmospheric pressure as the depth increases.

4.4 Volatility

Volatility of an organic compound or of the organic components of a mixture is an important loss mechanism in the overall materials balance. The key environmental factors affecting volatilization are the reaction constant (surface transfer rate of dissolved oxygen per mixed depth of the water body), wind speed, and the mixed depth of the water. Furthermore, when evaluating the volatilization of complex mixtures vs. single chemicals, only physicochemical properties can be affected differently by mixtures. Thus, only by altering either aqueous solubility or vapor pressure of a chemical by interactions with other chemicals can volatilization rates be altered. Also, if the composition of a given mixture is known as well as the solubility of each constituent (from knowledge of the physiochemical properties of each constituent), it is possible to anticipate that rate of change for volatilization due to chemical interactions.

Organic chemicals, whether they are produced by the chemical industry or by the pharmaceutical industry or the petroleum industry, can be considered as environmentally transportable materials, the character of which is determined by several chemical and physical properties (i.e., solubility, vapor pressure, and propensity to bind with soil and organic particles). These properties are the basis of measures of leachability and volatility of individual hydrocarbons. Thus, the transport or organic chemicals either as individual chemicals or as mixtures (such as the various crude oil-derived products) can be considered by equivalent carbon number to be grouped into thirteen different fractions. The analytical fractions are then set to match these transport fractions, using specific *n*-alkane derivatives to mark the analytical results for aliphatic compounds and selected aromatic compounds to delineate hydrocarbons containing benzene rings.

Although organic chemicals grouped by transport fraction generally have similar toxicological properties, this is not always the case. For example, benzene is a carcinogen but many alkyl-substituted benzenes do not fall under this classification. However, it is more appropriate to group benzene with compounds that have similar environmental transport properties than to group it with other carcinogens such as benzo(a)pyrene that have very different environmental transport properties. Nevertheless, consultation of any reference work that lists the properties of chemicals will show the properties and hazardous nature of the types of chemicals that are found in petroleum. In addition, petroleum is used to make petroleum products, which can contaminate the environment.

The range of chemicals produced by the organic chemicals industry and by the petroleum refining industry is so vast that summarizing the properties and/or the toxicity or general hazard of crude oil in general or even for a specific crude oil is a difficult task. However, petroleum and some petroleum products, because of the hydrocarbon content, are at least theoretically biodegradable but large-scale spills can overwhelm the ability of the ecosystem to break the oil down. The toxicological implications from petroleum occur primarily from exposure to or biological metabolism of aromatic structures. These implications change as an oil spill ages or is weathered.

4.4.1 Low-Boiling Organic Chemicals

Many of the gaseous organic chemicals and liquid organic chemicals that are low-boiling materials (including crude oil and crude oil-delved products) fall into the class of chemicals which have one or more of the following characteristics that are considered to be hazardous by the Environmental Protection Agency: (1) ignitability, (2) flammability, (3) corrosivity, and (4) reactivity. In summary, many of the specific chemicals in crude oil and crude oil-derived products are hazardous because of their chemical reactivity, fire hazard, toxicity, and other properties. In fact, a simple definition of a hazardous chemical (or hazardous waste) is that it is a chemical substance (or chemical waste) that has been inadvertently released, discarded, abandoned, neglected, or designated as a waste material and has the potential to be detrimental to the environment. Alternatively, a hazardous chemical may be a chemical that may interact with other (chemical) substances to give a product that is hazardous to the environment. Low-boiling organic chemicals (whether crude oil-derived or derived from another source) fit very well into this definition; examples of some commonly encountered volatile hydrocarbons are

Aliphatic Hydrocarbons	Aromatic Hydrocarbons
Pentane derivatives	Benzene
Hexane derivatives	Toluene
Heptane derivatives	Ethylbenzene
Octane derivatives	Xylene isomers and derivatives
Nonane derivatives	Naphthalene derivatives
Decane derivatives	Phenanthrene derivatives
	Anthracene derivatives
	Acenaphthylene derivatives

For example, a liquid that has a flash point of <60°C (140°F) is considered *ignitable*. Some examples are: benzene, hexane, heptane, benzene, pentane, petroleum ether (low boiling), toluene, and the xylene isomers. An organic chemical is classed as *flammable* if the chemical has the ability of a substance to burn or ignite, causing fire or combustion. The degree of difficulty required to cause the combustion of a substance is quantified through standard test methods (Speight, 2001, 2014, 2015). The data from such test methods are used in regulations that govern the storage and handling of highly flammable substances inside and outside of structures and in surface and air transportation.

An aqueous solution that has a pH of ≤2, or ≥12.5 is considered *corrosive*. Most organic chemicals, crude oils, and crude oil-derived products are not corrosive but many of the chemicals used in refineries are corrosive—corrosive materials include inorganic chemicals such as sodium hydroxide as well as other acids or bases. The term *reactivity* applies to chemicals that react violently with air or water and, as a result, are considered to be hazardous chemicals. Reactive organic chemicals include chemicals capable of detonation (TNT) when subjected to an initiating source.

Gas condensate also falls within the VOC category and condensate release can be equated to the release of volatile constituents but are often named as such because of the specific constituents of the condensate, often with some reference to the gas condensate that is produced by certain crude oil wells and natural gas wells. However, the condensate is often restricted to the low-boiling alkane derivative as well as benzene, toluene, ethyl benzene, and the xylene isomers (BTEX).

4.4.2 High-Boiling Organic Chemicals

In almost all cases of hydrocarbon contamination, attention must be directed to the presence of semivolatile hydrocarbon derivatives and nonvolatile hydrocarbon derivatives.

Among the polynuclear aromatic hydrocarbons, the toxicity of many hydrocarbon liquids (especially crude oils and crude oil-derived products) is a function of its di- and tri-aromatic hydrocarbon content. Like the single aromatic ring variations, including benzene, toluene, and the xylene isomers, all are relatively volatile compounds with varying degrees of water solubility. However, in the higher boiling hydrocarbons liquids (particular products designed as fuel oil), the two-ring condensed aromatic hydrocarbons, naphthalene and the various homologs, are less acutely toxic than benzene but are more prevalent for a longer period after a spill or discharge. The toxicity of different crude oils and refined products (such as naphtha and naphtha-derived solvents and fuels) depends not only on the total concentration of hydrocarbons but also on the hydrocarbon composition in the water-soluble fraction (WSF) as well as on the (1) the degree of water solubility, (2) the concentrations of the individual component, and (3) the toxicity of the components either individually or collectively. The WSFs prepared from different crude oils will vary in terms of these three parameters. WSFs of the refined products (such as naphtha and naphtha-derived solvents, and fuels such as No. 2 fuel oil and Bunker C oil) are more toxic to the floral and faunal inhabitants of many ecosystems than the typical WSF of crude. Organic chemical either having a higher number of condensed rings or with methyl substituents on the rinds are typically more toxic than the less substituted derivatives but tend to be less water soluble and thus less plentiful in the WSF. There are also indications that pure naphthalene (a constituent of moth balls that are, by definition, toxic to moths) and alkyl naphthalene derivatives are from three-to-ten times more toxic to test animals than are benzene and alkylbenzene derivatives. In addition, and because of the low water-solubility of tricyclic and polycyclic (polynuclear) aromatic hydrocarbons (that is, those aromatic hydrocarbons heavier than naphthalene), these compounds are generally present at very low concentrations in the WSF of oil. Therefore, the results of this study and others conclude that the soluble aromatics of crude oil (such as benzene, toluene, ethylbenzene, xylene isomers, and naphthalene derivatives) produce the majority of toxic effects of crude oil in the environment.

The higher molecular weight aromatic structures (with four to five condensed aromatic rings), which are the more persistent in the environment, have the potential for chronic toxicological effects. Since these compounds are nonvolatile and are relatively insoluble in water, their main routes of exposure are through ingestion and epidermal contact. Some of the compounds in this classification are considered possible human carcinogens; these include benzo(a)pyrene, benzo(e)pyrene, benzo(a)anthracene, benzo(b, j, and k)fluorene, benzo(ghi)perylene, chrysene, dibenzo(ah)anthracene, and pyrene.

4.5 Water Solubility

The water solubility of an organic compound is the solubility of the organic compound in water (typically in grams per liter) at a designated temperature and pressure. At the molecular level, solubility is controlled by the energy balance of intermolecular forces between solute-solute, solvent-solvent, and solute-solvent molecules. Intermolecular forces vary in strength from very weak induced dipole to the much stronger dipole-dipole forces (such as hydrogen bonding). Most organic molecules are relatively nonpolar and are usually soluble in organic solvents (e.g., diethyl ether, dichloromethane, chloroform, organic chemicals ether, hexanes, etc.) but not in polar solvents such as water. However, some organic molecules are more polar and therefore soluble in water, is generally indicative of a high ratio of polar group(s) to the nonpolar hydrocarbon chain, i.e., a low molecular weight compound containing a hydroxyl group ($-OH$) or an amino group ($-NH_2$) or a carboxylic acid group ($-CO_2H$ group) or a higher molecular weight compound containing several polar groups. The presence of an acidic carboxylic acid group or a basic amino group in a water-soluble compound can be detected by the low or high pH, respectively, of the solution.

Water solubility is one of the most important properties for evaluating the fate and direct measure of chemical since chemicals with high water solubility will partition readily and rapidly into the aqueous phase and will often remain in solution and be available for degradation. Chemicals that are sparingly soluble in water will often dissolve slowly into solution and partition more readily into other phases including air, solids, and the surface of solid particles including soil (Boethling and Mackay, 2000). Water solubility is used to determine the maximum theoretical concentration of a chemical in the soil pore water and is also important for estimating the air-water partitioning coefficient (K_{aw}), used to determine the partitioning behavior of an organic chemical in the soil.

There is a simple, very useful, and practical empirical rule that is quite reliable—*like dissolves like*—and it is based on the polarity of the systems insofar as polar molecules dissolve in polar solvents (such as water, alcohols) and nonpolar molecules dissolve in nonpolar solvents (such as liquid hydrocarbons).

The solubility of organic chemicals is an important property when assessing toxicity since the water solubility of an organic chemical determines the routes of exposure that are possible. In fact, an easy way of assessment of the water solubility of organic chemicals is that the solubility of an organic chemical is approximately inversely proportional to molecular weight—lower molecular weight hydrocarbon derivatives (excluding the presence of polar functional groups) are typically more soluble in water than the higher molecular weight compounds. Thus, lower molecular weight hydrocarbons (specifically, the C_4 to C_8 alkanes, including the aromatic compounds) are relatively soluble, up to approximately 2000 ppm, while the higher molecular weight hydrocarbon derivatives are much less soluble in water. Typically, the most soluble components are also the most toxic but whether this toxicity is due to the chemical and structural aspects of the lower molecular weight derivatives or whether it is noticed more frequently because of the relative ease of transportation is subject to debate.

Finally, each functional group has a particular set of chemical properties that allow it to be identified. Some of these properties can be demonstrated by observing solubility behavior, while others can be seen in chemical reactions that are accompanied by color changes, precipitate formation, or other visible effects. The identification and characterization of the organic chemicals is an important aspect of environmental organic chemistry. Although it is often possible to establish the structure of spectroscopic test methods (Chapter 5), the spectroscopic data should be supplemented with additional information such as (1) the physical state and (2) the relevant properties such as melting point, boiling point, solubility, odor, elemental analysis, and confirmatory tests for the presence of functional groups. The later test methods usually involve simple laboratory tests for solubility from which conclusions can be drawn which, when combined with other test data, can provide valuable information about the nature of the sample, but whether or not this can be applied to complex mixtures is another issue. Nevertheless, the solubility of an organic compound in water, dilute acid, or dilute base can provide useful information about the presence or absence of certain functional groups. For example, (1) solubility in water, (2) solubility in sodium hydroxide, and (3) solubility in hydrochloric acid (Fig. 6.1).

Most organic compounds are insoluble in water, except for low molecular weight amines and oxygen-containing compounds. Low molecular weight compounds are generally limited to those with fewer than five carbon atoms. Carboxylic acids (RCO_2H) with fewer than five carbon atoms in the molecule are soluble in water and form solutions that give an acidic response (pH <7) when tested with litmus paper. Amines (RNH_2) with fewer than five carbon atoms in the molecule are also soluble in water, and their amine solution gives a basic

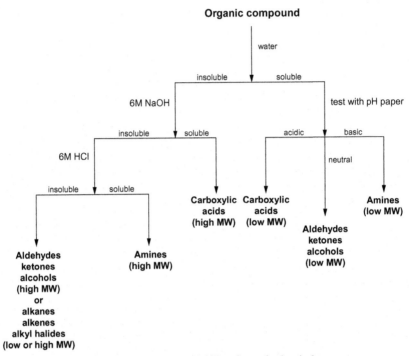

FIG. 6.1 General schematic for testing the solubility of organic chemicals.

response (pH >7) when tested with litmus paper. Ketones, aldehydes, and alcohols with fewer than five carbon atoms are soluble in water and form neutral solutions (pH 7).

In addition, solubility in sodium hydroxide solution (usually 6 M NaOH, 6 molar NaOH) is a positive identification test for acids and acid derivatives. A carboxylic acid that is insoluble in pure water will be soluble in base due to the formation of the sodium salt of the acid as the acid is neutralized by the base. Also, solubility in hydrochloric acid (usually 6 M HCl, 6 molar HCl) is a positive identification test for bases. Amines that are insoluble in pure water will be soluble in hydrochloric acid due to the formation of an ammonium chloride-type salt ($RNH_3^+ Cl^-$).

4.6 Total Petroleum Hydrocarbons

Within the environmental arena, there is often reference to *total petroleum hydrocarbons* (TPHs) which is used because; given the wide variety of chemicals in petroleum and in petroleum products, it is not practical to consider each constituent separately. After an incident, it is more usual to measure the amount of TPHs at the site. The term TPHs is used environmentally to describe the family of several hundred chemical compounds that originally come from petroleum (Speight, 2001, 2014, 2015). Chemicals that may be included in the TPHs are hexane, heptane (and higher molecular weight homologs), benzene, toluene, xylene isomers (and the higher molecular weight homologs), and naphthalene as well as the constituents of other petroleum products such as gasoline and diesel fuel. It is likely that samples of the TPHs collected at a specific site will contain only some, or a mixture, of these chemicals but samples from different sites cannot be (should not be) expected to be the same in terms of composition and content of chemicals because of the variations in the composition of the starting crude oil feedstocks.

Petroleum products, themselves, are the source of many components, but do not adequately define TPHs. However, the composition of petroleum products assists in understanding the hydrocarbon derivatives that become environmental contaminants, but any ultimate exposure is determined also by how the product changes with use, by the nature of the release, and by the hydrocarbon's environmental fate. When petroleum products are released into the environment, changes occur that significantly affect their potential effects. Physical, chemical, and biological processes change the location and concentration of hydrocarbons at any particular site.

Hydrocarbons (in this context, petroleum-derived hydrocarbons) may enter the environment through accidents, from industrial releases, or as by-products from commercial or private uses such as direct release into water through spills or leaks. When release into water occurs, some of the hydrocarbons float on the water and form surface films while others may sink and form bottom sediments. Bacteria and microorganisms in the water have the potential to break down some of the hydrocarbons over varying periods of time that are dependent on the ambient conditions. On the other hand, hydrocarbon derivatives that are spilled on to the soil may remain for a long time, subject to the molecular size and volatility.

In addition, the amount of TPHs is the measurable amount of petroleum-based hydrocarbon in an environmental medium, whether it is air, water, or land. It is, thus, dependent on analysis of the medium in which it is found and since it is a measured, gross quantity without identification of its constituents, the TPHs' data still represent a mixture. Thus, the data derived from measurement of the petroleum hydrocarbons in a particular environment is not a direct indicator of risk to humans or to the environment. The data may even be the results from one of several analytical methods, some of which have been used for decades and others developed in the past several years.

Analysis for TPHs (EPA Method 418.1, 2016) provides a *one number* value of the petroleum hydrocarbons in a given environmental medium. It does not, however, provide information on the composition (i.e., individual constituents) of the hydrocarbon mixture. The amount of hydrocarbon contaminants measured by this method depends on the ability of the solvent used to extract the hydrocarbon from the environmental media and the absorption of infrared light (infrared spectroscopy) by the hydrocarbons in the solvent extract. The method is not specific to hydrocarbon derivatives and does not always indicate petroleum contamination since humic acid, a nonpetroleum material and a constituent of many soils, can be detected by this method. Another analytical method commonly used for TPHs (EPA Method 8015 Modified, 2016) gives the concentration of purgeable and extractable hydrocarbons. These are sometimes referred to as gasoline range organics and diesel range organics because the boiling point ranges of the hydrocarbon in each roughly correspond to those of gasoline (C_6 to C_{10-12}) and diesel fuel (C_{8-12} to C_{24-26}), respectively. Purgeable hydrocarbons are measured by purge-and-trap gas chromatography analysis using a flame ionization detector, while the extractable hydrocarbons are extracted and concentrated prior to analysis. The results are most frequently reported as single numbers for purgeable and extractable hydrocarbon derivatives. Another method (based on EPA Method 8015 Modified, 2016) gives a measure of the aromatic and aliphatic content of the hydrocarbon in each of several carbon number ranges (fractions). An important feature of the analytical methods for the TPHs is the use of an *equivalent carbon number index*. This index represents equivalent boiling points for hydrocarbons and is the physical characteristic that is the basis for separating petroleum (and other) components in chemical analysis and for identifying the ratios of specific hydrocarbon derivatives in the sample from which a source might be deduced.

REFERENCES

Andrews, J.E., Brimblecombe, P., Jickells, T.D., Liss, P.S., 1996. An Introduction to Environmental Chemistry. Blackwell Science Publications, Oxford.

Bergen, B.J., Nelson, W.G., Pruell, R.J., 1993. Bioaccumulation of PCB congeners by blue mussels (*Mytilus edulis*) deployed in New Bedford Harbor, Massachusetts. Environ. Toxicol. Chem. 12, 1671–1681.

Boethling, R.S., Mackay, D., 2000. Handbook of Property Estimation Methods. Lewis Publishers, Boca Raton, FL.

Calvert, J. (Ed.), 1994. The Chemistry of the Atmosphere: Its Impact on Global Change. Blackwell Scientific Publications, Oxford.

Cole, J.G., Mackay, D., 2000. Correlating environmental partitioning properties of organic compounds: the three solubility approach. Environ. Toxicol. Chem. 19 (2), 265–270.

Corvalán, C., Kjellström, T., 1996. Health and environment analysis for decision-making. In: Briggs, D., Corvalán, C., Nurminen, M. (Eds.), Linkage Methods for Environment and Health Analysis—General Guidelines. United Nations Environment Program (in cooperation with the World Health Organization and the United States Environmental Protection Agency), Geneva, pp. 1–18.

Crompton, T.R., 2012. Organic Compounds in Soils, Sediments and Sludge: Analysis and Determination. CRC Press, Taylor & Francis Group, Boca Raton, FL.

Dean, J.R., 1998. Extraction Methods for Environmental Analysis. John Wiley & Sons, Inc., New York, NY.

Delle Site, A., 2001. Factors affecting sorption of organic compounds in natural sorbent/water systems and sorption coefficients for selected pollutants. A review. J. Phys. Chem. Ref. Data 30 (1), 187–439.

EPA Method 418.1, 2016. Petroleum Hydrocarbons: Total Recoverable. United States Environmental Protection Agency, Washington, DC.

EPA Method 8015 Modified, 2016. Purgeable and Extractable Hydrocarbons. United States Environmental Protection Agency, Washington, DC.

Firor, J., 1990. The Changing Atmosphere. Yale University Press, New Haven, CT.

Goody, R., 1995. Principles of Atmospheric Physics and Chemistry. Oxford University Press, Oxford.

Lyman, W.J., Reehl, W.F., Rosenblatt, D.H. (Eds.), 1982. Handbook of Chemical Property Estimation Methods. McGraw-Hill, New York, NY.

Lyman, W.J., Reehl, W.F., Rosenblatt, D.H., 1990. Handbook of Chemical Property Estimation Methods. American Chemical Society, Washington, DC.

Mackay, D., 1982. Correlation of Bioconcentration Factors. Environ. Sci. Technol. 16, 274–278.

Mackay, D., Shiu, W.-Y., Ma, K.-C., Lee, S., 2006. Handbook of Physical-Chemical Properties and Environmental Fate of Organic Chemicals, second ed. CRC Press, Taylor & Francis Group, Boca Raton, FL.

Manahan, S.E., 2010. Environmental Chemistry, ninth ed. CRC Press, Taylor & Francis Group, Boca Raton, FL.

McCall, P.J., Laskowski, D.A., Swann, R.L., Dishburger, H.J., 1983. Estimation of environmental partitioning of organic chemicals in model ecosystems. In: Gunther, F.A., Gunther, J.D. (Eds.), ResidueReviews, vol. 85. Springer, New York, NY, pp. 231–244.

Patnaik, P. (Ed.), 2004. Dean's Analytical Chemistry Handbook, second ed. McGraw-Hill, New York, NY.

Schwarzenbach, R.P., Gschwend, P.M., Imboden, D.M., 2003. Environmental Organic Chemistry, second ed. John Wiley & Sons Inc., Hoboken, NJ.

Smith, R.K., 1999. Handbook of Environmental Analysis, fourth ed. Genium Publishing, Schenectady, NY.

Speight, J.G., 2001. Handbook of Petroleum Analysis. John Wiley & Sons Inc., Hoboken, NJ.

Speight, J.G., 2005. Environmental Analysis and Technology for the Refining Industry. John Wiley & Sons Inc., Hoboken, NJ.

Speight, J.G., 2014. The Chemistry and Technology of Petroleum, fifth ed. CRC Press, Taylor & Francis Group, Boca Raton, FL.

Speight, J.G., 2015. Handbook of Petroleum Products Analysis, second ed. John Wiley & Sons Inc., Hoboken, NJ.

Speight, J.G., Islam, M.R., 2016. Peak Energy—Myth or Reality. Scrivener Publishing, Salem, MA.

Spellman, F.R., 2016. Handbook of Environmental Engineering. CRC Press, Taylor & Francis Group, Boca Raton, FL.

WMO, 1992. Scientific Assessment of Stratospheric Ozone 1991. WMO Global Ozone Research and Monitoring Project, Report No. 25; World Meteorological Organization, Geneva.

Chapter 7

Chemical Transformations in the Environment

1 INTRODUCTION

The major groups of organic pollutants, the principal reactions for their degradation or transformation require (1) the knowledge of the type of pollutants and (2) the reactivity of the pollutants so that the design of the relevant method to remove the pollutants from the environment can be achieved. Although much emphasis has been, and continue to be, placed on biotic reactions carried out by the various biota (such as bacteria), important transformation reactions of the pollutants in the environment must not be omitted. Some of these chemical reactions will be beneficial—in terms of pollutant removal—while other chemical reactions may have adverse effects on pollutant removal by converting the pollutant to a product that is more capable of remaining in the environment and may even prove to be persistent. Thus, emphasis must be placed on the occurrence of partial degradation of pollutants and the role of the intermediate products that are toxic to floral and faunal organisms, inhibit further degradation, or have adverse effects on the environment.

From the standpoint of environmental organic chemistry, it is essential to know the important groups of potential organic pollutants:

- Refined petrochemicals, which include gasoline, diesel fuel, fuel oil, and lubricating oil, especially those products that contain monocyclic aromatic hydrocarbons.
- Bulk chemicals, which include a wide range of compounds including nitrobenzene, aniline, solvents, and monomers used in plastics manufacture.
- Agrochemicals including herbicides and pesticides.
- Chemicals used in plastics, mining, metal-working, wood preservation, paints, textiles, pigments, flame retardants, and household products.
- Pharmaceuticals.

To complicate matters even further, these diverse groups of chemicals belong to a variety of chemical functional groups, which include: (1) hydrocarbons, (2) organo-halogen compounds, (3) substituted aromatic hydrocarbons, (4) nitrogen compounds, (5) sulfur compounds, and (6) phosphorus compounds (Table 7.1).

Environmental Organic Chemistry for Engineers. http://dx.doi.org/10.1016/B978-0-12-804492-6.00007-1

TABLE 7.1 Types of Organic Chemical Groups and the Related Products

Hydrocarbons

Products of degradation of higher plants including the simplest hydrocarbon (methane), aliphatic hydrocarbons (alkanes and alkenes), alicyclic, and aromatic structures with one or more rings

Organo-halogen compounds

Halogen (fluorine, chlorine, bromine, or iodine) bound to aliphatic, alicyclic, and aromatic structures including halogenated alkane derivatives and alkene derivatives, chlorinated alkanoate derivatives, halogenated aromatic derivatives, halogenated aniline derivatives, halogenated phenol derivatives, halogenated phenoxy-acetate derivatives, polybrominated phenol derivatives, polybrominated diphenyl ether derivatives

Substituted aromatic hydrocarbons

Phenol derivatives, aniline derivatives, benzoate derivatives, and phthalate derivatives, nitroaromatic derivatives, sulfonated aromatic derivatives

Nitrogen compounds

Aliphatic and aromatic amine derivatives, amide derivatives, carbamate derivatives, nitrile derivatives, nitro- and azarene derivatives, nitrophenol derivatives, heteroaromatic derivatives, nitrate esters, nitramine derivatives

Oxygenated compounds

Aliphatic diol derivatives, aliphatic and aromatic ester derivatives, ether derivatives, dibenzofuran derivatives, dibenzo-*p*-dioxin derivatives

Sulfur compounds

Aliphatic and aromatic sulfate ester derivatives, sulfonate derivatives, dibenzothiophene derivatives

Phosphorus compounds

Organo-phosphate derivatives, organo-thiophosphate derivatives, aliphatic phosphonate derivatives, phosphoro-fluoridate derivatives

The group containing oxygen, nitrogen, and phosphorus compounds is very diverse but as a general rule it contains compounds with relatively high solubility in water, low solubility, and relatively low persistence in the environment. This is due to the presence of bonds with relatively high levels of polarity due to carbon and other atoms being attached to oxygen, nitrogen, or phosphorus conferring a high level of polarity onto the related compounds. In addition, important plastics are prepared from monomeric intermediates of which examples include those produced by polymerization of substituted ethylene derivatives (such as polyethylene, polystyrene, poly-tetrafluoroethylene, polyvinyl chloride, polyvinyl acetate, and polyacrylonitrile) and condensation polymers (such as polyurethane from aliphatic diol derivatives and di-isocyanate derivatives nylon from diamino-hexanoate and aliphatic dicarboxylic acid derivatives, and polyethylene terephthalate from terephthalic acid and ethylene glycol).

Thus, the chemical transformation of an organic chemical contaminant in the environment is an issue that needs to be given serious consideration because of the changes (often nonbenign) that can occur to the chemical (Manzetti et al., 2014). It would be unusual if the chemical transformation did not show some effect on the properties of the discharged chemical. For example, using oxidation as the example, the conversion of a totally hydrocarbon derivative to an oxygenated derivative can have a major effect of the properties and behavior of the product vis-à-vis the starting material.

Hydrocarbon	Oxygenated Hydrocarbon
Nonpolar	Polar
Some solubility in water	Enhanced solubility in water
Little (if any) activity with minerals	Enhanced tendency to bond to minerals
Little (if any) activity with alkali	Enhanced reactivity with alkali
No intramolecular interactions	Enhanced intramolecular interactions
No intermolecular interactions	Enhanced intermolecular interactions

Thus, the incorporation of an oxygen function (or oxygen functions) has a major effect on the behavior of the product, whether the oxygenated product (assuming that there is little is any change in the molecular size other than the change accounting for the incorporation of the oxygen function) contains a (1) hydroxyl function or (2) a phenolic function or (3) a ketone function or (4) a carboxylic acid function (Table 7.2).

Thus, chemical transformations of organic chemicals released into the environment are, in the context of this book, considered to be the transformation of the released chemical into a product that is still of concern in terms of toxicity. Furthermore, knowledge of the relative amounts of each species present is critical because of the potential for differences in behavior and toxicity (including the possibility of enhanced toxicity) which are of concern because of the potential fate of such chemicals.

As used here, the term *fate* refers to the ultimate disposition of the organic chemical in the ecosystem, either by chemical or biological transformation to a new form which (hopefully) is nontoxic (degradation) or, in the case of an ultimately persistent organic pollutants (POPs), by conversion to a less offensive chemical or even by sequestration in a sediment or other location which is expected to remain undisturbed. Thus latter option—the sequestration in a sediment or other location—is not a viable option as for safety reasons the chemical must be dealt with at some stage of its environmental life cycle. Using the old adage *bad pennies always turn up* can also be applied to a hidden-away chemical and it is likely to manifest its presence at some future date. In summary, hiding the chemical away on paper (a note in a file giving the written reason why the chemical is considered to be of limited danger) is not an effective way of protecting the environment. However, for organic chemicals that are

TABLE 7.2 Comparison of the Relative Polarity of Hydrocarbon Derivatives and Oxygen-Containing Compounds

Solvent	Chemical Formula	Boiling Point (°C)	Dielectric Constant[a]	Density (g/mL)	Dipole Moment[b] (D)
Hydrocarbon derivatives					
Pentane	$CH_3CH_2CH_2CH_2CH_3$	36	1.84	0.626	0.00
Hexane	$CH_3CH_2CH_2CH_2CH_2CH_3$	69	1.88	0.655	0.00
Benzene	C_6H_6	80	2.3	0.879	0.00
Toluene	$C_6H_5CH_3$	111	2.38	0.867	0.36
Oxygen derivatives					
Diethyl ether	$CH_3CH_2OCH_2CH_3$	35	4.3	0.713	1.15
Acetone	CH_3COCH_3	56	21	0.786	2.88
n-Butanol	$CH_3CH_2CH_2CH_2OH$	118	18	0.810	1.63
Isopropanol	$CH_3CH(OH)CH_3$	82	18	0.785	1.66
n-Propanol	$CH_3CH_2CH_2OH$	97	20	0.803	1.68
Ethanol	CH_3CH_2OH	79	24.55	0.789	1.69
Methanol	CH_3OH	65	33	0.791	1.70
Acetic acid	CH_3CO_2H	118	6.2	1.049	1.74
For comparison					
Water	H_2O	100	80	1.000	1.85

[a]Also known by the modern term relative permittivity which is a relative measure of the chemical polarity of the molecule but it is not an indicator of the solubility of the molecule in water but may be used as a guide to the relative reactivity of the molecule.
[b]A general indicator of the relative polarity of a molecule—higher values indicate more polar molecules; the bond dipole uses the concept of the electric dipole moment as an indicator of the polarity of a chemical bond within a molecule.

effectively degraded, whether by hydrolysis, photolysis, microbial degradation, or other chemical transformation in the ecosystem, it would seem necessary to collect, tabulate, and store any information related to the chemical reaction parameters which can serve as indicators of the processes and the rates at which transformation (i.e., degradation) would occur.

In fact, organic chemicals are subject to two processes that determine the fate of the chemical in the environment: (1) the potential for transportation of the chemical and (2) the chemical changes that can occur once the chemical has been released to the environment and which depend upon the physical and chemical properties:

Chemical Transport Processes

- runoff
- erosion
- wind
- leaching
- movement in streams or in groundwater

Chemical Fate Processes

- transport
- transformation/degradation
- sorption
- volatilization
- biological processes

Transformation and Degradation Processes

- biological transformations due to microorganism
  (1) aerobic
  (2) anaerobic processes

Physical Properties	Chemical Changes
Melting point	Reaction with acids
Boiling point	Reaction with alkalis (bases)
Vapor pressure	Reaction with oxygen (oxidation)
Color	Reaction with oxygen (combustion)
State (gas, liquid, and solid)	Ability to act as an oxidizing agent
Density	Ability to act as a reducing agent
Electrical conductivity	Reaction with other chemicals
Solubility	Decomposition into lower molecular weight chemicals
Adsorption to a surface	Decomposition into lower molecular weight chemicals
Hardness	Can cause corrosion

Thus, release into the environment of a POP leads to an exposure level which ultimately depends on the length of time the chemical remains in circulation, and how many times it is recirculated in some sense, before ultimate termination of the environmental life cycle of the chemical—the same rationale applied to product formed from the pollutant by any form of chemical transformation. In addition, the potential for transportation and chemical change (either before or after transportation) raises the potential for the chemical to behave in an unpredictable manner.

A particular question which needs to be addressed more often for POPs relates to the fraction that remains in circulation (until the end of the life cycle) and the means by which the environmental existence of the organic chemical can be terminated as expeditiously as possible and without further harm to the environment. The findings may not always be positive but must be given serious consideration in terms of as near-as-possible complete removal of the organic chemical and any products of chemical transformation.

Thus, in the present context, organic chemicals in the environment are of particular concern because of the high potential for toxicity to a wide variety of floral and faunal species. Some chemicals are well known for their adverse effects on flora and fauna at high levels of exposure. These chemicals typically have no known essential role in the human body and for these nonessential chemicals, at very low exposure, the chemicals are tolerated with little, if any, adverse effect, but at higher exposure their toxicity is exerted and health consequences become obvious. These can have consequences as severe as the ones which result from excessive intakes. In between, there may be an acceptable range of exposures within which the body is able to regulate an optimum level of the element. Generally, it is safer to assume that organic chemicals (other than those prescribed by a physician in regulated dosages) are harmful to humans.

When an organic chemical (or a mixture of organic chemicals) is released into the environment, the issues that need to be considered are: (1) the toxicity of the organic chemical, (2) the concentration of the released organic chemical, (3) the concentration of the toxic organic chemical in the released material, (4) the potential of the organic chemical to migrate to other sites, (5) the potential of the organic chemical to produce a toxic degradation product, whether or not the toxicity is lower or higher than the toxicity of the released chemical, (6) the potential of the toxic degradation product or products to migrate to other sites, (7) the persistence of the organic chemical in an ecosystem, (8) the persistence of any toxic degradation product in an ecosystem, (9) the potential for the toxic degradation product to degrade even further into harmful or nonharmful constituents and the rate of degradation, and (10) the degree to which the constituent or any degradation product of the constituent can accumulate in an ecosystem. Other factors that may be appropriate may also be considered—this list *is not meant to be complete* but does serve to indicate the types of issues that must be given serious consideration preferably before a spill or discharge of an organic chemical into the environment.

Thus, in order to complete such a list and monitor the behavior and effects of an organic chemical in an ecosystem, an understanding of chemical transformation processes in which a disposed or discharged chemical might particulate is valuable to any study of the effects on the environment. Chemical transformation processes change the chemical composition and structure of the discharged chemical which can change the properties (and possibly the toxicity) of the chemical and influence behavior and life cycle of the chemical in the environment.

As an example of chemical transformation of organic chemicals in the environment, weathering processes are ever-present and include such phenomena as: (1) evaporation, (2) leaching, which is transfer to the aqueous phase through dissolution, (3) entrainment, which is physical transport along with the aqueous phase, (4) chemical oxidation, and (5) microbial degradation. The rate of transformation of the chemical is highly dependent on environmental conditions. For example, a product such as low-boiling naphtha solvent (boiling range 30–90°C, 86–194°F) will evaporate readily when spilled on to the surface of the earth (specifically a water surface or land surface) and will give the appearance of a reception in the amount that remains. But the low-boiling constituents have not merely *disappeared* or *gone away* but have transferred from the land or from the water into the atmosphere. On the other hand, naphtha that has been inadvertently released into a formation that lies below a formation of clay minerals will tend to evaporate slowly (the clay can act as a formation trap) and may not be readily detectable. Unfortunately, the database on such transformations and that available on the composition of spilled chemicals that have been transformed in the environment is limited.

However, the various chemical transformation processes, which influence the presence and the analysis of organic chemicals at a particular site, although often represented by simple (and convenient) chemical equations, can be very complex (Neilson and Allard, 2012) and the true nature of the chemical transformation process is difficult to elucidate. The extent of transformation is dependent on many factors including the (1) the properties of the chemical, (2) the geology of the site, (3) the climatic conditions, such as temperature, oxygen levels, and moisture, (4) the type of microorganisms present, and (5) any other environmental conditions that can influence the life cycle of the chemical. In fact, the primary factor controlling the extent of chemical transformation is the molecular composition of the organic chemicals contaminant. Multiple ring cycloalkanes are more difficult to degrade than alkane derivative or single ring cycloalkane derivatives—straight-chain alkanes biodegrade rapidly with branched alkanes and single saturated ring compounds degrading more slowly—and polynuclear aromatic hydrocarbon derivatives display varying degrees of degradation.

Polycyclic aromatic hydrocarbons are ubiquitous pollutants which derive from various sources but more particularly from source such as crude oil and crude oil-based products, coal combustion, volcanic eruptions, incineration of various types of waste, and combustion of biomass. Their mutagenic character (the character of a chemical or physical agent that has the ability to change genetic material) and carcinogenic character (the character of a chemical or physical agent that has the ability to cause cancer) has been studied extensively; however the environmental fate and chemical transformation processes that affect these organic compounds are still open to discussion and represent a central part in environmental chemistry frontiers (Manzetti, 2011, 2012a,b, 2013). Furthermore, environmental exposure to organic chemicals, particularly exposure to the mutagens or to the carcinogens, is very dangerous despite the minuscule risks associated with many such exposures at typical environmental concentrations. Examples are benzene (largely from vehicle emissions into the atmosphere) and polynuclear aromatic hydrocarbons which are generated by combustion of fossil fuels and biomass. Examples of polynuclear systems that are hazardous to the flora and fauna of a variety of ecosystems are benzo(*a*)pyrene and the equally hazardous dioxin derivative 2,3,7,8-tetrachlorodibenzodioxin:

(A) Benzene (B) Benzo(a)pyrene   (C) 2,3,7,8-Tetrachlorodibenzodioxin

However, it must be re-emphasized that an organic chemical deposited into the environment has the potential to undergo transformation to another chemical form which is still of concern in terms of toxicity. Moreover, when they are released into the environment, the fate of organic compounds depends on the physical and chemical properties of the compound(s) and the ability of these chemicals (i.e., the chemical reactivity) to undergo transformation to products. In addition, it is not only the structure of the chemical deposited into the environment but also the chemical forms that can result of the chemical transformation that are the result of the chemicals undergoing weathering (oxidation) and other environmental effects that cause change to the chemical structure. Thus, organic chemicals that are not directly toxic to environmental flora and faunas (including humans) at current environmental concentrations can become capable of causing environmental damage after chemical transformation has occurred.

In chemistry a chemical transformation is the conversion of a substrate to a product. In the environment, a chemical transformation is the same principle as in the laboratory or in the chemical process industries—the transformation of a substrate to a product—but whether or not the product is benign and less likely to harm the environment (relative to the substrate) or is more detrimental by exerting a greater impact on the environment depends upon the origin, properties, and reactivity pathways of the starting substrate. Thus, a chemical transformation requires a chemical reaction to lead to the transformation of one chemical substance to another. Typically, chemical reactions encompass changes that only involve the positions of electrons in the forming and breaking of chemical bonds between atoms, with no change to the nuclei (no change to the elements present), and can often be described by a chemical equation. However, recall that the various chemical transformation processes which influence the presence and the analysis of organic chemicals at a particular site and are often represented by simple (and convenient) chemical equations, can be very complex (Neilson and Allard, 2012).

The substance (or substances) initially involved in a chemical reaction (the reactants or the substrates) are usually characterized by a chemical change, and yield one or more products, which usually have properties different from the original substrates. Reactions often consist of a sequence of individual (and often complex) substeps and the information on the precise course of action is part of the reaction mechanism is not always clear. Chemical reactions typically occur under a specific set of parameters (temperature, chemical concentration, and time) and (under theses parameters) at a characteristic reaction rate. Typically, reaction rates increase with increasing temperature because there is more thermal energy available to reach the activation energy necessary for breaking bonds between atoms. The general rule of thumb is that for every 10°C (18°F) increase in temperature the rate of an organic chemical reaction is doubled.

Reactions may proceed in the forward direction and processed to completion as well as in the reverse direction until they reach equilibrium.

$$A + B \rightarrow C + D$$
$$C + D \rightarrow A + B$$

Thus,

$$A + B \leftrightarrow C + D$$

Reactions that proceed in the forward direction to approach equilibrium are often described as spontaneous, requiring no input of free energy to go forward. Nonspontaneous reactions require input of free energy to go forward (e.g., application of heat for the reaction to proceed). In organic chemical synthesis, different chemical reactions are used in combinations during chemical synthesis in order to obtain a desired product. Also, in organic chemistry, a consecutive series of chemical reactions (where the product of one reaction is the reactant of the next reaction) are often catalyzed by a variety of catalysts which increase the rates of biochemical reactions, so that syntheses and decompositions impossible under ordinary conditions can occur at the temperatures, pressures, and reactant concentrations present within a reactor and, by inference, within the environment.

$$A + B \rightarrow C$$
$$C \rightarrow D + E$$

This simplified equation illustrates the potential complexity of organic chemical reaction and such complexity must be anticipated when an organic chemical is transformed in an environmental ecosystem. Mother Nature can be quite complex!

Thus, the focus in this chapter is upon developing a fundamental understanding of the nature of these chemical processes, so that activities that have an effect on the environment the chemistry can be presented.

2 ORGANIC REACTIONS

Organic reactions are chemical reactions involving organic compounds and the basic organic chemical reaction types are (1) addition reactions, (2) elimination reactions, substitution reactions, (3) redox reactions, and (4) rearrangement reactions (Table 7.3) (March, 1992; Morrison et al., 1992). In organic synthesis, organic reactions are used in the construction of new organic molecules but in the discipline known as *environmental organic chemistry* these reactions often occur and cause chemical transformation of an organic pollutant in the environment. Factors governing organic reactions (hence, organic transformations) in the environment are essentially the same as that of any chemical reaction and these are factors that determine the stability of reactants and products.

Furthermore, while organic reactions can be organized into several basic reaction types (Table 7.3), some reactions fit into more than one category. Organic reactions can also be categorized on the basis of the type of functional group involved in the reaction as a reactant and the functional group that is formed as a result of this reaction. A functional group confers specific reactivity patterns on the molecules of which it is a part. Although the properties of each of the several million organic molecules whose structure is known are unique in some way, all molecules that contain the same functional group have a similar pattern of reactivity at the functional group site. Thus, functional groups are a key organizing feature of organic chemistry. By focusing on the functional groups present in a molecule (most molecules have more than one functional group), several of the reactions that the molecule will undergo can be predicted and understood. Thus, functional group transformation/inter conversion is the process of converting one functional group into another by several type of reactions like, substitution, addition, elimination, reduction, or oxidation, by the use of reagents and different reaction conditions.

Organic reactions can also be classified by the type of bond to carbon with respect to the element involved. In heterocyclic chemistry, organic reactions are classified by the type of heterocycle formed with respect to ring-size and type of heteroatom. Examples are ring expansion and ring contraction, polymerization reactions, insertion reactions, ring-opening reactions, and ring-closing reactions.

TABLE 7.3 Types of Organic Reactions

Reaction Type	Subtype	Examples
Addition reactions	Electrophilic addition	Halogenation, hydrohalogenation, hydration
	Nucleophilic addition	
	Radical addition	
Elimination reaction		Example: condensation in which a molecule of water is eliminated from the reactants
		Also: dehydration—removal of water
Substitution reactions	Nucleophilic aliphatic substitution	
	Nucleophilic aromatic substitution	
	Nucleophilic acyl substitution	
	Electrophilic substitution	
	Electrophilic aromatic substitution	
	Radical substitution	
Organic redox reactions		Oxidation-reduction reactions specific to organic compounds
Rearrangement reactions	1,2-Rearrangements	*1,2-Rearrangement*: a reaction in which a substituent can move from one atom to another atom, such as movement to an adjacent atom
	Pericyclic reactions	*Pericyclic reaction*: rearrangement reaction where the intermediate is cyclic
	Metathesis	*Metathesis*: a reaction involving the exchange of bonds between two reacting chemical species, such as $A - B + C - D \rightarrow A - D + C - B$

In fact, there is no limit to the number of possible organic reactions and mechanisms. However, certain general patterns are observed that can be used to describe many common or useful reactions. Each reaction typically has a stepwise reaction mechanism that can be used to explain or visualize the mean by which the reaction occurs, although a detailed description of steps may not always be evident from a list of reactants alone.

2.1   Addition and Elimination Reactions

An addition reaction is an organic reaction where two or more molecules combine to form a product (the adduct) (March, 1992; Morrison et al., 1992). Addition reactions are limited to chemical compounds that have multiple bonds, such as molecules with carbon-carbon double bonds (alkenes, $>C=C<$), or with triple bonds (alkynes, $-C\equiv C-$). Molecules containing carbon-heteroatom double bonds such as the carbonyl group ($>C=O$) groups, or the imine group ($C=N$) groups, can also participate in addition since they also have double bond character. An addition reaction is the reverse of an elimination reaction, such as the hydration of an alkene to an alcohol which is reversed by dehydration:

$$CH_2 = CH_2 + H_2O \rightarrow CH_3CH_2OH$$
$$CH_3CH_2OH \rightarrow CH_2 = CH_2 + H_2O$$

The main driver behind addition reactions to alkene is that alkenes contain the unsaturated $>C=C<$ functional group which characteristically undergoes addition reactions which is the conversion of the weaker π bond into two new, stronger σ bonds. In addition to addition reactions being typical of the unsaturated hydrocarbon derivatives (alkenes and alkynes) and aldehydes and ketones, which have a carbon-to-oxygen double bond, an addition reaction may be visualized as a process by which the double or triple bonds are fully or partially broken in order to accommodate additional atoms or groups of atoms in the molecule. Addition reactions to alkenes and alkynes are sometimes called saturation reactions because the reaction causes the carbon atoms to become saturated with the maximum number of attached groups.

In addition, reactions to aldehydes and ketones, the sequence of events is reversed insofar as the initial step is addition of the negatively charged component of the reagent to the carbon atom, followed by addition of the positively charged component to the oxygen atom.

2.2   Substitution Reactions

A substitution reaction (sometime referred to as a single displacement reaction or single replacement reaction) is a reaction in which a functional group in an organic chemical is replaced by another functional group (March, 1992; Morrison et al., 1992). Substitution reactions are of prime importance in environmental organic chemistry because of the simplicity of the reaction which is accompanied by a substantial change in the chemical and physical properties of the product vis-à-vis compared to the chemical and physical properties of the starting chemical. Substitution reactions in organic chemistry are classified either as (1) nucleophilic substitution or (2) electrophilic substitution depending upon the reagent involved.

A nucleophile is a chemical species (an ion or a molecule) which is strongly attracted to a region of positive charge in something else. Nucleophiles are either fully negative ions or have a strongly (δ^-) charge somewhere on a molecule. Common nucleophiles are hydroxide ions (OH^-), cyanide ions (CN^-), water (H_2O), and ammonia (NH_3). An example of a simple nucleophilic substitution reaction is the halogenation of an alkane such as when methane (CH_4) is reacted with chlorine (Cl_2) to form methyl chloride and hydrogen chloride:

$$CH_4 + Cl_2 \rightarrow CH_3Cl + HCl$$

Thus, a nucleophilic substitution reaction (which is common in organic chemistry) is a reaction in which an electron-rich nucleophile selectively bonds with or attacks the positive or partially positive charge of an atom or a group of atoms to replace the leaving electrophile:

$$(\text{Nucleophile}) + R - (\text{leaving group}) \rightarrow R - (\text{Nucleophile}) + (\text{leaving group})$$

On the other hand, electrophilic substitution involves electrophiles and an example is electrophilic aromatic substitution. In this type of substitution, the benzene ring (which has a π-electron cloud above and below the plane of the ring of carbon atoms) is attacked by an electrophile (shown as E^+). The π-electron cloud is disturbed and a carbocation resonating structure results after which a proton (H^+) is ejected from the intermediate and a new aromatic compound is formed:

$$C_6H_6 + E^+ \rightarrow C_6H_5E + H^+$$

A relevant reaction in the environment (because of the ever-presence of water in many ecosystems) is the hydrolysis reaction in which an organic chemical compound is decomposed by reaction with water—also the hydrolysis reaction should not be confused with the *hydrogenolysis* reaction which is a reaction of hydrogen as practiced widely in the petroleum industry to produce liquid fuels (Speight, 2014, 2016). Hydrolysis is an example of a larger class of reactions referred to as nucleophilic displacement reactions in which a nucleophile (an electron-rich species containing an unshared pair of electrons) attacks an electrophilic atom (an electron-deficient reaction center). Hydrolytic processes encompass several types of reaction mechanisms that can be defined by the type of reaction center (i.e., the atom bearing the leaving group, X) where hydrolysis occurs. The reaction mechanisms encountered most often are direct and indirect nucleophilic substitution and nucleophilic addition-elimination.

This type of reaction can be used to predict the persistency of a chemical in the environment, the chemical's physical-chemical properties and its reactivity in the environment need to be known or at least estimated (Rahm et al., 2005). The chemicals that can undergo elimination reactions are rapidly transformed, as are perhalogenated chemicals that can undergo substitution reactions. These chemicals are not likely to persist in the environment, while those that do not show any observable reactivity under similar hydrolytic conditions are likely to be POPs (Chapter 1), the fate of which are intimately linked to the cycling or organic chemicals in the environment (deBruyn and Gobas, 2004).

Typically, the hydrolysis reaction is a chemical transformation in which an organic molecule, RX, reacts with water, resulting in the formation of a new covalent bond with the hydroxy function (OH) and cleavage of the covalent bond with halogen function (e.g., Cl) (the leaving group) in the original molecule. The net reaction is the displacement of the halogen group (X) by the hydroxyl group (OH) (Table 7.4). In fact, for many types of organic contaminants, hydrolysis may be the dominant pathway for their transformation in aquatic ecosystems. Hydrolytic processes are not limited to the bodies of water such as rivers, streams, lakes, and oceans usually associated with the term aquatic ecosystems. Hydrolysis of organic chemicals can also occur in groundwater systems and the aqueous environment in solids and sediments.

In the aqueous hydrolysis reaction, the reacting water molecules are split into hydrogen (H^+) and hydroxide (OH^-) ions, which react with and break up (or "lyse") the other reacting compound. The term *hydrolysis* is also applied to the electrolysis of water (i.e., breaking up of water molecules by an electric current) to produce hydrogen and oxygen. The hydrolysis reaction is distinct from a *hydration reaction*, in which water molecules attach to molecules of the other reacting compound without breaking up the latter compound, such as:

TABLE 7.4 Examples of Hydrolysis Reactions[a]

1. The hydrolysis of a primary amide forms a carboxylic acid and ammonia:

$RCONH_2 + H_2O \rightarrow RCOOH + NH_3$

2. The hydrolysis of a secondary amide forms a carboxylic acid and primary amine:

$RCONHR' + H_2O \rightarrow RCOOH + R'NH_2$

3. The hydrolysis of an ester forms a carboxylic acid and an alcohol:

$RCOOR' + H_2O \rightarrow RCOOH + R'OH$

4. The hydrolysis of a halogenoalkane forms an alcohol:

$RBr + H_2O \rightarrow ROH + H^+ + Br^-$

[a]*The chemical equations presented above illustrate the hydrolysis by reaction with water.*

$$CH_2 = CH_2 + H_2O \rightarrow CH_3CH_2OH$$

The hydrolysis reaction mainly occurs between an ion and water molecules and often changes the pH (acidity or alkalinity) of a solution. In chemistry, there are three main types of hydrolysis: (1) salt hydrolysis, (2) acid hydrolysis, and (3) base hydrolysis.

In water, salts will dissociate to form ions (either completely or incompletely depending on the respective solubility constant, K_{sp}. For example:

$$NH_4Br(s) \rightarrow NH_4^+(aq) + Br^-(aq)$$

In this equation, the salt (ammonium bromide, NH4Br) is dissolved in water upon which the salt dissociates into ammonium ions (NH_4^+) and bromide ions (Br^-).

In the hydrolysis reaction, water can act as an acid or a base: (1) if that water acts as an acid, the water molecule would donate a proton (H^+), also written as a hydronium ion (H_3O^+) or (2) if the water acts as a base, the water molecule would accept a proton (H^+). An acid hydrolysis reaction is very much the same as an acid-dissociation reaction.

$$CH_3COOH + H_2O \rightleftharpoons H_3O^+ + CH_3COO^-$$

In the above reaction, the proton H^+ from CH_3COOH (acetic acid) is donated to water, producing the hydronium ion (H_3O^+) and an acetate ion (CH_3COO^-). The bonds between proton and the acetate ion are dissociated by the addition of water molecules. A reaction with acetic acid (CH_3COOH) a weak acid, is similar to an acid-dissociation reaction, and water forms a conjugate base and a hydronium ion. When a weak acid is hydrolyzed, a hydronium ion is produced.

A base hydrolysis reaction will resemble the reaction for base dissociation. A common weak base that dissociates in water is ammonia:

$$NH_3 + H_2O \rightleftharpoons NH_4^+ + OH^-$$

In the hydrolysis of ammonia, the ammonia molecule accepts a proton from the water (i.e., water acts as an acid), producing a hydroxide anion (OH^-). Similar to a basic dissociation reaction, ammonia forms ammonium and a hydroxide from the addition of a water molecule.

Generally, in organic chemistry, hydrolysis can be considered the reverse (or opposite) of condensation, a reaction in which two molecular fragments are joined for each water molecule produced. Since hydrolysis may be a reversible reaction, condensation, and hydrolysis can take place at the same time, with the position of equilibrium determining the amount of each product, such as the hydrolysis of an ester to an acid and an alcohol:

$$R^1CO_2R^2 + H_2O \rightleftharpoons R^1CO_2H + R^2OH$$

Thus, hydrolysis reactions involving organic compounds may be illustrated by the reaction of water with an ester of a carboxylic acid; all such esters have the general formula R^1COOR^2, in which R^1 and R^2 are combining groups (e.g., if R^1 and R^2 are both methyl groups, CH_3, the ester is methyl acetate). The hydrolysis involves several steps, of which the slowest is the formation of a covalent bond between the oxygen atom of the water molecule and the carbon atom of the ester. In succeeding steps, which are very rapid, the carbon-oxygen bond of the ester breaks and hydrogen ions become detached from the original water molecule and attached to the nascent alcohol molecule. The whole reaction is represented by the equation:

$$\underset{\text{Ester}}{R^1COOR^2} + H_2O \rightarrow \underset{\text{acid}}{R^1COOH} + \underset{\text{alcohol}}{R^2OH}$$

Thus, the products of hydrolysis depend very much upon that substrate that is to be hydrolyzed (Table 7.4).

More pertinent to the present text, the hydrolysis of a pesticide is basically a reaction with a water molecule involving specific catalysis by proton or hydroxide, and sometimes inorganic ions such as phosphate ion, present in the aquatic environment that play a role in general acid-base catalysis (Katagi, 2002). However, the hydrolytic profiles depend on the chemical structure and functional group(s) in the pesticide molecule, which are not always consistent within a chemical class of pesticides (Stoytcheva, 2011). For example, pesticides that are composed of organophosphorus derivatives are primarily susceptible to alkaline hydrolysis with less acidic catalysis, but some of phosphorodithioate derivatives are found to be acid labile. Various instrumental techniques have been applied to chemical identification of degraded products, leading to clarification of the reaction mechanisms involved. Moreover, pesticides are usually applied as a suitable formulation, and thus the effects of surfactants and other formulation reagents on hydrolysis should be examined in more detail. To assess the fate and impact of pesticides and their degraded products in real aquatic environments, these concerns should be further examined using the various analytical techniques together with simulation models.

When substituted benzene compounds undergo electrophilic substitution reactions it is necessary to compare the relative reactivity of the compound compared with benzene itself. Experiments have shown that substituents on a benzene ring can influence reactivity in a profound manner. For example, a hydroxy substituent ($-OH$) or methoxy ($-OCH^3$) substituent increases the rate of electrophilic substitution about 10,000-fold, as illustrated by the case of anisole in the virtual demonstration (above). In contrast, a nitro substituent decreases the reactivity of the ring substantially. This activation or deactivation of the benzene ring toward electrophilic substitution may be correlated with the electron donating or electron withdrawing influence of the substituents, as measured by molecular dipole moments. Electron donating substituents activate the benzene ring toward electrophilic attack, and electron withdrawing substituents deactivate the benzene ring and render it less reactive to electrophilic attack.

Activating substituents

$\ddot{N}H_2$	$\ddot{O}H$	$\ddot{O}CH_3$	CH_3
1.52	1.45	1.20	0.40

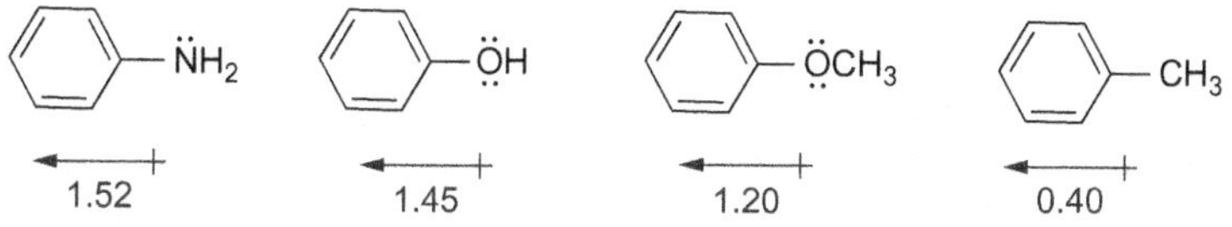

Deactivating substituents

NO_2	$C{\equiv}N$	CO_2CH_3	Cl
3.97	3.90	1.91	1.56

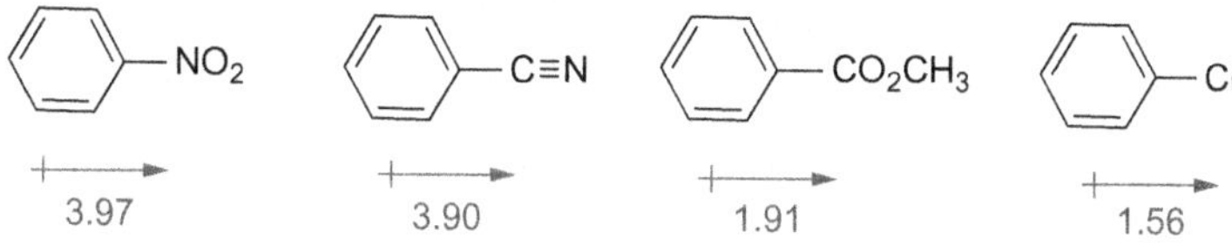

The influence a substituent exerts on the reactivity of a benzene ring may be explained by the interaction of two effects: (1) an inductive effect and (2) a conjugative effect. The *inductive effect* arises because most elements other than metals and carbon have a significantly greater electronegativity than hydrogen. Consequently, substituents in which nitrogen, oxygen, and halogen atoms form sigma-bonds to the aromatic ring exert an inductive electron withdrawal, which deactivates the ring (left-hand diagram below). The second effect (the conjugative effect) is the result of a conjugation of a substituent function with the aromatic ring which facilitates electron pair donation or withdrawal, to or from the benzene ring, in a manner different from the inductive shift. If the atom bonded to the ring has one or more nonbonding valence shell electron pairs, as do nitrogen, oxygen, and the halogens, electrons may flow into the aromatic ring by π conjugation (resonance), as in the middle diagram. Finally, polar double and triple bonds conjugated with the benzene ring may withdraw electrons, as in the right-hand diagram. The charge distribution in the benzene ring is greatest at sites *ortho* and *para* to the substituent. In the case of the nitrogen and oxygen activating groups displayed in the top row of the previous diagram, electron donation by resonance dominates the inductive effect and these compounds show exceptional reactivity in electrophilic substitution reactions. Although halogen atoms have nonbonding valence electron pairs that participate in p-π conjugation, their strong inductive effect predominates, and compounds such as chlorobenzene are less reactive than benzene.

2.3 Redox Reactions

Redox reactions (reduction-oxidation reactions) are reactions in which one species is reduced and another is oxidized. Therefore, the oxidation state of the species involved must change. The word *reduction* originally referred to the loss in weight upon heating a metallic ore such as a metal oxide to extract the metal—the ore was *reduced* to the metal. However, the meaning of *reduction* has become generalized to include all processes involving gain of electrons. The term *hydrogenation* could be used instead of *reduction*, since hydrogen is the reducing agent in a large number of reactions, especially in organic chemistry and biochemistry. But, unlike oxidation, hydrogenation has maintained its specific connection to reactions that *add* hydrogen to another substance such as the hydrogenation processes used in a petroleum refinery (Speight, 2014, 2016). On the other hand, the word *oxidation* originally implied reaction with oxygen to form an oxide but the word has been expanded to encompass oxygen-like substances that accomplished parallel chemical reactions and ultimately, the meaning was generalized to include all processes involving loss of electrons. For example, the production of iron from the iron oxide ore:

$$Fe_2O_3 + 3CO \rightarrow 2Fe + 3CO_2$$

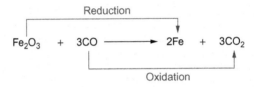

Similarly, in the context of organic chemistry, there is the oxidation of ethyl alcohol (CH_3CH_2OH) where it is oxidized to acetaldehyde (CH_3CHO) and the reverse reaction in which acetaldehyde is reduced to ethyl alcohol:

$$CH_3CH_2OH \rightarrow CH_3CHO + H_2 \rightarrow CH_3CH_2OH$$

This equation can be subdivided as follows into oxidation by loss of hydrogen and reduction by gain of hydrogen. Thus:

$$CH_3CH_2OH \longrightarrow CH_3CHO$$

Oxidation by loss of hydrogen

Reduction by gain of hydrogen
$$CH_3CHO \longrightarrow CH_3CH_2OH$$

The two species that exchange electrons in a redox reaction are given special names. The ion or molecule that accepts electrons is the *oxidizing agent* which, by accepting electrons causes the oxidation of another species. Conversely, the species that donates electrons is the *reducing agent* which; when the reaction occurs, reduces the other species. Thus, the species that is oxidized is the reducing agent and the species that is reduced is the oxidizing agent. To complicate matters even further, the oxidizing and reducing agents can be the same element or compound, as is the case when disproportionation of the reactive species occurs. For example:

$$2A \rightarrow (A + n) + (A - n)$$

In this equation, n is the number of electrons transferred. Disproportionation reactions do not need to commence with a neutral molecule and can involve more than two species with differing oxidation states.

Redox reactions are important for a number of applications, including energy storage devices (batteries), photographic processing, and energy production and utilization in living systems including humans. For example, a *reduction reaction* is a reaction in which an atom gains an electron and therefore decreases (or reduces) its oxidation number. The result is that the positive character of the species is reduced. On the other hand, an oxidation reaction is a reaction in which an atom loses an electron and therefore increases its oxidation number. The result is that the positive character of the species is increased.

Although oxidation reactions are commonly associated with the formation of oxides from oxygen molecules, these are only specific examples of a more general concept of reactions involving electron transfer. Redox reactions are a matched set, that is, there cannot be an oxidation reaction without a reduction reaction happening simultaneously. The oxidation reaction and the reduction reaction always occur together to form a whole reaction. Although oxidation and reduction properly refer to *a change in the oxidation state*, the actual transfer of electrons may never occur. The oxidation state of an atom is the fictitious charge that an atom would have if all bonds between atoms of different elements were 100% ionic. Thus, oxidation is best defined as an *increase in oxidation state*, and reduction as a *decrease in oxidation state*. In practice, the transfer of electrons will always cause a change in oxidation state, but there are many reactions that are classed as redox reactions even though no electron transfer occurs (such as those involving covalent bonds). There are simple redox processes, such as the oxidation of carbon to carbon dioxide (CO_2) or the reduction of carbon by hydrogen to methane (CH_4).

$$CH_4 + 2O_2 \rightarrow CO_2 + 2H_2O$$

The key to identifying oxidation-reduction reactions is recognizing when a chemical reaction leads to a change in the oxidation number of one or more atoms.

2.4   Rearrangement Reactions

Organic reactions typically yield products that are in accordance with the generally accepted mechanism of the reactions. However, in some instances, organic reactions do not give exclusively and solely the anticipated products but may lead to other product that arise from unexpected and mechanistically different reaction path. These unexpected products are often referred to as rearranged products and, while such a product may not be the expected product it may be the major product of the reaction. Thus, the reaction has involved a rearrangement of the expected product to an unexpected product—a rearrangement reaction has occurred. More than likely, this may have resulted from a plausible rearrangement occurring during the mechanistic course of the reaction to fulfill the principle of the minimum energy state of the whole system, that is, of the transition state which assumed another configuration to maintain a minimum energy balance to the system. In many cases, the rearrangement affords products of an isomerization, coupled with some stereochemical changes. An energetic requirement is also observed in order for a rearrangement to take place; that is, the rearrangement usually involves an evolution of energy (typically in the form of heat, i.e., the reaction is overall an exothermic reaction) to be able to yield a more stable compound (Moulay, 2002).

Thus, a rearrangement reaction falls into a class of organic reactions where the carbon skeleton of an organic molecule is rearranged to form a structural isomer of the original molecule that assumes the minimal energy content of the product—i.e., the most stable product is formed (March, 1992; Morrison et al., 1992; Moulay, 2002). As a result of the reaction, a substituent group typically moves from one atom to another atom in the same molecule to yield an isomer of the original reactant:

$$-\overset{|}{\underset{R}{C}}-C-C- \qquad \longrightarrow \qquad -C-\overset{|}{\underset{R}{C}}-C-$$

Carbon atom: 1 2 3   1 2 3

In the above equation, there has been movement of the substituents group (represented by R) from carbon atom number 1 to carbon atom number 2.

Thus, a rearrangement reaction is a reaction in which an atom or a group (or in some cases a bond) is caused to move or migrate to another part of the molecule. The atom, having been initially located at one site in a reactant molecule ultimately becomes located at a different site in the product molecule. A rearrangement reaction may involve several steps, but the defining feature is that the atom or a group or the bond shifts from one site of attachment to another. The simplest (perhaps, the most common) types of rearrangement reactions are intramolecular rearrangements insofar as the reactions occur in one reactant molecule and the product of the reaction is a structural isomer of the reactant. In summary, molecular rearrangement reactions occur in many organic reactions and the rearranged product usually results from the thermodynamic stability aspect of the compound or the reaction (Moulay, 2002).

2.5 Hydrolysis Reactions

Many organic compounds can be altered by a direct reaction of the chemical with water (hydrolysis) in which a chemical bond is cleaved and two new bonds are formed, each one having either the hydrogen component (H) or the hydroxyl component (OH) of the water molecule. Typically, the hydroxyl replaces another chemical group on the organic molecule and hydrolysis reactions are usually catalyzed by hydrogen ions or hydroxyl ions. This produces the strong dependence on the acidity or alkalinity (pH) of the solution often observed but, in some cases, hydrolysis can occur in a neutral (pH 7) environment. Adsorption on to a mineral sediment (such as a clay sediment that has strong adsorptive powers) generally reduces the rates of hydrolysis for acid- or base-catalyzed reactions. Neutral reactions appear to be unaffected by adsorption although there is always the possibility that the mineral sediment can cause catalyzed chemical transformation reactions.

The rate of a hydrolysis reaction is typically rate expressed in terms of the acid-catalyzed, neutral-catalyzed, and base-catalyzed hydrolysis rate constants. Furthermore, the hydrolysis of organic compounds is influenced by the composition of the solvent (Lyman et al., 1982) and the rate constants may be much higher in water than in organic solvents. In fact, the introduction of a complex mixture of chemicals into a water body can be expected to produce a significant shift in acidity or alkalinity of the medium and, therefore, it would not be surprising to anticipate that hydrolysis would be affected in complex mixtures.

2.6 Photolysis Reactions

Photolysis is a chemical process by which chemical bonds are broken as the result of transfer of light energy (direct photolysis) or radiant energy (indirect photolysis) to these bonds. The rate of photolysis depends upon numerous chemical and environmental factors including the light adsorption properties and reactivity of the chemical, and the intensity of solar radiation (Lyman et al., 1982). In the process, the photochemical mechanism of photolysis is divided into three stages: (1) the adsorption of light which excites electrons in the molecule, (2) the primary photochemical processes which transform or de-excite the excited molecule, and (3) the secondary ("dark") thermal reactions which transform the intermediates produced in the previous step (step 2).

In addition, before photolysis can occur, the photochemically excited state must be deactivated, such as a radiative process (fluorescence) in which energy (usually in the form of light) is emitted during the transition to ground electronic state and some residual vibrational excitation is rapidly lost via collision processes. Quenching of a photochemical process occurs when the excitation energy in the target organic molecule is transferred to some other chemical species in solution. This process results in net deactivation of the organic substance of concern via energy transfer. Energy can be transferred to any chemical species with a lower triplet energy. A very important and effective quencher (acceptor) is molecular and other chemicals in a complex mixture could act as acceptors and thereby reduce the photolytic degradation rate of a given compound to below that expected.

Indirect photolysis or sensitized photolysis occurs when the light energy captured (absorbed) by one molecule is transferred to the organic molecule of concern. The donor species (the sensitizer) undergoes no net reaction in the process but has an essentially catalytic effect. Moreover, the probability of a sensitized molecule donating its energy to an acceptor molecule is proportional to the concentration of both chemical species. Thus, complex mixtures may, in some cases, produce enhancement of photolysis rates of individual constituents through sensitized reactions.

3   CATALYSIS

In catalysis reactions, the reaction does not proceed directly, but through reaction with a third substance (the catalyst) and, although the catalyst takes part in the reaction, it is (in theory) returned to its original state by the end of the reaction and so is not consumed. However, the catalyst is not immune to being inhibited, deactivated, or destroyed by secondary processes. Catalysts can be used in a different phase (heterogeneous catalysis) or in the same phase (homogeneous catalysis) as the reactants.

In heterogeneous catalysis, typical secondary processes include coking (coke production from organic starting materials) where the catalyst becomes covered by ill-defined high-molecular weight by-side products as is observed in the petroleum refining industry during the production of fuels and other products (Speight, 2014, 2016). Heterogeneous catalysis is used in automobile exhaust systems to decrease nitrogen oxide, carbon monoxide, and unburned hydrocarbon emissions. The exhaust gas is vented through a high-surface area chamber lined with platinum, palladium, and rhodium. For example, the carbon monoxide is catalytically converted to carbon dioxide by reaction with oxygen.

Additionally, heterogeneous catalysts can dissolve into the solution in a solid-liquid system or evaporate in a solid-gas system. Catalysts can only speed up the reaction—chemicals that slow down the reaction are called inhibitors and there are chemicals that increase the activity of catalysts (catalyst promoters) as well as chemicals that deactivate catalysts (catalytic poisons). With a catalyst, a reaction which is kinetically inhibited by a high activation energy can take place in circumvention of this activation energy. Heterogeneous catalysts are usually solids, powdered in order to maximize their surface area. Of particular importance in heterogeneous catalysis are the platinum metals and other transition metals, which are used in crude oil refining processes such as hydrogenation and catalytic reforming.

Homogeneous catalysis involves a reaction in which the soluble catalyst is in solution—as long as the catalyst is in the same phase as the reactants.

Although the term is used almost exclusively to describe reactions (and catalysts) in solution, it often implies catalysis by organometallic compounds but can also apply to phase reactions and solid-phase reactions. Homogeneous catalysis differs from heterogeneous catalysis insofar as the catalyst is in a different phase than the reactants. The advantage of homogeneous catalysts is the ease of mixing them with the reactants, but they may also be difficult to separate from the products. Therefore, heterogeneous catalysts are preferred in many industrial processes for the production and transformation (conversion) of the starting compound(s). However, heterogeneous catalysis offers the advantage that products are readily separated from the catalyst, and heterogeneous catalysts are often more stable and degrade much slower than homogeneous catalysts. However, heterogeneous catalysts are difficult to study, so their reaction mechanisms are often unknown.

Catalysts in solution with the reactants usually provide fast reaction paths by allowing reactants to form an unstable intermediate that quickly decomposes into products. For example, the substitution reaction is catalyzed by acid because the ethanol is converted into unstable $CH_3CH_2OH_2^+$, which quickly reacts with Cl^- to produce the products:

$$CH_3CH_2OH(g) + HCl(g) \rightarrow CH_3CH_2Cl + H_2O$$

Other similar examples also exist which indicate that catalysts could be a major reaction path in environmental organic chemistry for which a variety of catalysts are under development (Janssen and Van Santen, 1999).

4 ADSORPTION AND ABSORPTION

Organic chemicals interact with the environment in different ways—two such ways are adsorption and absorption—both are important phenomena with differences in the outcomes (Table 7.5). The structure (physical and electronic) of the organic molecules play a role in both phenomena as well as such properties as water solubility and (in the case of mixtures) the composition is particularly important. Evaluation of adsorption or absorption can be obtained through either laboratory measurements or by use of several property correlations. In addition, any deductions from laboratory measurement must also take into account the potential for transformation of the organic chemicals in the environment as well as degradation of the chemicals.

4.1 Adsorption

Adsorption is the physical accumulation of material (usually a gas or liquid) on the surface of a solid adsorbent and is a *surface phenomenon* (Calvet, 1989). Typically, adsorption processes remove solutes from liquids based on their mass transfer from liquids to porous solids. Ion exchange is the exchange of dissolved ions for ions on solid media. The process can be used to remove water

TABLE 7.5 Comparison of Adsorption and Absorption

	Adsorption	Absorption
Definition	Accumulation of the molecular species at the surface rather than in the bulk of the solid or liquid	Assimilation of molecular species throughout the bulk of the solid or liquid
Characteristic	A surface phenomenon	A bulk phenomenon
Reaction type	Exothermic process	Endothermic process
Temperature	Unaffected by temperature	Not affected by temperature
Reaction rate	Increases to equilibrium	Occurs at a uniform rate
Concentration	Different at surface to bulk	Same throughout

hardness and toxic metals during wastewater treatment. Disinfection is the removal or inactivation of pathogenic organisms in wastewater prior to discharge to the receiving body of water.

The adsorption process creates a film of the *adsorbate* on the surface of the *adsorbent* and the process differs from the absorption process in which a fluid (the *absorbate*) is dissolved by a liquid or permeates into a solid (the *absorbent*), respectively. Thus, adsorption is a surface-based process while absorption involves the whole volume of the material. The term *sorption* encompasses both processes, while *desorption* is the reverse of sorption. In the environment, organic compounds will collect on the surfaces of particles, such as soil or suspended sediment. Most of these particles are covered with a layer of organic material; thus, the adsorption results from the attraction of two organic materials for one another.

In an industrial setting, adsorption processes are used to remove certain components from a mobile phase (i.e., a gas phase or a liquid phase) or to separate mixtures. The applications of adsorption can be production-related or abatement-related and may include the removal of water from natural gas or the removal of organic constituents from flue gas, such as is often witnessed in refinery processes and/or in natural gas processing operations and/or coal gas processing operations (Mokhatab et al., 2006; Speight, 2007, 2014, 2016). The most preferential adsorbents are characterized by a wide distribution of a large number of varying-sized pores and, accordingly, activated carbon, zeolites, silica gel, and aluminum oxide are the most commercially important adsorbent. This enable adsorbent to accommodate types and sizes of the various molecular species that occur in gas or liquid streams. Zeolites (molecular sieves) have a very narrow distribution of micropores and preferentially adsorb polar or polarizable materials (e.g., water or carbon dioxide). By contrast, activated carbon has a hydrophobic character and is especially suitable for the removal of organic substances.

In nature, it is different—a variety of potential natural adsorbents exits in the soil—adsorption occurs in many natural, physical, biological, and chemical systems (especially in the environment) where organic molecules can adsorb on to minerals (such as clay) or on to charred wood that remains after a forest fire. In fact, clay minerals are particularly good adsorbents and have a high adsorption capacity for organic chemicals that have been released into the environment.

Clay minerals are typically ultrafine-grained [normally considered to be less than 2 μm (<2 μm, $<2 \times 10^{-6}$ m) in size on standard particle size classifications]. In the present context, clay minerals, which can be classified into various chemical groups, such as the silicate clay mineral groups (Table 7.6) are an important part of many soils thus rendering the soil capable of having a high adsorption capacity for organic chemicals. Generally, no two clay minerals are the same and the adsorption capacity will vary accordingly.

Adsorption of an organic chemical on to a solid adsorbent is measured by a partition coefficient, which is the ratio of the concentration of the organic chemical on the solid to the concentration of the chemical in the fluid (usually water) surrounding the solid:

$$K_d = C_{solid}/C_{water}$$

The concentration on the solid has units of mol/kg, and the concentration in the water is mol/L and, thus, the adsorption coefficient (K_d) has units of L/kg. Assuming a solid density of 1 kg/L, these units are often ignored. The adsorption coefficient will often depend on how much of the total mass of the particle is organic material. Thus, the adsorption coefficient can be corrected by the fraction of organic material (f_{om}) in the particles:

$$K_{om} = K_d/f_{om}$$

TABLE 7.6 Illustration of Various Clay Mineral Groups

Group	Layer Type	Layer Charge (x)	Type of Chemical Formula
Kaolinite	1:1	<0.01	$[Si_4]Al_4O_{10}(OH)_8 \cdot nH_2O$ ($n=0$ or 4)
Illite	2:1	1.4–2.0	$M_x[Si_{6.8}Al_{1.2}]$ $Al_3Fe_{0.25}Mg_{0.75}O_{20}(OH)_4$
Vermiculite	2:1	1.2–1.8	$M_x[Si_7Al]AlFe_{0.5}Mg_{0.5}O_{20}(OH)_4$
Smectite	2:1	0.5–1.2	$M_x[Si_8]Al_{3.2}Fe_{0.2}Mg_{0.6}O_{20}(OH)_4$
Chlorite	2:1:1	Variable	$(Al(OH)_{2.55})_4[Si_{6.8}Al0_{1.2}]$ $Al_{3.4}(Mg_{0.6})_{20}(OH)_4$

Adsorbed molecules are those that are resistant to washing with the same solvent medium in the case of adsorption from solutions. The washing conditions can thus modify the measurement results, particularly when the interaction energy is low. The exact nature of the bonding depends on the details of the chemical species involved, but the adsorption process is generally classified as physisorption (which is characteristic of weak van der Waals forces) or chemisorption (which is characteristic of covalent bonding). It may also occur due to electrostatic attraction.

4.2 Absorption

Absorption is another phenomenon that can be a beneficial or adverse influence of the environment and involves the uptake of one substance into the inner structure of another; most typically a gas into a liquid solvent. Furthermore, absorption is a physical or chemical phenomenon or a process in which atoms, molecules, or ions enter some bulk phase—gas, liquid, or solid material. This is a different process from *adsorption*, since molecules undergoing absorption are taken up by the volume, not by the surface (as in the case for adsorption). A more general term is *sorption*, which covers absorption and adsorption—the former (absorption) is a condition in which something takes in another substance. In many processes important in technology, the chemical absorption is used in place of the physical process. It is possible to extract from one liquid phase to another a solute without a chemical reaction. The process of absorption means that a substance captures and transforms energy and distributes the material it captures throughout whole absorbent whereas an adsorbent only distributes it on the surface.

In industry, absorption is a unit operation not only for chemical production but also for environmental protection in the abatement of gaseous emissions (where it may be known as washing or scrubbing), such as course in gas processing operations (Mokhatab et al., 2006; Speight, 2007, 2014, 2016). The interaction of absorbed materials with the solvent can be physical or chemical in nature. In physical absorption, the gas molecules are physically changed (polarized) but remain chemically unchanged. The concentration of dissolved gases in the absorbing solvent increases in relation to the partial pressure of the gases.

In chemical absorption (sometimes referred to in the shortened word form as *chemisorption*), the absorbed material is generally converted to a product different to the starting material. Thus, chemical absorption or *reactive absorption* involves a chemical reaction between the absorbent (the absorbing substance) and the absorbate (the absorbed substance) and may be combined with the physical absorption phenomenon. This type of absorption depends upon the stoichiometry of the reaction and the concentration of the potential reactants.

Reactions and conversions between gaseous and liquid phases are much slower than those between one-phase mixtures, and so relatively large reaction volumes are required in gas absorption installations. Absorption equipment generally consists of a column with internals for heat and material exchange in which the feed gas is brought into counter-current contact with the regenerated absorbent (Mokhatab et al., 2006). The equipment internals (which may be absorption plates, randomly poured packing, or structured packing) direct the liquid and gas streams into close contact and also serve to maintain the contact area between the two phases.

Physical absorption or nonreactive absorption is made between two phases of matter: a liquid absorbs a gas, or a solid absorbs a liquid. When a liquid solvent absorbs a gas mixture or part of it, a mass of gas moves into the liquid. For example, water may absorb oxygen from the air. This mass transfer takes place at the interface between the liquid and the gas, at a rate depending on both the gas and the liquid. This type of absorption depends on the solubility of gases, the pressure and the temperature. The rate and amount of absorption also depend on the surface area of the interface and its duration in time. For example, when the water is finely divided and mixed with air, as may happen in a waterfall or a strong ocean surf, the water absorbs more oxygen. When a solid absorbs a liquid mixture or part of it, a mass of liquid moves into the solid. This mass transfer takes place at the interface between the solid and the liquid, at a rate depending on both the solid and the liquid. Absorption is essentially molecules attaching themselves to a substance and will not be attracted from other molecules.

On the other hand, chemical absorption or reactive absorption is a chemical reaction between the absorbed and the absorbing substances. Sometimes it combines with physical absorption. This type of absorption depends upon the stoichiometry of the reaction and the concentration of its reactants.

5 BIODEGRADATION

Biodegradation is one of the most important environmental processes that cause the breakdown of organic chemicals (Speight and Arjoon, 2012) and, for some organic chemicals may be the only process by which decomposition may occur. In the presence of microbially produced biological enzymes, chemical reactions may proceed at high rates of reaction. Some microorganisms can utilize some organic chemicals as food sources to provide energy and carbon for growth and cell maintenance of the microbial population (growth metabolism). On the other hand, some organic chemicals may be transformed by microbes without the microbial population being able to derive energy from the chemical reactions (cometabolism).

Growth metabolism, the use of the pollutant as a food source, requires that the microbial community adapt to the chemical. Usually a lag phase is associated with this adaptation during which the microbial population develops sufficiently large numbers to be effective in rapidly degrading the chemical. The time required for adaptation depends upon (1) prior exposure of the community to the pollutant, (2) the initial numbers of a suitable species, (3) the presence of more easily degraded carbon sources, and (4) the concentration of the pollutant in the water. Growth metabolism frequently results in complete mineralization of the pollutant.

Cometabolism, the degradation of compounds that cannot be used as growth substances, is believed to occur when some microbially produced enzymes alter the compound to form products which other enzymes cannot utilize. The resulting metabolites are structurally similar to the parent molecule and frequently retain their toxicity. In some instances, the metabolites may be more toxic than the parent compounds and these metabolites often accumulate in the environment but may be used as food sources by other organisms. Several environmental conditions, many of which may be modified by the presence of other chemicals in solution, have been shown to influence the rate of biodegradation of organic chemicals. Important parameters include temperature, nutrient availability, sorption to substrates, solubility, pH, and dissolved oxygen. Microbial degradation has been shown to be a major source of uncertainty for the prediction of the fate of a complex mixture, such as is often the case with crude oil-derived products and coal-derived products.

Chemically, biodegradation involves molecular transformations mediated by microorganisms that: (1) satisfy nutritional requirements, (2) satisfy energy requirements, (3) detoxify the immediate environment, or (4) occur fortuitously such that the organism receives no nutritional or energy benefit (Stoner, 1994; Obire and Nwaubeta, 2001; Obire and Anyanwu, 2009). In addition, *mineralization* is the complete biodegradation of organic materials to inorganic products, and often occurs through the combined activities of microbial consortia rather than through a single microorganism (Shelton and Tiedje, 1984). *Cometabolism* is the partial biodegradation of organic compounds that occurs fortuitously and that does not provide energy or cell biomass to the microorganisms. Cometabolism can result in partial transformation to an intermediate that can serve as a carbon and energy substrate for microorganisms, as with some hydrocarbons, or can result in an intermediate that is toxic to the transforming microbial cell, as with trichloroethylene (TCE) and methanotrophic organisms (organisms that can grow aerobically—with oxygen—or anaerobically—without oxygen—and able to metabolize methane as their only source of carbon and energy).

5.1 Chemical Reactions

Biodegradation of hydrocarbons can occur under both aerobic (oxic) and anaerobic (anoxic) conditions (Zengler et al., 1999), albeit by the action of different consortia of organisms. In the subsurface, biodegradation of chemicals occurs primarily under anoxic conditions, mediated by sulfate-reducing bacteria (e.g., Holba et al., 1996) or other anaerobes using a variety of other electron acceptors as the oxidant. Thus, two classes of biodegradation reactions are: (1) aerobic biodegradation and (2) anaerobic biodegradation.

Aerobic biodegradation involves the use of molecular oxygen (O_2), where oxygen (the "terminal electron acceptor") receives electrons transferred from an organic contaminant:

Organic substrate $+ O_2 \rightarrow$ biomass $+ CO_2 + H_2O +$ other inorganic products

For example, hydrocarbons are major components of crude oil and of refined products such as naphtha, kerosene, gas oil, and lubricating oils. These compounds may enter the environment as a result of accident and bioremediation has been attempted in the aquatic and the terrestrial environments (Speight and Arjoon, 2012). They may enter groundwater in which anaerobic degradation is significant and bacterial degradation under aerobic conditions is initiated by either of two reactions: (1) terminal hydroxylation followed by successive dehydrogenation. The resulting carboxylates are further degraded by oxidation to yield ultimately acetate from even-membered alkanes or propionate from odd-membered alkanes or (2) subterminal hydroxylation followed by oxidation to ketones.

Terminal oxidation of alkane derivatives:

$$RCH_2CH_2CH_2CH_3 \rightarrow RCH_2CH_2CH_2CH_2OH \rightarrow RCH_2CH_2CH_2CHO$$
$$\text{alkane} \qquad\qquad \text{alcohol} \qquad\qquad\qquad \text{aldehyde}$$

Subterminal oxidation of alkane derivatives:

$$RCH_2CH_2CH_2CH_3 \rightarrow RCH_2CH_2CH(OH)CH_3 \rightarrow RCH_2CH_2COCH_3$$
$$\text{alkane} \qquad\qquad \text{alcohol} \qquad\qquad\qquad \text{ketone}$$

Further reaction will produce a wider range of oxygenated products and, in the case of long-chain alkane derivatives, hydroxylation may occur at both ends of the chain (Neilson and Allard, 2008).

Thus, the organic substrate is oxidized (addition of oxygen), and the oxygen is reduced (addition of electrons and hydrogen) to water (H_2O). In this case, the organic substrate serves as the source of energy (electrons) and the source of cell carbon used to build microbial cells (biomass). Some microorganisms (chemoautotrophic aerobes or litho-trophic aerobes) oxidize reduced inorganic compounds (NH_3, Fe^{2+}, or H_2S) to gain energy and fix carbon dioxide to build cell carbon:

$$NH_3 \left(\text{or } Fe^{2+} \text{or } H_2S\right) + CO_2 + H_2 + O_2 \rightarrow \text{biomass} + NO_3 \left(\text{or Fe or } SO_4\right) + H_2O$$

At some contaminated sites, as a result of consumption of oxygen by aerobic microorganisms and slow recharge of oxygen, the environment becomes anaerobic (lacking oxygen), and mineralization, transformation, and cometabolism depend upon microbial utilization of electron acceptors other than oxygen (anaerobic biodegradation). Nitrate (NO_3), iron (Fe^{3+}), manganese (Mn^{4+}), sulfate (SO_4), and carbon dioxide (CO_2) can act as electron acceptors if the organisms present have the appropriate enzymes (Sims, 1990).

Anaerobic biodegradation is the microbial degradation of organic substances in the absence of free oxygen. While oxygen serves as the electron acceptor in aerobic biodegradation processes forming water as the final product, degradation processes in anaerobic systems depend on alternative acceptors such as sulfate, nitrate, or carbonate yielding, in the end, hydrogen sulfide, molecular nitrogen, and/or ammonia and methane (CH_4), respectively.

In the absence of oxygen, some microorganisms obtain energy from fermentation and anaerobic oxidation of organic carbon. Many anaerobes use nitrate, sulfate, and salts of iron (III) as practical alternates to oxygen acceptor. The anaerobic reduction process of nitrates, sulfates, and salts of iron is an example:

$$2NO_3^- + 10e^- + 12H^+ \rightarrow N_2 + 6H_2O$$

$$SO_4^{2-} + 8e^- + 10H^+ \rightarrow H_2S + 4H_2O$$

$$Fe(OH)_3 + e^- + 3H^+ \rightarrow Fe^{2+} + 3H_2O$$

Anaerobic biodegradation is a multistep process performed by different bacterial groups. It involves hydrolysis of polymeric substances like proteins or carbohydrates to monomers and the subsequent decomposition to soluble acids, alcohols, molecular hydrogen, and carbon dioxide. Depending on the prevailing environmental conditions, the final steps of ultimate anaerobic biodegradation are performed by denitrifying, sulfate-reducing or methanogenic bacteria.

In contrast to the strictly anaerobic sulfate-reducing and methanogenic bacteria, the nitrate-reducing microorganisms as well as many other decomposing bacteria are mostly facultative anaerobic insofar as these microorganisms are able to grow and to degrade organic substances (under aerobic as well as anaerobic conditions). Thus, aerobic and anaerobic environments represent the two extremes of a continuous spectrum of environmental habitats which are populated by a wide variety of microorganisms with specific biodegradation abilities.

Anaerobic conditions occur where vigorous decomposition of organic matter and restricted aeration result in the depletion of oxygen. Anoxic conditions may represent an intermediate stage where oxygen supply is limited, still allowing a slow (aerobic) degradation of organic compounds. In a digester the various bacteria also have different requirements to the surrounding environment. For example, acidogenic bacteria need pH values from 4 to 6, whilst methanogenic bacteria from 7 to 7.5. In batch tests the dynamic equilibrium is often interrupted because of an enrichment of acidogenic bacteria as a consequence of lacking substrate in- and outflow.

The inherent biodegradability of these individual components is a reflection of their chemical structure, but is also strongly influenced by the physical state and toxicity of the compounds. As an example, while the *n*-alkane derivatives (Chapter 2) are the most biodegradable hydrocarbon derivatives, the $C_5–C_{10}$ homologs have been shown to exhibit the occasional inhibitory action to the majority of hydrocarbon degrading microbes (Speight and Arjoon, 2012).

As solvents, these homologs tend to disrupt lipid membrane structures of micro-organisms. Similarly, alkanes in the $C_{20}-C_{40}$ range are hydrophobic solids at physiological temperatures. Apparently, it is this physical state that strongly influences their biodegradation (Bartha and Atlas, 1977).

Primary attack on intact hydrocarbons requires the action of oxygenase organisms and, therefore, requires the presence of free oxygen. In the case of alkanes, mono-oxygenase attack results in the production of alcohol. Most microorganisms attack alkanes terminally whereas some perform subterminal oxidation. The alcohol product is oxidized finally into an aldehyde. Extensive methyl branching interferes with the beta-oxidation process and necessitates terminal attack or other bypass mechanisms. Therefore, *n*-alkanes are degraded more readily than iso-alkanes.

Cycloalkanes are transformed by an oxidase system to a corresponding cyclic alcohol, which is dehydrated to ketone after which a mono-oxygenase system forms a lactose-type ring, which is subsequently opened by a lactone hydrolase. These two oxygenase systems usually never occur in the same organisms and hence, the frustrated attempts to isolate pure cultures that grow on cycloalkanes (Bartha, 1986b). However, synergistic actions of microbial communities are capable of dealing with degradation of various cycloalkanes quite effectively.

As in the case of alkanes, the monocyclic compounds, cyclopentane (C_5H_{10}), cyclohexane (C_6H_{12}), and cycloheptane (C_7H_{14}) have a strong solvent effect on lipid membranes, and are toxic to the majority of hydrocarbon degrading microorganisms. Highly condensed cycloalkane compounds resist biodegradation due to their relatively complex structure and physical state (Bartha, 1986a).

Condensed polycyclic aromatics are degraded, one ring at a time, by a similar mechanism, but biodegradability tends to decline with the increasing number of rings and degree of condensation. Aromatics with more than four condensed rings are generally not suitable as substrates for microbial growth, though, they may undergo metabolic transformations. The biodegradation process also declines with the increasing number of alkyl substituents on the aromatic nucleus. In fact, some iso-alkanes are apparently spared as long as *n*-alkanes are available as substrates, while some condensed aromatics are metabolized only in the presence of more easily utilizable hydrocarbons, a process referred to as cometabolism (Wackett 1996).

Finally, a word on the issue of adhesion as it affects biodegradation and, hence, bioremediation. Adhesion to hydrophobic surfaces is a common strategy used by microorganisms to overcome limited bioavailability of hydrocarbons. Intuitively, it may be assumed that adherence of cells to a hydrocarbon would correlate with the ability to utilize it as a growth substrate and conversely that cells able to utilize hydrocarbons would be expected to be able to adhere to them. However, species like *Staphylococcus aureus* and *Serratia marcescens*, which are unable to grow on hydrocarbons, adhere to them (Rosenberg et al., 1980). Thus, adherence to hydrocarbons does not necessarily predict utilization (Abbasnezhad et al., 2011).

Biodegradation of poorly water-soluble liquid hydrocarbons is often limited by low availability of the substrate to microbes. Adhesion of microorganisms to a hydrocarbon-water interface can enhance this availability, whereas detaching cells from the interface can reduce the rate of biodegradation. The capability of microbes to adhere to the interface is not limited to hydrocarbon degraders, nor is it the only mechanism to enable rapid uptake of hydrocarbons, but it represents a common strategy. The general indications are that microbial adhesion can benefit growth on and biodegradation of very poorly water-soluble hydrocarbons such as n-alkanes and large polycyclic aromatic hydrocarbons dissolved in a nonaqueous phase. Adhesion is particularly important when the hydrocarbons are not emulsified thereby giving limited interfacial area between the two liquid phases.

When mixed communities are involved in biodegradation, the ability of cells to adhere to the interface can enable selective growth and enhance bioremediation with time. The critical challenge in understanding the relationship between growth rate and biodegradation rate for adherent bacteria is to accurately measure and observe the population residing at the interface of the hydrocarbon phase.

5.2 Kinetics

The kinetics for modeling the bioremediation of contaminated soils can be extremely complicated. This is largely due to the fact that the primary function of microbial metabolism is not for the remediation of environmental contaminants. Instead the primary metabolic function, whether bacterial or fungal in nature, is to grow and sustain more of the microorganism. Therefore, the formulation of a kinetic model must start with the active biomass and factors, such as supplemental nutrients, oxygen source, that are necessary for subsequent biomass growth (Cutright, 1995; Rončević et al., 2005; Pala et al., 2006).

Studies of the kinetics of the bioremediation process proceed in two directions: (1) the first is concerned with the factors influencing the amount of transformed compounds with time and (2) the other approach seeks the types of curves describing the transformation and determines which of them fits the degradation of the given compounds by the microbiologic culture in the laboratory microcosm and sometimes, in the field. However, studies of biodegradation kinetics in the natural environment are often empiric, reflecting only a basic level of knowledge about the microbiologic population and its activity in a given environment (Maletić et al., 2009).

One such example of the empirical approach is the simple (perhaps oversimplified) model:

$$dC/dt = kC^n$$

C is the concentration of the substrate, t is time, k is the degradation rate constant of the compound, and n is a fitting parameter (most often taken to be unity) (Hamaker 1972; Wethasinghe et al. 2006). Using this model, the curve of substrate removal can be fitted by varying n and k until a satisfactory outcome is obtained. It is evident from this equation that the rate is proportional to the exponent of substrate concentration. First order kinetics are the most often used equation for representation of the degradation kinetics (Heitkamp et al. 1987; Heitkamp and Cerniglia 1987; Venosa et al. 1996; Seabra et al. 1999; Holder et al. 1999; Winningham et al. 1999; Namkoonga et al. 2002; Grossi et al. 2002; Hohener et al. 2003; Collina et al. 2005; Rončević et al. 2005; Pala et al. 2006).

However, researchers involved in kinetic studies do not always report whether the model they used was based on theory or experience and whether the constants in the equation have a physical meaning or if they just serve as fitting parameters (Rončević et al., 2005).

5.3   Effect of Salt

Salt is a common cocontaminant that can adversely affect the bioremediation potential at sites such as flare pits and drilling sites (*upstream sites*) contaminated with saline produced formation water, or at natural oil and crude oil processing facilities contaminated by refinery wastes containing potassium chloride (KCl) and sodium chloride (NaCl) salts (Pollard et al., 1994). Because of increasing emphasis and interest in the viability of intrinsic bioremediation as a remedial alternative, the impact of salt on these processes is of interest.

The effect of salinity on microbial cells varies from disrupted tertiary protein structures and denatured enzymes to cell dehydration (Pollard et al., 1994), with different species having different sensitivities to salt (Tibbett et al., 2011). A range of organic pollutants, including hydrocarbons, has been shown to be mineralized by marine or salt-adapted terrestrial microorganisms that are able to grow in the presence of salt (Margesin and Schinner, 2001; Oren et al., 1992; Nicholson and Fathepure, 2004). In naturally saline soils, it has been shown that bioremediation of diesel fuel is possible at salinities up to 17.5% (w/v) (Riis et al., 2003; Kleinsteuber et al. (2006).

However, an inverse relationship between salinity and the biodegradation of hydrocarbons by halophilic enrichment cultures from the Great Salt Lake (Utah) has been observed (Ward and Brock, 1978). These cultures were unable to metabolize hydrocarbons at salt concentrations above 20% (w/v) in this hyper-saline environment. An inhibitory effect of salinity at concentrations above 2.4% (w/v) sodium chloride was found to be greater for the biodegradation of aromatic and polar fractions than for saturated hydrocarbons incubated with marine sediment (Mille et al., 1991). This represents exsitu hydrocarbon degradation by salt-adapted terrestrial microorganisms.

Furthermore, the effects of salt as a cocontaminant on hydrocarbon degradation in naturally nonsaline systems has been described (De Carvalho and daFonseca, 2005). The results showed that in the degradation of C_5–C_{16} hydrocarbons at 28°C (82°F) in the presence of 1.0%, 2.0%, or 2.5% (w/v) NaCl by the isolate *Rhodococcus erythropolis* DCL14 the lag phase of the cultures increased and growth rates decreased with increasing concentrations of sodium chloride. In a similar study (Rhykerd et al., 1995), soils were fertilized with inorganic nitrogen and phosphorus, and amended with sodium chloride at 0.4%, 1.2%, or 2% (w/w). After 80 days at 25°C (77°F), the highest salt concentration had inhibited hydrocarbon mineralization.

However, investigation of the combinations of factors limiting biodegradation of hydrocarbon contamination at upstream natural and crude oil production facilities have received relatively little attention. A laboratory solid-phase bioremediation study reported that high salinity levels reduced the degradation rate of flare pit hydrocarbons (Amatya et al., 2002), and more recently it has been observed that addition of sodium chloride to a contaminated soil decreased hexadecane mineralization rates in the initial stages of bioremediation and increased lag times, but that the final extent of mineralization was comparable over a narrow range of salinity from 0% to 0.4% (w/w) (Børresen and Rike, 2007). However, before embarking on anaerobic microcosm tests, field evidence of indicators of anaerobic biodegradation including changes in terminal electron acceptors, presence of metabolites, and isotopic analysis would be a reasonable way to initiate the investigation (Ulrich et al., 2009).

6 CHEMISTRY IN THE ENVIRONMENT

In terms of organic chemicals, a chemical transformation is the conversion of a substrate (or reactant) to a product. In more general terms, a chemical transformation involves (or is) a chemical reaction which is characterized by a chemical change, and yields one or more products, which usually have properties substantially different from the properties of the individual reactants. Reactions often consist of a sequence of individual substeps that can be described by means of chemical equations, which symbolically present the starting materials, end products, and sometimes intermediate products and reaction conditions.

Chemical reactions occur at a characteristic rate (the reaction rate) at a given temperature and chemical concentration. Typically, reaction rates increase with increasing temperature because there is more thermal energy available to reach the activation energy necessary for breaking bonds between atoms. The general rule of thumb (see above) is that for every 10°C (18°F) increase in temperature the rate of an organic chemical reaction is doubled and there is no reason to doubt that this would not be the case for organic chemicals discharged into the environment (Jury et al., 1987).

The chemical industry involves physical, thermal, and manufacture of chemical intermediates and end-product organic chemicals. The product slate is varied but includes fuels, petrochemicals, fertilizers, pesticides, paints, waxes, thinners, solvents, cleaning fluids, detergents, refrigerants, antifreeze, resins, sealants, insulations, latex, rubber compounds, hard plastics, plastic sheeting, plastic foam, and synthetic fibers. The composition of the chemicals is varied and there are very few indications of how these chemical will behave once they are discharged into the environment either as a single chemical or as a mixture. It is at this stage that a knowledge of chemical properties can bring some knowledge of predictability about chemical behavior.

These organic chemicals vary from simple hydrocarbons of low-to-medium molecular weight to higher molecular weight organic compounds containing sulfur, oxygen, and nitrogen, as well as compounds containing metallic constituents, particularly vanadium nickel, iron, and copper and contain one or more functional groups that dictate the behavior of the chemical. However, the behavior of an organic chemical on the basis of functional groups depends upon (1) the type of functional group, (2) the number of functional groups, (3) the position of the functional groups within the molecule, and (4) the ecosystem into which the chemical is discharged.

Chemicals can enter the environment (air, water, and soil) when they are produced, used, or disposed (Chapter 6) and the impact on the environment is determined by the amount of the chemical that is released, the type and concentration of the chemical, and where it is found, as well as through any chemical transformation that occur after the chemical has entered the environment whether it is in the atmosphere, the aquasphere, or the terrestrial biosphere (Jury et al., 1987). Some chemicals can be harmful if released to the environment even when there is not an immediate, visible impact. Some chemicals are of concern as they can work their way into the food chain and accumulate and/or persist in the environment for prolonged periods, including years (Chapters 1 and 4–6) which is in direct contradiction of the earlier *conventional wisdom* (or unbridled optimism) that assumed that organic chemicals would either (1) degrade into harmless byproducts as a result of microbial or chemical reactions, (2) immobilize completely by binding to soil solids, or (3) volatilize to the atmosphere where dilution to harmless levels was assured. This false assurance led to years of agricultural chemical use and chemical waste disposal with no monitoring of atmosphere, or groundwater (the aquasphere), or soil (the terrestrial biosphere) in the vicinity of discharge (Jury et al., 1987). Thus, the volatility of an organic chemical is of concern predominantly for surface-located chemicals and is affected by (1) temperature of the soil, (2) the water content of the soil, (3) the adsorptive interaction of the chemical and the soil, (4) the concentration of the chemical in the soil, (5) the vapor pressure of the chemical, and (6) the solubility of the chemical in water, which is the predominant liquid in the soil.

However, before delving into the realm of chemicals in the environment, it is necessary for any investigator to recognize that there are chemicals that exist naturally in the environment and which must be taken into account before accurate assessment of chemicals in the environment can be made. These naturally occurring organic chemicals are often grouped under the umbrella name *natural organic matter* (NOM) which is an inherently complex mixture of polyfunctional organic molecules (Macalady and Walton-Day, 2011). Because of their universality and chemical reversibility, oxidation/reductions (redox) reactions of NOM have an especially interesting and important role in geochemistry. Variabilities in NOM composition and chemistry make studies of its redox chemistry particularly challenging, and details of NOM-mediated redox reactions are only partially understood. This is in large part due to the analytical difficulties associated with NOM characterization and the wide range of reagents and experimental systems used to study NOM redox reactions.

When dealing with organic chemicals that have been released (advertently or inadvertently, dispensing upon the circumstances), there are several types of chemical transformations of organic chemical transformation reactions that can occur in the environment. These reactions can be grouped into four major categories: (1) oxidation-reduction reactions, also known as redox reactions, (2) carbon-carbon bond formation, (3) carbon-heteroatom bond formation in which a carbon atom of one molecule forms a bond with the nitrogen atom or oxygen atom or sulfur atom of another molecule, (4) carbon-carbon bond cleavage, (5) carbon-heteroatom bond cleavage, and (6) organic-inorganic interactions.

Redox reactions would include the hydrogenation of olefin derivatives and acetylene derivatives, the loss of hydrogen through aromatization reactions, the oxidation or reduction of alcohols, aldehydes and ketones, and the oxidative cleavage of olefins. Examples of chemical transformations involving bond formation are polymerization or condensation reactions, esterification or amide ($-CONH_2$) formation, and cyclization (ring formation) reactions. Several types of bond cleavage reactions which might affect the fate or longevity of organic chemicals discharged into the environment are the formation of amino acids from peptides and proteins, and the hydrolysis of esters and amides to form carboxylic acids, as well as other forms of chemical degradation (Wham et al., 2005). Organic-inorganic interactions include the formation of organometallic compounds and organo-mineral phase interactions. There are at least two types of organometallic complexes that are found in the environment: (1) a compound that contains covalently bound metals such as metallo-enzymes as well as anthropogenic alkyl metal compounds, illustrated as RM or R^-M^+, where R is the alkyl groups and M is the metal and (2) the more abundant chelate-type complexes such as metal humate derivatives, where the humate derivatives are formed from humic acid derivatives (produced as a collection of organic acids by the biodegradation of dead organic matter). Some algal products form complexes with metals and there is always the potential for metal detoxification or making the metals otherwise available to the phytoplankton cells as micronutrients. Organo-mineral phase interactions involve the adsorption of highly surface active dissolved organic matter to ocean particulate matter. The mechanisms by which this takes place include ion exchange (such materials such as calcium carbonate, $CaCO_3$), interlayering of organic compounds in clay minerals, formation of clathrates, hydrogen bonding, and van der Waals interactions. By way of explanation, van der Waals' forces are the residual attractive or repulsive forces between molecules or atomic (functional) groups that do not arise from a covalent bond, or electrostatic interaction of ions or of ionic groups with one another or with neutral molecules. The resulting van der Waals' forces can be attractive or repulsive.

Long-term trends of chemical species in the environment are determined by emissions from anthropogenic and natural sources as well as by transport of the organic chemical, physical, and chemical processes that affect the behavior of the chemical, and deposition. While continually increasing emissions of such trace species as carbon dioxide (CO_2), nitrous oxide (N_2O), and methane (CH_4) that can arise from transformation occurring during the life cycle of organic chemicals are predicted to raise global temperatures via the *greenhouse effect*, growing emissions of sulfur dioxide (SO_2), which forms sulfate ($-SO_4$) aerosol through oxidation most likely will have a cooling effect by reflecting solar radiation back to space. However, these postulates do not take into account the fact that the earth is in an interglacial period during which time there will be an overall rise in climatic temperature as the natural order of climatic variation. Therefore, the extent of the anthropogenic contributions to temperature rise (climate change) cannot be accurately assessed (Speight and Islam, 2016). Complicating matters is the fact that the organic chemical reactions are sensitive to climatic conditions, being functions of temperature, the presence of water vapor, as well as a variety of other physical parameters.

Thus, environmental organic chemistry—the study of organic chemical processes occurring in the environment—is impacted by a variety of external activities (including anthropogenic activities and climatic variations) and these impacts may be felt on a local scale (through the presence of urban air pollutants or toxic substances arising from a chemical waste site) or on a global scale (through depletion of stratospheric ozone or the phenomenon that has become known as global climate change).

6.1 Chemistry in the Atmosphere

Organic chemicals can be emitted directly into the atmosphere or formed by chemical conversion or through chemical reactions of precursors species. In these reactions, highly toxic organic chemicals can be converted into less toxic products but the result of the reactions can also be products having a higher toxicity than the starting chemicals. In order to understand these reactions, it is also necessary to understand the chemical composition of the natural atmosphere, the way gases, liquids, and solids in the atmosphere interact with each other and with the earth's surface and associated biota, and how human activities may be changing the chemical and physical characteristics of the atmosphere.

There are a number of critical environmental issues associated with a changing atmosphere, including photochemical smog, global climate change, toxic air pollutants, acidic deposition, and stratospheric ozone depletion (Gouin et al., 2013). A great deal of research and development activity aimed at understanding and hopefully solving some of these problems is underway. Much of the anthropogenic (human) impact on the atmosphere is associated with our increasing use of fossil fuels as an energy source—for things such as heating, transportation, and electric power production. Photochemical smog/tropospheric ozone is a serious environmental problem associated with burning fossil fuels. In fact, the combustion of fossil fuels (which are in fact, organic chemicals) is one of the most common sequences of chemistry that causes pollution in the atmosphere. This phenomenon may not be classed as direct pollution (in the sense of organic chemistry) but is certainly an indirect form of pollution (again, in the sense of organic chemistry). The result is the formation and deposition of acid rain.

Acid rain is formed when sulfur dioxide and nitrogen oxides react with water vapor and other chemicals in the presence of sunlight to form various acidic compounds in the air. The principle source of acid rain-causing pollutants, sulfur dioxide and nitrogen oxides, are from fossil fuel combustion and from the combustion of fossil fuel-derived fuels:

$$2[C]_{\text{fossil fuel}} + O_2 \rightarrow 2CO$$

$$[C]_{\text{fossil fuel}} + O_2 \rightarrow CO_2$$

$$2[N]_{\text{fossil fuel}} + O_2 \rightarrow 2NO$$

$$[N]_{\text{fossil fuel}} + O_2 \rightarrow NO_2$$

$$[S]_{\text{fossil fuel}} + O_2 \rightarrow SO_2$$

$$2SO_2 + O_2 \rightarrow 2SO_3$$

Hydrogen sulfide and ammonia are produced from processing sulfur-containing and nitrogen containing feedstocks:

$$[S]_{\text{fossil fuel}} + H_2 \rightarrow H_2S + \text{hydrocarbons}$$

$$2[N]_{\text{fossil fuel}} + 3H_2 \rightarrow 2NH_3 + \text{hydrocarbons}$$

$$SO_2 + H_2O \rightarrow H_2SO_3 \, (\text{sulfurous acid})$$

$$SO_3 + H_2O \rightarrow H_2SO_4 \, (\text{sulfuric acid})$$

$$NO + H_2O \rightarrow HNO_2 \, (\text{nitrous acid})$$

$$3NO_2 + 2H_2O \rightarrow HNO_3 \, (\text{nitric acid})$$

Two of the pollutants that are emitted are hydrocarbons (e.g., unburned fuel) and nitric oxide (NO). When these pollutants build up to sufficiently high levels, a chain reaction occurs from their interaction with sunlight in which the NO is converted to nitrogen dioxide (NO_2)—a brown gas and at sufficiently high levels can contribute to urban haze. However, a more serious problem is that nitrogen dioxide (NO_2) can absorb sunlight and break apart to produce oxygen atoms that combine with the oxygen in the air to produce ozone (O_3), a powerful oxidizing agent, and a toxic gas.

In addition, as a result of a variety of human activities (e.g., agriculture, transportation, industrial processes) a large number of different toxic organic chemical pollutants are emitted into the atmosphere. Among the chemicals that may pose a human health risk are pesticides, polychlorobiphenyl derivatives (PCBs), polycyclic aromatic hydrocarbon derivatives (PAHs), dioxin derivatives, and volatile organic compounds (e.g., benzene, carbon tetrachloride).

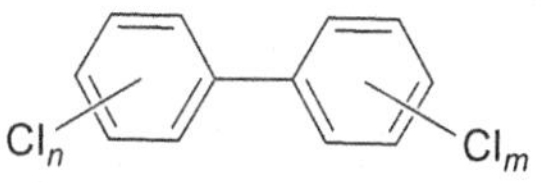

Polychlorobiphenyl derivatives;
n and m can by any number from 1 to 5

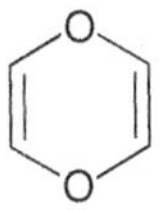

1,4-Dioxin

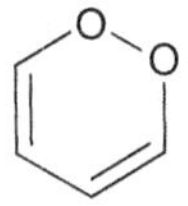

1,2-Dioxin

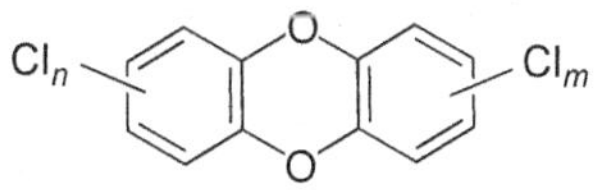

Polychlorinated dibenzo-*p*-dioxin derivatives;
n and m can by any number from 1 to 5

Many of the more environmentally persistent compounds (such as the PCBs) have been measured in various floral and faunal species.

6.2 Chemistry in the Aquasphere

Water pollution has become a widespread phenomenon and has been known for centuries, particularly the pollution of rivers and groundwater (Samin and

Janssen, 2012). By way of example, in ancient time up to the early part of the 20th century, many cities deposited waste into the nearby river or even into the ocean. It is only very recently (because of serious concerns for the condition of the environment) that an understanding of the behavior and fate of chemicals, which are discharged to the aquatic environment as a result of these activities, is essential to the control of water pollution. In rivers the basic physical movement of pollutant molecules is the result of advection, but superimposed upon this are the effects of dispersion and mixing with tributaries and other discharges. Some of the chemicals discharged are relatively inert, so their concentration changes only due to advection, dispersion, and mixing. However, many substances are not conservative in their behavior and undergo changes due to chemical or bio-chemical processes, such as oxidation.

In addition, there are many indications that the organic chemicals materials in the aquasphere (also called, when referring to the sea, the marine aquasphere) are subject to intense chemical transformations and physical recycling processes imply that a total organic-carbon approach is not sufficient to resolve the numerous processes occurring. The transport of anthropogenically produced or distributed organic compounds such as petroleum hydrocarbons and halogenated hydrocarbons, including the PCBs the DDT family, and the Freon derivatives and the chemistry of these organic chemicals in water is not fully understood.

The effects of an organic chemical released into the marine environment (or any part of the aquasphere) depends on several factors such as (1) the toxicity of the chemical, (2) the quantity of the chemical, (3) the resulting concentration of the chemical in the water column, (4) the length of time that floral and faunal organisms are exposed to that concentration, and (5) the level of tolerance of the organisms, which varies greatly among different species and during the life cycle of the organism. Even if the concentration of the chemical is below what would be considered as the lethal concentration, a sublethal concentration of an organic chemical can still lead to a long-term impact within the aqueous marine environment. For example, chemically induced stress can reduce the overall ability of an organism to reproduce, grow, feed or otherwise function normally within a few generations. In addition, the characteristics of some organic chemicals can result in an accumulation of the chemical within an organism (*bioaccumulation*) and the organism may be particularly vulnerable to this problem. Furthermore, subsequent bio-magnification may also occur if the organic chemical (or a toxic product produced by one or more transformation reactions) can be passed on, following the food chain up to higher flora or fauna.

In terms of the marine environment and a spill of crude oil, complex processes of crude oil transformation start developing almost as soon as the oil contacts the water although the progress, duration, and result of the transformations depend on the properties and composition of the oil itself, parameters of the actual oil spill, and environmental conditions. The major operative processes are (1) physical transport, (2) dissolution, (3) emulsification,

(4) oxidation, (5) sedimentation, (6) microbial degradation, (7) aggregation, and (8) self-purification.

In terms of *physical transport*, the distribution of oil spilled on the sea surface occurs under the influence of gravitation forces and is controlled by the viscosity of the crude oil as well as the surface tension oil and water. In addition, during the first several days after the spill, a part of is lost through evaporation of oil (into the gaseous phase) and any water-soluble constituents disappear into the sea. The portion of the crude oil that remains is the more viscous fraction. Further changes take place under the combined impact of meteorological and hydrological factors and depend mainly on the power and direction of wind, waves, and currents. A considerable part of oil disperses in the water as fine droplets that can be transported over large distances away from the place of the spill.

Crude oil is not particularly *soluble* in water although some constituents may be water-soluble to a certain degree, especially low-molecular-weight aliphatic and aromatic hydrocarbons. Polar compounds formed as a result of oxidation of some oil fractions in the marine environment also dissolve in seawater. Compared to evaporation process, the dissolution of cured oil constituents in water is a slow process. However, the *emulsification* of crude oil constituents in the marine environment does occur but depends predominantly on the presence of functional groups in the oil, which can increase with time due to oxidation. Emulsions usually appear when heavy oil is spilled into the ocean because of the higher proportion of polar constituents compared to conventional (lighter) crude oil (Speight, 2014). The rate of emulsification process can be decreased by use of emulsifiers—surface-active chemicals with strong hydrophilic properties used to eliminate oil spills—which help to stabilize oil emulsions and promote dispersing oil to form microscopic (invisible) droplets that accelerates the decomposition of the crude oil constituents in the water.

Oxidation is a complex process that ultimately results in the destruction of the crude oil constituents. The final products of oxidation (such as hydroperoxide derivatives, phenol derivatives, carboxylic acid derivatives, ketone derivatives, and aldehyde derivatives) usually have increased water solubility. This can result in the apparent disappearance of the crude oil from the surface of the water. What is actually happening is the incorporation of functional groups into the oil constituents which results in a change in density with an increase in the ability of the constituents to become miscible (or emulsify) and sink to various depths of the ocean as these changes intensify. These chemical changes also result in an increase in the viscosity of the crude oil which promotes the formation of solid oil aggregates. The reactions of photo-oxidation, photolysis in particular, also initiate transformation of the more complex (polar) constituents in the crude oil (Speight, 2014).

As these processes occur, some of the crude oil constituents are adsorbed on any suspended material and deposited on the ocean floor (sedimentation), the rate of which is dependent upon the ocean depth—in deeper areas remote from the shore, sedimentation of oil (except for the heavy fractions) is a slow process.

Simultaneously, the process of *biosedimentation* occurs—in this process, plankton and other organisms absorb the emulsified oil—and the crude oil constituents are sent to the bottom of the ocean as sediment with the metabolites of the plankton and other organisms. However, this situation radically changes when the suspended oil reaches the sea bottom—the decomposition rate of the oil on the ocean bottom abruptly ceases—especially under the prevailing anaerobic conditions—and any crude oil constituents accumulated inside the sediments can be preserved for many months and even years. These products can be swept to the edge of the ocean (the beach) by turbulent condition at some later time.

The fate of most of the constituents of crude oil in the marine environment is ultimately defined by their transformation and degradation due to *microbial degradation*. The degree and rates of biodegradation depend, first of all, upon the structure of the crude oil constituents—alkanes biodegrade faster than aromatic constituents and naphthenic constituents and, with increasing complexity of molecular structure as well as with increasing molecular weight—the rate of microbial decomposition usually decreases. Besides, this rate depends on the physical state of the oil, including the degree of its dispersion as well as environmental factors such as temperature, availability of oxygen, and the abundance of oil-degrading microorganisms.

Aggregation occurs when crude oil forms lumps or tar balls which are produced from crude oil after the evaporation and dissolution of its relatively low-boiling fractions, emulsification of oil residuals, and chemical and microbial transformation. The chemical composition of oil aggregates is changeable but typically includes asphaltene constituents (up to 50%) and other high-molecular-weight constituents of the oil (Speight, 2014). These tar balls have an uneven shape and vary from 1 mm to 10 cm in size (sometimes reaching up to 50 cm) and complete their life cycle by slowly degrading in the water column, on the shore (if they are washed there by currents), or on the sea bottom (if they lose their floating ability).

Self-purification is a result of the processes previously described above in which crude oil in the marine environment rapidly loses its original properties and disintegrates into various fractions. These fractions have different chemical composition and structure and exist in different migrational forms and they undergo chemical transformations that slow after reaching thermodynamic equilibrium with the environmental parameters. Eventually, the original and intermediate compounds disappear, and carbon dioxide and water form. This form of self-purification inevitably happens in water ecosystems if the toxic load does not exceed acceptable limits.

As an example of chemical transformation that can occur in a water system, the chemistry of methyl iodide (which is thermodynamically unstable in seawater) is known and its chemical fate is kinetically controlled. The equations showing the fate of methyl iodide are as follows (Gacosian and Lee, 1981):

$$CH_3I + Cl^- = CH_3Cl + I$$
$$CH_3I + Br^- = CH_3Br + I^-$$
$$CH_3Br + Cl^- = CH_3Cl + Br^-$$
$$CH_3X + H_2O = CH_3OH + X^-$$

In this equation, $X = Cl^-$, Br^-, I^-.

Chloride ion was theoretically predicted to be the most kinetically reactive species, with water second, and other anions of lesser importance. This suggested that methyl iodide in seawater would react predominantly via a nucleophilic substitution reaction with chloride ion to yield methyl chloride. Methyl iodide and the methyl chloride produced by would also react with water, although more slowly, to yield methanol and halide ions. According to these experiments, substantial amounts of methyl chloride should be formed in seawater. Methyl chloride has a long half-life for decomposition by known reactions in seawater. Hence, its presence could be a useful label for some surface-derived water masses. Methyl chloride is in fact found in the atmosphere, where compared to methyl iodide, it is less stable to photo-degradation reactions.

Steroids are a class of biogenic compounds which may serve as an indicator of certain processes transforming organic matter in seawater and sediments. The steroid hydrocarbon structure forms a relatively stable nucleus which may incorporate functional groups such as alcohols (sterol derivatives and stanol derivatives), ketone derivatives (stanone derivatives), and olefin linkages (sterene derivatives) either in the four ring system or on the side chain originating at C-17.

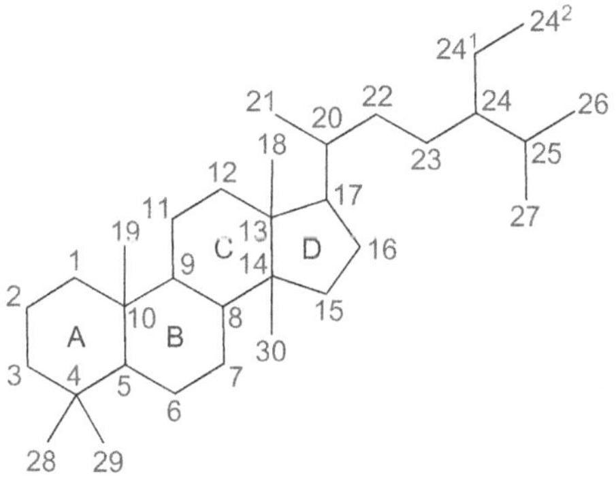

The hydrocarbon framework of the steroid system (ring lettering and atom numbering are shown).

These compounds are produced by a wide variety of marine and terrestrial organisms and often have specific species sources. Diagenetic alteration of steroids by geochemical and biochemical processes can lead to the accumulation of transformed products in seawater and sediments.

Within the group of chlorinated compounds, chlorinated ethylene derivatives are the most often detected groundwater pollutants. Tetrachloroethylene (PCE) is the only chlorinated ethylene derivative that resists aerobic biodegradation. TCE, all three isomers of dichloroethylene ($CCl_2=CH_2$ and the *cis/trans* isomers of $CHCl=CHCl$), and vinyl chloride ($CH_2=CHCl$) are mineralized in aerobic cometabolic processes by methanotropic or phenol-oxidizing bacteria. Oxygenase derivatives with broad substrate spectra are responsible for the cometabolic oxidation. Vinyl chloride is furthermore utilized by certain bacteria as carbon and electron source for growth. All chlorinated ethylene derivatives are reductively dechlorinated under anaerobic conditions with possibly ethylene or ethane as harmless end-products.

PCE ($CCl_2=CCl_2$) is dechlorinated to trichloroethylene ($CCl_2=CHCl$) in a cometabolic process by methanogens, sulfate reducers, homoacetogen derivatives and others. Furthermore, PCE and TCE serve in several bacteria as terminal electron acceptors in a respiration process. The majority of these isolates dechlorinate PCE and TCE to *cis*-1,2-dichloroethene although they have been isolated from systems where complete dechlorination to ethene occurred.

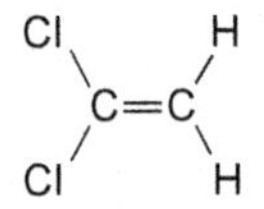

$$Cl \diagdown \quad \diagup H$$
$$C=C$$
$$Cl \diagup \quad \diagdown H$$

1,1-Dichloroethylene

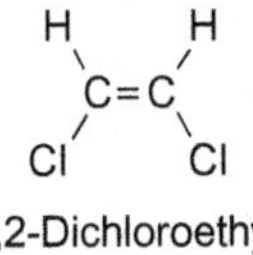

$$H \diagdown \quad \diagup H$$
$$C=C$$
$$Cl \diagup \quad \diagdown Cl$$

cis-1,2-Dichloroethylene

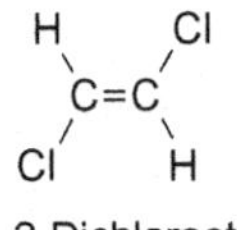

$$H \diagdown \quad \diagup Cl$$
$$C=C$$
$$Cl \diagup \quad \diagdown H$$

trans-1,2-Dichloroethylene

The natural organic subsurface products coal and crude oil have been and are still used to cover the tremendous energy demand of industrialized countries and to produce almost innumerable synthetic organic chemicals. Due to leakage of underground storage tanks and pipelines, due to spills at production wells, refineries and distribution terminals, and due to improper disposal and accidents during transport, organic compounds have become subsurface contaminants that threaten important drinking water resources. One strategy to remediate such polluted subsurface environments is with the help of the degradative capacity of bacteria.

6.3 Chemistry in the Terrestrial Biosphere

The terrestrial biosphere is predominantly the soil that is on the surface of the earth and which can house many different types of organic chemicals. A prime exposure pathway, either directly or indirectly, for soil-borne organic chemical contaminants is via transport in the pore-water solution though the structured and chemically reactive medium of our soils. Soils are also home to plant roots and a myriad of floral and faunal species. In predicting pollutant transport, it is important to distinguish between whether the fate is in the soil itself, or in the receiving water (Clothier et al., 2010).

The monocyclic aromatic compounds benzene, toluene, ethylbenzene, and the xylene isomers and the PAHs belong to the most often encountered subsurface contaminants and they are the most threatening compounds within the hydrocarbons. Aerobic bacteria able to degrade aromatic hydrocarbons are widespread. However, several reasons make the application of an aerobic treatment in the subsurface difficult. The limited availability of oxygen due to its low solubility restricts not only the respiration process, but also the degradation itself. Oxygen is needed by aerobic bacteria to activate and cleave the aromatic ring by the action of oxygenase derivatives. In contrast to the oxidative attack of the ring during aerobic degradation, aromatic compounds are reductively activated under anaerobic conditions.

Nitroaromatic compounds are widespread in the environment and are mainly of anthropogenic origin. One of the most problematic is 2,4,6-trinitrotoluene, a munition compound that is found wherever munition is produced, loaded, handled, or packed. Aerobic bacteria can use nitroaromatic compounds as growth substrates and derive carbon, nitrogen, and energy from their degradation.

REFERENCES

Abbasnezhad, H., Gray, M., Foght, J.M., 2011. Influence of adhesion on aerobic biodegradation and bioremediation of liquid hydrocarbons. Appl. Microbiol. Biotechnol. 92, 653–675.

Amatya, P.L., Hettiaratchi, J.P.A., Joshi, R.C., 2002. Biotreatment of flare pit waste. J. Can. Pet. Technol. 41, 30–36.

Bartha, R., 1986a. Microbial Ecology: Fundamentals and Applications. Addisson-Wesley Publishers, Reading, MA.

Bartha, R., 1986b. Biotechnology of petroleum pollutant biodegradation. Microb. Ecol. 12, 155–172.

Bartha, R., Atlas, R.M., 1977. The microbiology of aquatic oil spills. Adv. Appl. Microbiol. 22, 225–266.

Børresen, M.H., Rike, A.G., 2007. Effects of nutrient content, moisture content and salinity on mineralization of hexadecane in an arctic soil. Cold Reg. Sci. Technol. 48, 129–138.

Calvet, R., 1989. Adsorption of organic chemicals in soils. Environ. Health Perspect. 83, 145–177.

Clothier, B., Green, S., Deurer, M., Smith, E., Robinson, B., 2010. Transport and fate of contaminants in soils: challenges and development. In: Proceedings of 19th World Congress of Soil Science, Soil Solutions for a Changing World, August 1–6, Brisbane, Australia, pp. 73–76.

Collina, E., Bestetti, G., Di Gennaro, P., Franzetti, A., Gugliersi, F., Lasagni, M., Pitea, D., 2005. Naphthalene biodegradation kinetics in an aerobic slurry-phase bioreactor. Environ. Int. 31 (2), 167–171.

Cutright, T.J., 1995. Polycyclic aromatic hydrocarbon biodegradation and kinetics using *Cunninghamella echinulatu var. elegans*. Int. Biodeterior. Biodegrad. 35 (4), 397–408.

De Carvalho, C.C.C.R., daFonseca, M.M.R., 2005. Degradation of hydrocarbons and alcohols at different temperatures and salinities by *Rhodococcus erythropolis* DCL14. FEMS Microbiol. Ecol. 51, 389–399.

deBruyn, A.M.H., Gobas, F.A.P.C., 2004. Modelling the diagenetic fate of persistent organic pollutants in organically enriched sediments. Ecol. Model. 179, 405–416.

Gacosian, R.B., Lee, C., 1981. Processes controlling the distribution of biogenic organic compounds in seawater. In: Duursma, E.K., Dawson, R. (Eds.), Marine Organic Chemistry: Evolution, Composition, Interactions, and Chemistry of Organic Matter in Seawater. Amsterdam, Netherlands, pp. 91–123 (chapter 5).

Gouin, T., James, Y., Armitage, M., Cousins, I.T., Muir, D.C.G., Ng, C.A., Reid, L., Tao, S., 2013. Influence of global climate change on chemical fate and bioaccumulation: the role of multimedia models. Environ. Toxicol. Chem. 32 (1), 20–31.

Grossi, V., Massias, D., Stora, G., Bertrand Burial, J.C., 2002. Exportation and degradation of acyclic petroleum hydrocarbons following simulated oil spill in bioturbated mediterranean coastal sediments. Chemosphere 48 (9), 947–954.

Hamaker, W., 1972. Decomposition: quantitative aspects. In: Goring, C.A.I., Hamaker, J.W., Thomson, J. (Eds.), Organic Chemicals in the Soil Environment. Marcel Dekker Inc., New York.

Heitkamp, M.A., Cerniglia, C.E., 1987. Effects of chemical structure and exposure on the microbial degradation of polycyclic aromatic hydrocarbons in freshwater and estuarine ecosystems. Environ. Toxicol. Chem. 6 (7), 535–546.

Heitkamp, M.A., Freeman, J.P., Cerniglia, C.E., 1987. Naphthalene biodegradation in environmental microcosms: estimates of degradation rates and characterization of metabolites. Appl. Environ. Microbiol. 53 (1), 129–136.

Hohener, P., Duwig, C., Pasteris, G., Kaufmann, K., Dakhel, N., Harms, H., 2003. Biodegradation of petroleum hydrocarbon vapors: laboratory studies on rates and kinetics in unsaturated alluvial sand. J. Contam. Hydrol. 66 (1–2), 93–115.

Holba, A.G., Dzou, I.L., Hickey, J.J., Franks, S.G., May, S.J., Lenney, T., 1996. Reservoir geochemistry of South Pass 61 Field, Gulf of Mexico: compositional heterogeneities reflecting filling history and biodegradation. Org. Geochem. 24, 1179–1198.

Holder, E.L., Miller, K.M., Haines, J.R., 1999. Crude oil component biodegradation kinetics by marine and freshwater consortia. In: Alleman, B.C., Leeson, A. (Eds.), In Situ Bioremediation of Polycyclic Hydrocarbons and Other Organic Compounds. Battelle, Columbus, OH, pp. 245–250.

Janssen, F.J.J.G., Van Santen, R.A., 1999. Environmental catalysts. Catalytic Science Series, vol. 1. World Scientific Publishing Co. Inc., Hackensack, NJ.

Jury, W.A., Winer, A.M., Spencer, W.F., Focht, D.D., 1987. Transport and transformations of organic chemicals in the soil-air-water ecosystem. In: Ware, G.W. (Ed.), In: Reviews of Environmental Contamination and Toxicology, vol. 99. Springer, New York, pp. 119–164.

Katagi, T., 2002. Abiotic hydrolysis of pesticides in the aquatic environment. Rev. Environ. Contam. Toxicol. 175, 79–261.

Kleinsteuber, S., Riis, V., Fetzer, I., Harms, H., Müller, S., 2006. Population dynamics within a microbial consortium during growth on diesel fuel in saline environments. Appl. Environ. Microbiol. 72, 3531–3542.

Lyman, W.J., Reehl, W.F., Rosenblatt, D.H. (Eds.), 1982. Handbook of Chemical Property Estimation Methods. McGraw-Hill, New York.

Macalady, D.L., Walton-Day, K., 2011. Redox chemistry and natural organic matter (NOM): geochemists' dream, analytical chemists' nightmare. In: Tratnyek, P.G., Grundl, T.J., Haderlein, S.B. (Eds.), Aquatic Redox Chemistry. ACS Symposium Series No. 1071, American Chemical Society, Washington, DC, pp. 85–111 (chapter 5).

Maletić, S., Dalmacija, B., Rončević, S., Agbaba, J., Petrović, O., 2009. Degradation kinetics of an aged hydrocarbon-contaminated soil. Water Air Soil Pollut. 202, 149–159.

Manzetti, S., 2011. Are polycyclic aromatic hydrocarbons from fossil emissions potential hormone-analogue sources for modern man? Pathophysiology 19, 65–67.

Manzetti, S., 2012a. Chemical and electronic properties of polycyclic aromatic hydrocarbons: a review. In: Bandeira, G.C., Meneses, H.E. (Eds.), Handbook of Polycyclic Aromatic Hydrocarbons: Chemistry, Occurrence and Health Issues. In: Chemistry Research and Applications Series. Environmental Remediation Technologies, Regulations and Safety, Nova Science Publishers, Hauppauge, NY.

Manzetti, S., 2012b. Ecotoxicity of polycyclic aromatic hydrocarbons, aromatic amines and nitroarenes from molecular properties. Environ. Chem. Lett. 10, 349–361.

Manzetti, S., 2013. Polycyclic aromatic hydrocarbons in the environment: environmental fate and transformation. Polycycl. Aromat. Compd. 33 (4), 311–330.

Manzetti, S., Van der Spoel, E.R., Van der Spoel, D., 2014. Chemical properties, environmental fate, and degradation of seven classes of pollutants. Chem. Res. Toxicol. 27 (5), 713–737.

March, J., 1992. Advanced Organic Chemistry: Reactions, Mechanisms, and Structure, fourth ed. John Wiley & Sons Inc., Hoboken, NJ.

Margesin, R., Schinner, F., 2001. Biodegradation and bioremediation of hydrocarbons in extreme environments. Appl. Microbiol. Biotechnol. 56, 650–663.

Mille, G., Almallah, M., Bianchi, M., Van Wambeke, F., Bertrand, J.C., 1991. Effect of salinity on petroleum biodegradation. Fresenius J. Anal. Chem. 339, 788–791.

Mokhatab, S., Poe, W.A., Speight, J.G., 2006. Handbook of Natural Gas Transmission and Processing. Elsevier, Amsterdam, Netherlands.

Morrison, R.T., Boyd, R.N., Boyd, R.K., 1992. Organic Chemistry, sixth ed. Benjamin Cummings, Pearson Publishing, San Francisco, CA.

Moulay, S., 2002. The most well-known rearrangements in organic chemistry at hand. Chem. Educ. 3 (1), 33–64.

Namkoonga, W., Hwangb, E.Y., Parka, J.S., Choic, J.Y., 2002. Bioremediation of diesel-contaminated soil with composting. Environ. Pollut. 119 (1), 23–31.

Neilson, A.H., Allard, A.-S., 2008. Environmental Degradation and Transformation of Organic Chemicals. Taylor and Francis Group, Boca Raton, FL.

Neilson, A.H., Allard, A.-S., 2012. Organic Chemicals in the Environment: Mechanisms of Degradation and Transformation, second ed. CRC Press, Taylor & Francis Group, Boca Raton, FL.

Nicholson, C.A., Fathepure, B.Z., 2004. Biodegradation of benzene by halophilic and halotolerant bacteria under aerobic conditions. Appl. Environ. Microbiol. 70, 1222–1225.

Obire, O., Anyanwu, E.C., 2009. Impact of various concentrations of crude oil on fungal populations of soil. Int. J. Environ. Sci. Technol. 6 (2), 211–218.

Obire, O., Nwaubeta, O., 2001. Biodegradation of refined petroleum hydrocarbons in soil. J. Appl. Sci. Environ. Manage. 5 (1), 43–46.

Oren, A., Gurevich, P., Azachi, M., Henis, Y., 1992. Microbial degradation of pollutants at high salt concentrations. Biodegradation 3, 387–398.

Pala, D.M., de Carvalho, D.D., Pinto, J.C., Sant'Anna Jr., G.L., 2006. A suitable model to describe bioremediation of a petroleum-contaminated soil. Int. Biodeterior. Biodegrad. 58 (3-4), 254–260.

Pollard, S.J.T., Hrudey, S.E., Fedorak, P.M., 1994. Bioremediation of petroleum- and creosote-contaminated soils: a review of constraints. Waste Manage. Res. 12, 173–194.

Rahm, S., Green, N., Norrgran, J., Bergman, A., 2005. Hydrolysis of environmental contaminants as an experimental tool for indication of their persistency. Environ. Sci. Technol. 39 (9), 3128–3133.

Rhykerd, R.L., Weaver, R.W., McInnes, K.J., 1995. Influence of salinity on bioremediation of oil in soil. Environ. Pollut. 90, 127–130.

Riis, V., Kleinsteuber, S., Babel, W., 2003. Influence of high salinities on the degradation of diesel fuel by bacterial consortia. Can. J. Microbiol. 49, 713–721.

Rončević, S., Dalmacija, B., Ivančev-Tumbas, I., Petrović, O., Klašnja, M., Agbaba, J., 2005. Kinetics of degradation of hydrocarbons in the contaminated soil layer. Arch. Environ. Contam. Toxicol. 49 (1), 27–36.

Rosenberg, M., Gutnick, D., Rosenberg, E., 1980. Adherence of bacteria to hydrocarbons: a simple method for measuring cell-surface hydrophobicity. FEMS Microbiol. Lett. 9, 29–33.

Samin, G., Janssen, D.B., 2012. Transformation and biodegradation of 1,2,3-trichloropropane (TCP). Environ. Sci. Pollut. Res. Int. 19 (8), 3067–3078.

Seabra, P.N., Linhares, M.M., Santa Anna, L.M., 1999. Laboratory study of crude oil remediation by bioaugmentation. In: Alleman, B.C., Leeson, A. (Eds.), Situ Bioremediation of Polycyclic Hydrocarbons and Other Organic Compounds. Battelle, Columbus, OH, pp. 421–426.

Shelton, D.R., Tiedje, J.M., 1984. Isolation and partial characterization of bacteria in an anaerobic consortium that mineralizes 3-chlorobenzoic acid. Appl. Environ. Microbiol. 48, 840–848.

Sims, R.C., 1990. Soil remediation techniques at uncontrolled hazardous waste sites. J. Air Waste Manage. Assoc. 40 (5), 703–732.

Speight, J.G., 2007. Natural Gas: A Basic Handbook. GPC Books, Gulf Publishing Company, Houston, TX.

Speight, J.G., 2014. The Chemistry and Technology of Petroleum, fifth ed. CRC Press, Taylor & Francis Group, Boca Raton, FL.

Speight, J.G., 2016. Handbook of Petroleum Refining. CRC Press, Taylor & Francis Group, Boca Raton, FL.

Speight, J.G., Arjoon, K.K., 2012. Bioremediation of Crude Oil and Crude Oil Products. Scrivener Publishing, Salem, MA.

Speight, J.G., Islam, M.R., 2016. Peak Energy—Myth or Reality. Scrivener Publishing, Salem, MA.

Stoner, D.L., 1994. Biotechnology for the Treatment of Hazardous Waste. CRC Press, Boca Raton, FL.

Stoytcheva, M., 2011. Pesticides—Formulations, Effects, Fate. InTech Publishers, Rijeka, Croatia.

Tibbett, M., George, S.J., Davie, A., Barron, A., Milton, N., Greenwood, P.F., 2011. Just add water and salt: the optimization of petrogenic hydrocarbon biodegradation in soils from semi-arid Barrow Island, Western Australia. Water Air Soil Pollut. 216, 513–525.

Ulrich, A.C., Guigard, S.E., Foght, J.M., Semple, K.M., Pooley, K., Armstrong, J.E., Biggar, K.W., 2009. Effect of salt on aerobic biodegradation of petroleum hydrocarbons in contaminated groundwater. Biodegradation 20, 27–38.

Venosa, A., Suidan, M., Wrenn, B., Strohmeier, K., Haines, J., Eberhart, B., 1996. Bioremediation of an experimental oil spill on the shoreline of Delaware bay. Environ. Sci. Technol. 30 (5), 1764–1775.

Wackett, L.P., 1996. Co-metabolism: is the emperor wearing any clothes? Curr. Opin. Biotechnol. 7, 321–325.

Ward, D.M., Brock, T.D., 1978. Hydrocarbon biodegradation in hypersaline environments. Appl. Environ. Microbiol. 35, 353–359.

Wethasinghe, C., Yuen, S.T.S., Kaluarachchi, J.J., Hughes, R., 2006. Uncertainty in biokinetic parameters on bioremediation: health risks and economic implications. Environ. Int. 32 (3), 312–323.

Wham, R.M., Fisher, J.F., Forrester III, R.C., Irvine, A.R., Salmon, R., Singh, S.P.N., Ulrich, Zhang, H.L., He, Y., Li, S., Liu, X.B., 2005. Synthesis and hydrolytic degradation of aliphatic polyester amides. Polym. Degrad. Stab. 88 (2), 309–316.

Winningham, J., Britto, R., Patel, M., McInturff, F., 1999. A land farming field study of creosote-contaminated soil. In: Leeson, A., Alleman, B.C. (Eds.), Bioremediation Technologies for PAH Compounds. Battelle, Columbus, OH, pp. 421–426.

Zengler, K., Richnow, H.H., Rossello-Mora, R., Michaelis, W., Widdel, F., 1999. Methane formation from long-chain alkanes by anaerobic microorganisms. Nature 401, 266–269.

Chapter 8

Environmental Regulations

1 INTRODUCTION

The latter part of the 20th century started with the important realization that all chemicals, especially organic chemicals, can act as environmental pollutants depending upon the ecosystem into which the chemicals are discharged and the amount of chemical discharged. In addition, there came the realization that emissions of carbon dioxide (CO_2), methane (CH_4), nitrous oxide (N_2O), and chlorofluorocarbons (CFCs) to the atmosphere had either a direct impact or even an indirect impact on the global climate as well as on depletion of the ozone layer. As a result, unprecedented efforts were then made to reduce all global emissions of all chemicals in order to maintain a *green perspective*. Through the evolution of environmental-oriented thinking followed the formulation and passage of environmental regulations (Table 8.1).

Notwithstanding early efforts at environmental protection and control, the concept of environmentally oriented laws and regulations as a separate and distinct body of law is a 20th-century development (Lazarus, 2004). The recognition that the environment is (collectively) a fragile organism that is in need of legal protection did not occur until the 1960s (Chapter 1). At that time, numerous influences including (1) a growing awareness of the unity and fragility of the biosphere, (2) increased public concern over the impact of industrial activity on natural resources and human health, (3) the increasing strength of the regulatory state, and (4) the success of various movement to protect the environment led to a collection of laws in a relatively short period of time (Tables 8.1 and 8.2). While the modern history of environmental law is one of continuing (political) discussion and evolution, environmental laws had (by the end of the 20th century) been established as a component of the legal landscape in all of the developed and industrialized countries of the world as well as in many developing countries.

In concert with the evolution of laws related to the protection of the environment, commercial processes have been designed to reduce the direct emissions of organic chemical products as well as emissions of organic chemical by-products into the air (*the atmosphere*), water (*the aquasphere*), and soil (*the terrestrial biosphere*), and to recycle and reuse these chemicals and chemical waste as much as possible. Advanced technologies for the rapid, economical, and effective elimination of industrial and domestic chemical wastes

Environmental Organic Chemistry for Engineers. http://dx.doi.org/10.1016/B978-0-12-804492-6.00008-3

TABLE 8.1 Chronology of Environmental Events and Regulations in the United States (Not Necessarily Related to The Organic Chemicals Industry or The Crude Oil-Natural Gas Industry)

1906	The Pure Food and Drug Act established the Food and Drug Administration (FDA) that now oversees the manufacture and use of all foods, food additives, and drugs; amendments (1938, 1958, and 1962) strengthened the law considerably
1924	The Oil Pollution Act
1935	The Chemical Manufacturers Association (CMA), a private group of people working in the chemical industry and especially involved in the manufacture and selling of chemicals, established a Water Resources Committee to study the effects of their products on water quality
1948	The Chemical Manufacturers Association established an Air Quality Committee to study methods of improving the air quality that could be implemented by chemical manufacturers
1953	The Delaney Amendment to the Food and Drug Act defined and controlled food additives; any additives showing an increase in cancer tumors in rats, even if extremely large doses were used in the animal studies, had to be outlawed in foods; recent debates have focused on a number of additives, including the artificial sweetener cyclamate
1959	Just before Thanksgiving the government announced that it had destroyed cranberries contaminated with a chemical aminotriazole that produced cancer in rats; the cranberries were from a lot frozen from 2 years earlier when the chemical was still an approved weed killer
1960	Diethylstilbestrol (DES), taken in the late 1950s and early 1960s to prevent miscarriages and also used as an animal fattener, was reported to cause vaginal cancer in the daughters of these women as well as premature deliveries, miscarriages, and infertility
1962	Thalidomide, a prescription drug used as a tranquilizer and flu medicine for pregnant women in Europe to replace dangerous barbiturates that caused 2000–3000 deaths per year by overdoses, was found to cause birth defects. Thalidomide had been kept off the market in America because of the insistence that more safety data be produced for the drug
1962	The Kefauver-Harris Amendment to the Food and Drug Act began that required drugs be proven safe before being put on the market
1962	The publication of Silent Spring (authored by Rachel Carson) that outlined many environmental problems associated with chlorinated pesticides, especially DDT and its use was banned in 1972
1965	Nonlinear, nonbiodegradable synthetic detergents made from propylene tetramer were banned after these materials were found in large amounts in rivers, so much as to cause soapy foam in many locations. Phosphates in detergents were banned in detergents by many states in the 1970s

TABLE 8.1 Chronology of Environmental Events and Regulations in the United States (Not Necessarily Related to The Organic Chemicals Industry or The Crude Oil-Natural Gas Industry)—cont'd

1965	Mercury poisoning from concentration in the food chain recognized
1966	Polychlorinated biphenyls (PCBs) were first found in the environment and in contaminated fish; banned in 1978 except in closed systems
1968	TCDD (a dioxin derivative) tested positive as a teratogen in rats
1969	The artificial sweetener cyclamate was banned because of its link to bladder cancer in rats fed with large doses; many 20 subsequent studies have failed to confirm this result but cyclamate remains banned
1972	Federal Water Pollution Control Act
1974	Safe Drinking Water Act
1977	Saccharin found to cause cancer in rats; banned by the FDA temporarily but Congress placed a moratorium on this ban because of public pressure; saccharin is still available
1970	Earth Day recognized because of concern about with the effects of many substances on the environment
1970	The Clean Air Act
1971	TCDD (see above) outlawed by the Environmental Protection Agency
1971	The Chemical Manufacturers Association established the Chemical Emergency Transportation System (CHEMTREC) to provide immediate information on chemical transportation emergencies
1972	The Clean Water Act
1974	Vinyl chloride investigated as a possible carcinogen
1976	The Toxic Substances Control Act (TSCA or TOSCA); Environmental Protection Agency developed rules to limit manufacture and use of PCBs
1976	The Resource Conservation and Recovery Act (RCRA)
1977	Dibromochloropropane (DBCP) investigated for causes leading to sterility; now banned
1977	Benzene was linked to an abnormally high rate of leukemia; increased concern with benzene use in industry
1978	Ban on chlorofluorocarbons (CFCs) as aerosol propellants; react with ozone in the stratosphere causing an increase in the penetration of ultraviolet sunlight and increase in the risk of skin cancer
1978	Love Canal, Niagara Falls, New York
1980	CHEMTREC (see above) recognized by the Department of Transportation as the central service to provide immediate information on chemical transportation emergencies

Continued

TABLE 8.1 Chronology of Environmental Events and Regulations in the United States (Not Necessarily Related to The Organic Chemicals Industry or The Crude Oil-Natural Gas Industry)—cont'd

1980	The Comprehensive Environmental Response, Compensation, Liability Act
1986	The Safe Drinking Water Act Amendments
1986	The Emergency Planning and Community-Right-to-Know Act; companies must also report inventories of specific chemicals kept in the workplace and annual release of hazardous materials into the environment
1986	The Superfund Amendments and Reauthorization Act
1989	Pasadena, TX: explosion caused by leakage of ethylene and iso-butane when leaked from a pipeline
1990	Channelview, TX: explosion in a petrochemical treatment tank of wastewater and chemicals
1991	Sterlington, LA: explosion at a nitro-paraffin plant
1991	Charleston, SC: explosion at a plant manufacturing Antiblaze 19, a phosphonate ester and flame retardant used in textiles and polyurethane foam; manufactured from trimethyl phosphite, dimethyl methylphosphonate, and trimethyl phosphate

TABLE 8.2 Federal Regulations Relevant to Chemicals (Including Organic Chemicals)[a]

- Atomic Energy Act (AEA)
- Beaches Environmental Assessment and Coastal Health (BEACH) Act
- Chemical Safety Information, Site Security, and Fuels Regulatory Relief Act
- Clean Air Act (CAA)
- Clean Water Act (CWA; original title: Federal Water Pollution Control Amendments of 1972); also contains the Federal Water Pollution Control Amendments
- Comprehensive Environmental Response, Compensation and Liability Act (CERCLA, also known as Superfund or the Superfund Act); also contains the Superfund Amendments and Reauthorization Act (SARA)
- Emergency Planning and Community Right-to-Know Act (EPCRA)
- Endangered Species Act (ESA)
- Energy Independence and Security Act (EISA)
- Energy Policy Act
- Federal Food, Drug, and Cosmetic Act (FFDCA)
- Federal Insecticide, Fungicide, and Rodenticide Act (FIFRA)
- Food Quality Protection Act (FQPA)
- Marine Protection, Research, and Sanctuaries Act (MPRSA, also known as the Ocean Dumping Act)

TABLE 8.2 Federal Regulations Relevant to Chemicals (Including Organic Chemicals) — cont'd

- National Environmental Policy Act (NEPA)
- National Technology Transfer and Advancement Act (NTTAA)
- Nuclear Waste Policy Act (NWPA)
- Occupational Safety and Health (OSHA)
- Oil Pollution Act (OPA)
- Pesticide Registration Improvement Act (PRIA)
- Pollution Prevention Act (PPA)
- Resource Conservation and Recovery Act (RCRA)
- Safe Drinking Water Act (SDWA)
- Shore Protection Act (SPA)
- Toxic Substances Control Act (TSCA)

[a]*The various Acts are listed alphabetically and not in the order of importance to the chemical and refining industries. Executive Orders signed by the President are not included in this list.*

have been developed and employed on a large scale and, in fact, advanced technologies for the control and monitoring of chemical pollutants on regional and global scales continue to be developed and implemented. Satellite-based instruments are able to detect, quantify, and monitor a wide range of chemical pollutants. In addition, an understanding of the fate and consequences of chemicals in the environment (Chapters 6 and 7) has increased dramatically and there are now available the means of predicting many of the environmental, ecological, and biochemical consequences of the inadvertent introduction of organic chemicals into the environment, with much greater precision.

Organic chemicals used in a range of products and industrial processes are an intrinsic part of modern life. Many products which have contributed to making our lifestyles more comfortable, including domestic appliances, detergents, pharmaceuticals, and personal computers, involve the use of chemicals to varying degrees. Chemicals are also an essential component of industrial production and are used in sectors ranging from agriculture and mining to manufacturing. Besides the benefits chemicals bring for the economy, trade and employment, the rapid increase in their use and build up in our environment also comes at a price for human health and wildlife, when not managed effectively. Potential adverse impacts on humans include acute poisoning and even long-term effects such as cancers, neurological disorders, and birth defects. For the environment, harmful chemicals can trigger eutrophication of water bodies, ozone depletion, and pose a threat to sensitive ecosystems and biodiversity.

Industrial processes and rising consumption have also led to a rapid increase in generation of hazardous wastes. These wastes not only pose risks and hazards because of their nature but also have the potential to contaminate large quantities of otherwise nonhazardous wastes if allowed to get mixed. It is increasingly recognized that the sound management of chemicals throughout their lifecycle as well as the proper segregation, treatment, and disposal of hazardous wastes are critical to the protection of vulnerable ecosystems, biodiversity, and the livelihoods and health of communities. To address concerns related to harmful organic chemicals and waste, it is necessary to promote chemical safety and provide ready access to information on toxic chemicals and also to promote chemical safety by providing technical guidance on the regulations that govern the manufacture, use, and disposal of organic chemicals. Although it may be felt that the effect and activity of organic chemicals in the environment must, by definition focus on the chemistry of these chemicals, knowledge of the various environmental regulations (Table 8.2) is always helpful in determining the analyses that must be performed. These laws are designed to help protect human health and the environment—the Environmental Protection Agency is charged with administering all or a part of each law.

The definitions of hazardous substances (40 CFR 300.5) and pollutants or contaminants (40 CFR 300.5) specifically exclude *organic chemicals, including crude oil or any fraction thereof* unless specifically listed. Although there is no definition of *organic chemicals* in Superfund, the Environmental Protection Agency interprets the *organic chemicals exclusion* provision to include organic chemicals and fractions of organic chemicals, including the hazardous substances, such as benzene, that are indigenous in organic chemicals and are, therefore, included in the term *organic chemicals*. The term also includes hazardous substances that are normally mixed with or added to crude oil or crude oil fractions during the refining process, including hazardous substances whose levels are increased during refining. These substances are also part of *organic chemicals* because their addition is part of the normal oil separation and processing operations at refineries that produce the product commonly understood to be organic chemicals. However, hazardous substances that are added to organic chemicals (e.g., mixing of solvents with used oil) or that increase in concentration solely as a result of contamination of the organic chemicals during use are not part of the organic chemicals and thus are not excluded from Superfund (Wagner, 1999).

Furthermore, as a result of the evolving environmental awareness, organic chemicals refinery operators face more stringent regulations in the treatment, storage, and disposal (TSD) of hazardous wastes. Waste management laws govern the transport, treatment, storage, and disposal of all manner of chemical waste, including municipal solid waste, hazardous waste, and nuclear waste. Waste laws are generally designed to minimize or eliminate the uncontrolled dispersal of waste materials into the environment in a manner that may cause ecological or biological harm, and include laws designed to reduce the generation of waste and promote or mandate waste recycling. Regulatory efforts include identifying and categorizing waste types and mandating transport, treatment, storage, and disposal practices. Thus, under recent regulations, a larger number of organic chemical compounds have been, and are being, studied.

Long-time methods of disposal, such as land farming of chemical waste, are being phased out. New regulations are becoming even more stringent, and they encompass a broader range of chemical constituents and processes.

However, it is not the purpose of this chapter here to enter into any political discussion and the levy of fines for infringement of the environmental laws. The purpose of this chapter is to introduce the reader to an overview of a selection of the many and varied regulations instituted in the United States that regulate the disposition of organic chemicals into the environment.

2   ENVIRONMENTAL IMPACT OF PRODUCTION PROCESSES

Organic chemicals can enter the air, water, and soil when they are produced, used, or disposed. Their impact on the environment is determined by the amount of the chemical that is released, the type, and concentration of the chemical, and where it is found. Some chemicals can be harmful if released to the environment even when there is not an immediate, visible impact. Some chemicals are of concern as they can work their way into the food chain and accumulate and/ or persist in the environment for many years, even for decades.

The broad category of "environmental regulations" may be broken down into a number of more specific regulatory subjects. While there is no single agreed-upon taxonomy, the core environmental regulations address the environmental impact of the production and use of organic chemicals on the environment. A related but distinct set of regulatory regimes, now strongly influenced by environmental legal principles, focus on the management of specific natural resources, such as forests, minerals, or fisheries. Other areas, such as environmental impact assessment, may not fit neatly into either category, but are nonetheless important components of environmental law.

Environmental impact assessment (EA) is the term used for the assessment of the environmental consequences (positive and/or negative) of a plan, policy, program, or project prior to the decision to move forward with the proposed action. In this context, the term "environmental impact assessment" (EIA) is usually used when applied to concrete projects and the term "strategic environmental assessment" applies to policies, plans, and programs. Environmental assessments may be governed by rules of administrative procedure regarding public participation and documentation of decision making, and may be subject to judicial review.

The organic chemicals industry is one of the largest industries in the United States and potential environmental hazards have caused increased concern for communities in close proximity to them. This update provides a general overview of the processes involved and some of the potential environmental hazards associated with organic chemicals (Chapters 3, 4 and 7). Briefly, production of organic chemicals involves a series of steps that includes separation and blending of organic chemicals. Production facilities are generally considered a major source of pollutants in areas where they are located and are regulated by a number of environmental laws related to air, land, and water (Table 8.2). Thus, the organic chemicals industry (correctly or incorrectly, without any form of condemnation here) has been considered to be a major source of pollutants in areas where they are located and are regulated by a number of environmental laws related to air (the atmosphere), water (the aquasphere), and land (the terrestrial biosphere).

2.1 Air Pollution

Air quality laws govern the emission of air pollutants into the atmosphere while a specialized subset of air quality laws regulates the quality of air inside buildings. Air quality laws are often designed specifically to protect human health by limiting or eliminating airborne pollutant concentrations. Other initiatives are designed to address broader ecological problems, such as limitations on chemicals that affect the ozone layer, and emissions trading programs to address acid rain or climate change. Regulatory efforts include identifying and categorizing air pollutants, setting limits on acceptable emissions levels, and dictating necessary or appropriate mitigation technologies.

Organic chemicals are a source of hazardous and toxic air pollutants such as BTEX compounds (benzene, toluene, ethylbenzene, and the xylene isomers). They are also a major source of criteria air pollutants: particulate matter (PM), nitrogen oxides (NOx), carbon monoxide (CO), hydrogen sulfide (H_2S), and sulfur oxides (SOx). Currently, organic chemicals production (including refinery operations) releases less toxic hydrocarbons than in prior decades.

Air emissions can come from a number of sources within the industry including: equipment leaks (from valves or other devices); high-temperature combustion processes in the actual burning of fuels for electricity generation; the heating of steam and process fluids; and the transfer of products. These pollutants are typically emitted into the environment over the course of a year through normal emissions, fugitive releases, accidental releases, or plant upsets. The combination of volatile hydrocarbons and oxides of nitrogen also contribute to ozone formation, one of the most important air pollution problems.

2.2 Water Pollution

Water quality laws govern the release of pollutants into water resources, including surface water, groundwater, and stored drinking water. Some water quality laws, such as drinking water regulations, may be designed solely with reference to human health. Many others, including restrictions on the alteration of the chemical, physical, radiological, and biological characteristics of water resources, may also reflect efforts to protect aquatic ecosystems more broadly. Regulatory efforts may include identifying and categorizing water pollutants, dictating acceptable pollutant concentrations in water resources, and limiting pollutant discharges from effluent sources. Regulatory areas include sewage treatment and disposal, industrial waste, and agricultural waste water management, and control of surface runoff from construction sites and urban environments.

Production facilities are also potential contributors to ground water and surface water contamination. Some companies have used (or continue to use, with the appropriate permits) deep-injection wells to dispose of wastewater generated inside the plants, and some of these wastes end up in aquifers and groundwater. These wastes are then regulated under the Safe Drinking Water Act (SDWA). Industrial wastewater may be highly contaminated and may arise from various processes (such as wastewaters from desalting, water from cooling towers, storm water, distillation, or cracking). This water is recycled through many stages during the production process and goes through several treatment processes, including a wastewater treatment plant, before being released (through government-issued permits) into surface waters.

The wastes discharged into surface waters are subject to state discharge regulations and are regulated under the Clean Water Act (CWA). These discharge guidelines limit the amounts of sulfides, ammonia, and suspended solids and other compounds that may be present in the wastewater. Although these guidelines are in place, contamination from past discharges may remain in surface water bodies.

2.3   Soil Pollution

Contamination of soils from the various processes is generally a less significant problem when compared to contamination of the atmosphere and the various water systems. Past (pre-1970) production practices may have led to spills on company property that now need to be cleaned up. Natural bacteria that may use the organic chemical products as food are often effective at cleaning up organic chemicals spills and leaks compared to many other pollutants. Many waste materials are produced during the production of organic chemicals, and some of them are recycled through other stages in the process. Other wastes are collected and disposed of in landfills, or they may be recovered by other facilities. Soil contamination including by some hazardous wastes, spent catalysts or coke dust, tank bottoms, and sludge from the treatment processes can occur from leaks as well as accidents or spills on or off site during the transport process.

2.4   Wastewater

A number of wastewater issues face the organic chemicals industry (which also includes the fossil fuel industries). These issues include chemicals in waste process waters. However, efforts by the industry are being continued to eliminate any water contamination that may occur, whether it be from inadvertent leakage of petroleum or petroleum products or leakage of contaminated water from one or more processes. In addition to monitoring organics in the water, metals concentration must be continually monitored since heavy metals tend to concentrate in the body tissues of fishes and animals and increase in concentration as they go up the food chain. General sewage problems face every municipal sewage treatment facility regardless of size.

Primary treatment (solid settling and removal) is required and secondary treatment (use of bacteria and aeration to enhance organic degradation) is becoming more routine, tertiary treatment (filtration through activated carbon, applications of ozone, and chlorination) have been, or are being, implemented by all chemical production companies. Wastewater pretreaters that discharge water into sewer systems have new requirements. Pollutant standards for sewage sludge have been set. Toxics in the water must be identified and plans must be developed to alleviate any problems. In addition, regulators have established, and continue to establish, water quality standards for priority toxic pollutants.

3 ENVIRONMENTAL REGULATIONS IN THE UNITED STATES

The broad category of *environmental regulations* (or *environmental laws*) that cover the discharge or organic chemicals into the environment may be broken down into a number of more specific regulatory subjects (Table 8.2). While there is no single agreed-upon classification of these regulations, the core environmental regulation addresses environmental pollution and many do specify pollution by chemical or by organic chemicals such as pesticides. A related but distinct set of regulatory regulations, focuses on the management of specific natural resources, such as forests, minerals, or fisheries. Other areas, such as environmental impact assessment, may not fit neatly into either category, but are nonetheless important components of the protection of the environment.

The toxic chemicals found within the organic chemicals industry are not necessarily unique and although general air pollution, water pollution, and land pollution controls are affected by the organic chemicals defining, these problems and solutions are not unique to the industry. In fact, because the issues are so diverse, the organic chemicals industry (and because a segment of the industry—the refining industry—is an industrial complex consisting of many integrated unit processes) may be looked upon as a series of complex pollution-prevention issues, each one unique to the unit processes from which the effluent originates. Therefore, there may be many examples of laws and controls that have been enacted by governments with input from the producers of chemicals that address pollution prevention and control and discharge of hazardous chemicals into the environment.

Thus, air quality laws govern the emission of chemical pollutants into the atmosphere (Chapter 6). Other initiatives are designed to address broader ecological problems, such as limitations on chemicals that affect the ozone layer, and emissions trading programs that address the formation and disposition of acid rain. Regulatory efforts include identifying and categorizing air pollutants, setting limits on acceptable emissions levels, and dictating necessary or appropriate mitigation technologies.

Similarly, water quality laws govern the release of chemical pollutants into water resources, including surface water, groundwater, and drinking water. Some water quality laws, such as drinking water regulations, may be designed solely with reference to human health but many other water quality laws, including restrictions on the alteration of the chemical, physical, radiological, and biological characteristics of water resources are designed to protect the aquatic ecosystems and are often broader in their respective application to ecosystem. Regulatory efforts may also include identifying and categorizing water pollutants, dictating acceptable pollutant concentrations in water resources, and limiting pollutant discharges from effluent sources. Regulatory areas include sewage treatment and disposal, industrial wastewater management and agricultural wastewater management, and control of surface runoff from construction sites and urban environments, especially where there is a high likelihood that chemicals are dissolved in the waste water.

Waste management regulations govern the transport, treatment, storage, and disposal of all manner of waste, including municipal solid waste, hazardous waste, and nuclear waste, among many other types of chemical-containing waste. These laws are typically designed to minimize or eliminate the uncontrolled disposal of waste (chemical) materials into the environment in a manner that may cause ecological or biological harm, and the regulations also include laws that are designed to reduce the generation of waste and promote or mandate waste recycling. Regulatory efforts include identifying and categorizing waste types and mandating transport, treatment, storage, and disposal practices.

Environmental cleanup laws govern the removal of (chemical) pollutants or (chemical) contaminants from environmental ecosystems such as soil, sediment in aqueous ecosystems, surface water, or groundwater. Unlike pollution control laws (which are designed to be followed before-the-fact), cleanup laws are designed to respond after-the-fact to environmental contamination, and consequently must often define not only the necessary response actions, but also the parties who may be responsible for undertaking the actual cleanup as well as bearing the cost of the cleanup actions. Regulatory requirements may include rules for emergency response, liability allocation, site assessment, remedial investigation, feasibility studies, remedial action, postremedial monitoring, and site reuse.

Thus pollution prevention and control of hazardous chemical materials is an issue not only for the organic chemicals industry but also for many industries and has been an issue for decades (Table 8.1) (Noyes, 1993). In this context, there are specific definitions for terms such as *hazardous substances*, *toxic substances*, and *hazardous waste* (Chapter 1). These are all terms of art and must be fully understood in the context of their statutory or regulatory meanings and not merely limited to their plain English or dictionary meanings. It is absolutely imperative from a legal sense that each statute or regulation promulgated be read in conjunction with terms defined in that specific statute or regulation (Majumdar, 1993).

In order to combat any threat to the environment, it is necessary to understand the nature and magnitude of the problems involved. It is in such situations that environmental technology has a major role to play. Environmental issues even arise when outdated laws are taken to task. Thus the concept of what seemed to be a good idea at the time the action occurred, no longer holds when the law influences the environment.

Finally, it is worth of note that regulatory disincentives to voluntary reductions of emissions from organic chemicals also exist. Many environmental statutes define a baseline period and measure progress in pollution reductions from that baseline. Any reduction in emissions before it is required could lower a facility's baseline emissions. Consequently, future regulations requiring a specified reduction from the baseline could be more difficult (and, consequently, have a much greater effect on the economic bottom line) to achieve because the most easily-applied and, hence, the most cost-effective reductions would already have been made and establishing the environmental base case may be no longer realistic. With no credit given for voluntary reductions, those facilities that do the minimum may be in fact be rewarded when emission reductions are required.

3.1 Clean Air Act

The Clean Air Act Amendments (CAAA) of 1990 have made significant changes in the basic Clean Air Act (CAA) enacted in 1970. The Clean Air allowed the establishment of air quality standards and provisions for their implementation and enforcement. This law was strengthened in 1977 and the CAAA of 1990 imposed many new standards that included controls for industrial pollutants.

The CAA of 1970 and the 1977 amendments that followed consist of three titles. Title I deals with stationary air emission sources, Title II deals with mobile air emission sources, and Title III includes definitions of appropriate terms, provisions for citizen suits, and applicable standards for judicial review. However, in contrast to the previous clean air statutes, the 1990 Amendments contained extensive provisions for control of the accidental release of air toxics from storage or transportation (TPG, 1995) as well as the formation of acid rain. At the same time, the 1990 Amendments provided new and added requirements for such original ideas as state implementation plans (SIPs) for attainment of the national ambient air quality standards and permitting requirements for the attainment and nonattainment areas. Title III now calls for a vastly expanded program to regulate *hazardous air pollutants* (HAPs) or the so-called *air toxics*.

Under the CAAA of 1990, the mandate is to establish, during the first phase, technology-based maximum achievable control technology (MACT) emission standards that apply to the major categories or subcategories of sources of the listed HAPs (EPA, 1997). In addition, Title III provides for health-based standards that address the issue of residual risks due to air toxic emissions from the sources equipped with MACT and to determine whether the MACT standards can protect health with an *ample margin of safety*.

Section 112 of the original CAA that dealt with HAPs has been greatly expanded by the 1990 Amendments. The list of HAPs has been increased many fold. In addition, the standards for emission control have been tightened and raised to a very high level, referred to as the *best of the best*, in order to reduce the risk of exposure to various HAPs.

Thus, the 1990 CAAA aimed to encourage voluntary reductions above the regulatory requirements by allowing facilities to obtain emission credits for voluntary reductions in emissions. These credits would serve as offsets against any potential future facility modifications resulting in an increase in emissions. Other regulations established by the amendments, however, will require the construction of major new units within existing chemicals producers to reduce emissions even further and these new operations will require emission offsets in order to be permitted. This will consume many of the credits available for existing facility modifications. A shortage of credits for facility modifications will make it difficult to receive credits for emission reductions through pollution-prevention projects.

Thus, under this CAA, the Environmental Protection Agency sets limits on how much of a pollutant can be in the air anywhere in the United States. The law does allow individual states to have stronger pollution controls, but states are not allowed to have weaker pollution controls than those set for the whole country. The law recognizes that it makes sense for states to take the lead in carrying out the CAA, because pollution control problems often require special understanding of local industries and geography as well as housing developments near to industrial sites.

In addition, Title IV of the 1990 Amendments to the CAA (Title IV—Acid Deposition Control) mandates requirements for the control of acid deposition acid rain. The purpose of this title is to reduce the adverse effects of acid deposition through reductions in annual emissions of sulfur dioxide and also, in combination with other provisions of this Act, through reductions of nitrogen oxides emissions in the 48 contiguous States and the District of Columbia. It is the intent of this title to effectuate such reductions by requiring compliance by affected sources with prescribed emission limitations by specified deadlines, which limitations may be met through alternative methods of compliance provided by an emission allocation and transfer system. It is also the purpose of this title to encourage energy conservation, use of renewable and clean alternative technologies, and pollution prevention as a long-range strategy, consistent with the provisions of this title, for reducing air pollution and other adverse impacts of energy production and use. Furthermore, individual states are required to develop SIPs that is a collection of the regulations a state will use to clean up polluted areas. The states must involve the public, through hearings and opportunities to comment, in the development of each SIP. The Environmental Protection Agency must approve each plan and if a state implementation is not acceptable, the Environmental Protection Agency can take over enforcing the CAA in that state.

Air pollutant often travels from its source in one state to another state. In many metropolitan areas, people live in one state and work or shop in another; air pollution from cars and trucks may spread throughout the interstate area. The 1990 CAAA provide for interstate commissions on air pollution control, which are to develop regional strategies for cleaning up air pollution. The 1990 Amendments also cover pollution that originates in nearby countries, such as Mexico and Canada, and drifts into the United States as well as pollution from the United States that reaches Canada and Mexico.

In the current context, the 1990 Amendments provide economic incentives for reducing pollution. For instance, producers of organic chemicals can get credits if they produce cleaner products than required, and use the credits when the product falls short of the requirements. Furthermore, organic chemicals (like many industrial products) can be extremely toxic to the environment. As an example, to combat such effects refiners have started to reformulate gasoline sold in the formerly smog-prone areas. This gasoline contains less volatile organic chemicals such as benzene (which is also a hazardous air pollutant that causes cancer and aplastic anemia, a potentially fatal blood disease). The reformulated gasoline also contains detergents, which, by preventing build-up of engine deposits, keep engines working smoothly and burning fuel cleanly.

3.2 Clean Water Act

The CWA started life as a regulatory act The Federal Water Pollution Control Act (FWPCA) of 1948 was the first major US law to address water pollution. Growing public awareness and concern for controlling water pollution led to amendments in 1972 and, as amended in 1972, the law became commonly known as the Clean Water Act (CWA). The 1972 amendments: (1) established the basic structure for regulating pollutant discharges into the waters of the United States, (2) gave the Environmental Protection Agency the authority to implement pollution control programs such as setting wastewater standards for industry, (3) maintained existing requirements to set water quality standards for all contaminants in surface waters, (4) made it unlawful for any person to discharge any pollutant from a point source into navigable waters, unless a permit was obtained under its provisions, (5) funded the construction of sewage treatment plants under the construction grants program, and (6) recognized the need for planning to address the critical problems posed by nonpoint source pollution.

Subsequent amendments modified some of the earlier provisions of the CWA. For example, revisions in 1981 streamlined the municipal construction grants process, improving the capabilities of treatment plants built under the program. Further changes to the Act in 1987 phased out the construction grants program, replacing it with the State Water Pollution Control Revolving Fund (the Clean Water State Revolving Fund) which addressed water quality needs by building on EPA-state partnerships.

Over the years, many other laws have changed parts of the CWA. Title I of the Great Lakes Critical Programs Act of 1990, for example, put into place parts of the Great Lakes Water Quality Agreement of 1978, signed by the United States and Canada, where the two nations agreed to reduce certain toxic pollutants in the Great Lakes. That law required the Environmental Protection Agency to establish water quality criteria for the Great Lakes addressing 29 toxic pollutants with maximum levels that are safe for humans, wildlife, and aquatic life. It also required the Environmental Protection Agency to help the States implement the criteria on a specific schedule.

Thus, the Clean Water Act (CWA or the Water Pollution Control Act) is the cornerstone of surface water quality protection in the United States and employs a variety of regulatory and nonregulatory tools to sharply reduce direct pollutant discharges into waterways and manage polluted runoff. The objective of the CWA is to restore and maintain the chemical, physical, and biological integrity of water systems. The Act established the basic structure for regulating discharges of pollutants into the waters of the United States and regulating quality standards for discharge of pollutants into the waters of the United States and gave the Environmental Protection Agency the authority to implement pollution control programs such as setting wastewater standards for industry. The CWA also continued establishing requirements to set water quality standards for all contaminants in surface waters. The Act is credited with the first comprehensive program for controlling and abating water pollution. In addition, the Act made it unlawful to discharge any pollutant from a point source into navigable waters, unless a permit was obtained. Point sources are discrete conveyances such as pipes or man-made ditches. Individual homes that are connected to a municipal system, use a septic system, or do not have a surface discharge do not need an National Pollutant Discharge Elimination System (NPDES) permit; however, industrial, municipal, and other facilities must obtain permits if their discharges go directly to surface waters.

The statute makes a distinction between conventional and toxic pollutants. As a result, two standards of treatment are required prior to their discharge into the navigable waters of the nation. For conventional pollutants that generally include degradable nontoxic organic chemicals and inorganic chemicals, the applicable treatment standard is best conventional technology (BCT). For toxic pollutants, on the other hand, the required treatment standard is best available technology, which is a higher standard than BCT.

The statutory provisions of CWA have five major sections that deal with specific issues: (1) nationwide water quality standards; (2) effluent standards from certain specific industries; (3) permit programs for discharges into receiving water bodies based on the NPDES; (4) discharge of toxic chemicals, including oil spills; and (5) construction grant program for publicly owned treatment works. In addition, Section 311 of CWA includes elaborate provisions for regulating intentional or accidental discharges of oil and hazardous substances. Included there are response actions required for oil spills and the release or discharge of toxic and hazardous substances. Pursuant to this, certain elements and compounds are designated as hazardous substances and an appropriate list has been developed (40 CFR 116.4). The person in charge of a vessel or an onshore or offshore facility from which any designated hazardous substances is discharged, in quantities equal to or exceeding its reportable quality, must notify the appropriate federal agency as soon as such knowledge is obtained. Such notice should be provided in accordance with the designated procedures (33 CFR 153.203).

Under the CWA, discharge of water-borne pollutants is limited by NPDES permits. Chemical-producing companies that easily meet their permit requirements may find that the permit limits will be changed to lower values and be less stringent. However, because occasional system upsets do occur resulting in significant excursions above the normal performance values, many companies may feel that they must maintain a large operating margin below the permit limits to ensure continuous compliance. Those companies that can significantly reduce water-borne emissions may find the risk of having their permit limits lowered to be a substantial disincentive.

3.3 Comprehensive Environmental Response, Compensation, and Liability Act

The Comprehensive Environmental Response, Compensation, and Liability Act (CERCLA), commonly known as Superfund, 1980, created a tax on the chemical and organic chemicals industries and provided broad federal authority to respond directly to releases or threatened releases of hazardous substances that may endanger public health or the environment. The Act was amended by the Superfund Amendments and Reauthorization Act (SARA) in 1986 and stressed the importance of permanent remedies and innovative treatment technologies in cleaning up hazardous waste sites. Thus the Act provides a Federal *superfund* to clean up uncontrolled or abandoned hazardous waste sites as well as accidents, spills, and other emergency releases of pollutants and contaminants into the environment. Through CERCLA, the Environmental Protection Agency was given power to seek out those parties responsible for any release and assure their cooperation in the cleanup.

A CERCLA response or liability will be triggered by an actual release or the threat of a *hazardous substance or pollutant or contaminant* being released into the environment. A hazardous substance [CERCLA 101(14)] is any substance requiring special consideration due to its toxic nature under the CAA, the CWA, or the Toxic Substances Control Act (TSCA) and as defined under

Resource Conservation and Recovery Act (RCRA). Additionally, a pollutant or contaminant can be any other substance not necessarily designated or listed but that "will or may reasonably" be anticipated to cause any adverse effect in organisms and/or their offspring [CERCLA 101(33)].

The central purpose of CERCLA is to provide a response mechanism for cleanup of any hazardous substance released, such as an accidental spill, or of a threatened release of a hazardous substance (Nordin et al., 1995). Section 102 of CERCLA is a catchall provision because it requires regulations to establish *that quantity of any hazardous substance the release of which shall be reported pursuant to Section 103* of CERCLA. Thus, under CERCLA, the list of potentially responsible parties can include all direct and indirect culpable parties who have either released a hazardous substance or violated any statutory provision. In addition, responsible private parties are liable for cleanup actions and/or costs as well as for reporting requirements for an actual or potential release of a hazardous substance, pollutant, or contaminant.

CERCLA (Superfund) legislation deals with actual or potential releases of hazardous materials that have the potential to endanger people or the surrounding environment at uncontrolled or abandoned hazardous waste sites. The Act requires responsible parties or the government to clean up waste sites. Among CERCLA's major purposes are the following: (1) site identification; (2) evaluation of danger from waste sites; (3) evaluation of damages to natural resources; (4) monitoring of release of hazardous substances from sites; and (5) removal or cleanup of wastes by responsible parties or government.

The SARA addresses closed hazardous waste disposal sites that may release hazardous substances into any environmental medium. Title III of SARA also requires regular review of emergency systems for monitoring, detecting, and preventing releases of extremely hazardous substances at facilities that produce, use, or store such substances.

The most revolutionary part of SARA is the Emergency Planning and Community Right-to-Know Act (EPCRA) that is covered under Title III of SARA. EPCRA includes three subtitles and four major parts: emergency planning, emergency release notification, hazardous chemical reporting, and toxic chemical release reporting. Subtitle A is a framework for emergency planning and release notification. Subtitle B deals with various reporting requirements for *hazardous chemicals* and *toxic chemicals*. Subtitle C provides various dimensions of civil, criminal, and administrative penalties for violations of specific statutory requirements.

Other provisions of SARA basically reinforce and/or broaden the basic statutory program dealing with the releases of hazardous substances (CERCLA Section 313). It requires owners and operators of certain facilities that manufacture, process, or otherwise use one of the listed chemicals and chemical categories to report all environmental releases of these chemicals annually. This information about total annual releases of chemicals from the industrial facilities can be made available to the public.

The Act also requires (under Section 4) testing of chemicals by manufacturers, importers, and processors where risks or exposures of concern are found and (under Section 5) issuance of significant new use rules when a significant new use is identified for a chemical that could result in exposures to, or releases of, a substance of concern. In addition, companies or persons importing or exporting chemicals are required to comply with certification reporting and/ or other requirements which also requires (under Section 8) reporting and record-keeping by persons who manufacture, import, process, and/or distribute chemical substances in commerce as well as that any person who manufactures (including imports), processes, or distributes commercially a chemical substance or mixture and who obtains information which reasonably supports the conclusion that such substance or mixture presents a substantial risk of injury to health or the environment to immediately inform the Environmental Protection Agency, except where the Agency has been adequately informed of such information.

The TSCA is also authorized (under Section 8) to maintain the TSCA Inventory which contains more than 83,000 chemicals and, as new chemicals are commercially manufactured or imported, they are placed on the list.

3.4 Hazardous Materials Transportation Act

The Hazardous Materials Transportation Act, passed in 1975, is the law governing transportation of chemicals and hazardous materials. It is the principal federal law in the United States that regulates the transportation of hazardous materials (such as, in the current context, organic chemicals). The purpose of the Act is to *protect against the risks to life, property, and the environment that are inherent in the transportation of hazardous material in intrastate, interstate, and foreign commerce* under the authority of the United States Secretary of Transportation.

The Act was passed as a means to improve the uniformity of existing regulations for transporting hazardous materials and to prevent spills and illegal dumping endangering the public and the environment, a problem exacerbated by uncoordinated and fragmented regulations. Regulations are enforced through four key provisions that encompass federal standards under Title 49 of the United States Code (a code that regards the role of transportation in the United States of America): (1) procedures and policies, (2) material designations and labeling, (3) packaging requirements, and (4) operational rules. Violation of the Act regulations can result in civil or criminal penalties, unless a special permit is granted under the discretion of the Secretary of Transportation.

Thus, the basic purpose of Hazardous Materials Transportation Act is to ensure safe transportation of hazardous materials through the nation's highways, railways, and waterways. The basic theme of the Act is to prevent any person from offering or accepting a hazardous material for transportation anywhere within this nation if that material is not properly classified, described, packaged, marked, labeled, and properly authorized for shipment pursuant to the regulatory requirements. In addition, the Act includes a comprehensive assessment of the regulations, information systems, container safety, and training for emergency response and enforcement. The regulations apply to

any person who transports, or causes to be transported or shipped, a hazardous material; or who manufactures, fabricates, marks, maintains, reconditions, repairs, or tests a package or container which is represented, marked, certified, or sold by such person for use in the transportation in commerce of certain hazardous materials.

Under this statutory authority, the Secretary of Transportation has broad authority to determine what is a hazardous material, using the dual tools of quantity and type. By this two-part approach, any material that may pose an unreasonable risk to human health or the environment may be declared a hazardous material. Such a designated hazardous material obviously includes both the quantity and the form that make the material hazardous. Furthermore, under the Department of Transportation (DOT) regulations, a hazardous material is *any substance or material, including a hazardous substance and hazardous waste that is capable of posing an unreasonable risk to health, safety, and property when transported in commerce.* DOT thus has broad authority to regulate the transportation of hazardous materials that, by definition, include hazardous substances as well as hazardous wastes.

3.5 Occupational Safety and Health Act

The objective of the Occupational Safety and Health Act (OSHA) Hazard Communication Standard is to inform workers of potentially dangerous substances in the work place and to train them on how to protect themselves against potential dangers. The Act is formulated to

to assure safe and healthful working conditions for working men and women; by authorizing enforcement of the standards developed under the Act; by assisting and encouraging the States in their efforts to assure safe and healthful working conditions; by providing for research, information, education, and training in the field of occupational safety and health; and for other purposes.

The goal of OSHA is to ensure that *"no employee will suffer material impairment of health or functional capacity"* due to a lifetime occupational exposure to chemicals and hazardous substances. The statute imposes a duty on the employers to provide employees with a safe workplace environment, free of known hazards that may cause death or serious bodily injury. Thus, the Act is entrusted with the major responsibility for workplace safety and worker's health (Wang, 1994). It is responsible for the means by which chemicals are contained (TPG, 1995) through the inspection of workplaces to ensure compliance and enforcement of applicable standards under OSHA. It is also the means by which guidelines have evolved for the destruction of chemicals used in chemical laboratories.

The statute covers all employers and their employees in all the states and federal territories with certain exceptions (Lunn and Sanstone, 1994). Generally, the statute does not cover self-employed persons, farms solely employing family members, and those workplaces covered under other federal statutes. Chemicals producers must evaluate whether the chemicals that the company manufactures and sell are hazardous. Under the General Duty Clause of OSHA, employers are required to provide an environment that is free from recognized hazards that could cause physical harm or death.

All employers are required to develop, implement, and maintain at the work place a written hazard communication program. The program must include the following components: (1) a list of hazardous chemicals in the work place, (2) the methods the employer will use to inform employees of the hazards associated with these chemicals, and (3) a description of how the labeling, Material Safety Data Sheet, and employee training requirements will be met.

The following information must be included in the program for employers who produce, use, or store hazardous chemicals in the workplace: (1) the means by which manufacturer's safety data sheets will be made available to the outside contractor for each hazardous chemical, (2) the means by which the employer will inform the outside contractor of precautions necessary to protect the contractor's employees both during normal operating conditions and in foreseeable emergencies, and (3) the methods that the employer will use to inform contractors of the labeling system used in the workplace.

3.6 Oil Pollution Act

The Oil Pollution Act (OPA) of 1924 was the first federal statute prohibiting pollution of waters strictly by crude oil and crude oil-derived products. The FWPCA of 1972 provided a comprehensive plan for the cleanup of waters polluted by oil spills and intentional or accidental release of oil into the water. The subsequent laws, including the CWA of 1977 and its later amendments, provide for regulation of pollution of waters by oil spills and other forms of discharges. These legislations also incorporate certain provisions of the Rivers and Harbors Act of 1899, which was intended to prevent any obstruction to the use of navigable waters for interstate commerce.

One of top priorities of the Environmental Protection Agency is to prevent, prepare for, and respond to spills of crude oil that occur in and around inland waters of the United States. As such, the Environmental Protection Agency is the lead federal response agency for oil spills occurring in inland waters. The United States Coast Guard is the lead response agency for spills in coastal waters and deepwater ports. The Environmental Protection Agency oil spill prevention program includes the Spill Prevention, Control, and Countermeasure

(SPCC) and the Facility Response Plan (FRP) rules. The SPCC rule helps facilities prevent a discharge of oil into navigable waters or adjoining shorelines. In addition, the FRP rule requires certain facilities to submit a response plan and prepare to respond to a worst case oil discharge or threat of a discharge.

The OPA of 1990 streamlined and strengthened the ability of the Environmental Protection Agency to prevent and respond to catastrophic oil spills. A trust fund financed by a tax on oil is available to clean up spills when the responsible party is incapable or unwilling to do so. The Act requires oil storage facilities and vessels to submit to the Federal government plans detailing how they will respond to large discharges. The Act also requires the development of Area Contingency Plans to prepare and plan for oil spill response on a regional scale.

3.7 Resource Conservation and Recovery Act

The hazardous waste regulatory program, as we know it today, began with the RCRA in 1976. Since its enactment in 1976, the Resource Conservation and Recovery Act has been amended several times, to promote safer solid and hazardous waste management programs (Dennison, 1993). The Used Oil Recycling Act of 1980 and the Hazardous and Solid Waste Amendments of 1984 (HSWA) were the major amendments to the original law. The 1984 amendments also brought the owners and operators of underground storage tanks under the Resource Conservation Recovery Act umbrella. This can have a significant effect on refineries that store organic chemicals in underground tanks. Now, in addition to the hazardous waste being controlled, the Resource Conservation Recovery Act Subtitle I regulates the handling and storage of organic chemicals.

The Resource Conservation Recovery Act controls disposal of solid waste and requires all wastes destined for land disposal be evaluated for their potential hazard to the environment. Solid waste includes liquids, solids, and containerized gases and is divided into nonhazardous waste and hazardous waste. The various amendments are aimed at preventing the disposal problems that lead to a need for the CERCLA, or Superfund, as it is known.

Subtitle C of the original Resource Conservation Recovery Act lists the requirements for the management of hazardous waste. This includes criteria for identifying hazardous waste, and the standards for generators, transporters, and companies that treat, store, or dispose of the waste. The Resource Conservation Recovery Act regulations also provide standards for design and operation of such facilities. However, before any action under the Act is planned, it is essential to understand what constitutes a solid waste and what constitutes a hazardous waste. The first step to be taken by a generator of waste is to determine whether that waste is hazardous. Waste may be hazardous by being listed in the regulations or by meeting any of the four characteristics: ignitability, corrosivity, reactivity, and extraction procedure (EP) toxicity.

Section 1004(27) of the Resource Conservation Recovery Act defines *solid waste* as garbage, refuse, sludge, from a waste treatment plant, water supply treatment plant, or air pollution control facility and other discarded material, including solid, liquid, semisolid, or contained gaseous material resulting from industrial, commercial, mining and agricultural operations and from community activities, but does not include solid or dissolved materials in domestic sewage, or solid or dissolved materials in irrigation return flows or industrial discharges which are point sources subject to permits under Section 402 of the Federal Water Pollution Control Act, as amended (86 Stat. 880), or source, special nuclear, or by-product material as defined by the Atomic Energy Act of 1954, as amended (68 Stat. 923).

This statutory definition of solid waste is pursuant to the regulations of the Environmental Protection Agency insofar as a solid waste is a hazardous waste if it exhibits any one of four specific characteristics: (1) ignitability, (2) reactivity, (3) corrosivity, and (4) toxicity. However, a waste listed solely for the characteristic of ignitability, reactivity, and/or corrosivity is (may be) excluded from regulation as a hazardous waste once it no longer exhibits a characteristic of hazardous waste [Section 261.3(g)(1)].

In terms of *ignitability*, a waste is an *ignitable hazardous waste* if it has a flash point of less than 140 degrees (40 CFR 261.21) Fahrenheit as determined by the Pensky-Martens closed cup flash point test; readily causes fires and burns so vigorously as to create a hazard; or is an ignitable compressed gas or an oxidizer (as defined by the Department of Transport regulations). A simple method of determining the flash point of a waste is to review the material. Ignitable wastes carry the waste code D1001. Naphtha is an example of an ignitable hazardous waste.

On the other hand, a *corrosive waste* is a liquid waste that has a pH of less than or equal to 2 (a highly acidic waste) or greater than or equal to 12.5 (a highly alkaline waste) and is considered to be a corrosive hazardous waste (40 CFR 261.22). Also, a corrosive waste may be a liquid and corrodes steel (SAE 1020) at a rate greater than 6.35 mm (0.250 in.) per year at a test temperature of 55°C (130 F). Corrosivity testing is conducted using the standard test method as formulated by the National Association of Corrosion Engineers (NACE)—Standard TM-01-69 or, in place of this method, an EPA-approved equivalent test method.

For example, sodium hydroxide (caustic soda), with a high pH, is often used by the organic chemicals production industry (especially by the refining industry and the natural gas industry) in the form of a caustic wash to remove sulfur compounds or acid gases. When these caustic solutions become contaminated and must be disposed of, the waste would be a corrosive hazardous waste. Corrosive wastes carry the waste code D002. Acid solutions also fall under this category.

A chemical waste material is considered to be a *reactive* hazardous waste if it is normally unstable, reacts violently with water, generates toxic gases when exposed to water or corrosive materials, or if it is capable of detonation or explosion when exposed to heat or a flame (40 CFR 261.23). Materials that are defined as forbidden explosives or Class A or B explosives by the DOT are also considered reactive hazardous waste.

Typically, reactive wastes are solid wastes that exhibit any of the following properties as defined at 40 CFR 61.23(a): (1) it is normally unstable and readily undergoes violent change without detonating, (2) it reacts violently with water, (3) it forms potentially explosive mixtures with water, (4) when mixed with water, it generates toxic gases, vapors, or fumes in a quantity sufficient to present a danger to human health or the environment, (5) it is a cyanide or sulfide-bearing waste which, when exposed to pH conditions between 2 and 12.5, can generate toxic gases, vapors or fumes in a quantity sufficient to present a danger to human health or the environment, (6) it is capable of detonation or explosive reaction if it is subjected to a strong initiating source or if heated under confinement, (7) it is readily capable of detonation or explosive decomposition or reaction at standard temperature and pressure, and (8) it is a forbidden explosive as defined in 49 CFR 173.51, or a Class A explosive as defined in 49 CFR 173.53 or a Class B explosive as defined in 49 CFR 173.88.

The Environmental Protection Agency has assigned the Hazardous Waste Number D003 to reactive characteristic waste.

The fourth characteristic that could make a waste a hazardous waste is *toxicity* (40 CFR 261.24). To determine if a waste is a toxic hazardous waste, a representative sample of the material must be subjected to a test conducted in a certified laboratory using a test procedure (Toxicity Characteristic Leaching Procedure, TCLP). Under federal rules (40 CFR 261), all generators are required to use the TCLP test when evaluating wastes.

Wastes that fail a toxicity characteristic test are considered hazardous under the Resource Conservation and Recovery Act. There is less incentive for a company to attempt to reduce the toxicity of such waste below the toxicity characteristic levels because, even though such toxicity reductions may render the waste nonhazardous, it may still have to comply with new Land Disposal treatment standards under subtitle C of the Resource Conservation and Recovery Act before disposal. Similarly, there is little positive incentive to reduce the toxicity of listed hazardous wastes because, once listed, the waste is subject to subtitle C regulations without regard to how much the toxicity levels are reduced.

Besides the four characteristics of hazardous wastes, the Environmental Protection Agency has established three hazardous waste lists: (1) hazardous wastes from nonspecific sources, such as spent nonhalogenated solvents, (2) hazardous wastes from specific sources, such as bottom sediment sludge from the treatment of wastewaters from wood preserving, and (3) discarded commercial chemical products and off-specification species, containers, and spill residues. However, under regulations of the Environmental Protection Agency, certain types of solid wastes (e.g., household waste) are not considered to be hazardous wastes irrespective of their characteristics. Additionally, the

Environmental Protection Agency has provided certain regulatory exemptions based on very specific criteria. For example, hazardous waste generated in a product or raw material storage tank, transport vehicle, or manufacturing processes, and samples collected for monitoring and testing purposes are exempt from the regulations.

Finally, in terms of waste classification, the Environmental Protection Agency has also designated certain wastes as *incompatible wastes*, which are hazardous wastes that, if placed together, could result in potentially dangerous consequences. As defined in 40 CFR 260.10, an *incompatible waste* is a hazardous waste which is unsuitable for: (1) placement in a particular device or facility because it may cause corrosion or decay of containment materials, such as container inner liners or tank walls, or (2) commingling with another waste or material under uncontrolled conditions because the commingling might produce heat or pressure, fire or explosion, violent reaction, toxic dusts, mists, fumes, or gases, or flammable fumes or gases.

Once the physical and chemical properties of a hazardous waste have been adequately characterized, hazardous waste compatibility charts can be consulted to identify other types of wastes with which it is potentially incompatible. For example, Appendix V to both 40 CFR 264 and 265 presents examples of potentially incompatible wastes and the potential consequences of their mixture.

Under the Resource Conservation Recovery Act, the hazardous waste management program is based on a *cradle-to-grave* concept so that all hazardous wastes can be traced and fully accounted for. Section 3010(a) of the Act requires all generators and transporters of hazardous wastes as well as owners and operators of all TSD facilities to file a notification with the Environmental Protection Agency within 90 days after the promulgation of the regulations. The notification should state the location of the facility and include a general description of the activities as well as the identified and listed hazardous wastes being handled.

Submission of the Part A permit application for existing facilities prior to Nov. 19, 1980, qualified a refinery for interim status. This meant that the refinery was allowed to continue operation according to certain regulations during the permitting process. The HSWA represented a strong bias against land disposal of hazardous waste. Some of the provisions that affect the companies (especially refineries) involved in the production of organic chemicals are:

- A ban on the disposal of bulk or noncontainerized liquids in landfills. The prohibition also bans solidification of liquids using absorbent material including absorbents used for spill cleanup.
- Five hazardous wastes from specific sources come under scheduled for disposal prohibition and/or treatment standards. These five are: dissolved air flotation (DAF) float, slop oil emulsion solids, heat exchanger bundle cleaning sludge, API separator sludge, and leaded tank bottoms (EPA waste numbers: K047–K051).

- Producers of organic chemicals (such as refineries) must retrofit surface impoundments that are used for hazardous waste management. Retrofitting must involve the use of double liners and leak detection systems.

Under the Resource Conservation Recovery Act, the Environmental Protection Agency has the authority to require a company to clean up releases of hazardous waste or waste constituents. The regulation provides for cleanup of hazardous waste released from active TSD facilities. Superfund was expected to handle contamination that had occurred before that date.

3.8 Safe Drinking Water Act

The Safe Drinking Water Act, was enacted in 1974 to assure high-quality water supplies through public water system. It is the federal law that protects public drinking water supplies throughout the nation. Under the Safe Drinking Water Act, the Environmental Protection Agency is authorized to set the standards for drinking water quality and with its partners implements various technical and financial programs to ensure drinking water safety.

The Act is truly the first federal intervention to set the limits of contaminants in drinking water. The 1986 amendments came 2 years after the passage of the HSWA or the so-called Resource Conservation Recovery Act Amendments of 1984. As a result, certain statutory provisions were added to these 1986 amendments to reflect the changes made in the underground injection control systems. In addition, the SARA of 1986 set the groundwater standards the same as the drinking water standards for the purpose of necessary cleanup and remediation of an inactive hazardous waste disposal site.

The law was amended in 1986 and 1996 and requires many actions to protect drinking water and its sources—rivers, lakes, reservoirs, springs, and ground water wells (the SDWA does not regulate private wells which serve fewer than 25 individuals). For example, the 1986 amendments of SDWA included additional elements to establish maximum contaminant level goals (MCLGs) and national primary drinking water standards. The MCLGs must be set at a level at which no known or anticipated adverse effects on human health occur, thus providing an *adequate margin of safety*. Establishment of a specific MCLG depends on the evidence of carcinogenicity in drinking water or a reference dose that is individually calculated for each specific contaminant. The MCLGs, an enforceable standard, however, must be set to operate as the nation's primary drinking water standard (NPDWS). The 1996 amendments greatly enhanced the existing law by recognizing source water protection, operator training, funding for water system improvements, and public information as important components of safe drinking water. This approach ensures the quality of drinking water by protecting it from source to tap.

The SDWA calls for regulations that (1) apply to public water systems, (2) specify contaminants that may have any adverse effect on the health of persons, and (3) specify contaminant levels. The difference between primary and secondary drinking water regulations are defined, as well as, other applicable terms. Information concerning national drinking water regulations and protection of underground sources of drinking water is given.

In the context of the organic chemicals industry, the priority list of drinking water contaminants is very important since it includes the contaminants known for their adverse effect on public health. Furthermore, most if not all are known or suspected to have hazardous or toxic characteristics that can compromise human health.

3.9   Toxic Substances Control Act

The TSCA, enacted in 1976, was designed to understand the use or development of chemicals and to provide controls, if necessary, for those chemicals that may threaten human health or the environment (Ingle, 1983; Sittig, 1991). The Act provides the Environmental Protection Agency with the authority to require reporting, record-keeping and testing requirements, and restrictions relating to chemical substances and/or mixtures.

This Act has probably had more effect on the producers of chemicals, and the reining industry, than any other Act. It has caused many changes in the industry and may even create further modifications in the future. The basic thrust of the Act is (1) to develop data on the effects of chemicals on our health and environment, (2) to grant authority to the Environmental Protection Agency to regulate substances presenting an unreasonable risk, and (3) to assure that this authority is exercised so as not to impede technological innovation. In short, the Act calls for regulation of *chemical substances* and *chemical mixtures* that present an unreasonable risk or injury to health or the environment. Furthermore, the introduction and evolution of this Act has led to a central bank of information on existing commercial chemical substances and chemical mixtures, procedures for further testing of hazardous chemicals, and detailed permit requirements for submission of proposed new commercial chemical substances and chemical mixtures.

As used in the Act, the term *chemical substance* means any organic or inorganic substance of a particular molecular identity, including any combination of such substances occurring in whole or in part as a result of a chemical reaction or occurring in nature, and any element or uncombined radical. Items not considered *chemical substances* are listed in the definition section of the Act. The term *mixture* means any combination of two or more chemical substances if the combination does not occur in nature and is not, in whole or in part, the result of a chemical reaction; except that such term does include any combination which occurs, in whole or in part, as a result of a chemical reaction if none of the chemical substances comprising the combination is a new chemical substance and if the combination could have been manufactured for commercial purposes without a chemical reaction at the time the chemical substances comprising the combination were combined.

For many, familiarity with the TSCA generally stems from its specific reference to polychlorinated biphenyls, which raise a vivid, deadly characterization of the harm caused by them. But the Act is not a statute that deals with a single chemical or chemical mixture or product. In fact, under the TSCA, the Environmental Protection Agency is authorized to institute testing programs for various chemical substances that may enter the environment. Under Tosca's broad authorization, data on the production and use of various chemical substances and mixtures may be obtained to protect public health and the environment from the effects of harmful chemicals. In actuality, the Act supplements the appropriate sections dealing with toxic substances in other federal statutes such as the Clean Water Act (Section 307) and the Occupational Safety and Health Act (Section 6).

At the heart of the TSCA is a premanufacture notification requirement under which a manufacturer must notify the Environmental Protection Agency at least 90 days prior to the production of a new chemical. In this context, a *new chemical* is a chemical that is not listed in the Act-based Inventory of Chemical Substances or is an unlisted reaction product of two or more chemicals. For chemicals already on this list, a notification is required if there is a new use that could significantly increase human or environmental exposure. No notification is required for chemicals that are manufactured in small quantities solely for scientific research and experimentation.

The Chemical Substances Inventory of the TSCA is a comprehensive list of the names of all existing chemical substances and currently contains over 70,000 existing chemicals. Information in the inventory is updated every 4 years. A facility must submit a premanufacture notice prior to manufacturing or importation for any chemical substances not on the list and not excluded by the Act. Examples of regulated chemicals include lubricants, paints, inks, fuels, plastics, and solvents.

4 OUTLOOK

A good deal is already known about the effects of chemical substances on man during their production, transport and use, from the study of occupational medicine, toxicology, pharmacology, etc., and to a lesser extent their effects on resource-organisms, from veterinary science and plant pathology. Much less is known about the effects of a chemical upon wildlife species, following its disposal in the environment. This last area of knowledge needs improvement because ecological cycles and food chains may deliver the potentially hazardous chemical from affected wildlife back to man. Moreover, wildlife can often be used as indicators of environmental states and trends for a potentially harmful substance, giving an early warning of future risks to man. We are also frequently ignorant of how far wildlife may be supportive to human well-being:

as a food-base for an important resource-species, e.g., fisheries or grazing animals; as a key species maintaining the stability of economically valuable ecosystems; as predators of crop or livestock pests; as a species involved in mineral recycling or biodegradation; as an important amenity.

Failure to recognize the mutually interactive roles of man, resource-species, wildlife organisms and climate in the biosphere and their different tolerances to chemical substances has hindered the development of a unitary environmental management policy embracing all four biosphere components. Although a good deal is already known about the influence of molecular structure on the toxicity to human beings of drugs and certain other chemicals, much less is known about the influence of molecular structure on the environmental persistence of a chemical. For wildlife, persistence is probably the most important criterion for predicting potential harm because there is inevitably some wild species or other which is sensitive to any compound and any persistent chemical, apparently harmless to a limited number of toxicity-test organisms, will eventually be delivered by biogeochemical cycles to a sensitive target-species in nature. This means that highly toxic, readily biodegradable substances may pose much less of an environmental problem, than a relatively harmless persistent chemical which may well damage a critical wild species. The study of chemical effects in the environment resolves itself into a study of (a) the levels of a substance accumulating in air, water, soils (including sediments) and biota (including man), and (b) when the threshold action-level has been reached, effects produced in biota which constitute a significant adverse response (i.e., environmental dose-response curve). In order to predict trends in levels of a chemical, much more information is needed about rates of injection, flow and partitioning between air, water, soils and biota; and loss via degradation (environmental balance-sheets).

These dynamic phenomena are governed by the physico-chemical properties of the molecule. Fluid mechanics and meteorology may in future provide the conceptual and technical tools for producing predictive models of such systems. Most of our knowledge of effects is derived from acute toxicology and medical studies on man, but since environmental effects are usually associated with chronic exposure, studies are being increasingly made of long-term continuous exposure to minute amounts of a chemical. The well-known difficulty of recognizing such effects when they occur in the field is aggravated by the fact that many of the effects are nonspecific and are frequently swamped by similar effects derived from exposure to such natural phenomena as famines, droughts, cold spells, etc. Even when a genuine effect is recognized, a candidate causal agent must be found and correlated with it. This process must be followed by experimental studies, unequivocally linking chemical cause and adverse biological effect. All three stages are difficult and costly, and it is not surprising that long delays are often experienced between the recognition of a significant adverse effect and a generally agreed chemical cause. There is often ample uncertainty to allow underreaction as well as overreaction to potential hazards, both backed up by "scientific" evidence.

4.1 Hazardous Waste Regulations

Operators of organic chemicals production companies face stringent regulation of the TSD of hazardous wastes. Under recent regulations, a larger number of compounds have been, and are being, studied and long-time methods of disposal, such as land farming of chemical waste, are being phased out. As a result, many companies are changing their waste management practices.

New regulations are becoming even more stringent, and they encompass a broader range of chemical constituents and processes. Continued pressure from the US Congress has led to more explicit laws allowing little leeway for industry, the US Environmental Protection Agency (Environmental Protection Agency), or state agencies. A summary of the current regulations and what they mean is given in the following.

4.2 Regulatory Background

The hazardous waste regulatory program, as we know it today, began with the RCRA in 1976. The Used Oil Recycling Act of 1980 and Hazardous and Solid Waste Amendments of 1984 (HSWA) were the major amendments to the original law.

The Resource Conservation Recovery Act provides for the tracking of hazardous waste from the time it is generated, through storage and transportation, to the treatment or disposal sites. The Act and the various Amendments are aimed at preventing the disposal problems that lead to a need for the CERCLA, or Superfund, as it is known. Subtitle C of the original Resource Conservation Recovery Act lists the requirements for the management of hazardous waste. This includes the Environmental Protection Agency criteria for identifying hazardous waste, and the standards for generators, transporters, and companies that treat, store, or dispose of the waste. The Resource Conservation Recovery Act regulations also provide standards for design and operation of such facilities.

4.3 Requirements

The first step to be taken by a generator of waste is to determine whether that waste is hazardous. Waste may be hazardous by being listed in the regulations, or by meeting any of the four characteristics: ignitability, corrosivity, reactivity, and EP toxicity.

Generally: (1) if the material has a flash point less than 140°F it is considered ignitable; (2) if the waste has a pH less than 2.0 or above 12.5, it is considered corrosive. It may also be considered corrosive if it corrodes stainless steel at a certain rate; (3) a waste is considered reactive if it is unstable and produces toxic materials, or it is a cyanide or sulfide-bearing waste which generates toxic gases or fumes; (4) a waste which is analyzed for EP toxicity and fails is also considered a hazardous waste. This procedure subjects a sample of the waste to an acidic environment. After an appropriate time has elapsed, the liquid portion of the sample (or the sample itself if the waste is liquid) is analyzed for certain metals and pesticides. Limits for allowable concentrations are given in the regulations. The specific analytical parameters and procedures for these tests are referred to in 40CRF 261.

The 1984 amendments also brought the owners and operators of underground storage tanks under the umbrella of the Resource Conservation Recovery Act. This can have a significant effect on chemicals production companies (such as refineries) that store products in underground tanks. In addition, organic chemicals are also regulated by the Resource Conservation Recovery Act, Subtitle I.

5 MANAGEMENT OF PROCESS WASTE

The chemicals production industry will increasingly feel the effects of the land bans on their hazardous waste management practices. Current practices of land disposal must change along with management attitudes for waste handling. The way companies handle the waste products in the future depends largely on the ever-changing regulations. Waste management is the focus and reuse/recycle options must be explored to maintain a balanced waste management program. This requires that a waste be recognized as either *nonhazardous* or *hazardous*.

However, before a company can determine if its waste is hazardous, it must first determine that the waste is indeed a solid waste. In 40 CFR 261.2, the definition of solid waste can be found. If a waste material is considered a solids waste, it may be a hazardous waste in accordance with 40 CFR 261.3. There are two ways to determine whether a waste is hazardous. These are to see if the waste is listed in the regulations or to test the waste to see if it exhibits one of the characteristics (40 CFR 261).

There are five lists of hazardous wastes in the regulations. These are wastes from nonspecific sources (F list) (Appendix: Table A2), wastes from specific sources (K list) (Appendix: Table A3), acutely toxic wastes (P list) (Appendix: Table A4), toxic wastes (U list) (Appendix: Table A5), and the D-Code List (Appendix: Table A7). And there are the four characteristics mentioned before: ignitability, corrosivity, reactivity, and EP toxicity. Certain waste materials are excluded from regulation under the Resource Conservation Recovery Act. The various definitions and situations that allow waste to be exempted can be confusing and difficult to interpret. One such case is the interpretation of the *mixture* and *derived-from* rules. According to the mixture rule, mixtures of solid waste and listed hazardous wastes are, by definition, considered hazardous. Likewise, the *derived-from* rule defines solid waste resulting from the management of hazardous waste to be hazardous (40 CFR 261.3a and 40 CFR. 261.1c).

There are five specific listed hazardous wastes (K list) generated in crude oil refineries—K048–K052. According to 40 CFR 261.32, these waste codes are defined and listed as follows: (1) K048: DAF float, (2) K049: slop oil emulsion solids, (3) K050: heat exchanger bundle cleaning sludge, (4) K051: API separator sludge, and (5) K052: tank bottoms, leaded. Additional listed wastes include those from nonspecific sources (F list), those from the commercial chemical product lists (P and U), and the D list of hazardous waste, which can apply to those wastes generated during chemical production (Appendix). Because of the mixture and derived-from rules, special care must be taken to ensure that hazardous wastes do not *contaminate* nonhazardous waste. Under the mixture rule, adding one drop of hazardous waste in a container of nonhazardous materials makes the entire container contents a hazardous waste.

As an example of the problems such mixing can cause is the API separator sludge that is a listed hazardous waste (K051). The wastewater from a properly operating API separator is not hazardous unless it exhibits one of the characteristics of a hazardous waste. That is, the derived-from rule does not apply to the wastewater. However, if the API separator is not functioning properly, solids carry over in the wastewater can occur. In this case, the wastewater contains a listed hazardous waste, the solids from the API sludge, and the wastewater would be considered a hazardous waste because it is a mixture of a nonhazardous waste and a hazardous waste.

This wastewater is often further cleaned by other treatment systems (filters, impoundments, etc.). The solids separating in these systems continue to be API separator sludge, a listed hazardous waste. Therefore, all downstream wastewater treatment systems are receiving and treating a hazardous waste and are considered hazardous waste management units that are subject to regulation.

Oily wastewater is often treated or stored in unlined wastewater treatment ponds in refineries. These wastes appear to be similar to API separator waste.

REFERENCES

Dennison, M.S., 1993. RCRA Regulatory Compliance Guide. Noyes Data Corp., Park Ridge, NJ.

EPA, 1997. Organic Chemicals Refinery MACT Standard Guidance. Environmental Protection Agency, Washington, DC.

Ingle, G.W. (Ed.), 1983. TSCA's Impact on Society and the Chemical Industry. ACS Symposium Series, American Chemical Society, Washington, DC.

Lazarus, R., 2004. The Making of Environmental Law. Cambridge University Press, Cambridge.

Lunn, G., Sanstone, E.B., 1994. Destruction of Hazardous Chemicals in the Laboratory, second ed. McGraw-Hill, New York, NY.

Majumdar, S.B., 1993. Regulatory Requirements for Hazardous Materials. McGraw-Hill, New York, NY.

Nordin, J.S., Sheesley, D.C., King, S.B., Routh, T.K., 1995. Environ. Solut. 8 (4), 49.

Noyes, R. (Ed.), 1993. Pollution Prevention Technology Handbook. Noyes Data Corp., Park Ridge, NJ.

Sittig, M., 1991. Handbook of Toxic and Hazardous Chemicals and Carcinogens, third ed. Noyes Data Corp., Park Ridge, NJ.

TPG, 1995. Aboveground Storage Tank Guide. vols. 1 and 2 Thompson Publishing Group, Washington, DC.

Wagner, T.P., 1999. The Complete Guide to Hazardous Waste Regulations. John Wiley & Sons Inc., New York, NY.

Wang, C.C.K., 1994. OSHA Compliance and Management Handbook. Noyes Data Corp., Park Ridge, NJ.

Chapter 9

Removal of Organic Compounds From the Environment

1 INTRODUCTION

One of the major and continuing environmental problems is contamination resulting from the activities related to industrial processes and products. Contamination of the air, land, and water causes extensive damage of local ecosystems since accumulation of pollutants in animals and plant tissue may cause death or mutations. As a result, once a spill has occurred, every effort must be made to rid the environment of the toxins. The chemicals of known toxicity range in degree of toxicity from low to high and represent considerable danger to human health and must be removed (Frenzel et al., 2009). Many of these chemical substances come in contact with, and are sequestered by, soil or water systems. While conventional methods to remove, reduce, or mitigate the effects of toxic chemical in nature are available include (1) pump and treat systems, (2) soil vapor extraction, (3) incineration, and (4) containment, each of these conventional methods of treatment of contaminated soil and/or water suffers from recognizable drawbacks and may involve some level of risk. In short, these methods, depending upon the chemical constituents of the spilled material, may limited effectiveness and can be expensive (Speight, 1996, 2005; Speight and Lee, 2000).

Although the effects of bacteria (microbes) on chemicals, especially hydrocarbons, have been known for decades, this technology biodegradation (also known as *bioremediation* in the sense of applied cleanup of a site by other than natural means) has shown promise and, in some cases, high degrees of effectiveness for the treatment of contaminated sites since it is cost-effective and will lead to complete mineralization. The concept of biodegradation may also refer to complete *mineralization* of the organic contaminants into carbon dioxide, water, inorganic compounds to other simpler organic compounds that are not detrimental to the environment. In fact, unless they are overwhelmed by the amount of the spilled material or it is toxic, many indigenous microorganisms in soil and/or water are capable of degrading hydrocarbon contaminants (Speight and Arjoon, 2012).

Environmental Organic Chemistry for Engineers. http://dx.doi.org/10.1016/B978-0-12-804492-6.00009-5

The United States Environmental Protection Agency (US EPA) uses biodegradation because it takes advantage of natural processes and relies on microbes that occur naturally or can be laboratory cultivated; these consist of bacteria, fungi, actinomycetes, cyanobacteria, and, to a lesser extent, plants (US EPA, 2006). These microorganisms either consume and convert the contaminants or assimilate within them all harmful compounds from the surrounding area, thereby rendering the region virtually contaminant-free. Generally, the substances that are consumed as an energy source are organic compounds, while those, which are assimilated within the organism, are heavy metals. Biodegradation harnesses this natural process by promoting the growth and/or rapid multiplication of these organisms that can effectively degrade specific contaminants and convert them to nontoxic by-products.

The capabilities of microorganisms and plants to degrade and transform contaminants provide benefits in the cleanup of pollutants from spills and storage sites. These remediation ideas have provided the foundation for many ex situ waste treatment processes (including sewage treatment) and a host of in situ biodegradation methods that are currently in practice.

Thus, biodegradation—*the ability of living organisms to reduce or eliminate environmental hazards resulting from accumulations of toxic chemicals and other hazardous wastes*—is an option that offers the possibility to destroy or render harmless various contaminants using natural biological activity (Gibson and Sayler, 1992). In addition, biodegradation can also be used in conjunction with a wide range of traditional physical and chemical technology to enhance the effectiveness of these technologies (Vidali, 2001).

In the current context, biodegradation at contaminated sites is the natural or stimulated cleanup of spills of organic chemicals by the use of microbes to breakdown the organic contaminants of the spill into less harmful (usually lower molecular weight) and easier-to-remove products (biodegradation). The microbes transform the contaminants through metabolic or enzymatic processes, which vary greatly, but the final product is usually harmless and includes carbon dioxide, water, and cell biomass. Thus, the emerging science and technology of biodegradation offers an alternative method to detoxify soil and water from chemical contaminants. Furthermore, and by means of clarification, *biodegradation (biotic degradation, biotic decomposition) is the chemical degradation of contaminants by bacteria or other biological means.* Organic material can be degraded aerobically (in the presence of oxygen) or anaerobically (in the absence of oxygen). Most biodegradation reactions operate run under aerobic conditions, but a system under anaerobic conditions may permit microbial organisms to degrade chemical species that are otherwise nonresponsive to aerobic treatment, and vice versa.

Thus, biodegradation is a natural process (or a series of processes) by which spilled organic chemicals are broken down (degraded) into nutrients that can be used by other organisms. As a result, the ability of a chemical to be biodegraded is an indispensable element in understanding the risk posed by that chemical on the environment.

Biodegradation is a key process in the natural attenuation (reduction or disposal) of chemical compounds at hazardous waste sites, but the success of the process depends on the ability to determine these conditions and establish them in the contaminated environment. Thus, important site factors required for success include (1) the presence of metabolically capable and sustainable microbial populations; (2) suitable environmental growth conditions, such as the presence of oxygen; (3) temperature, which is an important variable—keeping a substance frozen or below the optimal operating temperature for microbial species, can prevent biodegradation—most biodegradation occurs at temperatures between 10°C and 35°C (50°F and 95°F); (4) the presence of water; (5) appropriate levels of nutrients and contaminants; and (6) favorable acidity or alkalinity (Table 9.1). In regard to the last parameter, soil pH is extremely important because most microbial species can survive only within a certain pH range—generally the biodegradation of hydrocarbons is optimal at a pH 7 (neutral) and the acceptable (or optimal) pH range is on the order of 6–8. Furthermore, soil (or water) pH can affect availability of nutrients.

Thus, through biodegradation processes, living microorganisms (primarily bacteria, but also yeasts, molds, and filamentous fungi) can alter and/or metabolize various classes of chemical compounds. Furthermore, biodegradation also alters subsurface accumulations of chemicals Winters and Williams, 1969; Speight, 2014a,b).

Temperature influences rate of biodegradation by controlling rate of enzymatic reactions within microorganisms. Generally, the rate of an enzymatic

TABLE 9.1 Essential Factors for Microbial Bioremediation

Factor	Optimal Conditions
Microbial population	Suitable kinds of organisms that can biodegrade all of the contaminants
Oxygen	Enough to support aerobic biodegradation (c.2% oxygen in the gas phase or 0.4 mg/L in the soil water)
Water	Soil moisture should be from 50% to 70% (w/w) of the water holding capacity of the soil
Nutrients	Nitrogen, phosphorus, sulfur, and other nutrients to support good microbial growth
Temperature	Appropriate temperatures for microbial growth (0-40°C)
pH	Best range is from 6.5 to 7.5

reaction approximately doubles for each 10°C (18°F) rise in temperature (Nester et al., 2001). However, there is an upper limit to the temperature that microorganisms can withstand. Most bacteria found in soil, including many bacteria that degrade hydrocarbons, are mesophile organisms which have an optimum working temperature range on the order of 25–45°C (77–113°F) (Nester et al., 2001). Thermophilic bacteria (those which survive and thrive at relatively high temperatures) which are normally found in hot springs and compost heaps exist indigenously in cool soil environments and can be activated to degrade hydrocarbons with an increase in temperature to 60°C (140°F). This indicates the potential for natural attenuation in cool soils through thermally enhanced biodegradation reactions (Perfumo et al., 2007).

In order to enhance and make favorable the parameters presented above to ensure microbial activity, there are two other enhanced biodegradation methods that offer useful options for cleanup of spills of chemicals: (1) fertilization and (2) seeding. *Fertilization* (*nutrient enrichment*) is the method of adding nutrients such as phosphorus and nitrogen to a contaminated environment to stimulate the growth of the microorganisms capable of biodegradation. Limited supplies of these nutrients in nature usually control the growth of native microorganism populations. When more nutrients are added, the native microorganism population can grow rapidly, potentially increasing the rate of biodegradation. *Seeding* is the addition of microorganisms to the existing native degrading population. Some species of bacteria that do not naturally exist in an area will be added to the native population. As with fertilization, the purpose of seeding is to increase the population of microorganisms that can biodegrade the spilled chemical. Thus, biodegradation is an environmentally acceptable naturally occurring process that takes place when all of the nutrients and physical conditions involved are suitable for growth. The process allows for the breakdown of a compound to either fully oxidized or reduced simple molecules such as carbon dioxide/methane, nitrate/ammonium, and water. However, in some cases, where the process is not complete, the products of biodegradation can be more harmful than the substance degraded.

Intrinsic biodegradation is the combined effect of natural destructive and nondestructive processes to reduce the mobility, mass, and associated risk of a contaminant. Nondestructive mechanisms include sorption, dilution, and volatilization. Destructive processes are aerobic and anaerobic biodegradation. *Intrinsic aerobic biodegradation* is well documented as a means of remediating soil and groundwater contaminated with fuel hydrocarbons. In fact, intrinsic aerobic degradation should be considered an integral part of the remediation process (McAllister et al., 1995; Barker et al., 1995). There is growing evidence that natural processes influence the immobilization and biodegradation of chemicals such as aromatic hydrocarbons, mixed hydrocarbons, and chlorinated organic compounds (Ginn et al., 1995; King et al., 1995).

Phytoremediation is the use of living green plants for the removal of contaminants and metals from soil and is, essentially, an in situ treatment of pollutant-contaminated soils, sediments, and water—terrestrial, aquatic, and wetland plants and algae can be used for the phytoremediation process under specific cases and conditions of hydrocarbon contamination (Brown, 1995; Nedunuri et al., 2000; Radwan et al., 2000; Magdalene et al., 2009). It is best applied at sites with relatively shallow contamination of pollutants that are amenable to the various subcategories of phytoremediation: (1) phytotransformation—the breakdown of organic contaminants sequestered by plants; (2) rhizosphere biodegradation—the use of rhizosphere microorganisms to degrade organic pollutants; (3) phytostabilization—a containment process using plants, often in combination with soil additives to assist plant installation, to mechanically stabilizing the site and reducing pollutant transfer to other ecosystem compartments and the food chain; (4) phytoextraction—the ability of some plants to accumulate metals/metalloids in their shoots; (5) rhizofiltration; and/or (6) phytovolatilization/rhizovolatilization—processes employing metabolic capabilities of plants and associated rhizosphere microorganisms to transform pollutants into volatile compounds that are released to the atmosphere (Korade and Fulekar, 2009).

These technologies are especially valuable where the contaminated soils are fragile, and prone to erosion. The establishment of a stable vegetation community stabilizes the soil system and prevents erosion. This aspect is especially relevant to certain types of soil where removal of large volumes of soil destabilizes the soil system, which leads to extensive erosion. However, when the above parameters are not conducive to bacterial activity, the bacteria (1) grow too slowly, (2) die, or (3) create more harmful chemicals.

Phytotransformation and *rhizosphere biodegradation* are applicable to sites that have been contaminated with organic pollutants, including pesticides. It is a technology that should be considered for remediation of contaminated sites because of its cost-effectiveness, esthetic advantages, and long-term applicability (Brown, 1995).

Plants have shown the capacity to withstand relatively high concentrations of organic chemicals without toxic effects, and they can uptake and convert chemicals quickly to less toxic metabolites in some cases. In addition, they stimulate the degradation of organic chemicals in the rhizosphere by the release of root exudates, enzymes, and the buildup of organic carbon in the soil.

Microorganisms degrade or transform contaminants by a variety of mechanisms. Hydrocarbons (particularly alkanes) for example are converted to carbon dioxide and water:

$$2C_{12}H_{26} + 37O_2 \rightarrow 24CO_2 + 26H_2O$$

Or the hydrocarbon may be used as a primary food source by the bacteria, which use the energy to generate new cells.

Some contaminants, such as chlorinated organic or high aromatic hydrocarbons, are generally resistant to microbial attack. They are degraded either slowly or not at all, hence it is not easy to predict the rates of cleanup for biodegradation; there are no rules to predict if a contaminant can be degraded.

When the hydrocarbons are chlorinated, degradation takes place as a secondary or cometabolic process rather than a primary metabolic process. In such a case, enzymes, which are produced during aerobic utilization of carbon sources such as methane, degrade the chlorinated compounds. Under aerobic conditions, a chlorinated solvent such as trichloroethylene ($CHCl=CCl_2$) can be degraded through a sequence of metabolic steps, where some of the intermediary by-products may be more hazardous than the parent compound (e.g., vinyl chloride, $CH_2=CHCl$).

Over the past two decades, opportunities for applying biodegradation to a much broader set of contaminants have been identified. Indigenous and enhanced organisms have been shown to degrade industrial solvents, polychlorinated biphenyls, explosives, and many different agricultural chemicals. Pilot, demonstration, and full-scale applications of biodegradation have been carried out on a limited basis. However, the full benefits of biodegradation have not been realized because processes and organisms that are effective in controlled laboratory tests are not always equally effective in full-scale applications. The failure to perform optimally in the field setting stems from a lack of predictability due, in part, to inadequacies in the fundamental scientific understanding of how and why these processes work.

2 BIODEGRADATION

Biodegradation is looked upon as an environmentally friendly technique used to restore soil and water to its original state by using indigenous microbes to break down and eliminate contaminants. Biological technologies are often used as a substitute to chemical or physical cleanup of chemical spills because biodegradation does not require as much equipment or labor as other methods; therefore it is usually cheaper. It also allows cleanup workers to avoid contact with polluted soil and water.

The microorganisms used for biodegradation may be indigenous to a contaminated area, or they may be isolated from elsewhere and brought to the contaminated site. Contaminants are transformed by living organisms through reactions that take place as a part of their metabolic processes. Biodegradation of a compound is often a result of the actions of multiple organisms. When microorganisms are imported to a contaminated site to enhance degradation, we have a process known as bioaugmentation.

For biodegradation to be effective, microorganisms must convert the pollutants and convert them to harmless products. As biodegradation can be effective only where environmental conditions permit microbial growth and activity, its application often involves the manipulation of environmental parameters to allow microbial growth and degradation to proceed at a faster rate. However,

TABLE 9.2 Advantages and Disadvantages of Bioremediation

Advantages	Disadvantages
Remediates contaminants that are adsorbed onto or trapped within the geologic materials of which the aquifer is composed along with contaminants dissolved in groundwater	Injection wells and/or infiltration galleries may become plugged by microbial growth or mineral precipitation
Application involves equipment that is widely available and easy to install	High concentrations (TPH greater than 50,000 ppm) of low solubility constituents may be toxic and/or not bioavailable
Creates minimal disruption and/or disturbance to ongoing site activities	Difficult to implement in low-permeability aquifers
Time required for subsurface remediation may be shorter than other approaches (e.g., pump-and-treat)	Reinjection wells or infiltration galleries may require permits or may be prohibited. Some states require permit for air injection
Generally recognized as being less costly than other remedial options	May require continuous monitoring and maintenance
Can be combined with other technologies (e.g., bioventing, SVE) to enhance site remediation	Remediation may only occur in more permeable layer or channels within the aquifer
In many cases this technique does not produce waste products that must be disposed	

as is the case with other technologies, biodegradation has its limitations, and there are several disadvantages that must be recognized (Table 9.2).

The control and optimization of biodegradation processes is a complex system of many factors. These factors include: the existence of a microbial population capable of degrading the pollutants; the availability of contaminants to the microbial population; and the environment factors (type of soil, temperature, pH, and the presence of oxygen or other electron acceptors, and nutrients).

One of the important factors in biological removal of hydrocarbons from a contaminated environment is their bioavailability to an active microbial population, which is *the degree of interaction of chemicals with living organisms* or the degree to which a contaminant can be readily taken up and metabolized by a bacterium (Harms et al., 2010). Moreover, the bioavailability of a contaminant is controlled by factors such as the physical state of the hydrocarbon in situ, its hydrophobicity, water solubility, sorption to environmental matrices such as soil, and diffusion out of the soil matrix. When contaminants have very low solubility in water, as in the case of *n*-alkanes and polynuclear aromatic hydrocarbons, the organic phase components will not partition efficiently into the aqueous phase supporting the microbes.

In the case of soil, the contaminants will also partition to the soil organic matter and become even less bioavailable. Two-phase bioreactors containing an aqueous phase and a nonaqueous phase liquid (NAPL) have been developed and used for biodegradation of hydrocarbon-contaminated soil to address this very problem, but the adherence of microbes to the NAPL-water interface can still be an important factor in reaction kinetics. Similarly, two-phase bioreactors, sometimes with silicone oil as the nonaqueous phase, have been proposed for biocatalytic conversion of hydrocarbons like styrene (Osswald et al., 1996) to make the substrate more bioavailable to microbes in the aqueous phase. When the carbon source is in limited supply, then its availability will control the rate of metabolism and hence biodegradation, rather than catabolic capacity of the cells or availability of oxygen or other nutrients.

In the case of the biomediation of waterways, similar principles apply. Under enhanced conditions (1) certain fuel hydrocarbons can be removed preferentially over others, but the order of preference is dependent upon the geochemical conditions and (2) augmentation and enhancement via electron acceptors to accelerate the biodegradation process. For example, with regard to the aromatic benzene-toluene-ethylbenzene-xylenes (BTEX): (1) toluene can be preferentially removed under intrinsic biodegradation conditions, (2) biodegradation of benzene is relatively slow, (3) augmentation with sulfate can preferentially stimulated biodegradation of *o*-xylene, and (4) ethylbenzene may be recalcitrant under sulfate-reducing conditions but readily degradable under denitrifying conditions (Cunningham et al., 2000).

In the current context, biodegradation is a collection of chemical reactions for dealing with chemicals contaminants, and the process typically occurs through the degradation of the chemical through the action of microorganisms (biodegradation). The method utilizes indigenous bacteria (microbes) compared to the customary (physical and chemical) remediation methods. Also, the microorganisms engaged are capable of performing almost any detoxification reaction. Biodegradation studies provide information on the fate of a chemical or mixture of chemicals (such as crude oil spills or process wastes) in the environment, thereby opening the scientific doorway to develop further methods of cleanup by (1) analyzing the contaminated sites, (2) determining the best method suited for the environment, and (3) optimizing the cleanup techniques which lead to the emergence of new processes.

2.1 Natural Biodegradation

Natural biodegradation typically involves the use of molecular oxygen (O_2), where oxygen (the *terminal electron acceptor*) receives electrons transferred from an organic contaminant:

Organic substrate + O_2 → biomass + CO_2 + H_2O + other products

In the absence of oxygen, some microorganisms obtain energy from fermentation and anaerobic oxidation of organic carbon. Many anaerobic organisms (*anaerobes*) use nitrate, sulfate, and salts of iron (III) as practical alternates to oxygen acceptor as, for example, in the anaerobic reduction process of nitrates, sulfates, and salts of iron (III):

$$2NO_3^- + 10e^- + 12H^+ \rightarrow N_2 + 6H_2O$$
$$SO_4^{2-} + 8e^- + 10H^+ \rightarrow H_2S + 4H_2O$$
$$Fe(OH)_3 + e^- + 3H^+ \rightarrow Fe^{2+} + 3H_2O$$

2.2 Traditional Biodegradation Methods

Methods for the cleanup of pollutants have usually involved removal of the polluted materials, and their subsequent disposal by land filling or incineration (so-called *dig, haul, bury, or burn* methods) (Speight, 1996, 2005; Speight and Lee, 2000). Furthermore, available space for landfills and incinerators is declining. Perhaps one of the greatest limitations to traditional cleanup methods is the fact that in spite of their high costs, they do not always ensure that contaminants are completely destroyed.

Conventional biodegradation methods that have been, and are still, used are (1) composting, (2) land farming, (3) biopiling, and (4) use of a bioslurry reactor (Speight, 1996; Speight and Lee, 2000; Semple et al., 2001).

Composting is a technique that involves combining contaminated soil with nonhazardous organic materials such as manure or agricultural wastes; the presence of the organic materials allows the development of a rich microbial population and elevated temperature characteristic of composting. *Land farming* is a simple technique in which contaminated soil is excavated and spread over a prepared bed and periodically tilled until pollutants are degraded. *Biopiling* is a hybrid of land farming and composting, it is essentially engineered cells that are constructed as aerated composted piles. A *bioslurry reactor* can provide rapid biodegradation of contaminants due to enhanced mass transfer rates and increased contaminant-to-microorganism contact. These units are capable of aerobically biodegrading aqueous slurries created through the mixing of soils or sludge with water. The most common state of bioslurry treatment is batch; however, continuous-flow operation is also possible.

The technology selected for a particular site will depend on the limiting factors present at the location. For example, where there is insufficient dissolved oxygen, bioventing or sparging is applied, and biostimulation or bioaugmentation is suitable for instances where the biological count is low. On the other hand, application of the composting technique, if the operation is unsuccessful, it will result in a greater quantity of contaminated materials. Land farming is only effective if the contamination is near the soil surface or else bed preparation is required. The main drawback with slurry bioreactors is that high-energy input is required to maintain suspension and the potential needed for volatilization.

Other techniques are also being developed to improve the microbe-contaminant interactions at treatment sites so as to use biodegradation technologies at their fullest potential. These biodegradation technologies consist of monitored natural attenuation, bioaugmentation, biosimulation, surfactant addition, anaerobic bioventing, sequential anaerobic/aerobic treatment, soil vapor extraction, air sparging, enhanced anaerobic dechlorination, and bioengineering (Speight, 1996; Speight and Lee, 2000).

2.3   Enhanced Biodegradation Treatment

Enhanced biodegradation is a process in which indigenous or inoculated microorganisms (e.g., fungi, bacteria, and other microbes) degrade (metabolize) organic contaminants found in soil and/or groundwater and convert the contaminants to innocuous end products. The process relies on general availability of naturally occurring microbes to consume contaminants as a food source or as an electron acceptor (chlorinated solvents, which may be waste materials from chemical processes). In addition to microbes being present, in order to be successful, these processes require nutrients such as carbon, nitrogen, and phosphorus.

Enhanced biodegradation involves the addition of microorganisms (e.g., fungi, bacteria, and other microbes) or nutrients (e.g., oxygen, nitrates) to the subsurface environment to accelerate the natural biodegradation process.

2.4   Biostimulation and Bioaugmentation

Biostimulation is the method of adding nutrients such as phosphorus and nitrogen to a contaminated environment to stimulate the growth of the microorganisms that break down chemicals. Additives are usually added to the subsurface through injection wells although injection well technology for biostimulation purposes is still emerging. Limited supplies of these necessary nutrients usually control the growth of native microorganism populations. Thus, addition of nutrients causes rapid growth of the indigenous microorganism population, thereby increasing the rate of biodegradation.

It is to be anticipated that the success of biostimulation is case specific and site specific, depending on the properties of the chemicals, the nature of the nutrient products, and the characteristics of the contaminated environments. When oxygen is not a limiting factor, one of keys for the success of biostimulation is to maintain an optimal nutrient level in the interstitial pore water. Several types of commercial biostimulation agents are available for use in biodegradation (Zhu et al., 2004).

Bioaugmentation is the addition of pregrown microbial cultures to enhance microbial populations at a site to improve contaminant clean up and reduce clean up time and cost. Indigenous or native microbes are usually present in very small quantities and may not be able to prevent the spread of the contaminant. In some cases, native microbes do not have the ability to degrade a particular contaminant. Therefore, bioaugmentation offers a way to provide specific microbes in sufficient numbers to complete the biodegradation (Atlas, 1991).

Mixed cultures have been most commonly used as inocula for seeding because of the relative ease with which microorganisms with different and complementary biodegradative capabilities can be isolated (Atlas, 1977). Different commercial cultures were reported to degrade various hydrocarbons (Compeau et al., 1991; Leavitt and Brown, 1994; Chhatre et al., 1996; Mishra et al., 2001; Vasudevan and Rajaram, 2001).

Microbial inocula (the microbial materials used in an inoculation) are prepared in the laboratory from soil or groundwater either from the site where they are to be used or from another site where the biodegradation of the chemicals of interest is known to be occurring. Microbes from the soil or groundwater are isolated and are added to media containing the chemicals to be degraded. Only microbes capable of metabolizing the chemicals will grow on the media. This process isolates the microbial population of interest. One of the main environmental applications for bioaugmentation is at sites with chlorinated solvents. Microbes (such as *Dehalococcoides ethenogenes*) usually perform reductive dechlorination of solvents such as perchloroethylene and trichloroethylene.

Bioaugmentation adds highly concentrated and specialized populations of specific microbes to the contaminated area, while biostimulation is dependent on appropriate indigenous microbial population and organic material being present at the site.

2.5 In Situ and Ex Situ Biodegradation Methods

Biodegradation can be used as a cleanup method for both contaminated soil and water. Its applications fall into two broad categories: in situ or ex situ. In situ biodegradation treats the contaminated soil or groundwater in the location in which it was found, while ex situ biodegradation processes require excavation of contaminated soil or pumping of groundwater before they can be treated.

In situ technologies do not require excavation of the contaminated soils so may be less expensive, create less dust, and cause less release of contaminants than ex situ techniques. Also, it is possible to treat a large volume of soil at once. In situ techniques, however, may be slower than ex situ techniques, may be difficult to manage, and are only most effective at sites with permeable soil.

The most effective means of implementing in situ biodegradation depends on the hydrology of the subsurface area, the extent of the contaminated area, and the nature (type) of the contamination. In general, this method is effective only when the subsurface soils are highly permeable, the soil horizon to be treated falls within a depth of 8–10 m, and shallow groundwater is present at 10 m or less below ground surface. The depth of contamination plays an important role in determining whether or not an in situ biodegradation project should be employed. If the contamination is near the groundwater but the groundwater is not yet contaminated, then it would be unwise to set up a hydrostatic system. It would be safer to excavate the contaminated soil and apply an on-site method of treatment away from the groundwater.

The typical time frame for an in situ biodegradation project can be in the order of 12–24 months depending on the levels of contamination and depth of contaminated soil. Due to the poor mixing in this system it becomes necessary to treat for long periods of time to ensure that all the pockets of contamination have been treated.

In situ biodegradation is a very site specific technology that involves establishing a hydrostatic gradient through the contaminated area by flooding it with water carrying nutrients and possibly organisms adapted to the contaminants. Water is continuously circulated through the site until it is determined to be clean.

In situ biodegradation of groundwater speeds the natural biodegradation processes that take place in the water-soaked underground region that lies below the water table. One limitation of this technology is that differences in underground soil layering and density may cause reinjected conditioned groundwater to follow certain preferred flow paths. On the other hand, ex situ techniques can be faster, easier to control, and used to treat a wider range of contaminants and soil types than in situ techniques. However, they require excavation and treatment of the contaminated soil before and, sometimes, after the actual biodegradation step.

In situ biodegradation is the preferred method for large sites and is used when physical and chemical methods of remediation may not completely remove the contaminants, leaving residual concentrations that are above regulatory guidelines. This method has the potential to provide advantages such as complete destruction of the contaminant(s), lower risk to site workers, and lower equipment/operating costs. In situ biodegradation can be used as a cost-effective secondary treatment scheme to decrease the concentration of contaminants to acceptable levels or as a primary treatment method, which is followed by physical or chemical methods for final site closure.

Finally, evidence for the effectiveness of biodegradation should include: (1) faster disappearance of chemicals in treated areas than in untreated areas and (2) a demonstration that biodegradation was the main reason for the increased rate of disappearance of the chemical(s). To obtain such evidence, the analytical procedures must be chosen carefully and careful data interpretation is essential, but there are disadvantages and errors when the method is not applied correctly (Speight, 2005; Speight and Arjoon, 2012).

3 BIODEGRADATION METHODS

Biodegradation technology exploits various naturally occurring mitigation processes: (1) *natural attenuation*, (2) *biostimulation*, and (3) *bioaugmentation*.

Biodegradation which occurs without human intervention other than monitoring is often called *natural attenuation*. This natural attenuation relies on natural conditions and behavior of soil microorganisms that are indigenous to soil. *Biostimulation* also utilizes indigenous microbial populations to remediate contaminated soils and consists of adding nutrients and other substances to soil to catalyze natural attenuation processes. *Bioaugmentation* involves introduction of exogenic microorganisms (sourced from outside the soil environment) capable of detoxifying a particular contaminant, sometimes employing genetically altered microorganisms.

In recent years, in situ biodegradation concepts have been applied in treating contaminated soil and groundwater. Removal rates and extent vary based on the contaminant of concern and site-specific characteristics. Removal rates also are affected by variables such as contaminant distribution and concentration; cocontaminant concentrations; indigenous microbial populations and reaction kinetics; and parameters such as pH, moisture content, nutrient supply, and temperature. Many of these factors are a function of the site and the indigenous microbial community and, thus, are difficult to manipulate. Specific technologies may have the capacity to manipulate some variables and may be affected by other variables as well (US EPA, 2006).

During biodegradation, microbes utilize chemical contaminants in the soil as an energy source and, through oxidation-reduction reactions, metabolize the target contaminant into useable energy for microbes. By-products (metabolites) released back into the environment are typically in a less toxic form than the parent contaminants. For example, hydrocarbons can be degraded by microorganisms in the presence of oxygen through aerobic respiration. The hydrocarbon loses electrons and is oxidized, while oxygen gains electrons and is reduced. The result is formation of carbon dioxide and water (Nester et al., 2001).

When oxygen is limited in supply or absent, as in saturated or anaerobic soils or lake sediment, anaerobic (without oxygen) respiration prevails. Generally, inorganic compounds such as nitrate, sulfate, ferric iron, manganese, or carbon dioxide serve as terminal electron acceptors to facilitate biodegradation. Generally, a contaminant is more easily and quickly degraded if it is a naturally occurring compound in the environment, or chemically similar to a naturally occurring compound, because microorganisms capable of its biodegradation are more likely to have evolved. Development of biodegradation technologies of synthetic chemicals such chlorocarbons or chlorohydrocarbons is dependent on outcomes of research that seeks to develop natural or genetically improved strains of microorganisms to degrade such contaminants into less toxic forms.

In summary, biodegradation is increasingly viewed as an appropriate remediation technology for hydrocarbon-contaminated polar soils. As for all soils, the successful application of biodegradation depends on appropriate biodegradative microbes and environmental conditions in situ. Laboratory studies have confirmed that hydrocarbon-degrading bacteria typically assigned to the genera *Rhodococcus*, *Sphingomonas*, or *Pseudomonas* are present in contaminated polar soils. However, as indicated by the persistence of spilled hydrocarbons, environmental conditions in situ are suboptimal for biodegradation in polar soils.

Therefore, it is likely that ex situ biodegradation will be the method of choice for ameliorating and controlling the factors limiting microbial activity, i.e., low and fluctuating soil temperatures, low levels of nutrients, and possible alkalinity and low moisture. Care must be taken when adding nutrients to the coarse-textured, low-moisture soils prevalent in continental Antarctica and the high Arctic because excess levels can inhibit hydrocarbon biodegradation by decreasing soil water potentials. Biodegradation experiments conducted on-site in the Arctic indicate that land farming and biopiles may be useful approaches for biodegradation of polar soils (Aislabie et al., 2006; Nugroho et al., 2010).

Several factors that affect the decision of which method is chosen are (1) the nature of the contaminants; (2) the location of contaminated site, cost of cleanup; (3) the time allotted to the cleanup; (4) effects on humans, animals, and plants; and last but by no means least (5) the cost of the cleanup. Sometimes when one method is no longer effective and efficient, another remediation method can be introduced into the contaminated soil.

Oil spills introduce large amounts of toxic compounds into the environment and though different methods of biodegradation have been successful in remediating soils and water contaminated with the low-density organic compounds, the high-viscosity crude oil constituents (high-boiling constituents) are less susceptible to these techniques.

Conventional biodegradation methods used are biopiling, composting, land farming, bioslurry reactors, but there are limitations affecting the applicability and effectiveness of these methods (Speight and Lee, 2000). With the application of the composting technique, if the operation is unsuccessful, it will result in a greater quantity of contaminated materials. Land farming is only effective if the contamination is near the soil surface or else bed preparation needs to take place. The main drawback with slurry bioreactors is that high-energy inputs are required to maintain suspension and the potential needed for volatilization.

For a biodegradation method to be successful in soil and water cleanup, the physical, chemical, and biological environment must be feasible. Parameters that affect the biodegradation process are (1) low temperatures, (2) preferential growth of microbes obstructive to biodegradation, (3) high concentrations of chlorinated organics, (4) preferential flow paths severely decreasing contact between injected fluids and contaminants throughout the contaminated zones, and (5) the soil matrix prohibiting contaminant-microorganism contact.

The bioventing process combines an increased oxygen supply with vapor extraction. A vacuum is applied at some depth in the contaminated soil which draws air down into the soil from holes drilled around the site and sweeps out any volatile organic compounds. The development and application of venting and bioventing for in situ removal of hydrocarbons from soil have been shown to remediate hydrocarbons by venting and biodegradation (van Eyk, 1994). Even though a particular technology may have reports of improving biodegradation efficiency (e.g., surfactant addition), this may not be the case at times depending on the sample.

3.1 In Situ and Ex Situ Biodegradation

Biodegradation applications fall into two broad categories: (1) in situ or (2) ex situ. In situ biodegradation *processes* treats the contaminated soil or groundwater in the location in which it was found. *Ex situ biodegradation processes* require excavation of contaminated soil or pumping of groundwater before they can be treated. In situ techniques do not require excavation of the contaminated soils so may be less expensive, create less dust, and cause less release of contaminants than ex situ techniques. Also, it is possible to treat a large volume of soil at once. In situ techniques, however, may be slower than ex situ techniques, difficult to manage, and are most effective at sites with permeable soil.

In situ biodegradation of groundwater speeds the natural biodegradation processes that take place in the water-soaked underground region that lies below the water table. One limitation of this technology is that differences in underground soil layering and density may cause reinjected conditioned groundwater to follow certain preferred flow paths. On the other hand, ex situ techniques can be faster, easier to control, and used to treat a wider range of contaminants and soil types than in situ techniques. However, they require excavation and treatment of the contaminated soil before and, sometimes, after the actual biodegradation step.

In situ biodegradation is used when physical and chemical methods of remediation may not completely remove the contaminants, leaving residual concentrations that are above regulatory guidelines. Biodegradation can be used as a cost-effective secondary treatment scheme to decrease the concentration of contaminants to acceptable levels. In other cases, biodegradation can be the primary treatment method and followed by physical or chemical methods for final site closure. Also, it is the preferred method for very large sites.

3.2 Biostimulation and Bioaugmentation

Biostimulation is the method of adding nutrients such as phosphorus and nitrogen to a contaminated environment to stimulate the growth of the microorganisms that break down chemicals. Additives are usually added to the subsurface through injection wells although injection well technology for biostimulation purposes is still emerging. Limited supplies of these necessary nutrients usually control the growth of native microorganism populations. Thus, when nutrients are added, the indigenous microorganism population grows rapidly, potentially increasing the rate of biodegradation.

The primary advantage of biostimulation is that biodegradation will be undertaken by already present native microorganisms that are well suited to the subsurface environment and are well distributed spatially within the subsurface, but the main disadvantage is that the delivery of additives in a manner that allows the additives to be readily available to subsurface microorganisms is based on the local geology of the subsurface.

Bioaugmentation is the addition of pregrown microbial cultures to enhance microbial populations at a site to improve contaminant clean up and reduce clean up time and cost. Indigenous or native microbes are usually present in very small quantities and may not be able to prevent the spread of the contaminant. In some cases, native microbes do not have the ability to degrade a particular contaminant. Therefore, bioaugmentation offers a way to provide specific microbes in sufficient numbers to complete the biodegradation.

Microbial inocula are prepared in the laboratory from soil or groundwater either from the site where they are to be used or from another site where the biodegradation of the chemicals of interest is known to be occurring. Microbes from the soil or groundwater are isolated and are added to media containing the chemicals to be degraded. Only microbes capable of metabolizing the chemicals will grow on the media. This process isolates the microbial population of interest. One of the main environmental applications for bioaugmentation is at sites with chlorinated solvents and specific microbes (*D. ethenogenes*) usually perform reductive dechlorination of solvents such as perchloroethylene and trichloroethylene.

Bioaugmentation adds highly concentrated and specialized populations of specific microbes to the contaminated area, while biostimulation is dependent on appropriate indigenous microbial population and organic material being present at the site. Therefore, it might be that bioaugmentation is more effective than biostimulation, but most cleanup programs have a site specificity that is not able to be matched form one site to another.

However, results suggest that the success of biostimulation is case specific, depending on (1) chemical properties, (2) the nature of the nutrient products, and (3) the characteristics of the contaminated environments. When oxygen is not a limiting factor, one of keys for the success of the biostimulation process is to maintain an optimal nutrient level in the interstitial pore water.

3.3 Monitored Natural Attenuation

The term *monitored natural attenuation* refers to the reliance on natural attenuation to achieve site-specific remedial objectives within a time frame that is reasonable compared to that offered by other more active methods.

The *natural attenuation processes* that are at work in such a remediation approach include a variety of physical, chemical, or biological processes that, under favorable conditions, act without human intervention to reduce the mass, toxicity, mobility, volume, or concentration of contaminants in soil or groundwater. These in situ processes include biodegradation, dispersion, dilution, sorption, volatilization, and chemical or biological stabilization, transformation, or destruction of contaminants. A study of any contaminated site must first be performed to decide whether natural attenuation would make a positive input, and though it has degraded lighter chain hydrocarbons quite extensively; the heavier chain hydrocarbons are less susceptible.

As with any technique, there are disadvantages—the disadvantages of *monitored natural attenuation* method are the need for longer time frames to achieve remediation objectives, compared to active remediation, the site characterization may be more complex and costly and long-term monitoring will generally be necessary.

3.4   Soil Vapor Extraction, Air Sparging, and Bioventing

Soil vapor extraction removes harmful chemicals, in the form of vapors, from the soil above the water table. The vapors are extracted from the ground by applying a vacuum to pull it out.

Like Soil vapor extraction, *air sparging* uses a vacuum to extract the vapors. Air sparging uses air to help remove harmful vapors like the lighter gasoline constituents (i.e., BTEX), because they readily transfer from the dissolved to the gaseous phase, but is less applicable to diesel fuel and kerosene. When air is pumped underground, the chemicals evaporate faster, which makes them easier to remove. Methane can be used as an amendment to the sparged air to enhance cometabolism of chlorinated organics. Soil vapor extraction and air sparging are often used at the same time to clean up both soil and groundwater.

Biosparging is used to increase the biological activity in soil by increasing the oxygen supply via sparging air or oxygen into the soil. In some instances, air injections are replaced by pure oxygen to increase the degradation rates. However, in view of the high costs of this treatment in addition to the limitations in the amount of dissolved oxygen available for microorganisms, hydrogen peroxide (H_2O_2) was introduced as an alternative, and it was used on a number of sites to supply more oxygen (Schlegel, 1977) and more efficient in enhancing microbial activity during the biodegradation of contaminated soil and groundwater (Brown and Norris, 1994; Flathman et al., 1991; Lee et al., 1988; Lu, 1994; Lu and Hwang, 1992; Pardieck et al., 1992), but it can be a disadvantage if the toxicity is sufficiently to microorganisms even at low concentrations (Brown and Norris, 1994; Scragg, 1999).

Soil vapor extraction requires drilling extraction wells within the polluted area. The necessary equipment to create a vacuum is attached to the well, which pulls air and vapors through the soil and up to the surface. Once the extraction wells pull the air and vapors out of the ground, special air pollution control equipment collects them. The equipment separates the harmful vapors from the clean air.

Air sparging works very much like soil vapor extraction. However, the wells that pump air into the ground are drilled into water-soaked soil below the water table. Air pumped into the wells disturbs the groundwater. This helps the pollution change into vapors. The vapors rise into the drier soil above the groundwater and are pulled out of the ground by extraction wells. The harmful vapors are removed in the same way as soil vapor extraction. The air used in soil vapor extraction and air sparging also helps clean up pollution by encouraging the growth of microorganisms. In general, the wells and equipment are simple to install and maintain and can reach greater depths than other methods that involve digging up soil. Soil vapor extraction and air sparging are effective at removing many types of pollution that can evaporate.

Air sparging should not be used if free products are present. Air sparging can create groundwater mounding which could potentially cause free product to migrate and contamination to spread. Also, it is not suitable around basements, sewers, or other subsurface confined spaces are present at the site. Potentially dangerous constituent concentrations could accumulate in basements unless a vapor extraction system is used to control vapor migration. If the contaminated groundwater is located in a confined aquifer system, air sparging is not advisable because the injected air would be trapped by the saturated confining layer and could not escape to the unsaturated zone. Anaerobic sparging, an innovative technique in biodegradation, depends on the delivery of an inert gas (nitrogen or argon) with low (<2%) levels of hydrogen. Cometabolic air sparging is the delivery of oxygen-containing gas with enzyme-inducing growth substrate (such as methane or propane).

Bioventing is a technology that stimulates the natural in situ biodegradation of any aerobically degradable compounds in soil by providing oxygen to existing soil microorganisms. In contrast to soil vapor vacuum extraction, bioventing uses low airflow rates to provide only enough oxygen to sustain microbial activity. Two basic criteria have to be satisfied for successful bioventing (1) the air must be able to pass through the soil in sufficient quantities to maintain aerobic conditions and (2) natural hydrocarbon-degrading microorganisms must be present in concentrations large enough to obtain reasonable biodegradation rates.

Bioventing is a medium to long-term technology—cleanup ranges from a few months to several years—and it is applicable to any chemical that can be aerobically biodegraded. The technique has been successfully used to remediate soils contaminated by hydrocarbons, nonchlorinated solvents, some pesticides, wood preservatives, and other organic chemicals. Though there are limitations, this technology does not require expensive equipment and relatively few personnel are involved operation and maintenance; therefore bioventing is receiving increased exposure in the remediation consulting community. Potential improvements on the current bioventing methods that have been taking place are the use of electrochemical oxygen gas sensors, detailed characterization of NAPL distribution, and neutron probe logging. Another bioventing enhancement is the use of this technique with bioslurping and soil vapor extraction (Baker, 1999).

Bioslurping is an in situ remediation technology that combines the two remedial approaches of bioventing and vacuum-enhanced free-product recovery.

It is faster than the conventional remedy of product recovery followed by bioventing. The system is made to minimize groundwater recovery and drawdown in the aquifer. Bioslurping was designed and is being tested to address contamination by hydrocarbons with a floating lighter nonaqueous phase liquids layer.

Bioslurping efficiently recovers free product and extracts less groundwater for treatment, which speeds up remediation and reduces water handling and treatment costs. It enhances natural in situ biodegradation of vadose zone (which extends from the top of the ground surface to the water table) soils and may be the only feasible remediation technology at low-permeability sites. But for bioslurping to even be considered at a contamination site, free product must be present and the product must be biodegradable, also the soil must respond to bioventing.

3.5   Use of Biosurfactants

Another common emerging technology is the use of *biosurfactants*, which are microbially produced surface-active compounds. They are amphiphilic molecules with both hydrophilic and hydrophobic regions, causing them to aggregate at interfaces between fluids with different polarities found in chemical spills. Many of the known biosurfactant producers are hydrocarbon-degrading organisms.

Biosurfactants have comparable solubilization properties to synthetic surfactants but have several additional advantages that make them superior candidates in biodegradation schemes. First, biosurfactants are biodegradable and are not a pollution threat. Furthermore, most studies indicate that they are nontoxic to microorganisms and therefore are unlikely to inhibit biodegradation of nonpolar organic contaminants. Biosurfactant production is less expensive, can be easily achieved ex situ at the contaminated site, and has the potential of occurring in situ.

Biosurfactants are also effective in many diverse geologic formations and are compatible with many existing remedial technologies (such as pump and treat rehabilitation, air sparging, and soil flushing) and significantly accelerate innovative approaches including microbial, natural attenuation-enhanced soil flushing, and bioslurping.

3.6   Rhizosphere Biodegradation

Rhizosphere biodegradation is the interaction between plants and microorganisms and is also known as phytostimulation or plant-assisted biodegradation. The plant root zone (the rhizosphere) has significantly larger numbers of microorganisms than soils which do not have plants growing in them, which appears to enhance the biodegradation of organic compounds (Wenzel, 2009). In the rhizosphere biodegradation process, plants provide oxygen, bacteria, and organic carbon to encourage the degradation of organics in the soil. The microorganisms in the environment created by the plants, together with the roots of the plants, can degrade more contaminants that could occur in a purely microbial system.

Plants release stimulants into the soil environment that help to motivate the degradation of organic chemicals by inducing enzyme systems of existing bacterial populations, stimulating growth of new species that are able to degrade the wastes, and/or increasing soluble substrate concentrations for all microorganisms. Plants help with microbial conversions where certain bacteria that metabolize pollutants are able to encourage degradation of chemicals in the soil, so allowing biodegradation to occur with less retardation.

Evaluation of the current efforts suggests that pollutant bioavailability in the rhizosphere of phytoremediation crops is decisive for designing phytoremediation technologies with improved, predictable remedial success. For phytoextraction, emphasis should be put on improved characterization of the bioavailable metal pools and the kinetics of resupply from less available fractions to support decision-making on the applicability of this technology to a given site. Limited pollutant bioavailability may be overcome by the design of plant-microbial consortia that are capable of mobilizing metals/metalloids by modification of rhizosphere pH and ligand exudation, or enhancing bioavailability of organic pollutants by the release of biosurfactants.

The complexity and heterogeneity polluted soils will require the design of integrated approaches of rhizosphere management such as (1) combining cocropping of phytoextraction and rhizodegradation crops, (2) inoculation of microorganisms, and (3) soil management.

3.7 Bioengineering in Biodegradation

In many cases, after a chemical spill, the natural microbial systems for degrading the chemical are overwhelmed. Therefore, molecular engineers are constructing starvation promoters to express heterologous genes needed in the field for survival and adding additional biodegradation genes that code for enzymes able to degrade a broader range of compounds present in the contaminated environments. Various bacterial strains are also being developed, where each strain is specific for a certain organic compound present in chemical spills. This will help increase the speed of biodegradation and allow detailed cleanup to take place where no organic contamination remains in the environment.

Thus, the decision to bioremediate a site is dependent on cleanup, restoration, and habitat protection objectives, and the factors that are present would have an impact on success. If the circumstances are such that no amount of nutrients will accelerate biodegradation, then the decision should be made on the need to accelerate disappearance of chemicals to protect a vital living resource or simply to speed up restoration of the ecosystem. These decisions are clearly influenced by the circumstances of the spill.

4 TEST METHODS FOR BIODEGRADATION

Various methods exist for the testing of biodegradability of substances. Biodegradability is assessed by following certain parameters which are considered to be indicative of the consumption of the test substance by microorganisms, or the production of simple basic compounds which indicate the mineralization of the test substance. Hence there are various biodegradability testing methods which measure the amount of carbon dioxide (or methane, for anaerobic cases) produced during a specified period; there are those which measure the loss of dissolved organic carbon for substances which are water soluble; those that measure the loss of hydrocarbon infrared bands, and there are yet others which measure the uptake of oxygen by the activities of microorganisms (biochemical oxygen demand).

However, when the reference is specifically to lubricants, there are two major methods of biodegradability testing, and these are outlined in the following paragraphs.

Standard test method ASTM D5864 is a method for the determination of the degradation of high-boiling hydrocarbon mixtures, such as lubricants. In the method, the rate and extent of aerobic aquatic biodegradation of lubricants is determined when the hydrocarbon mixture is exposed to an inoculum under laboratory conditions. The inoculum may be the activated sewage-sludge from a domestic sewage-treatment plant, or it may be derived from soil or natural surface waters, or any combination of the three sources. The degree of biodegradability is measured by calculating the rate of conversion of the lubricant to carbon dioxide. A lubricant, hydraulic fluid, or grease is classified as readily biodegradable when 60% or more of the test material carbon is converted to carbon dioxide in 28 days, as determined using this test method.

The most established test methods used by the lubricant industry for evaluating the biodegradability of their products are Method CEC-L-33-A-94 developed by the Coordinating European Council (CEC); Method OEC D 301B, the Modified Sturm Test, developed by the Organization for Economic Cooperation and Development (OECD); and Method EPA 560/6-82-003, number CG-2000, the Shake Flask Test, adapted by the U.S. Environmental Protection Agency (EPA). These tests also determine the rate and extent of aerobic aquatic biodegradation under laboratory conditions. The Modified Sturm Test and Shake Flask Test also calculate the rate of conversion of the lubricant to carbon dioxide. The CEC test measures the disappearance of the lubricant by analyzing test material at various incubation times through infrared spectroscopy. Laboratory tests have shown that the degradation rates may vary widely among the various test methods indicated earlier (US Army Corps of Engineers, 1999).

Biodegradability tests based on the CEC method described earlier has certain trends which indicate that alkylated benzenes and polyalkeleneglycols among others generally have poor biodegradability.

5 POLLUTION PREVENTION

A major aspect of pollution prevention is to determine the organic chemicals that are toxic and dangerous to the flora and fauna, especially organic chemicals that can influence the flora and fauna of waterways (Table 9.3). However, assessing the ability of organic chemicals to be toxic and dangerous to the flora and

TABLE 9.3 Harmful Effects of Selected Organic Chemicals in Waterways

PAHs (polycyclic aromatic hydrocarbons)

PAHs in the bottom sediments of a waterway can cause tumors in marine flatfish

PAHs from oil and fuel spills in water can cause heart defects in the developing embryos of herring and other fish species

Crude oil-related compounds

Crude oil-related compounds—including gasoline, motor oil, hydraulic fluids, diesel, and jet fuels—are mixtures of many different chemicals, including additives

Many petrochemicals are toxic to algae and invertebrates. And can cause changes in metabolism, reduced feeding, and poor shell formation

These compounds can poison fish at all life stages and kill their eggs and larva

Crude oil-related compounds can damage the skin, lungs, liver, and kidneys of birds and mammals as well as increase vulnerability to deadly infections by suppressing the immune systems of animals

Petrochemicals can reduce the reproductive success of invertebrates such as shellfish and insects, fish, birds, and mammals, leading to population declines

Petrochemicals can also damage plants and impair or stop seed germination

PBDE flame retardants (polybrominated diphenyl ethers)

PBDE flame retardants can affect the development, reproduction, and survival of many species

These compounds build up in the food chain and are found in people as well other organisms including fish and sea mammals, such as orcas

Phthalates (including DEHP or bis(2-ethylhexyl) phthalate)

Exposure to DEHP, a phthalate, is associated with developmental and reproductive harm, especially the male reproductive system in humans and animals

PCBs (polychlorinated biphenyls)

PCBs build up in the food chain and can cause adverse health effects in humans and wildlife, including cancer and harm to immune, nervous, and reproductive systems

PCBs disrupt thyroid hormone levels in animals and humans, hindering growth, and development

TABLE 9.3 Harmful Effects of Selected Organic Chemicals in Waterways— cont'd

DDT (dichlorodiphenyltrichloroethane)

The pesticide DDT builds up in the food chain and can last for decades in the environment

DDT is linked to the decline of the bald eagle, peregrine falcon, and other birds because it makes their egg shells too thin, decreasing the survival of chicks

PCDD/Fs dioxins (polychlorinated dibenzo-*p*-dioxins and polychlorinated dibenzofurans)

Even at very low concentrations, PCDD/F dioxins are toxic to humans and animals

Dioxins can cause cancer, disrupt the endocrine (hormone) system, and harm reproduction and development

Generally, birds and mammals are at greater risk than fish, dioxins build up in the food chain and may affect people and animals that eat fish

Triclopyr (3,5,6-Trichloro-2-pyridinyloxyacetic acid)

The herbicide/fungicide Triclopyr breaks down in soil with a half-life of between 30 and 90 days

One of the by-products of breakdown, trichloropyridinol, remains in the soil for up to a year. Triclopyr degrades rapidly in water

It remains active in decaying vegetation for about 3 months

If misapplied, the herbicide/fungicide Triclopyr can harm fish and other aquatic species

Nonylphenol

Nonylphenol, a chemical found in detergents

Nonylphenol has a potential role as an endocrine disruptor and xenoestrogen, due to its ability to act with estrogen-like activity

fauna is not always an easy task unless the properties of the organic chemicals are known and understood since different organic chemicals cause harm in different ways and to different organisms. In addition, the hazard posed any specific organic chemical depends on the degree of toxicity of the chemical and the amount of the chemical enters the environment. Furthermore, some toxic organic chemicals do not break down easily in the environment which allows these chemical to move up through the food chain. These persistent, bioaccumulative toxic organic chemicals can build up in the tissues of small organisms which are then eaten by larger animals which are then, in turn, eaten by even larger animals, sometimes by humans.

In order to combat such effects, including bioaccumulation, knowledge of organic chemistry which leads to knowledge of the structure, properties, composition, reactions, and preparation of carbon-containing compounds become advantageous (Chapters 1 and 2). This knowledge should not only include an understanding of the properties and behavior of hydrocarbons derivatives but also compounds with any number of other elements, including hydrogen (most compounds contain at least one carbon-hydrogen bond), nitrogen, oxygen, halogens, phosphorus, silicon, and sulfur (Chapter 1). This branch of chemistry was originally limited to compounds produced by living organisms but has been broadened to include human-made substances such as plastics.

The organic chemicals process industry plays an important role in the development of a country by providing a wide variety of products, which are being used in providing basic needs of rising demand. The organic chemicals process industries use raw materials derived from crude oil and natural gas, salt, oil and fats, biomass and energy from coal, natural gas, and a small percentage from renewable energy resources. Although initially manufacture of organic chemicals initially started with coal and alcohol from fermentation industry, however later due to availability of crude oil and natural gas dominated the scene and now more than 90% of organic chemicals are produced from crude oil and natural gas routes. However, variable costs of crude oil and natural gas and continuous decrease in the reserves have spurred the chemical industry for alternative feedstock like coal, biomass, coalbed methane, shale gas, sand oil as an alternate source of fuel, and chemical feedstock.

The range of application of organic compounds is extensive, and the fundamental organic chemicals (commodity organic chemicals) are a broad chemical category including polymers, bulk petrochemicals, and intermediates, as well as many other derivatives, some of which appear on various lists compiled by the US Environmental Protection Agency (Appendix). The production of polymers includes production of all categories of plastics and man-made fibers. Examples are:

- Polyethylene, which is used in packaging films and other markets such as milk bottles, containers, and pipe.
- Polyvinyl chloride, which is used to make pipe for construction markets as well as siding and, to a much smaller extent, transportation, and packaging materials.
- Polypropylene, which is used in markets ranging from packaging, appliances, and containers to clothing and carpeting.
- Polystyrene, which is used for appliances and packaging as well as toys and recreation.
- Man-made fibers, such as polyester, nylon, and acrylic fibers, which are used in applications such as apparel, home furnishings, and other industrial and consumer use.

The principal raw materials for polymers are bulk petrochemicals (Speight, 2014a,b, 2016). Organic chemicals in the bulk petrochemicals arena are primarily made from natural gas (predominantly methane), liquefied crude oil gas

(LPG, various mixtures of propane, C_3H_8, and butane, C_4H_{10}), and the gaseous products such as ethylene ($CH_2=CH_2$) from crude oil refining. Typical large-volume products include ethylene ($CH_2=CH_2$), propylene ($CH_3CH=CH_2$), butadiene ($CH_2=CHCH=CH_2$), benzene, (C_6H_6) toluene ($C_6H_5CH_3$), xylene isomers ($CH_3C_6H_4CH_3$), methanol (methyl alcohol, CH_3OH), vinyl chloride ($CH_2=CHCl$), styrene ($C_6H_5CH=CH_2$), and other intermediates. These basic or commodity chemicals are the starting materials used to manufacture many polymers and other more complex organic chemicals particularly those that are made for use in the production of specialty chemicals. Other derivatives and basic industrial chemicals include synthetic rubber, surfactants, dyestuffs, turpentine, resins, and carbon black. From the engineering perspective, the organic chemicals industry involves the use of reaction engineering to produce a wide variety of solid, liquid, and gaseous materials, most of which are used to manufacture the final products (Speight, 2002).

The organic chemicals industry includes manufacturers of organic industrial chemicals, ceramic products, petrochemicals, agrochemicals, polymers and rubber (elastomers), oleochemicals (oils, fats, and waxes), explosives, fragrances, and flavors (Table 9.4). In addition, the organic pharmaceuticals industry (i.e., not often included under the general umbrella of the *chemicals industry*) uses many different starting materials and products that are not always categorized as general organic chemicals. For example, the pharmaceutical industry develops, produces, and markets drugs licensed for use as medications for humans or animals. Some pharmaceutical companies deal in brand-name (i.e., has a trade name and can be produced and sold only by the company holding the patent) and/or generic (i.e., chemically equivalent, lower-cost version of a brand-name drug) medications and medical devices (agents that act on

TABLE 9.4 Examples of Products Produced by the Organic Chemicals Industry

Product Type	Examples
Organics	Acrylonitrile, phenol, ethylene oxide, urea
Petrochemicals	Ethylene, propylene, benzene, styrene
Agrochemicals	Fertilizers, insecticides, herbicides
Polymers	Polyethylene, Bakelite, polyester
Elastomers	Polyisoprene, neoprene, polyurethane
Oleochemicals	Lard, soybean oil, stearic acid
Explosives	Nitroglycerin, nitrocellulose
Fragrances and flavors	Benzyl benzoate, coumarin, vanillin

diseases without chemical interaction with the body). Pharmaceuticals (brand name and generic) and medical devices are subject to a large number of country-specific laws and regulations regarding patenting, testing, safety assurance, efficacy, monitoring, and marketing. Other closely industries include the crude oil industry and the natural gas industry (Mokhatab et al., 2006; Speight, 2007, 2014a,b, 2016).

The processes used to produce organic chemicals are usually tested during and after manufacture by dedicated instruments and on-site quality control laboratories to ensure safe operation and to assure that the product will meet the required specifications. More organizations within the chemicals industry are implementing chemical compliance software to maintain quality products and manufacturing standards. The products are packaged and delivered by many methods, including pipelines, tank-cars and tank-trucks (for both solids and liquids), cylinders, drums, bottles, and boxes. Chemical companies often have a research and development (R&D) laboratory for developing and testing products and processes. These facilities may include pilot plants, and such research facilities may be located at a site separate from the production plant(s).

In addition, the production of organic chemicals produces not only the desired chemical product but also organic chemical waste that is composed of by-products and other unwanted reaction materials that is, for the most part, composed of harmful (even hazardous) organic chemicals. This waste typically falls under one or more environmental regulations (Chapter 8) as well as a variety of state and local regulations also regulate chemicals use and disposal. Chemical waste may or may not be classed as hazardous waste which can be a gas, liquid, or solid that displays either a *hazardous characteristic* or is specifically listed by name as a hazardous waste. There are four characteristics chemical wastes may have to be considered as hazardous. These are ignitability, corrosivity, reactivity, and toxicity. This type of hazardous waste must be categorized as to its identity, constituents, and hazards so that it may be safely handled and managed. Organic chemical waste may not be a single compound but may be a composite of many types of organic compounds (with some inorganic materials included for good measure!). For clarification, it is recommended that the Material Safety Data Sheet (MSDS), product data sheet, or label be consulted for a list of the constituents of the wastes. These sources should state whether or not this chemical waste is a waste that needs special disposal protocols. However, in spite of the numerous safety regulations that are applied daily (in fact, hourly) and consistently, accidents do happen and pollution does occur.

Pollution prevention is the operational guideline for process operators, process engineers, process chemists, and, for that matter, anyone who handles organic chemicals. It is in this area that environmental observance plays a major role. Pollution prevention is, simply, reduction or elimination of discharges or emissions to the environment. The limits of pollutants emitted to the atmosphere, the land, and water are defined by various pieces of legislation that have been put into place over the past four decades (Chapter 8) (Speight, 1996;

Woodside, 1999; Speight and Arjoon, 2012). This includes all pollutants such as hazardous and nonhazardous wastes, regulated and unregulated chemicals from all sources.

Pollution associated with organic chemicals refining typically includes volatile organic compounds (volatile organic compounds), carbon monoxide (CO), sulfur oxides (SO_x), nitrogen oxides (NO_x), particulates, ammonia (NH_3), hydrogen sulfide (H_2S), metals, spent acids, and numerous toxic organic compounds (Speight, 2014a,b, 2016). Sulfur and metals result from the impurities in crude oil. The other wastes represent losses of feedstock and organic chemicals.

These pollutants may be discharged as air emissions, wastewater, or solid waste. All of these wastes are treated. However, air emissions are more difficult to capture than wastewater or solid waste. Thus, air emissions are the largest source of untreated wastes released to the environment.

Pollution prevention can be accomplished by reducing the generation of wastes at their source (source reduction) or by using, reusing, or reclaiming wastes once they are generated (environmentally sound recycling). However, environmental analysis plays a major role in determining if emissions-effluents (air, liquid, or solid) fall within the parameters of the relevant legislation. For example, issues to be addressed are the constituents of gaseous emissions, the sulfur content of liquid fuels, and the potential for leaching contaminants (through normal rainfall or through the agency of acid rain) from solid products such as coke.

Collecting site information is very important because chemicals could be used and discharged from industrial sources. For example, a battery manufacturing industry could be the source of metals such as mercury, cadmium, lead, nickel, manganese, iron, copper, and lithium. Certain chemicals listed in the appendices are widely used in industries; for example, degreasing agents (organic solvents) such as dichloromethane, 1,1-dichloroethane, 1,2-dichloroethane, 1,1,1-trichloroethane, 1,2-dichloroethene, trichloroethylene and tetrachloroethylene. Moreover, effluents from recycling industries contain a variety of chemicals depending on the type of raw and final products.

Finally, as part of the disposal protocol, it is necessary to know the compatibility of the constituents of the organic chemical waste. Many chemicals may react adversely when combined. It is recommended that incompatible chemicals are stored in separate areas of the facility and should be separated. The reason being when combined some incompatible compounds can undergo a violent exothermic reaction causing a gas emission that is highly flammable leading, in some cases, to an explosion. For example, oxidizers should be separated from organic chemicals because when combined oxidizers organic chemicals can produce volatile and flammable products as well as products that are highly toxic.

Furthermore, it is not only the compatibility of the individual constituents of an organic chemical waste that must be monitored but also the compatibility of the organic chemical(s) with the container. For example, chemical that should not be stored in *Nalgene* (a brand of shatterproof and lighter-than-glass plastic products) containers:

- reactive organic chlorides, such as amyl chloride ($C_5H_{12}Cl$) vinylidene chloride (1,1-dichloroethylene, $CH_2=CCl_2$);
- organic acids, such as butyric acid ($C_3H_8CO_2H$);
- solvents, such as carbon disulfide (CS_2); and
- aromatic liquids and solvents, such as nitrobenzene ($C_6H_5NO_2$).

Thus, proper management of organic chemicals and organic chemical waste is necessary to protect the environment. Federal and state regulations require all generators of organic chemicals, and organic chemical waste receive training and follow proper waste management and disposal procedures.

In fact, emission (gaseous, liquid, and solid emissions) abatement equipment represents one of the most important parts of site infrastructure. A wide variety of end-of-pipe pollution control techniques is available for gaseous, liquid, and solid wastes, and many are used in common ways across the chemical industry. The application of emission abatement technologies is highly dependent on site-specific situations and needs to be evaluated case by case. Where gaseous and liquid streams necessarily arise from a process (i.e., prevention techniques have been fully implemented), then the aim is to maximize the number of vents that are collected and diverted into appropriate treatment units. Many large sites make use of centralized environmental treatment facilities for waste water and waste gases (although waste gases are often harder to collect and so less suited to centralized treatment). Central treatment plants take advantage of economies of scale when installing and operating treatment equipment, and they damp hydraulic and chemical fluctuations in the effluent feeds thus improving the stability of performance. There may also be direct benefit from the combination of effluent streams (e.g., the combination of nitrogen-containing waste water streams with nitrogen-poor streams to aid their biological treatment). However, centralized treatment facilities should provide genuine benefits and not merely dilute pollutants prior to release.

The purpose of this section is to present a description of the methods by which organic chemicals are treated in an attempt to insure that pollution does not occur and any effluents and/or emissions fall within the legislative specifications. Indeed, as already noted, environmental compliance (Chapter 8) is the major discipline by which this potential for the production of effluents and emissions can be determined and, hence, monitored.

5.1 Chemical Wastes and Treatment

The manufacture of organic chemicals is a large source of pollution worldwide, and part of the reason for the expansive reach of chemical manufacturing is the diverse and varied types of sectors and activities that are included in it. The EPA defines chemical manufacturing as "creating products by transforming organic and inorganic raw materials with chemical processes." These are further broken up into commodity and specialty chemicals. Commodity chemicals are basic singular chemicals in ongoing production at industrial plants. Specialty chemicals are batches of combination chemicals made at the request of certain industries and produced on an as needed basis. New chemicals are introduced, and old chemicals are withdrawn constantly, changing the chemical manufacturing market frequently, making it difficult to monitor and evaluate. The sheer size of the industry makes it difficult to monitor as well since the industry accounts for substantial amounts of global income of international trade.

Organic chemicals can be released through the same pathways as other pollutants, including emissions from heating and processing, accidental release of dust or other particulates, accidental spills, and improper disposal of solid waste and wastewater. Once in the environment exposure media includes air, water, soil, and food. In the Blacksmith Institute's database, which focuses on chemical dumps and abandoned sites, the exposure pathways are evenly split between inhalation of contaminated dust and soil, ingestion of contaminated water and food, and inhalation of contaminated gases or vapor. The chemical manufacturing industry is the largest single consumer of water by sector in all OECD countries. The large amount of process water required provides many opportunities for pollutants to be released through wastewater.

The pollutants found in the largest quantities at chemical manufacturing sites include pesticides and volatile organic compounds. Furthermore, it is important to note that volatile organic compounds (VOCs), exposure to volatile organic compounds released from chemical manufacturing sites potentially puts more human health at risk at the sites. Volatile organic compounds are low-molecular-weight chemicals made from carbon and hydrogen, and often including oxygen, nitrogen, chlorine, and other elements. Because of their low molecular weight, volatile organic compounds convert to vapor easily, and vapors of volatile organic compounds are emitted from certain products and processes. There are thousands of volatile organic compounds, many of which are familiar compounds in everyday life, such as ethyl alcohol, propane, mineral spirits, and the chemicals in gasoline, kerosene, and oil.

While many volatile organic compounds are relatively nonhazardous (aside from their flammability), there are thousands of volatile organic compounds that are toxic, and some can cause eye, nose, and throat irritation and headaches, while others are known carcinogens. Some examples of toxic volatile organic compounds include benzene, formaldehyde, toluene, vinyl chloride, and chloroform. Volatile organic compounds come from a wide variety of products, most of which are used daily by society. The list includes most fuels, paints, stains and lacquers, cleaning supplies, pesticides, plastics, glues, adhesives, and refrigerants. Volatile organic compounds, including many more uncommon and toxic types, are very commonly used in manufacturing processes as solvents or raw materials in the production of plastics, chemicals, pharmaceuticals, and electronic products.

TABLE 9.5 Benefits of Waste Elimination

- Solve the waste disposal problems created by land bans
- Reduce waste disposal costs
- Reduce costs for energy, water, and raw materials
- Reduce operating costs
- Protect workers, the public, and the environment
- Reduce risk of spills, accidents, and emergencies
- Reduce vulnerability to lawsuits and improve its public image
- Generate income from wastes that can be sold

Waste elimination is common sense and provides several obvious benefits (Table 9.5). Yet waste elimination continues to elude many companies in every sector, including the process section, and activity from process waste (i.e., a function of their production system design). It may not matter how a producer categorizes the waste or how the producer chooses to pursue waste elimination, one thing remains constant and that is: once identified, waste can be eliminated. There are models and structures that allow a producer to identify and eliminate waste to increase productivity, and hence cost structures, that have a direct impact on process operations and, more than all else, profitability. Waste elimination though identification (by judicious analysis) and treatment subscribe to the smooth operation of a process.

Generally, process wastes (emissions) are categorized as gaseous, liquid, and solid. This does not usually include waste from accidental spillage of an organic chemical feedstock or from a product. Creating standards for the strategic and sound management of chemicals is essential to reducing the risk of exposure. Nationally and internationally both private and public organizations including the United Nations are working to create globally applied standards for the management of chemicals so that the need for chemicals and the hazardous effects of pollution can be balanced.

5.2 Air Emissions

Air emissions include point and nonpoint sources (Speight and Lee, 2000). Point sources are emissions that exit stacks and flares and, thus, can be monitored and treated. Nonpoint sources are *fugitive emissions* that are difficult to locate and capture. Fugitive emissions occur throughout production facilities and arise from the thousands of valves, pumps, tanks, pressure relief valves, and flanges. While individual leaks are typically small, the sum of all fugitive leaks from any process can be one of the largest emission sources in that process.

The numerous process heaters used in production facilities to heat process streams or to generate steam (boilers) for heating or steam stripping can be potential sources of SO_x, NO_x, CO, particulates, and hydrocarbons emissions. When operating properly and when burning cleaner fuels such as process fuel gas, fuel oil, or natural gas, these emissions are relatively low. If, however, combustion is not complete, or heaters are fired with process fuel pitch or residuals, emissions can be significant. As a result, there has been an increased interest in the application of control to combustion with the main objective to optimize combustor operation, monitor the process, and alleviate instabilities and their severe consequences. As combustion systems have to meet increasingly more demanding air pollution standards, their design and operation becomes more complex. The trend toward reduced emission levels has led to pollutant conversion (Docquier and Candel, 2002; Tecon and Van der Meer, 2008).

The majority of gas streams exiting each process contain varying amounts of process fuel gas, hydrogen sulfide, and ammonia. These streams are collected and sent to the gas treatment and sulfur recovery units to recover the process fuel gas and sulfur though a variety of add-on technologies (Speight and Lee, 2000; Speight, 2014a,b). Emissions from the sulfur recovery unit typically contain some hydrogen sulfide, sulfur oxides, and nitrogen oxides. Other emissions sources from various processes arise from periodic regeneration of catalysts. These processes generate streams that may contain relatively high levels of carbon monoxide, particulates, and volatile organic compounds. Before being discharged to the atmosphere, such off-gas streams may be treated first through a carbon monoxide boiler to burn carbon monoxide and any volatile organic compounds, and then through an electrostatic precipitator or cyclone separator to remove particulates.

Sulfur is removed from a number of process off-gas streams (sour gas) in order to meet the sulfur oxide emissions limits of the Clean Air Act and to recover saleable elemental sulfur. Process off-gas streams, or sour gas, from the coker, catalytic cracking unit, hydrotreating units, and hydroprocessing units can contain high concentrations of hydrogen sulfide mixed with light process fuel gases.

Before elemental sulfur can be recovered, the fuel gases (primarily methane and ethane) need to be separated from the hydrogen sulfide. This is typically accomplished by dissolving the hydrogen sulfide in a chemical solvent. Solvents most commonly used are amines, such as diethanolamine (DEA, $HOCH_2CH_2NHCH_2CH_2OH$). Dry adsorbents such as molecular sieves, activated carbon, iron sponge (Fe_2O_3), and zinc oxide (ZnO) are also used (Speight, 2014a,b). In the amine solvent processes, diethanolamine solution or similar ethanolamine solution is pumped to an absorption tower where the gases are contacted and hydrogen sulfide is dissolved in the solution. The fuel gases are removed for use as fuel in process furnaces in other process operations. The amine-hydrogen sulfide solution is then heated and steam stripped to remove the hydrogen sulfide gas.

Current methods for removing sulfur from the hydrogen sulfide gas streams are typically a combination of two processes in which the primary process is the Claus Process followed by either the Beavon Process or the SCOT Process or the Wellman-Lord Process.

In the Claus process (Speight, 2014a,b, 2016), the hydrogen sulfide, after separation from the gas stream using amine extraction, is fed to the Claus unit, where it is converted in two stages. The first stage is a thermal step: in which the hydrogen sulfide is partially oxidized with air in a reaction furnace at high temperatures (1000–1400°C, 1830–2550°F). Sulfur is formed, but some hydrogen sulfide remains unreacted, and some sulfur dioxide is produced. The second stage is a catalytic stage in which the remaining hydrogen sulfide is reacted with the sulfur dioxide at lower temperatures (200–350°C, 390–660°F) over a catalyst to produce more sulfur. The overall reaction is the conversion of hydrogen sulfide and sulfur dioxide to sulfur and water:

$$2H_2S + SO_2 \rightarrow 3S + 2H_2O$$

The catalyst is necessary to ensure that the components react with reasonable speed, but, unfortunately, the reaction does not always proceed to completion. For this reason, two or three stages are used, with sulfur being removed between the stages. For the analysts, it is valuable to know that carbon disulfide (CS_2) is a by-product from the reaction in the high-temperature furnace. The carbon disulfide can be destroyed catalytically before it enters the catalytic section proper.

Generally, the Claus process may only remove about 90% of the hydrogen sulfide in the gas stream and, as already noted, other processes such as the Beavon process, the SCOT process, or Wellman-Lord processes are often used to further recover sulfur.

In the Beavon process, the hydrogen sulfide in the relatively low concentration gas stream from the Claus process can be almost completely removed by absorption in a quinone solution. The dissolved hydrogen sulfide is oxidized to form a mixture of elemental sulfur and hydroquinone. The solution is injected with air or oxygen to oxidize the hydroquinone back to quinone. The solution is then filtered or centrifuged to remove the sulfur, and the quinone is then reused. The Beavon process is also effective in removing small amounts of sulfur dioxide, carbonyl sulfide, and carbon disulfide that are not affected by the Claus process. These compounds are first converted to hydrogen sulfide at elevated temperatures in a cobalt molybdate catalyst prior to being fed to the Beavon unit. Air emissions from sulfur recovery units will consist of hydrogen sulfide, sulfur oxides, and nitrogen oxides in the process tail gas as well as fugitive emissions and releases from vents.

The SCOT process is also widely used for removing sulfur from the Claus tail gas. The sulfur compounds in the Claus tail gas are converted to hydrogen sulfide by heating and passing it through a cobalt-molybdenum catalyst with the addition of a reducing gas. The gas is then cooled and contacted with a solution of diisopropanolamine (DIPA) that removes all but trace amounts of hydrogen sulfide. The sulfide-rich DIPA is sent to a stripper where hydrogen sulfide gas is removed and sent to the Claus plant. The DIPA is returned to the absorption column.

The Wellman-Lord process is divided into two main stages: (1) absorption and (2) regeneration. In the absorption section, hot flue gases are passed through a prescrubber where ash, hydrogen chloride, hydrogen fluoride, and sulfur trioxide are removed. The gases are then cooled and fed into the absorption tower. A saturated solution of sodium sulfite is then sprayed into the top of the absorber onto the flue gases; the sodium sulfite reacts with the sulfur dioxide forming sodium bisulfite ($NaHSO_3$). The concentrated bisulfate solution is collected and passed to an evaporation system for regeneration. In the regeneration section, sodium bisulfite is converted, using steam, to sodium sulfite that is recycled back to the flue gas. The remaining product, the released sulfur dioxide, is converted to elemental sulfur, sulfuric acid, or liquid sulfur dioxide.

Most process units and equipment are sent into a collection unit, called the blowdown system. Blowdown systems provide for the safe handling and disposal of liquid and gases that are either automatically vented from the process units through pressure relief valves or that are manually drawn from units. Recirculated process streams and cooling water streams are often manually purged to prevent the continued buildup of contaminants in the stream. Part or all of the contents of equipment can also be purged to the blowdown system prior to shut down before normal or emergency shutdowns. Blowdown systems utilize a series of flash drums and condensers to separate the blowdown into its vapor and liquid components. The liquid is typically composed of mixtures of water and hydrocarbons containing sulfides, ammonia, and other contaminants, which are sent to the wastewater treatment plant. The gaseous component typically contains hydrocarbons, hydrogen sulfide, ammonia, mercaptans, solvents, and other constituents and is either discharged directly to the atmosphere or is combusted in a flare. The major air emissions from blowdown systems are hydrocarbons in the case of direct discharge to the atmosphere and sulfur oxides when flared.

5.3 Wastewater

Wastewaters from the organic chemicals industry and the refining industry consist of process water, cooling water, storm water, and sanitary sewage water (Speight, 2005; Speight and Arjoon, 2012). In fact, water used in processing operations accounts for a significant portion of the total wastewater. Process wastewater arises from desalting crude oil, steam-stripping operations, pump gland cooling, product fractionator reflux drum drains, and boiler blowdown. Because process water often comes into direct contact with oil, it is usually highly contaminated. Most cooling water is recycled over and over. Cooling water typically does not come into direct contact with process oil streams and therefore contains less contaminants than process wastewater. However, it may contain some oil contamination due to leaks in the process equipment. Storm water (i.e., surface water runoff) is intermittent and will contain constituents from spills to the surface, leaks in equipment, and any materials that may have collected in drains. Runoff surface water also includes water coming from crude and product storage tank roof drains. Sewage water needs no further explanation of its origins but must be treated as opposed to discharge on to the land or into ponds.

Wastewater is treated in on-site wastewater treatment facilities and then discharged to publicly owned treatment works (POTWs) or discharged to surfaces waters under National Pollution Discharge Elimination System (NPDES) permits. Organic chemicals production facilities typically utilize primary and secondary wastewater treatment.

Primary wastewater treatment consists of the separation of oil, water, and solids in two stages. During the first stage, an API separator, a corrugated plate interceptor, or other separator design are used. Wastewater moves very slowly through the separator allowing free oil to float to the surface and be skimmed off, and solids to settle to the bottom and be scraped off to a sludge collection hopper. The second stage utilizes physical or chemical methods to separate emulsified oils from the wastewater. Physical methods may include the use of a series of settling ponds with a long retention time, or the use of dissolved air flotation (DAF). In DAF, air is bubbled through the wastewater, and both oil and suspended solids are skimmed off the top. Chemicals, such as ferric hydroxide or aluminum hydroxide, can be used to coagulate impurities into a froth or sludge that can be more easily skimmed off the top. Some wastes associated with the primary treatment of wastewater at organic chemicals production facilities may be considered hazardous and include API separator sludge, primary treatment sludge, sludge from other gravitational separation techniques, float from DAF units, and wastes from settling ponds.

After primary treatment, the wastewater can be discharged to a POTW or undergo *secondary treatment* before being discharged directly to surface waters under a NPDES permit. In secondary treatment, microorganisms may consume dissolved oil and other organic pollutants biologically. Biological treatment may require the addition of oxygen through a number of different techniques, including activated sludge units, trickling filters, and rotating biological contactors. Secondary treatment generates biomass waste that is typically treated anaerobically and then dewatered.

Some production facilities employ an additional stage of wastewater treatment called *polishing* to meet discharge limits. The polishing step can involve the use of activated carbon, anthracite coal, or sand to filter out any remaining impurities, such as biomass, silt, trace metals, and other inorganic chemicals, as well as any remaining organic chemicals.

Certain process wastewater streams are treated separately, prior to the wastewater treatment plant, to remove contaminants that would not easily be treated after mixing with other wastewater. One such waste stream is the sour water drained from distillation reflux drums. Sour water contains dissolved hydrogen sulfide and other organic sulfur compounds and ammonia which are stripped in a tower with gas or steam before being discharged to the wastewater treatment plant. Wastewater treatment plants are a significant source of process air emissions and solid wastes. Air releases arise from fugitive emissions from the numerous tanks, ponds, and sewer system drains. Solid wastes are generated in the form of sludge from a number of the treatment units.

Many production facilities unintentionally release, or have unintentionally released in the past, liquid hydrocarbons to groundwater and surface waters. At some production facilities, contaminated groundwater has migrated off-site and resulted in continuous *seeps* to surface waters. While the actual volume of hydrocarbons released in such a manner are relatively small, there is the potential to contaminate large volumes of groundwater and surface water possibly posing a substantial risk to human health and the environment.

5.4 Other Waste

Solid wastes are generated from many of the organic chemcials production processes and from refining processes, organic chemicals handling operations, as well as wastewater treatment (Chapter 4). Both hazardous and nonhazardous wastes are generated, treated, and disposed. Solid wastes in a process are typically in the form of sludge (including sludge from wastewater treatment), spent process catalysts, filter clay, and incinerator ash. Treatment of these wastes includes incineration, land treating off-site, land filling on-site, land filling off-site, chemical fixation, neutralization, and other treatment methods (Speight, 1996; Woodside, 1999; Speight and Lee, 2000; Speight and Arjoon, 2012).

A significant portion of the nonorganic chemicals product outputs of production facilities is transported off-site and sold as by-products. These outputs include sulfur, acetic acid, phosphoric acid, and recovered metals. Metals from catalysts and from the crude oil that have deposited on the catalyst during the production often are recovered by third-party recovery facilities.

Storage tanks are used throughout the refining process to store crude oil and intermediate process feeds for cooling and further processing. Finished organic chemicals are also kept in storage tanks before transport off-site. Storage tank bottoms are mixtures of iron rust from corrosion, sand, water, and emulsified oil and wax, which accumulate at the bottom of tanks. Liquid tank bottoms (primarily water and oil emulsions) are periodically drawn off to prevent their continued build up. Tank bottom liquids and sludge are also removed during periodic cleaning of tanks for inspection. Tank bottoms may contain amounts of tetraethyl or tetramethyl lead (although this is increasingly rare due to the phase out of leaded products), other metals, and phenols. Solids generated from leaded gasoline storage tank bottoms are listed as a hazardous waste.

5.5 Options

Pollution prevention is the responsibility of everyone and preventing pollution may be a new role for production-oriented managers and workers, but their cooperation is crucial. It will be the workers themselves who must make pollution prevention succeed in the workplace.

The best way to reduce pollution is to prevent it in the first place. Some companies have creatively implemented pollution prevention techniques that improve efficiency and increase profits while at the same time minimizing environmental impacts. This can be done in many ways such as reducing material inputs, reengineering processes to reuse by-products, improving management practices, and substituting benign chemicals for toxic ones. Some smaller facilities are able to actually get below regulatory thresholds just by reducing pollutant releases through aggressive pollution prevention policies. Furthermore, it is critical to emphasize that pollution prevention in the chemical industry is process specific and oftentimes constrained by site-specific considerations. As such, it is difficult to generalize about the relative merits of different pollution prevention strategies. The age, size, and purpose of the plant will influence the choice of the most effective pollution prevention strategy. Commodity chemical manufacturers redesign their processes infrequently so that redesign of the reaction process or equipment is unlikely in the short term. Here operational changes are the most feasible response. Specialty chemical manufacturers are making a greater variety of chemicals and have more process and design flexibility. Incorporating changes at the earlier research and development phases may be possible for them.

Several options have been identified that production facilities can undertake to reduce pollution. These include pollution prevention options, recycling options, and waste treatment options. Furthermore, pollution prevention options are often presented in four different categories, viz.: (1) pollution prevention options, (2) waste recycling, and (3) waste treatment. Either one or the other or any combination of the three options may be in operation in any given process.

Pollution prevention options are usually subdivided into four areas: (1) good operating practices, (2) processes modification, (3) feedstock modification, and (4) product reformulation (Lo, 1991). The options described here include only the first three of these categories since product reformulation is not an option that is usually available to the environmental analyst, scientist, or engineer.

5.5.1 Operating Practices

Good operating practices (Table 9.6) prevent waste by better handling of feedstocks and products without making significant modifications to current production technology. If feedstocks are handled appropriately, they are less likely to become wastes inadvertently through spills or outdating. If products are handled appropriately, they can be managed in the most cost-effective manner.

TABLE 9.6 A Selection of Good Operating Practices

- Specify sludge and water content for feedstock
- Minimize carryover to API separator
- Use recycled water for desalter
- Replace desalting with chemical treatment system
- Collect catalyst fines during delivery
- Recover coke fines

For example, a significant portion of process waste arises from oily sludge found in combined process/storm sewers. Segregation of the relatively clean rainwater runoff from the process streams can reduce the quantity of oily sludge generated. Furthermore, there is a much higher potential for recovery of oil from smaller, more concentrated process streams.

Solids released to the process wastewater sewer system can account for a large portion of a process's oily sludge. Solids entering the sewer system (primarily soil particles) become coated with oil and are deposited as oily sludge in the API oil/water separator. Because a typical sludge has a solids content of 5-30% by weight, preventing one pound of solids from entering the sewer system can eliminate several pounds 3-0 pounds of oily sludge.

Methods used to control solids include using a street sweeper on paved areas, paving unpaved areas, planting ground cover on unpaved areas, relining sewers, cleaning solids from ditches and catch basins, and reducing heat exchanger bundle cleaning solids by using antifoulant materials in cooling water. Benzene and other solvents in wastewater can often be treated more easily and effectively at the point at which they are generated rather than at the wastewater treatment plant after it is mixed with other wastewater.

5.5.2   Process Modifications

The organic chemicals industry requires very large, capital-intensive process equipment. Expected lifetimes of process equipment are measured in decades. This limits economic incentives to make capital-intensive process modifications to reduce wastes generation. However, some process modifications (Table 9.7) or process improvement (Table 9.8) reduce waste generation.

The organic chemicals industry has made many improvements in the design and modification of processes and technologies to recover product and unconverted raw materials. In the past, they pursued this strategy to the point that the cost of further recovery could not be justified. Now the costs of end-of-pipe treatment and disposal have made source reduction a good investment. Greater

TABLE 9.7 Options for Process Modifications

• Add coking operations

Certain process hazardous wastes can then be used as coker feedstock, reducing the quantity of sludge for disposal

• Install secondary seals on floating roof tanks

• Where appropriate, replace with fixed roofs to eliminate the collection of rainwater, contamination of crude oil or finished products, and oxidation of crude oil

• Where feasible,
  o Replace clay filtration with hydrotreating
  o Substitute air coolers or electric heaters for water heat exchangers to reduce sludge production
  o Install tank agitators. This can prevent solids from settling out
  o Concentrate similar wastewater streams through a common dewatering system

TABLE 9.8 Process Improvement

• Segregate oily wastes to reduce the quantity of oily sludge generated and increase the potential for oil recovery

• Reuse rinse waters where possible

• Use optimum pressures, temperatures, and mixing ratios

• Sweep or vacuum streets and paved process areas to reduce solids going to sewers

• Use water softeners in cooling water systems to extend the useful life of the water

reductions are possible when process engineers trained in pollution prevention plan to reduce waste at the design stage. For example, although barge loading is not a factor for all production facilities, it is an important emissions source for many facilities. One of the largest sources of volatile organic carbon emissions is the fugitive emissions from loading of tanker barges. These emissions could be reduced by more than 90% by installing a vapor loss control system that consists of vapor recovery or the destruction of the volatile organic carbon emissions in a flare.

Fugitive emissions are one of the largest sources of process hydrocarbon emissions. A leak detection and repair program consists of using a portable detecting instrument to detect leaks during regularly scheduled inspections of valves, flanges, and pump seals. Older process boilers may also be a significant source of emissions of sulfur oxides (SO$_x$), nitrogen oxides (NO$_x$), and particulate matter. It is possible to replace a large number of old boilers with a single new cogeneration plant with emissions controls.

Since storage tanks are one of the largest sources of VOC emissions, a reduction in the number of these tanks can have a significant impact. The need for certain tanks can often be eliminated through improved production planning and more continuous operations. By minimizing the number of storage tanks, tank bottom solids and decanted wastewater may also be reduced. Installing secondary seals on the tanks can significantly reduce the losses from storage tanks containing gasoline and other volatile products.

Solids entering the crude distillation unit are likely to eventually attract more oil and produce additional emulsions and sludge. The amount of solids removed from the desalting unit should, therefore, be maximized. A number of techniques can be used such as: using low shear mixing devices to mix desalter wash water and crude oil, using lower pressure water in the desalter to avoid turbulence, and replacing the water jets used in some production facilities with mud rakes which add less turbulence when removing settled solids.

Purging or blowing down a portion of the cooling water stream to the wastewater treatment system controls the dissolved solids concentration in the recirculating cooling water. Solids in the blowdown eventually create additional sludge in the wastewater treatment plant. However, minimizing the dissolved solids content of the cooling water can lower the amount of cooling tower blowdown. A significant portion of the total dissolved solids in the cooling water can originate in the cooling water makeup stream in the form of naturally occurring calcium carbonates. Such solids can be controlled either by selecting a source of cooling tower makeup water with less dissolved solids or by removing the dissolved solids from the makeup water stream. Common treatment methods include: cold lime softening, reverse osmosis, or electrodialysis.

In many production facilities, using high-pressure water to clean heat exchanger bundles generates and releases water and entrained solids to the process wastewater treatment system. Exchanger solids may then attract oil as they move through the sewer system and may also produce finer solids and stabilized emulsions that are more difficult to remove. Solids can be removed at the heat exchanger cleaning pad by installing concrete overflow weirs around the surface drains or by covering drains with a screen. Other ways to reduce solids generation are by using antifoulants on the heat exchanger bundles to prevent scaling and by cleaning with reusable cleaning chemicals that also allow for the easy removal of oil.

Surfactants entering the process wastewater streams will increase the amount of emulsions and sludge generated. Surfactants can enter the system from a number of sources including: washing unit pads with detergents; treating gasoline with an end point over 200°C (>392°F), thereby producing spent caustics; cleaning tank truck tank interiors; and using soaps and cleaners for miscellaneous tasks. In addition, the overuse and mixing of the organic polymers used to separate oil, water, and solids in the wastewater treatment plant can actually stabilize emulsions. The use of surfactants should be minimized by educating operators, routing surfactant sources to a point downstream of the DAF unit and by using dry cleaning, high pressure water or steam to clean oil surfaces of oil and dirt.

Replacing 55-gallon drums with bulk storage facilities can minimize the chances of leaks and spills. And, just as 55-gallon drums can lead to leaks, underground piping can be a source of undetected releases to the soil and groundwater. Inspecting, repairing or replacing underground piping with surface piping can reduce or eliminate these potential sources.

Finally, open ponds used to cool, settle out solids and store process water can be a significant source of volatile organic carbon emissions. Wastewater from coke cooling and coke volatile organic carbon removal is occasionally cooled in open ponds where volatile organic carbon easily escapes to the atmosphere. In many cases, open ponds can be replaced with closed storage tanks.

5.5.3 Material Substitution Options

Spent conventional degreaser solvents can be reduced or eliminated through substitution with less toxic and/or biodegradable products. In addition, chromate containing wastes can be reduced or eliminated in cooling tower and heat exchanger sludge by replacing chromates with less toxic alternatives such as phosphates.

Using catalysts of a higher quality will lead in increased process efficiency, while the required frequency of catalyst replacement can be reduced. Similarly, the replacement of ceramic catalyst support with activated alumina supports presents the opportunity for recycling the activated alumina supports with the spent alumina catalyst.

5.6 Recycling

Recycling is the use, reuse, or reclamation of a waste after it is generated. At present the organic chemicals industry is focusing on recycling and reuse as the best opportunities for pollution prevention (Table 9.9). Although pollution is reduced more if wastes are prevented in the first place, a next best option for reducing pollution is to treat wastes so that they can be transformed into useful products.

Caustic substances used to absorb and remove hydrogen sulfide and phenol contaminants from intermediate and final product streams can often be recycled. Spent caustics may be saleable to chemical recovery companies if concentrations of phenol or hydrogen sulfide are high enough. Process changes in the process may be needed to raise the concentration of phenols in the caustic to make recovery of the contaminants economical. Caustics containing phenols can also be recycled on-site by reducing the pH of the caustic until the phenols

TABLE 9.9 Options for Recycling

- Use phenols and caustics produced in the refining operations as chemical feeds in other applications

- Use oily waste sludge as feedstock in coking operations

- Regenerate catalysts. Extend useful life. Recover valuable metals from spent catalyst. Possibly use catalyst as a concrete admixture or as a fertilizer

- Maximize slop oil recovery. Agitate sludge with air and steam to recover residual oils

- Regenerate filtration clay. Wash clay with naphtha, dry by steam heating, and feed to a burning kiln for regeneration

- Recover valuable product from oily sludge with solvent extraction

become insoluble thereby allowing physical separation. The caustic can then be treated in the process wastewater system.

Oily sludge can be sent to a coking unit or the crude distillation unit where it becomes part of the process products. Sludge sent to the coker can be injected into the coke drum with the quench water, injected directly into the delayed coker, or injected into the coker blowdown contactor used in separating the quenching products. Use of sludge as a feedstock has increased significantly in recent years and is currently carried out by most production facilities. The quantity of sludge that can be sent to the coker is restricted by coke quality specifications that may limit the amount of sludge solids in the coke. Coking operations can be upgraded, however, to increase the amount of sludge that they can handle.

Significant quantities of catalyst fines are often present around the catalyst hoppers of fluid catalytic cracking reactors and regenerators. Coke fines are often present around the coker unit and coke storage areas. The fines can be collected and recycled before being washed to the sewers or migrating off-site via the wind. Collection techniques include dry sweeping the catalyst and coke fines and sending the solids to be recycled or disposed of as nonhazardous waste. Coke fines can also be recycled for fuel use. Another collection technique involves the use of vacuum ducts in dusty areas (and vacuum hoses for manual collection) that run to a small baghouse for collection.

An issue that always arises relates to the disposal of laboratory sample from any process control or even environmental laboratory that is associated with a process. Samples from such a laboratory can be recycled to the oil recovery system.

5.7   Treatment Options

When pollution prevention and recycling options are not economically viable, pollution can still be reduced by treating wastes so that they are transformed in

TABLE 9.10 Options for Chemicals Waste Reduction

- Segregate process (oily) waste streams from relatively clean rainwater runoff in order to reduce the quantity of oily sludge
- Generated and increased the potential for oil recovery. Significant portion of the process waste comes from oily sludge found in combined process/storm sewers
- Conduct inspection of organic chemicals process systems for leaks. For example, check hoses, pipes, valves, pumps, and seals. Make necessary repairs where appropriate
- Conserve water. Reuse rinse waters if possible. Reduce equipment-cleaning frequency where beneficial in reducing net waste generation
- Use correct pressures, temperatures, and mixing ratios for optimum recovery of product and reduction in waste produced
- Employ street sweeping or vacuuming of paved process areas to reduce solids to the sewers
- Pave runoff areas to reduce transfer of solids to waste systems. Use water softeners in cooling water systems to extend useful cycling time of the water

to less environmentally harmful wastes or can be disposed of in a less environmentally harmful media (Table 9.10). The toxicity and volume of some de-oiled and dewatered sludge can be further reduced through thermal treatment. Thermal sludge treatment units use heat to vaporize the water and volatile components in the feed and leave behind a dry solid residue. The vapors are condensed for separation into the hydrocarbon and water components. Noncondensable vapors are either flared or sent to the amine unit for treatment and use as process fuel gas.

Furthermore, because oily sludge makes up a large portion of process solid wastes, any improvement in the recovery of oil from the sludge can significantly reduce the volume of waste. There are a number of technologies currently in use to mechanically separate oil, water, and solids, including: belt filter presses, recessed chamber pressure filters, rotary vacuum filters, scroll centrifuges, disc centrifuges, shakers, thermal driers, and centrifuge-drier combinations.

Waste material such as tank bottoms from crude oil storage tanks constitute a large percentage of process solid waste and pose a particularly difficult disposal problem due to the presence of heavy metals. Tank bottoms are comprised of heavy hydrocarbons, solids, water, rust, and scale. Minimization of tank bottoms is carried out most cost effectively through careful separation of the oil and water remaining in the tank bottom. Filters and centrifuges can also be used to recover the oil for recycling.

Spent clay from process filters often contains significant amounts of entrained hydrocarbons and, therefore, must be designated as hazardous waste.

Back washing spent clay with water or steam can reduce the hydrocarbon content to levels so that it can be reused or handled as a nonhazardous waste. Another method used to regenerate clay is to wash the clay with naphtha, dry it by steam heating and then feed it to a burning kiln for regeneration. In some cases, clay filtration can be replaced entirely with hydrotreating process options.

Decant oil sludge from the fluidized bed catalytic cracking unit can (and often does) contain significant concentrations of catalyst fines. These fines often prevent the use of decant oil as a feedstock or require treatment which generates an oily catalyst sludge. Catalyst fines in the decant oil can be minimized by using a decant oil catalyst removal system. One system incorporates high voltage electric fields to polarize and capture catalyst particles in the oil. The amount of catalyst fines reaching the decant oil can be minimized by installing high efficiency cyclones in the reactor to shift catalyst fines losses from the decant oil to the regenerator where they can be collected in the electrostatic precipitator.

REFERENCES

Aislabie, J., Saul, D.J., Foght, J.M., 2006. Bioremediation of hydrocarbon-contaminated polar soils. Extremophiles 10, 171–179.

Atlas, R.M., 1977. Stimulated crude oil biodegradation. Crit. Rev. Microbiol. 5, 371–386.

Atlas, R.M., 1991. Bioremediation: using nature's helpers-microbes and enzymes to remedy mankind's pollutants. In: Lyons, T.P., Jacques, K.A. (Eds.), Proceedings Biotechnology in the Feed Industry. Alltech's Thirteenth Annual Symposium. Alltech Technical Publications, Nicholasville, KY, pp. 255–264.

Baker, R.S., 1999. Bioventing systems: a critical review. In: Adriana, D.C., Bollag, J.M., Frankenberger, W.T., Sims, R.C. (Eds.), Bioremediation of Contaminated Soils. American Society of Agronomy, Crop Science Society of America, and Soil Science Society of America, Madison, WI, pp. 595–630. Agronomy Monograph No. 37.

Barker, G.W., Raterman, K.T., Fisher, J.B., Corgan, J.M., Trent, G.L., Brown, D.R., et al., 1995. Assessment of natural hydrocarbon bioremediation at two gas condensate production sites. In: Hinchee, R.E., Wilson, J.T., Downey, D.C. (Eds.), Intrinsic Bioremediation. Battelle Press, Columbus, OH, pp. 181–188.

Brown, K.S., 1995. The green clean: the emerging field of phytoremediation takes root. Bioscience 45, 579–582.

Brown, R.A., Norris, R.D., 1994. The evolution of a technology: hydrogen peroxide in in situ bioremediation. In: Hinchee, R.E., Alleman, B.C., Hoeppel, R.E., Miller, R.N. (Eds.), Hydrocarbon Bioremediation. CRC Press, Boca Raton, FL, pp. 148–162.

Chhatre, S., Purohit, H., Shanker, R., Khanna, P., 1996. Bacterial consortia for crude oil spill remediation. Water Sci. Technol. 34, 187–193.

Compeau, G.C., Mahaffey, W.D., Patras, L., 1991. Full-scale bioremediation of a contaminated soil and water site. In: Sayler, G.S., Fox, R., Blackburn, J.W. (Eds.), Environmental Biotechnology for Waste Treatment. Plenum Press, New York, NY, pp. 91–110.

Cunningham, J.A., Hopkins, G.D., Lebron, C.A., Reinhard, M., 2000. Enhanced anaerobic bioremediation of groundwater contaminated by fuel hydrocarbons at Seal Beach, California. Biodegradation 11, 159–170.

Docquier, N., Candel, S., 2002. Combustion control and sensors: a review. Prog. Energy Combust. Sci. 28, 107–150.

Flathman, P.E., Carson Jr., J.H., Whitenhead, S.J., Khan, K.A., Barnes, D.M., Evans, J.S., 1991. Laboratory evaluation of the utilization of hydrogen peroxide for enhanced biological treatment of crude oil hydrocarbon contaminants in soil. In: Hinchee, R.E., Olfenbuttel, R.F. (Eds.), In Situ Bioreclamation: Applications and Investigations for Hydrocarbon and Contaminated Site Remediation. Butterworth-Heinemann, Stoneham, MA, pp. 125–142.

Frenzel, M., James, P., Burton, S.K., Rowland, S.J., Lappin-Scott, H.M., 2009. Towards bioremediation of toxic unresolved complex mixtures of hydrocarbons: identification of bacteria capable of rapid degradation of alkyltetralins. J. Soils Sediments 9, 129–136.

Gibson, D.T., Sayler, G.S., 1992. Scientific Foundation for Bioremediation: Current Status and Future Needs. American Academy of Microbiology, Washington, DC.

Ginn, J.S., Sims, R.C., Murarka, I.P., 1995. In situ bioremediation (natural attenuation) at a gas plant waste site. In: Hinchee, R.E., Wilson, J.T., Downey, D.C. (Eds.), Intrinsic Bioremediation. Battelle Press, Columbus, OH, pp. 153–162.

Harms, H., Smith, K.E.C., Wick, L.Y., 2010. Problems of hydrophobicity/bioavailability. In: Timmis, K.N. (Ed.), Handbook of Hydrocarbon and Lipid Microbiology. Springer, Berlin, pp. 1439–1450.

King, M.W.G., Barker, J.F., Hamilton, L.K., 1995. Natural attenuation of coal tar organics in groundwater. In: Hinchee, R.E., Wilson, J.T., Downey, D.C. (Eds.), Intrinsic Bioremediation. Battelle Press, Columbus, OH, pp. 171–180.

Korade, D.L., Fulekar, M.H., 2009. Development and evaluation of mycorrhiza for rhizosphere bioremediation. J. Appl. Biosci. 17, 922–929.

Leavitt, M.E., Brown, K.L., 1994. Bioremediation versus bioaugmentation—three case studies. In: Hinchee, R.E., Alleman, B.C., Hoeppel, R.E., Miller, R.N. (Eds.), Hydrocarbon Bioremediation. CRC Press, Inc., Boca Raton, FL, pp. 72–79.

Lee, M.D., Thomas, J.M., Borden, R.C., Bedient, P.B., Ward, C.H., 1988. Biorestoration of aquifers contaminated with organic compounds. CRC Crit. Rev. Environ. Control. 18, 29–89.

Lo, P., 1991. Waste Water and Solid Waste Management. County Sanitation District of Los Angeles County, Whittier, CA.

Lu, C.J., 1994. Effects of hydrogen peroxide on the in situ biodegradation of organic chemicals in a simulated groundwater system. In: Hinchee, R.E., Alleman, B.C., Hoeppel, R.E., Miller, R.N. (Eds.), Hydrocarbon Bioremediation. CRC Press, Inc., Boca Raton, FA, pp. 140–147.

Lu, C.J., Hwang, M.C., 1992. Effects of hydrogen peroxide on the in situ biodegradation of chlorinated phenols in groundwater. In: Proceedings of the Water Environ. Federation 65th Annual Conference, New Orleans, Louisiana, September 20–24.

Magdalene, O.E., Ufuoma, A., Gloria, O., 2009. Screening of four common nigerian weeds for use in phytoremediation of soil contaminated with spent lubricating oil. African J. Plant Sci. 3 (5), 102–106.

McAllister, P.M., Chiang, C.Y., Salanitro, J.P., Dortch, I.J., Williams, P., 1995. Enhanced aerobic bioremediation of residual hydrocarbon sources. In: Hinchee, R.E., Wilson, J.T., Downey, D.C. (Eds.), Intrinsic Bioremediation. Battelle Press, Columbus, OH, pp. 67–76.

Mishra, S., Jyot, J., Kuhad, R.C., Lal, B., 2001. In situ bioremediation potential of an oily sludge-degrading bacterial consortium. Curr. Microbiol. 43, 328–335.

Mokhatab, S., Poe, W.A., Speight, J.G., 2006. Handbook of Natural Gas Transmission and Processing. Elsevier, Amsterdam.

Nedunuri, K.V., Govundaraju, R.S., Banks, M.K., Schwab, A.P., Chen, Z., 2000. Evaluation of phytoremediation for field scale degradation of total crude oil hydrocarbons. J. Environ. Eng. 126, 483–490.

Nester, E.W., Anderson, D.G., Roberts Jr., C.E., Pearsall, N.N., Nester, M.T., 2001. Microbiology: A Human Perspective, third ed. McGraw-Hill, New York.

Nugroho, A., Effendi, E., Karonta, Y., 2010. Crude oil degradation in soil by thermophilic bacteria with biopile reactor. Makara, Teknologi 14 (1), 43–46.

Osswald, P., Baveye, P., Block, J.C., 1996. Bacterial influence on partitioning rate during the biodegradation of styrene in abiphasic aqueous-organic system. Biodegradation 7, 297–302.

Pardieck, D.L., Bouwer, E.J., Stone, A.T., 1992. Hydrogen peroxide use to increase oxidant capacity for in situ bioremediation of contaminated soils and aquifers: a review. J. Contam. Hydrol. 9, 221–242.

Perfumo, A., Banat, I.M., Marchant, R., Vezzulli, L., 2007. Thermally enhanced approaches for bioremediation of hydrocarbon-contaminated soils. Chemosphere 66, 179–184.

Radwan, S.S., Al-Mailem, D., El-Nemr, I., Salamah, S., 2000. Enhanced remediation of hydrocarbon contaminated desert soil fertilized with organic carbons. Int. Biodeter. Biodegr. 46, 129–132.

Schlegel, H.G., 1977. Aeration without air: oxygen supply by hydrogen peroxide. Biotechnol. Bioeng. 19, 413.

Scragg, A., 1999. Environmental Biotechnology. Pearson Education Limited, Harlow.

Semple, K.T., Reid, B.J., Fermor, T.R., 2001. Impact of composting strategies on the treatment of soils contaminated with organic pollutants. Environ. Pollut. 112, 269–283.

Speight, J.G., 1996. Environmental Technology Handbook. Taylor & Francis, Washington, DC.

Speight, J.G., 2002. Chemical Process and Design Handbook. McGraw-Hill, New York.

Speight, J.G., 2005. Environmental Analysis and Technology for the Refining Industry. John Wiley & Sons Inc., Hoboken, NJ.

Speight, J.G., 2007. Natural Gas: A Basic Handbook. GPC Books Gulf Publishing Company, Houston, TX.

Speight, J.G., 2014a. The Chemistry and Technology of Crude Oil, fifth ed. CRC Press, Taylor & Francis Group, Boca Raton, FL.

Speight, J.G., 2014b. Handbook of Crude Oil Refining. CRC Press, Taylor & Francis Group, Boca Raton, FL.

Speight, J.G., 2016. Deep Shale Oil and Gas. Gulf Professional Publishing, Elsevier, Oxford.

Speight, J.G., Arjoon, K.K., 2012. Bioremediation of Crude Oil and Crude Oil Products. Scrivener Publishing, Salem, MA.

Speight, J.G., Lee, S., 2000. Environmental Technology Handbook, second ed. Taylor & Francis, New York.

Tecon, R., Van der Meer, J.R., 2008. Bacterial biosensors for measuring availability of environmental pollutants. Sensors 8, 4062–4080.

US Army Corps of Engineers, 1999. US Army Manual EM1110-2-1424. www.usace.army.mil/usace-docs/engmanuals/em1110-2-1424/c-8.pdf. Chapter 8.

US EPA, 2006. In situ and ex situ biodegradation technologies for remediation of contaminated sites. In: Office of Research and Development National Risk Management Research Laboratory, United States Environmental Protection Agency, Cincinnati, OH Report No. EPA/625/R-06/015.

Van Eyk, J., 1994. Venting and bioventing for the in situ removal of crude oil from soil. In: Hinchee, R.E., Alleman, B.C., Hoeppel, R.E., Miller, R.N. (Eds.), Hydrocarbon Bioremediation. CRC Press, Boca Raton, FL, pp. 234–251.

Vasudevan, N., Rajaram, P., 2001. Bioremediation of oil sludge-contaminated soil. Environ. Int. 26, 409–411.

Vidali, M., 2001. Bioremediation: an overview. Pure Appl. Chem. 73 (7), 1163–1172.

Wenzel, W., 2009. Rhizosphere processes and management in plant-assisted bioremediation (phytoremediation) of soils. Plant Soil 321 (1–2), 385–408.

Winters, J.C., Williams, J.A., 1969. Microbiological alteration of crude oil in the reservoir. Preprints, division of crude oil chemistry. Am. Chem. Soc. 14 (4), E22–E31.

Woodside, G., 1999. Hazardous Materials and Hazardous Waste Management. John Wiley & Sons Inc., New York.

Zhu, X., Venosa, A.D., Suidan, M.T., 2004. Literature review on the use of commercial bioremediation agents for clean-up of oil contaminated estuarine environments. In: National Risk Management Research Laboratory, Environmental Protection Agency, Cincinnati, OH Report No. EPA/600/R-04/075.

Conversion Factors

1. Area
   1 square centimeter (1 cm^2) = 0.1550 square inches
   1 square meter (1 m^2) = 1.1960 square yards
   1 hectare = 2.4711 acres
   1 square kilometer (1 km^2) = 0.3861 square miles
   1 square inch (1 in.^{2}) = 6.4516 square centimeters
   1 square foot (1 ft^2) = 0.0929 square meters
   1 square yard (1 yd^2) = 0.8361 square meters
   1 acre = 4046.9 square meters
   1 square mile (1 mi^2) = 2.59 square kilometers
2. Concentration Conversions
   1 part per million (1 ppm) = 1 microgram per liter (1 μg/L)
   1 microgram per liter (1 μg/L) = 1 milligram per kilogram (1 mg/kg)
   1 microgram per liter (μg/L) × 6.243 × 10^8 = 1 lb per cubic foot (1 lb/ft^3)
   1 microgram per liter (1 μg/L) × 10^{-3} = 1 milligram per liter (1 mg/L)
   1 milligram per liter (1 mg/L) × 6.243 × 10^5 = 1 pound per cubic foot (1 lb/ft^3)
   I gram mole per cubic meter (1 g mol/m^3) × 6.243 × 10^5 = 1 pound per cubic foot (1 lb/ft^3)
   10,000 ppm = 1% w/w
   1 ppm hydrocarbon in soil × 0.002 = 1 lb of hydrocarbons per ton of contaminated soil
3. Nutrient Conversion Factor
   1 pound, phosphorus × 2.3 (1 lb P × 2.3) = 1 pound, phosphorous pentoxide (1 lb P$_2$O$_5$)
   1 pound, potassium × 1.2 (1 lb K × 1.2) = 1 pound, potassium oxide (1 lb K$_2$O)
4. Temperature Conversions
   °F = (°C × 1.8) + 32
   °C = (°F − 32)/1.8
   (°F − 32) × 0.555 = °C
   Absolute zero = −273.15°C
   Absolute zero = −459.67°F
5. Sludge Conversions
   1700 lbs wet sludge = 1 yd^3 wet sludge
   1 yd^3 sludge = wet tons/0.85

Wet tons sludge $\times\, 240 =$ gallons sludge

1 wet ton sludge $\times\, \%$ dry solids/100 $= 1$ dry ton of sludge

6. Various Constants

Atomic mass	$mu = 1.6605402 \times 10^{-27}$
Avogadro's number	$N = 6.0221367 \times 10^{23}\ \text{mol}^{-1}$
Boltzmann's constant	$k = 1.380658 \times 10^{-23}\ \text{J K}^{-1}$
Elementary charge	$e = 1.60217733 \times 10^{-19}\ \text{C}$
Faraday's constant	$F = 9.6485309 \times 104\ \text{C} \cdot \text{mol}^{-1}$
Gas (molar) constant	$R = k \cdot N \sim 8.314510\ \text{J} \cdot \text{mol}^{-1} \cdot \text{K}^{-1}$
	$= 0.08205783\ \text{L atm mol}^{-1}\ \text{K}^{-1}$
Gravitational acceleration	$g = 9.80665\ \text{m s}^{-2}$
Molar volume of an ideal gas at	$V_{\text{ideal gas}} = 24.465\ \text{L mol}^{-1}$
1 atm and 25°C	
Planck's constant	$h = 6.6260755 \times 10^{-34}\ \text{J s}$
Zero, Celsius scale	$0°C = 273.15°K$

7. Volume Conversion

Barrels (petroleum, US) to Cu feet multiply by 5.6146

Barrels (petroleum, US) to Gallons (US) multiply by 42

Barrels (petroleum, US) to Liters multiply by 158.98

Barrels (US, liq.) to Cu feet multiply by 4.2109

Barrels (US, liq.) to Cu inches multiply by 7.2765×103

Barrels (US, liq.) to Cu meters multiply by 0.1192

Barrels (US, liq.) to Gallons multiply by (US, liq.) 31.5

Barrels (US, liq.) to Liters multiply by 119.24

Cubic centimeters to Cu feet multiply by 3.5315×10^{-5}

Cubic centimeters to Cu inches multiply by 0.06102

Cubic centimeters to Cu meters multiply by 1.0×10^{-6}

Cubic centimeters to Cu yards multiply by 1.308×10^{-6}

Cubic centimeters to Gallons (US liq.) multiply by 2.642×10^{-4}

Cubic centimeters to Quarts (US liq.) multiply by 1.0567×10^{-3}

Cubic feet to Cu centimeters multiply by 2.8317×10^{4}

Cubic feet to Cu meters multiply by 0.028317

Cubic feet to Gallons (US liq.) multiply by 7.4805

Cubic feet to Liters multiply by 28.317

Cubic inches to Cu cm multiply by 16.387

Cubic inches to Cu feet multiply by 5.787×10^{-4}

Cubic inches to Cu meters multiply by 1.6387×10^{-5}

Cubic inches to Cu yards multiply by 2.1433×10^{-5}

Cubic inches to Gallons (US liq.) multiply by 4.329×10^{-3}

Cubic inches to Liters multiply by 0.01639

Cubic inches to Quarts (US liq.) multiply by 0.01732

Cubic meters to Barrels (US liq.) multiply by 8.3864

Cubic meters to Cu cm multiply by 1.0×10^{6}

Cubic meters to Cu feet multiply by 35.315

Cubic meters to Cu inches multiply by 6.1024×10^4
Cubic meters to Cu yards multiply by 1.308
Cubic meters to Gallons (US liq.) multiply by 264.17
Cubic meters to Liters multiply by 1000
Cubic yards to Bushels (Brit.) multiply by 21.022
Cubic yards to Bushels (US) multiply by 21.696
Cubic yards to Cu cm multiply by 7.6455×105
Cubic yards to Cu feet multiply by 27
Cubic yards to Cu inches multiply by 4.6656×10^4
Cubic yards to Cu meters multiply by 0.76455
Cubic yards to US Gallons (liquid) multiply by 201.97
Cubic yards to Liters multiply by 764.55
Cubic yards to Quarts multiply by 672.71
Cubic yards to Quarts multiply by 694.28
Cubic yards to Quarts multiply by 807.90
Gallons (US liq.) to Barrels (US liq.) multiply by 0.03175
Gallons (US liq.) to Barrels (petroleum, US) multiply by 0.02381
Gallons (US liq.) to Bushels (US) multiply by 0.10742
Gallons (US liq.) to Cu centimeters multiply by 3.7854×10^3
Gallons (US liq.) to Cu feet multiply by 0.13368
Gallons (US liq.) to Cu inches multiply by 231
Gallons (US liq.) to Cu meters multiply by 3.7854×10^{-3}
Gallons (US liq.) to Cu yards multiply by 4.951×10^{-3}
Gallons (US liq.) to Gallons (wine) multiply by 1.0
Gallons (US liq.) to Liters multiply by 3.7854
Gallons (US liq.) to Ounces (US fluid) multiply by 128.0
Gallons (US liq.) to Pints (US liq.) multiply by 8.0
Gallons (US liq.) to Quarts (US liq.) multiply by 4.0
Liters to Cu centimeters multiply by 1000
Liters to Cu feet multiply by 0.035315
Liters to Cu inches multiply by 61.024
Liters to Cu meters multiply by 0.001
Liters to Gallons (US liq.) multiply by 0.2642
Liters to Ounces (US fluid) multiply by 33.814

8. Weight Conversion
   1 ounce (1 ounce) = 28.3495 grams (18.2495 g)
   1 pound (1 lb) = 0.454 kilogram
   1 pound (1 lb) = 454 grams (454 g)
   1 kilogram (1 kg) = 2.20462 pounds (2.20462 lb)
   1 stone (English) = 14 pounds (14 lb)
   1 ton (US; 1 short ton) = 2000 lbs
   1 ton (English; 1 long ton) = 2240 lbs
   1 metric ton = 2204.62262 pounds
   1 ton = 2204.62262 pounds

9. Other Approximations

14.7 pounds per square inch (14.7 psi) – 1 atmosphere (1 atm)

1 kilopascal (kPa) $\times 9.8692 \times 10^{-3} = 14.7$ pounds per square inch (14.7 psi)

1 yd^3 = 27 ft^3

1 US gallon of water = 8.34 lbs

1 imperial gallon of water – 10 lbs

1 ft^3 = 7.5 gallon = 1728 cubic inches = 62.5 lbs.

1 yd^3 = 0.765 m^3

1 acre-inch of liquid = 27,150 gallons = 3.630 ft^3

1-foot depth in 1 acre (in-situ) = 1613 $\times$ (20–25% excavation factor) = ~2000 yd^3

1 yd^3 (clayey soils-excavated) = 1.1–1.2 tons (US)

1 yd^3 (sandy soils-excavated) = 1.2–1.3 tons (US)

Pressure of a column of water in psi = height of the column in feet by 0.434.

Appendix

TABLE A1 Uses of Select Organic Chemicals

Chemicals (in Alphabetical Order)	Structure and Properties	Typical Examples of Use
Acrylic materials: polymethyl methacrylate, man-made organic polymers	Large organic molecules formed by polymerizing unsaturated molecules (alkenes with the C=C double bond) such as methyl acrylate (methyl propanoate), methyl methyl acrylate (methyl 2-methyl propanoate), acrylonitrile, and acrylic.	Used to make tough flexible plastics or sticky resins that can solidify in context, e.g., paints, molded bone substitute. These resins and plastics have a huge range of uses. Poly(methyl acrylate) is used in emulsion form in textile and leather finishes, lacquers, paints, adhesives, and safety glass layers (can replace glass in many situations). Poly (methyl methacrylate) (pmma), is a clear plastic material (Perspex) which can be injection molded and extruded into a variety of shapes.
Alcohols (compounds)	Organic compounds of C, H, and O atoms forming a homologous series containing the hydroxyl functional group —OH; colorless molecular liquids at room temperature.	Uses of alcohols: wide range of uses, e.g., Fuels, solvents, esters, and starting molecule to make other molecules.
Alkanes	Covalent saturated hydrocarbon molecules consisting of combinations of carbon and hydrogen atoms. Flammable, smelly, colorless gases or liquids	Uses of alkanes: widely used as fuels from natural gas, petrol, central heating oil, paraffin/candle wax.

Continued

TABLE A1 Uses of Select Organic Chemicals—cont'd

Chemicals (in Alphabetical Order)	Structure and Properties	Typical Examples of Use
	or white waxy solids of little odor, depending on the value of *n*.	
Alkenes	They are small unsaturated covalent hydrocarbon molecules of carbon and hydrogen atoms; C=C double bond.	Uses of alkenes: alkenes are not used directly for anything but they are readily converted to other very useful organic molecules; can be polymerized to make useful polymer-plastic materials.
Antioxidants (usually organic compounds, naturally occurring and man-made)	These are usually organic molecules with quite variable molecular structure. They slow down oxidation rates in auto-oxidizable substances by removing highly reactive species like free radicals.	Uses of antioxidants: they are added to protect/preserve foods, particularly those containing fat. They are used to reduce the aging/deterioration of rubber and plastics. The body needs antioxidants to reduce the potential harm of free radicals. Vitamins c and e function as antioxidants and most fruits and vegetables contain antioxidants.
Antiseptics	Usually organic molecules, often based on phenol; often formulations of chlorinated phenol molecules.	Antiseptics kill microorganisms but are safe enough to use on the skin; not as strong as a disinfectant for surgical procedures; can be used safely on skin and other tissue cells.
Bakelite phenol-formaldehyde resin compounds	Hard brittle plastics or thick resins made from phenol and formaldehyde; solid has a cross-linked polymer structure which is an excellent electrical insulator and quite heat resistance; thermosetting polymer.	Uses of bakelite: these phenol-formaldehyde plastics/resins have a wide range of uses, e.g., Electrical fittings, saucepan handles, ... (but now replaced by pvc and poly(propene)).

TABLE A1 Uses of Select Organic Chemicals—cont'd

Chemicals (in Alphabetical Order)	Structure and Properties	Typical Examples of Use
Biofuels	Biofuels are organic fuel molecules made from some naturally grown crop.	Uses of biofuels: sugar beet to produce sugar which is fermented into ethanol; rape seed oil can be made into biodiesel for agricultural vehicles.
Butane (compound)	A colorless pungent petrol like smelling hydrocarbon gas belonging to the homologous series of alkanes.	Uses of butane: liquefied under pressure and stored in thick steel fuel gas tanks.
Carbohydrates (sugars and starches)	Covalent molecules of carbon, hydrogen, and oxygen; usually contain 6 or 12 carbon atoms, e.g., glucose, fructose, sucrose; starches are natural polymers where n is a very large number of repeating units.	Uses of carbohydrates: sugars are used in food preparation as a sweetening agent and provider of energy (high on calories) and starches like cornflower are used as thickening agents.
Carboxylic acids	Covalent molecules belonging to a homologous series of organic molecules.	Uses of carboxylic acids: usually converted into a more useful chemical form—combine with alcohols to form esters used in perfumes-fragrances and food flavorings; the analgesic/pain killer aspirin is a carboxylic acid.
Chloroethylene	Colorless gas; used to produce poly (chloroethylene).	Uses: to make poly (chloroethylene)= PVC=polyvinyl chloride; useful thermoplastic.
Citric acid (organic compound)	A naturally occurring carboxylic acid.	Variety of uses in sherbet powders; used in baking powders containing sodium bicarbonate.
Cotton	A natural organic polymer fiber.	Uses of cotton: textile industry for sheets and clothing, etc.

Continued

TABLE A1 Uses of Select Organic Chemicals—cont'd

Chemicals (in Alphabetical Order)	Structure and Properties	Typical Examples of Use
Detergents	Detergents are usually organic molecules with both a hydrophilic and a hydrophobic group.	Used to remove dirt from the surface of materials in combination with water.
Drugs and medicines	Externally administered substance which modifies or affects chemical reactions in the body.	Examples are: aspirin, ibuprofen, paracetamol, codeine.
Dyes	The original dyes came from plant or animal materials; most dyes are now synthesized organic molecules.	Used for dyeing fabrics and coloring plastics.
Emulsifiers	An emulsion is usually one liquid/solid dispersed (but not dissolved) in another liquid (often water) and an emulsifying agent inhibits the separation of the two main components.	Emulsion paints can be oil or water based and the pigment dispersed in the mixture; also used for cosmetic foundation creams and brushless shaving creams are oil-in-water emulsions, cold creams, and cleansing.
Epoxy resin	Polyethers (have a C—O—C linkage) formed by condensing together such as, e.g., 3-chloro-1,2-epoxypropane (epichlorhydrin) with polyols like bisphenol.	Used as adhesives, coatings and in composite materials; resins are thermosetting to tough adhesive materials which are chemical resistant and electrical insulating.
Esters	Prepared by reaction of a carboxylic acid (or derivative) and an alcohol.	Used in the cosmetics industry, e.g., perfumes-fragrances, enhancing the smell of household products, "air fresheners," food additives; also useful organic solvents; used as plasticizers.
Ethane	Colorless gas.	Used as a fuel gas; can be thermally decomposed (cracked) to make ethylene; flammable/explosive with air.

TABLE A1 Uses of Select Organic Chemicals—cont'd

Chemicals (in Alphabetical Order)	Structure and Properties	Typical Examples of Use
Ethanoic acid (acetic acid)	Strong smelling colorless liquid when pure.	Used in the food industry and food preparation, e.g., vinegar, pickling vegetables and eggs; combines with alcohols to produce acetate esters; used to manufacture cellulose ethanoate used for artificial fibers in the textile industry.
Ethanol	Highly toxic colorless liquid; in aqueous solution, the more concentrated, the more dangerous!	Used in alcoholic drinks and beverages, a solvent, a fuel such as methylated spirits; starting point for the manufacture of other organic chemicals, e.g., oxidized to acetaldehyde acetic acid which in turn converted to other useful products.
Ethylene	An unsaturated (C=C bond) organic covalent molecule belonging to the homologous series of hydrocarbons called alkenes.; a colorless gas.	Uses of ethene: it doesn't really have any uses as ethene itself but it is the starting molecule in the manufacture of a wide range of products. It is readily polymerized to make the plastic polyethylene (polythene); can be reacted with water to make ethanol (C_2H_5OH)
Fats	Saturated esters of glycerol and long chain fatty/ carboxylic acids.	Uses of fats: hydrogenated fats like margarine used in the food industry and in the home.
Fatty acids	See carboxylic acids.	Uses of fatty acids: see carboxylic acids.
Fibers	Apart from carbon fibers and glass fibers, they are usually long chain organic polymer molecules.	Glass fibers are used in insulation and fiber optics, carbon fiber composites are used in sports equipment; nylon fibers

Continued

TABLE A1 Uses of Select Organic Chemicals—cont'd

Chemicals (in Alphabetical Order)	Structure and Properties	Typical Examples of Use
		are in the clothing-textile industry.
Flavorings-flavor enhancers	Organic molecules (such as esters) from plant or animal extracts.	Used to enhance the appeal of food.
Fungicides	Can have a wide range of chemical structures; often compounds of sulfur, tin, mercury, nickel, and copper.	Used to inhibit fungal growth which attacks wood or plastic surfaces.
Glucose	A member of the carbohydrate family.	Widely used in the confectionery and food industry, e.g., sweetener in food, chocolate, etc. It can be fermented by yeast to make ethanol.
Herbicides	A range of chemical structures, though most are synthetic organic compounds.	Used as weed killers or growth regulators and are usually selective in their effect on plants.
Methane	A colorless hydrocarbon gas.	Used as a fuel (natural gas) and is reacted with chlorine to make chloromethane derivatives; also used as a source of hydrogen for the synthesis of ammonia.
Methanol	Colorless toxic liquid that causes blindness.	Can be oxidized to formaldehyde (methanol) which is used to make thermosetting polymers; a useful solvent.
Nylon	A polyamide class of synthetic fibers and thermo-softening plastics.	Used as: a synthetic fiber for the textile industry; fibers are strong and are used for fishing lines and climbing ropes.
Pesticides-insecticides	A wide range of chemical structures, some natural products but most are	Designed to kill insects by poisoning.

TABLE A1 Uses of Select Organic Chemicals—cont'd

Chemicals (in Alphabetical Order)	Structure and Properties	Typical Examples of Use
	synthetic organic molecules.	
Polyesters	Polymers made by condensing together a diol and a dicarboxylic acid; usually manufactured as fibers.	Use in the textile industry for clothing, curtain materials, fishing lines, parachutes, sleeping bags, and as Lycra for sports clothing.
Poly(ethylene)	Manufactured by polymerization of ethylene.	Used for plastic bags, buckets, bowls, and bottles.
Poly(propylene)	Manufactured by polymerization of propylene.	Used for making crates, fibers, and ropes.
Polystyrene	Manufactured by polymerization of styrene.	Used in packaging products, heat insulation material, toys, and models.
Poly (tetrafluoroethylene)	Manufactured by polymerization of tetrafluoroethylene.	Uses include coating for nonstick pans. Electrical fittings.
Polyurethane	Condensation polymers made from polyhydroxy compounds and polyisocyanate derivatives.	Uses of polyurethane: mainly used in foam form, e.g., Cushion fillers, thermal insulation.
Propane	A colorless hydrocarbon.	Used as bottled fuel gas, liquefies under the high pressure inside the steel cylinder.
Propylene	A colorless gas.	Used to make the plastic-polymer poly(propylene).
Vegetable oils	Unsaturated oils.	Used in cooking, conversion to margarine, biofuel, soap.

TABLE A2 The F-Code List of Hazardous Wastes

Solvent waste

F001	These spent halogenated solvents used in degreasing; spent solvent mixtures used in degreasing containing, before use, a total of 10% or more by volume of these solvents or the solvents listed in F002, F004, or F005, and still bottoms from the reclamation of these spent solvent and spent solvent mixtures used in degreasing. • Carbon tetrachloride • Chlorinated fluorocarbons • Methylene chloride • Tetrachloroethylene, also called perchloroethylene • 1,1,1-Trichloroethane • Trichloroethylene, also called TCE
F002	These spent halogenated solvents; spent solvent mixtures containing, before use, a total of 10% or more by volume of these solvents or the solvents listed in F001, F004, or F005, and still bottoms from the reclamation of these spent solvent and spent solvent mixtures. • Chlorobenzene • Methylene chloride • *ortho*-Dichlorobenzene • Tetrachloroethylene, also called "perchloroethylene" • 1,1,1-Trichloroethane • 1,1,2-Trichloroethane • Trichloroethylene, also called "TCE" • Trichlorofluoromethane • 1,1,2-Trichloro-1,2,2-trifluoroethane
F003	These spent nonhalogenated solvents; spent solvent mixtures containing, before use, either only these nonhalogenated solvents, or one or more of these nonhalogenated solvents and a total of 10% or more by volume of the solvents listed in F001, F002, F004, or F005, and still bottoms from the reclamation of these spent solvent and spent solvent mixtures. • Acetone • Cyclohexane • Ethyl acetate • Ethyl benzene • Ethyl ether • Methanol • Methyl isobutyl ketone • *n*-Butyl alcohol • Xylene
F004	These spent nonhalogenated solvents; spent solvent mixtures containing, before use, a total of 10% or more by volume of these solvents or the solvents listed in F001, F002, or F005, and still bottoms from the reclamation of these spent solvent and spent solvent mixtures. • Cresol derivatives and cresylic acid derivatives • Nitrobenzene

TABLE A2 The F-Code List of Hazardous Wastes—cont'd

F005	These spent nonhalogenated solvents; spent solvent mixtures containing, before use, a total of 10% or more by volume of these solvents or the solvents listed in F001, F002, or F004, and still bottoms from the reclamation of these spent solvent and spent solvent mixtures. • Benzene • Carbon disulfide • 2-Ethoxyethanol • Iso-butanol • Methyl ethyl ketone, also called "MEK" • 2-Nitropropane • Pyridine • Toluene

Metal treating waste

F006	All wastewater treatment sludge products from electroplating operations except those from these processes. However, these sludge products may still be hazardous for a hazardous waste characteristic. • Sulfuric acid anodizing of aluminum • Tin plating of carbon steel • Zinc plating (segregated basis) on carbon steel • Aluminum or zinc aluminum plating on carbon steel • Cleaning/stripping associated with tin, zinc, and aluminum plating on carbon steel • Chemical etching and milling of aluminum
F007	Spent cyanide plating bath solutions from electroplating operations.
F008	Plating bath sludge products from the bottom of plating baths from electroplating operations where cyanides are used in the process.
F009	Spent stripping and cleaning bath solutions from electroplating operations where cyanides are used in the process. Sludge formed in electroplating stripping and cleaning bath solution tanks where cyanides are used in the process is also included.
F010	Quenching bath residues from oil baths from metal heat-treating operations where cyanides are used in the process.
F011	Spent cyanide solutions from salt bath pot cleaning from metal heat-treating operations.
F012	Quenching waste water treatment sludge from metal heat-treating operations where cyanides are used in the process.
F019	Wastewater treatment sludge from the chemical conversion coating of aluminum except from zirconium phosphating in aluminum can washing when such phosphating is an exclusive conversion coating process.

Manufacturing and processing wastes

F020	Wastes (except wastewater and spent carbon from hydrogen chloride purification) from the production or manufacturing use (as a reactant,

Continued

TABLE A2 The F-Code List of Hazardous Wastes—cont'd

	chemical intermediate, or component in a formulating process) of trichlorophenol or tetrachlorophenol, or of intermediates used to produce their pesticide derivatives.
F021	Wastes (except wastewater and spent carbon from hydrogen chloride purification) from the production or manufacturing use (as a reactant, chemical intermediate, or component in a formulating process) of pentachlorophenol, or of intermediates used to produce its derivatives.
F022	Wastes (except wastewater and spent carbon from hydrogen chloride purification) from the manufacturing use (as a reactant, chemical intermediate, or component in a formulating process) of tetra-, penta-, or hexachlorobenzene derivatives under alkaline conditions.
F023	Wastes (except wastewater and spent carbon from hydrogen chloride purification) from the production of materials on equipment previously used for the production or manufacturing use (as a reactant, chemical intermediate, or component in a formulating process) of tri- and tetrachlorophenol derivatives.
F024	Process wastes from the production of chlorinated aliphatic hydrocarbons with carbon chain lengths from one through five by free radical catalyzed processes, with any amount and position of chlorine substitution. Process wastes include but are not limited to, distillation residues, heavy ends, tars, and reactor clean-out wastes, but do not include F025 wastes.
F025	Condensed light ends, spent filters and filter aids, and spent desiccant wastes from the production of chlorinated aliphatic hydrocarbons with carbon chain lengths from one through five by free radical catalyzed processes, with any amount and position of chlorine substitution.
F026	Wastes (except wastewater and spent carbon from hydrogen chloride purification) from the production of materials on equipment previously used for the manufacturing use (as a reactant, chemical intermediate, or component in a formulating process) of tetra-, penta-, or hexachlorobenzene under alkaline conditions.
Discarded unused products	
F027	Discarded unused formulations containing tri-, tetra-, or pentachlorophenol or discarded unused formulations containing compounds derived from these chlorophenol derivatives.
Contaminated soil treatment residues	
F028	Residues resulting from the incineration or thermal treatment of soil contaminated with hazardous waste codes F020, F021, F022, F023, F026, and F027.
Wood preserving wastes	
F032	Wastewaters (except those that have not come into contact with process contaminants), process residuals, preservative drippage, and spent

TABLE A2 The F-Code List of Hazardous Wastes—cont'd

	formulations from wood preserving processes generated at plants that currently use or have previously used chlorophenol formulations.
F034	Wastewaters (except those that have not come into contact with process contaminants), process residuals, preservative drippage, and spent formulations from wood preserving processes generated at plants that use creosote formulations.
F035	Wastewaters (except those that have not come into contact with process contaminants), process residuals, preservative drippage, and spent formulations from wood preserving processes generated at plants that use inorganic preservatives containing arsenic or chromium.

Petroleum refinery wastes

F037	Petroleum refinery primary oil/water/solids separation sludge—any sludge generated from the gravitational separation of oil/water/solids during the storage or treatment of process wastewaters and oily cooling wastewaters from petroleum refineries. Such types of sludge include, but not limited to, those generated in oil/water/solids separators; tanks and impoundments; ditches and other conveyances; sumps; and storm water units receiving dry weather flow.
F038	Petroleum refinery secondary (emulsified) oil/water/solids separation sludge—any sludge and/or float generated from the physical and/or chemical separation of oil/water/solids in process wastewaters and oily cooling wastewaters from petroleum refineries. Such wastes include, but are not limited to, all sludge and floats generated in: induced air flotation (IAF) units, tanks and impoundments, and all sludge generated in DAF units.

Landfill leachate

F039	Leachate (liquids that have percolated through land disposed wastes) resulting from the disposal of more than one restricted waste listed as a F-, K-, P-, or U hazardous waste. Leachate resulting from the disposal of one or more hazardous wastes bearing the following waste codes which is not mixed with any other hazardous wastes retains its original codes and is not F039: F020, F021, F022, F026, F027, and F028.

TABLE A3 The K-Code List of Hazardous Wastes

Waste Type and Number	Hazardous Waste
Wood preservation waste	
K001	Bottom sediment sludge from the treatment of wastewaters from wood preserving processes that use creosote and/or pentachlorophenol.

Continued

TABLE A3 The K-Code List of Hazardous Wastes—cont'd

Waste Type and Number	Hazardous Waste
Inorganic pigments	
K002	Wastewater treatment sludge from the production of chrome yellow and orange pigments.
K003	Wastewater treatment sludge from the production of molybdate orange pigments.
K004	Wastewater treatment sludge from the production of zinc yellow pigments.
K005	Wastewater treatment sludge from the production of chrome green pigments.
K006	Wastewater treatment sludge from the production of chrome oxide green pigments (anhydrous and hydrated).
K007	Wastewater treatment sludge from the production of iron blue pigments.
K008	Oven residue from the production of chrome oxide green pigments.
Organic chemical waste	
K009	Distillation bottoms from the production of acetaldehyde from ethylene.
K010	Distillation side cuts from the production of acetaldehyde from ethylene.
K011	Bottom stream from the wastewater stripper in the production of acrylonitrile.
K013	Bottom stream from the acetonitrile column in the production of acrylonitrile.
K014	Bottoms from the acetonitrile purification column in the production of acrylonitrile.
K015	Still bottoms from the distillation of benzyl chloride.
K016	Heavy ends or distillation residues from the production of carbon tetrachloride.
K017	Heavy ends (still bottoms) from the purification column in the production of epichlorohydrin.
K018	Heavy ends from the fractionation column in ethyl chloride production.
K019	Heavy ends from the distillation of ethylene dichloride in ethylene dichloride production.

TABLE A3 The K-Code List of Hazardous Wastes—cont'd

Waste Type and Number	Hazardous Waste
K020	Heavy ends from the distillation of vinyl chloride in vinyl chloride monomer production.
K021	Aqueous spent antimony catalyst waste from production of fluoromethanes.
K022	Distillation bottom tars from the production of phenol/acetone from cumene.
K02	Distillation light ends from the production of phthalic anhydride from naphthalene.
K024	Distillation bottoms from the production of phthalic anhydride from naphthalene.
K093	Distillation light ends from the production of phthalic anhydride from *ortho*-xylene.
K094	Distillation bottoms from the production of phthalic anhydride from *ortho*-xylene.
K025	Distillation bottoms from the production of nitrobenzene by the nitration of benzene.
K026	Stripping still tails from the production of methyl ethyl pyridines.
K027	Centrifuge and distillation residues from toluene diisocyanate production.
K028	Spent catalyst from the hydrochlorinator reactor in the production of 1,1,1-trichloroethane.
K029	Waste from the product steam stripper in the production of 1,1,1-trichloroethane.
K095	Distillation bottoms from the production of 1,1,1-trichloroethane.
K096	Heavy ends from the heavy ends column from the production of 1,1,1-trichloroethane.
K030	Column bottoms or heavy ends from the combined production of trichloroethylene and perchloroethylene.
K083	Distillation bottoms from aniline production.
K013	Process residues from aniline extraction from the production of aniline.
K104	Combined wastewater streams generated from nitrobenzene/aniline production.

Continued

TABLE A3 The K-Code List of Hazardous Wastes—cont'd

Waste Type and Number	Hazardous Waste
K085	Distillation or fractionation column bottoms from the production of chlorobenzenes.
K105	Separated aqueous stream from the reactor product washing step in the production of chlorobenzenes.
K107	Column bottoms from product separation from the production of 1,1-dimethylhydrazine (UDMH) from carboxylic acid hydrazines.
K108	Condensed column overheads from product separation and condensed reactor vent gases from the production of 1,1-dimethylhydrazine (UDMH) from carboxylic acid hydrazides.
K109	Spent filter cartridges from product purification from the production of 1,1-dimethylhydrazine (UDMH) from carboxylic acid hydrazides.
K110	Condensed column overheads from intermediate separation from the production of 1,1-dimethylhydrazine (UDMH) from carboxylic acid hydrazides.
K111	Product washwaters from the production of dinitrotoluene via nitration of toluene.
K112	Reaction by-product water from the drying column in the production of toluenediamine via hydrogenation of dinitrotoluene.
K113	Condensed liquid light ends from the purification of toluenediamine in the production of toluenediamine via hydrogenation of dinitrotoluene.
K114	Vicinals from the purification of toluenediamine in the production of toluenediamine via hydrogenation of dinitrotoluene.
K115	Heavy ends from the purification of toluenediamine in the production of toluenediamine via hydrogenation of dinitrotoluene.
K116	Organic condensate from the solvent recovery column in the production of toluene diisocyanate via phosgenation of toluenediamine.
K117	Wastewater from the reactor vent gas scrubber in the production of ethylene dibromide via bromination of ethylene.
K118	Spent absorbent solids from purification of ethylene dibromide in the production of ethylene dibromide via bromination of ethylene.

TABLE A3 The K-Code List of Hazardous Wastes—cont'd

Waste Type and Number	Hazardous Waste
K136	Still bottoms from the purification of ethylene dibromide in the production of ethylene dibromide via bromination of ethylene.
K149	Distillation bottoms from the production of alpha- (or methyl-) chlorinated toluenes, ring-chlorinated toluenes, benzoyl chlorides, and compounds with mixtures of these functional groups (this waste does not include still bottoms from the distillation of benzyl chloride).
K150	Organic residuals, excluding spent carbon adsorbent, from the spent chlorine gas and hydrochloric acid recovery processes associated with the production of alpha-(or methyl-) chlorinated toluenes, ring-chlorinated toluenes, benzoyl chlorides, and compounds with mixtures of these functional groups.
K151	Wastewater treatment sludges, excluding neutralization and biological sludges, generated during the treatment of wastewaters from the production of alpha- (or methyl-) chlorinated toluenes, ring-chlorinated toluenes, benzoyl chlorides, and compounds with mixtures of these functional groups.
K156	Organic waste (including heavy ends, still bottoms, light ends, spent solvents, filtrates, and decantates) from the production of carbamates and carbamoyl oximes (this listing does not apply to wastes generated from the manufacture of 3-iodo-2-propynyl *n*-butylcarbamate).
K157	Wastewaters (including scrubber waters, condenser waters, washwaters, and separation waters) from the production of carbamates and carbamoyl oximes (this listing does not apply to wastes generated from the manufacture of 3-iodo-2-propynyl *n*-butylcarbamate).
K158	Bag house dusts and filter/separation solids from the production of carbamates and carbamoyl oximes (this listing does not apply to wastes generated from the manufacture of 3-iodo-2-propynyl *n*-butylcarbamate).
K159	Organics from the treatment of thiocarbamate wastes.
K161	Purification solids (including filtration, evaporation, and centrifugation solids), bag house dust and floor sweepings from the production of dithiorcarbamate acids and their salts (this listing does not include K125 or K126).
K174	Wastewater treatment sludges from the production of ethylene dichloride or vinyl chloride monomer (including sludges that result from commingled ethylene dichloride or vinyl chloride

Continued

TABLE A3 The K-Code List of Hazardous Wastes—cont'd

Waste Type and Number	Hazardous Waste
	monomer wastewater and other wastewater), unless the sludge meets the following conditions: (1) they are disposed of in a RCRA subtitle C or in a nonhazardous landfill licensed or permitted by the state or federal government; (2) they are not otherwise placed on the land prior to final disposal; and (3) the generator maintains documentation demonstrating that the waste was either disposed of in an on-site landfill or consigned to a transporter or disposal facility that provided a written commitment to dispose of the waste in an off-site landfill. Respondents in any action brought to enforce the requirements of this division must, upon a showing by the government that the respondent managed wastewater treatment sludges from the production of vinyl chloride monomer or ethylene dichloride, demonstrate that they meet the terms of the exclusion set forth above. In doing so, they must provide appropriate documentation (e.g., contracts between the generator and the landfill owner/operator, invoices documenting delivery of waste to landfill, etc.) that the terms of the exclusion were met.
K175	Wastewater treatment sludges from the production of vinyl chloride monomer using mercuric chloride catalyst in an acetylene-based process
Inorganic chemicals	
K071	Brine purification muds from the mercury cell process in chlorine production, where separately prepurified brine is not used.
K073	Chlorinated hydrocarbon waste from the purification step of the diaphragm cell process using graphite anodes in chlorine production.
K106	Wastewater treatment sludge from the mercury cell process in chlorine production.
Pesticide wastes	
K031	By-product salts generated in the production of MSDMA and cacodylic acid.
K032	Wastewater treatment sludge from the production of chlordane.
K033	Wastewater and scrub water from the chlorination of cyclopentadiene in the production of chlordane; filter solids
K034	from the filtration of hexachlorocyclopentadiene in the production of chlordane.
K097	Vacuum stripper discharge from the chlordane chlorinator in the production of chlordane.

TABLE A3 The K-Code List of Hazardous Wastes—cont'd

Waste Type and Number	Hazardous Waste
K035	Wastewater treatment sludge generated in the production of creosote.
K036	Still bottoms from toluene reclamation distillation in the production of disulfoton.
K037	Wastewater treatment sludges from the production of disulfoton.
K038	Wastewater from the washing and stripping of disulfoton.
K039	Filter cake from the filtration of diethylphosphorodithioic acid in the production of phorate.
K040	Wastewater treatment sludge from the production of phorate.
K041	Wastewater treatment sludge from the production of toxaphene.
K098	Untreated process wastewater from the production of toxaphene.
K042	Heavy ends or distillation residues from the distillation of tetrachlorobenzene in the production of 2,4,5-T.
K043	2,6-Dichlorophenol waste from the production of 2,4-D.
K099	Untreated wastewater from the production of 2,4-D.
K123	Process wastewater (including supernates, filtrates, and washwaters) from the production of ethylenebisdithiocarbamic acid and its salt.
K124	Reactor vent scrubber water from the production of ethylenebisdithiocarbamic acid and its salts.
K125	Filtration, evaporation, and centrifugation solids from the production of ethylenebisdithiocarbamic acid and its salts.
K126	Baghouse dust and floor sweepings in milling and packaging operations from the production or formulation of ethylenebisdithiocarbamic acid and its salts.
K131	Wastewater from the reactor and spent sulfuric acid from the acid dryer resulting from the production of methyl bromide.
K132	Spent absorbent and wastewater separator solids from the production of methyl bromide.
Explosives waste	
K044	Wastewater treatment sludges from the manufacturing and processing of explosives.

Continued

TABLE A3 The K-Code List of Hazardous Wastes—cont'd

Waste Type and Number	Hazardous Waste
K045	Spent carbon from the treatment of wastewater containing explosives.
K046	Wastewater treatment sludges from the manufacturing, formulation, and loading of lead-based initiating compounds.
K047	Pink/red water from TNT operations.
Petroleum refining	
K048	Dissolved air flotation (DAF) float from the petroleum refining industry.
K049	Slop oil emulsion solids from the petroleum refining industry.
K050	Heat exchanger bundle cleaning sludge from the petroleum refining industry.
K051	API separator sludge from the petroleum refining industry.
K052	Tank bottoms (leaded) from the petroleum refining industry.
Iron and steel waste	
K061	Emission control dust/sludge from the primary production of steel in electric furnaces.
K062	Spent pickle liquor generated by steel finishing operations of facilities within the iron and steel industry (SIC Codes 331 and 332).
Primary aluminum waste	
K088	Spent potliners from primary aluminum reduction.
Secondary lead waste	
K069	Emission control dust/sludge from secondary lead smelting. (Note: this listing has been stayed administratively for sludge generated from secondary acid scrubber systems. The stay will remain in effect until further administrative action is taken. Further administrative action will be taken after the US EPA publishes a notice of action in the Federal Register and Department adopts regulations making this listing effective.)
K100	Waste leaching solution from acid leaching of emission control dust/sludge from secondary lead smelting.
Veterinary pharmaceuticals waste	
K084	Wastewater treatment sludges generated during the production of veterinary pharmaceuticals from arsenic or organo-arsenic compounds.

TABLE A3 The K-Code List of Hazardous Wastes—cont'd

Waste Type and Number	Hazardous Waste
K101	Distillation tar residues from the distillation of aniline-based compounds in the production of veterinary pharmaceuticals from arsenic or organo-arsenic compounds.
K102	Residue from the use of activated carbon for decolorization in the production of veterinary pharmaceuticals from arsenic or organo-arsenic compounds.
Ink formulation waste	
K086	Solvent washes and sludges, caustic washes and sludges, or water washes and sludges from cleaning tubs and equipment used in the formulation of ink from pigments, driers, soaps, and stabilizers containing chromium and lead.
Coking waste	
K060	Ammonia still lime sludge from coking operations.
K087	Decanter tank tar sludge from coking operations.
K141	Process residues from the recovery of coat tar, including, but not limited to, collecting sump residues from the production of coke from coal or the recovery of coke by-products produced from coal. This listing does not include K087 (decanter tank tar sludges from coking operations).
K142	Tar storage tank residues from the production of coke from the coal or from the recovery of coke by-products produced from coal.
K143	Process residues from the recovery of light oil, including, but not limited to, those generated in stills, decanters, and wash oil recovery units from the recovery of coke by-products produced from coal.
K144	Wastewater sump residues from light oil refining, including, but not limited to, intercepting or contamination sump sludges from the recovery of coke by-products produced from coal.
K145	Residues from naphthalene collection and recovery operations from the recovery of coke by-products produced from coal.
K147	Tar storage tank residues from coal tar refining.
K148	Residues from coal tar distillation, including but not limited to, still bottoms.

TABLE A4 The P-Code Wastes Listed Alphabetically by Chemical Name

Code	Chemical
P026	1-(o-Chlorophenyl)thiourea
P081	1,2,3-Propanetriol, trinitrate (R)
P042	1,2-Benzenediol, 4-[1-hydroxy-2-(methylamino)ethyl]-, (R)-
P067	1,2-Propylenimine
P185	1,3-Dithiolane-2-carboxaldehyde, 2,4-dimethyl-, O-[(methylamino)-carbonyl]oxime
P004	1,4,5,8-Dimethanonaphthalene, 1,2,3,4,10,10-hexachloro-1,4,4a,5,8, 8a,-hexahydro-(1alpha,4alpha, 4abeta,5alpha,8alpha,8abeta)
P060	1,4,5,8-Dimethanonaphthalene, 1,2,3,4,10,10-hexachloro-1,4,4a,5,8, 8a-hexahydro-(1alpha,4alpha, 4abeta, 5beta, 8beta, 8abeta)-
P002	1-Acetyl-2-thiourea
P048	2,4-Dinitrophenol
P051	2,7:3,6-Dimethanonaphth [2,3-b]oxirene, 3,4,5,6,9,9-hexachloro-1a,2,2a, 3,6,6a,7,7aoctahydro-(1aalpha,2beta,2abeta,3alpha,6alpha,6abeta,7 beta, 7aalpha)-, and metabolites
P037	2,7:3,6-Dimethanonaphth[2,3-b]oxirene, 3,4,5,6,9,9-hexachloro-1a,2,2a,3,6,6a,7,7aoctahydro-(1aalpha,2beta,2aalpha,3beta,6beta, 6aalpha,7 beta, 7aalpha)
P045	2-Butanone, 3,3-dimethyl-1-(methylthio)-, O-[(methylamino)carbonyl] oxime
P034	2-Cyclohexyl-4,6-dinitrophenol
P001	2H-1-Benzopyran-2-one, 4-hydroxy-3-(3-oxo-1-phenyl-butyl)- and salts, when present at concentrations greater than 0.3% (w/w)
P069	2-Methyllactonitrile
P017	2-Propanone, 1-bromo-
P005	2-Propen-1-ol
P003	2-Propenal
P102	2-Propyn-1-ol
P007	3(2H)-Isoxazolone, 5-(aminomethyl)-
P027	3-Chloropropionitrile
P047	4,6-Dinitro-o-cresol and salts
P059	4,7-Methano-1H-indene, 1,4,5,6,7,8,8-heptachloro-3a,4,7,7a-tetrahydro-
P008	4-Aminopyridine

TABLE A4 The P-Code Wastes Listed Alphabetically by Chemical
Name—cont'd

Code	Chemical
P008	4-Pyridinamine
P007	5-(Aminomethyl)-3-isoxazolol
P050	6,9-Methano-2,4,3-benzodioxathiepin, 6,7,8,9,10,10-hexachloro-1,5,5a,6,9,9a-hexahydro-, 3-oxide
P127	7-Benzofuranol, 2,3-dihydro-2,2-dimethyl-, methylcarbamate
P088	7-Oxabicyclo[2.2.1]heptane-2,3-dicarboxylic acid
P023	Acetaldehyde, chloro-
P057	Acetamide, 2-fluoro-
P002	Acetamide, N-(aminothioxomethyl)-
P058	Acetic acid, fluoro-, sodium salt
P003	Acrolein
P070	Aldicarb
P203	Aldicarb sulfone
P004	Aldrin
P005	Allyl alcohol
P046	alpha,alpha-Dimethylphenethylamine
P072	alpha-Naphthylthiourea
P006	Aluminum phosphide (R,T)
P009	Ammonium picrate (R)
P119	Ammonium vanadate
P099	Argentate(1-), bis(cyano-C)-, potassium
P010	Arsenic acid H_3AsO_4
P012	Arsenic oxide As_2O_3
P011	Arsenic oxide As_2O_5
P011	Arsenic pentoxide
P012	Arsenic trioxide
P038	Arsine, diethyl-
P036	Arsonous dichloride, phenyl-
P054	Aziridine

Continued

TABLE A4 The P-Code Wastes Listed Alphabetically by Chemical
Name—cont'd

Code	Chemical
P067	Aziridine, 2-methyl-
P013	Barium cyanide
P024	Benzenamine, 4-chloro-
P077	Benzenamine, 4-nitro-
P028	Benzene, (chloromethyl)-
P046	Benzeneethanamine, alpha,alpha-dimethyl-
P014	Benzenethiol
P188	Benzoic acid, 2-hydroxy-, compound with (3aS-*cis*)-1,2,3,3a,8, 8a-hexahydro-1,3a,8-trimethylpyrrolo [2,3-b]indol-5-yl methylcarbamate ester (1:1)
P028	Benzyl chloride
P015	Beryllium powder
P017	Bromoacetone
P018	Brucine
P021	Calcium cyanide Ca(CN)$_2$
P189	Carbamic acid, [(dibutylamino)-thio]methyl-, 2,3-dihydro-2,2-dimethyl-7-benzofuranyl ester
P191	Carbamic acid, dimethyl-, 1-[(dimethylamino) carbonyl]-5-methyl-1H-pyrazol-3-yl ester
P192	Carbamic acid, dimethyl-, 3-methyl-1-(1-methylethyl)-1H-pyrazol-5-yl ester
P190	Carbamic acid, methyl-, 3-methylphenyl ester
P127	Carbofuran
P022	Carbon disulfide
P095	Carbonic dichloride
P189	Carbosulfan
P023	Chloroacetaldehyde
P029	Copper cyanide
P029	Copper cyanide Cu(CN)
P030	Cyanides (soluble cyanide salts), not otherwise specified
P031	Cyanogen

TABLE A4 The P-Code Wastes Listed Alphabetically by Chemical Name—cont'd

Code	Chemical
P033	Cyanogen chloride
P033	Cyanogen chloride (CN)Cl
P016	Dichloromethyl ether
P036	Dichlorophenylarsine
P037	Dieldrin
P038	Diethylarsine
P041	Diethyl-p-nitrophenyl phosphate
P043	Diisopropylfluorophosphate (DFP)
P044	Dimethoate
P191	Dimetilan
P020	Dinoseb
P085	Diphosphoramide, octamethyl-
P111	Diphosphoric acid, tetraethyl ester
P039	Disulfoton
P049	Dithiobiuret
P050	Endosulfan
P088	Endothall
P051	Endrin
P051	Endrin and metabolites
P042	Epinephrine
P031	Ethanedinitrile
P194	Ethanimidothioc acid, 2-(dimethylamino)-N-[[(methylamino)carbonyl]oxy]-2-oxo-, methyl ester
P066	Ethanimidothioic acid, N-[[(methylamino)carbonyl]oxy]-, methyl ester
P101	Ethyl cyanide
P054	Ethyleneimine
P097	Famphur
P056	Fluorine
P057	Fluoroacetamide

Continued

TABLE A4 The P-Code Wastes Listed Alphabetically by Chemical Name—cont'd

Code	Chemical
P058	Fluoroacetic acid, sodium salt
P198	Formetanate hydrochloride
P197	Formparanate
P065	Fulminic acid, mercury(2+) salt (R,T)
P059	Heptachlor
P062	Hexaethyl tetraphosphate
P068	Hydrazine, methyl-
P116	Hydrazinecarbothioamide
P063	Hydrocyanic acid
P063	Hydrogen cyanide
P096	Hydrogen phosphide
P060	Isodrin
P192	Isolan
P196	Manganese dimethyldithiocarbamate
P196	Manganese, bis(dimethylcarbamodithioato-S,S')-
P202	m-Cumenyl methylcarbamate
P065	Mercury fulminate (R,T)
P092	Mercury, (acetato-O)phenyl-
P082	Methanamine, N-methyl-N-nitroso-
P064	Methane, isocyanato-
P016	Methane, oxybis[chloro-
P112	Methane, tetranitro- (R)
P118	Methanethiol, trichloro-
P198	Methanimidamide, N,N-dimethyl-N'-[2-methyl-4-[[(methylamino)carbonyl]oxy]phenyl]-
P199	Methiocarb
P066	Methomyl
P068	Methyl hydrazine
P064	Methyl isocyanate

TABLE A4 The P-Code Wastes Listed Alphabetically by Chemical Name—cont'd

Code	Chemical
P071	Methyl parathion
P190	Metolcarb
P128	Mexacarbate
P073	Nickel carbonyl
P073	Nickel carbonyl Ni(CO)$_4$, (T-4)-
P074	Nickel cyanide
P074	Nickel cyanide Ni(CN)$_2$
P075	Nicotine and salts
P076	Nitric oxide
P078	Nitrogen dioxide
P076	Nitrogen oxide NO
P078	Nitrogen oxide NO$_2$
P081	Nitroglycerine (R)
P082	N-Nitrosodimethylamine
P084	N-Nitrosomethylvinylamine
P040	o,o-Diethyl o-pyrazinyl phosphorothioate
P085	Octamethylpyrophosphoramide
P087	Osmium oxide OsO$_4$, (T-4)-
P087	Osmium tetroxide
P194	Oxamyl
P089	Parathion
P024	p-Chloroaniline
P199	Phenol, 3,5-dimethyl-4-(methylthio)-, methylcarbamate
P020	Phenol, 2-(1-methylpropyl)-4,6-dinitro-
P009	Phenol, 2,4,6-trinitro-, ammonium salt (R)
P048	Phenol, 2,4-dinitro-
P034	Phenol, 2-cyclohexyl-4,6-dinitro-
P047	Phenol, 2-methyl-4,6-dinitro- and salts
P202	Phenol, 3-(1-methylethyl)-, methyl carbamate

Continued

TABLE A4 The P-Code Wastes Listed Alphabetically by Chemical Name—cont'd

Code	Chemical
P201	Phenol, 3-methyl-5-(1-methylethyl)-, methyl carbamate
P128	Phenol, 4-(dimethylamino)-3,5-dimethyl-, methylcarbamate (ester)
P092	Phenylmercury acetate
P093	Phenylthiourea
P094	Phorate
P095	Phosgene
P096	Phosphine
P041	Phosphoric acid, diethyl 4-nitrophenyl ester
P094	Phosphorodithioic acid, O,O-diethyl S-[(ethylthio)methyl] ester
P039	Phosphorodithioic acid, O,O-diethyl S-[2-(ethylthio)ethyl] ester
P044	Phosphorodithioic acid, O,O-dimethyl S-[2-(methylamino)-2-oxoethyl] ester
P043	Phosphorofluoridic acid, bis(1-methylethyl) ester
P071	Phosphorothioic acid, O,O-dimethyl O-(4-nitrophenyl) ester
P089	Phosphorothioic acid, O,O-diethyl O-(4-nitrophenyl) ester
P040	Phosphorothioic acid, O,O-diethyl O-pyrazinyl ester 3
P097	Phosphorothioic acid, O-[4-[(dimethylamino)sulfonyl]phenyl] O, O-dimethyl ester
P204	Physostigmine
P188	Physostigmine salicylate
P110	Plumbane, tetraethyl-
P077	p-Nitroaniline
P098	Potassium cyanide
P098	Potassium cyanide K(CN)
P099	Potassium silver cyanide
P201	Promecarb
P203	Propanal, 2-methyl-2-(methyl-sulfonyl)-, O-[(methylamino)carbonyl] oxime
P070	Propanal, 2-methyl-2-(methylthio)-, O-[(methylamino)carbonyl]oxime
P101	Propanenitrile
P069	Propanenitrile, 2-hydroxy-2-methyl-

TABLE A4 The P-Code Wastes Listed Alphabetically by Chemical
Name—cont'd

Code	Chemical
P027	Propanenitrile, 3-chloro-
P102	Propargyl alcohol
P075	Pyridine, 3-(1-methyl-2-pyrrolidinyl)-, (S)- and salts
P204	Pyrrolo[2,3-b]indol-5-ol, 1,2,3,3a,8,8a-hexahydro-1,3a,8-trimethyl-, methylcarbamate (ester), (3aS-cis)-
P114	Selenious acid, dithallium(1 +) salt
P103	Selenourea
P104	Silver cyanide
P104	Silver cyanide Ag(CN)
P105	Sodium azide
P106	Sodium cyanide
P106	Sodium cyanide Na(CN)
P108	Strychnidin-10-one and salts
P018	Strychnidin-10-one, 2,3-dimethoxy-
P108	Strychnine and salts
P115	Sulfuric acid, dithallium(1 +) salt
P110	Tetraethyl lead
P111	Tetraethyl pyrophosphate
P109	Tetraethyldithiopyrophosphate
P112	Tetranitromethane (R)
P062	Tetraphosphoric acid, hexaethyl ester
P113	Thallic oxide
P113	Thallium oxide Tl_2O_3
P114	Thallium(I) selenite
P115	Thallium(I) sulfate
P109	Thiodiphosphoric acid, tetraethyl ester
P045	Thiofanox
P049	Thioimidodicarbonic diamide [(H2N)C(S)]2NH
P014	Thiophenol

Continued

TABLE A4 The P-Code Wastes Listed Alphabetically by Chemical Name—cont'd

Code	Chemical
P116	Thiosemicarbazide
P026	Thiourea, (2-chlorophenyl)-
P072	Thiourea, 1-naphthalenyl-
P093	Thiourea, phenyl-
P185	Tirpate
P123	Toxaphene
P118	Trichloromethanethiol
P119	Vanadic acid, ammonium salt
P120	Vanadium oxide V_2O_5
P120	Vanadium pentoxide
P084	Vinylamine, N-methyl-N-nitroso-
P001	Warfarin, and salts, when present at concentrations greater than 0.3% w/w
P121	Zinc cyanide
P121	Zinc cyanide $Zn(CN)_2$
P122	Zinc phosphide Zn_3P_2, when present at concentrations greater than 10% (R,T)
P205	Zinc, bis(dimethylcarbamodithioato-S,S')-,
P205	Ziram

TABLE A5 The U-Code Wastes Listed Alphabetically by Chemical Name

Code	Chemical
U021	[1,1'-Biphenyl]-4,4'-diamine
U073	[1,1'-Biphenyl]-4,4'-diamine, 3,3'-dichloro-
U091	[1,1'-Biphenyl]-4,4'-diamine, 3,3'-dimethoxy-
U095	q[1,1'-Biphenyl]-4,4'-diamine, 3,3'-dimethyl-
U208	1,1,1,2-Tetrachloroethane

TABLE A5 The U-Code Wastes Listed Alphabetically by Chemical Name—cont'd

Code	Chemical
U209	1,1,2,2-Tetrachloroethane
U227	1,1,2-Trichloroethane
U078	1,1-Dichloroethylene
U098	1,1-Dimethylhydrazine
U207	1,2,4,5-Tetrachlorobenzene
U085	1,2:3,4-Diepoxybutane (I,T)
U028	1,2-Benzenedicarboxylic acid, bis(2-ethylhexyl) ester
U069	1,2-Benzenedicarboxylic acid, dibutyl ester
U088	1,2-Benzenedicarboxylic acid, diethyl ester
U102	1,2-Benzenedicarboxylic acid, dimethyl ester
U107	1,2-Benzenedicarboxylic acid, dioctyl ester
U202	1,2-Benzisothiazol-3(2H)-one, 1,1-dioxide and salts
U066	1,2-Dibromo-3-chloropropane
U079	1,2-Dichloroethylene
U099	1,2-Dimethylhydrazine
U109	1,2-Diphenylhydrazine
U155	1,2-Ethanediamine, N,N-dimethyl-N'-2-pyridinyl-N'-(2-thienylmethyl)-
U193	1,2-Oxathiolane, 2,2-dioxide
U142	1,3,4-Metheno-2H-cyclobuta[cd]pentalen-2-one, 1,1a, 3,3a,4,5,5,5a,5b, 6-decachlorooctahydro-
U234	1,3,5-Trinitrobenzene (R,T)
U182	1,3,5-Trioxane, 2,4,6-trimethyl-
U201	1,3-Benzenediol
U364	1,3-Benzodioxol-4-ol, 2,2-dimethyl-,
U278	1,3-Benzodioxol-4-ol, 2,2-dimethyl-, methyl carbamate
U141	1,3-Benzodioxole, 5-(1-propenyl)-
U203	1,3-Benzodioxole, 5-(2-propenyl)-
U090	1,3-Benzodioxole, 5-propyl-
U128	1,3-Butadiene, 1,1,2,3,4,4-hexachloro-

Continued

TABLE A5 The U-Code Wastes Listed Alphabetically by Chemical Name—cont'd

Code	Chemical
U130	1,3-Cyclopentadiene, 1,2,3,4,5,5-hexachloro-
U084	1,3-Dichloropropene
U190	1,3-Isobenzofurandione
U186	1,3-Pentadiene (I)
U193	1,3-Propane sultone
U074	1,4-Dichloro-2-butene (I,T)
U108	1,4-Diethyleneoxide
U108	1,4-Dioxane
U166	1,4-Naphthalenedione
U166	1,4-Naphthoquinone
U172	1-Butanamine, N-butyl-N-nitroso-
U031	1-Butanol (I)
U011	1H-1,2,4-Triazol-3-amine
U186	1-Methylbutadiene (I)
U167	1-Naphthalenamine
U279	1-Naphthalenol, methylcarbamate
U194	1-Propanamine (I,T)
U111	1-Propanamine, N-nitroso-N-propyl-
U110	1-Propanamine, N-propyl- (I)
U235	1-Propanol, 2,3-dibromo-, phosphate (3:1)
U140	1-Propanol, 2-methyl- (I,T)
U243	1-Propene, 1,1,2,3,3,3-hexachloro-
U084	1-Propene, 1,3-dichloro-
U085	2,2'-Bioxirane
U237	2,4-(1H,3H)-Pyrimidinedione, 5-[bis(2-chloroethyl)amino]-
U240	2,4-D, salts and esters
U081	2,4-Dichlorophenol
U101	2,4-Dimethylphenol
U105	2,4-Dinitrotoluene

TABLE A5 The U-Code Wastes Listed Alphabetically by Chemical Name—cont'd

Code	Chemical
U197	2,5-Cyclohexadiene-1,4-dione
U147	2,5-Furandione
U082	2,6-Dichlorophenol
U106	2,6-Dinitrotoluene
U236	2,7-Naphthalenedisulfonic acid, 3,3'-[(3,3'-dimethyl[1,1'-biphenyl]-4,4'-diyl)]bis(azo)bis[5-amino-4-hydroxy]-, tetrasodium salt
U005	2-Acetylaminofluorene
U159	2-Butanone (I,T)
U160	2-Butanone, peroxide (R,T)
U053	2-Butenal
U074	2-Butene, 1,4-dichloro- (I,T)
U143	2-Butenoic acid, 2-methyl-, 7-[[2,3-dihydroxy-2-(1-methoxyethyl)-3-methyl-1-oxobutoxy]methyl]-2,3,5,7a-tetrahydro-1H-pyrrolizin-1-yl ester, [1S-[1alpha(Z),7(2S*,3R*),7aalpha]]-
U042	2-Chloroethyl vinyl ether
U125	2-Furancarboxaldehyde (I)
U058	2H-1,3,2-Oxazaphosphorin-2-amine, *N,N*-bis(2-chloroethyl)tetrahydro-, 2-oxide
U366	2H-1,3,5-Thiadiazine-2-thione, tetrahydro-3,5-dimethyl-
U248	2H-1-Benzopyran-2-one, 4-hydroxy-3-(3-oxo-1-phenyl-butyl)- and salts, when present at concentrations of 0.3% or less
U116	2-Imidazolidinethione
U168	2-Naphthalenamine
U171	2-Nitropropane (I,T)
U191	2-Picoline
U002	2-Propanone (I)
U007	2-Propenamide
U009	2-Propenenitrile
U152	2-Propenenitrile, 2-methyl- (I,T)
U008	2-Propenoic acid (I)
U118	2-Propenoic acid, 2-methyl-, ethyl ester

Continued

TABLE A5 The U-Code Wastes Listed Alphabetically by Chemical Name—cont'd

Code	Chemical
U162	2-Propenoic acid, 2-methyl-, methyl ester (I,T)
U113	2-Propenoic acid, ethyl ester (I)
U073	3,3'-Dichlorobenzidine
U091	3,3'-Dimethoxybenzidine
U095	3,3'-Dimethylbenzidine
U148	3,6-Pyridazinedione, 1,2-dihydro-
U375	3-Iodo-2-propynyl n-butylcarbamate
U157	3-Methylcholanthrene
U164	4(1H)-Pyrimidinone, 2,3-dihydro-6-methyl-2-thioxo-
U158	4,4'-Methylenebis(2-chloroaniline)
U036	4,7-Methano-1H-indene, 1,2,4,5,6,7,8,8-octachloro-2, 3,3a,4,7, 7a-hexahydro-
U030	4-Bromophenyl phenyl ether
U049	4-Chloro-o-toluidine, hydrochloride
U161	4-Methyl-2-pentanone (I)
U059	5,12-Naphthacenedione, 8-acetyl-10-[(3-amino-2,3,6-trideoxy-alpha-L-lyxohexopyranosyl) oxy]-7,8,9,10-tetrahydro-6,8,11-trihydroxy-1-methoxy-, (8S-cis)-
U181	5-Nitro-o-toluidine
U094	7,12-Dimethylbenz[a]anthracene
U367	7-Benzofuranol, 2,3-dihydro-2,2-dimethyl-
U394	A2213
U001	Acetaldehyde (I)
U034	Acetaldehyde, trichloro-
U187	Acetamide, N-(4-ethoxyphenyl)-
U005	Acetamide, N-9H-fluoren-2-yl-
U112	Acetic acid ethyl ester (I)
U240	Acetic acid, (2,4-dichlorophenoxy)-, salts and esters 3
U144	Acetic acid, lead(2+) salt
U214	Acetic acid, thallium(1+) salt

TABLE A5 The U-Code Wastes Listed Alphabetically by Chemical
Name—cont'd

Code	Chemical
U002	Acetone (I)
U003	Acetonitrile (I,T)
U004	Acetophenone
U006	Acetyl chloride (C,R,T)
U007	Acrylamide
U008	Acrylic acid (I)
U009	Acrylonitrile
U096	alpha,alpha-Dimethylbenzylhydroperoxide (R)
U167	alpha-Naphthylamine
U011	Amitrole
U012	Aniline (I,T)
U136	Arsinic acid, dimethyl-
U014	Auramine
U015	Azaserine
U010	Azirino[2',3':3,4]pyrrolo[1,2-a]indole-4,7-dio ne, 6-amino-8-[[(aminocarbonyl)oxy]methyl]-1,1a, 2,8,8a, 8b-hexahydro-8a-methoxy-5-methyl-, [1aS-(1aalpha, 8beta,8aalpha,8balpha)]-
U280	Barban
U364	Bendiocarb phenol
U278	Bendiocarb
U271	Benomyl
U018	Benz[a]anthracene
U094	Benz[a]anthracene, 7,12-dimethyl-
U016	Benz[c]acridine
U157	Benz[j]aceanthrylene, 1,2-dihydro-3-methyl-
U017	Benzal chloride
U192	Benzamide, 3,5-dichloro-N-(1,1-dimethyl-2-propynyl)-
U012	Benzenamine (I,T)
U328	Benzenamine, 2-methyl-
U222	Benzenamine, 2-methyl-, hydrochloride

Continued

TABLE A5 The U-Code Wastes Listed Alphabetically by Chemical Name—cont'd

Code	Chemical
U181	Benzenamine, 2-methyl-5-nitro-
U014	Benzenamine, 4,4'-carbonimidoylbis[N,N-dimethyl-
U158	Benzenamine, 4,4'-methylenebis[2-chloro-
U049	Benzenamine, 4-chloro-2-methyl-, hydrochloride
U353	Benzenamine, 4-methyl-
U093	Benzenamine, N,N-dimethyl-4-(phenylazo)-
U019	Benzene (I,T)
U055	Benzene, (1-methylethyl)- (I)
U017	Benzene, (dichloromethyl)-
U023	Benzene, (trichloromethyl)-
U061	Benzene, 1,1'-(2,2,2-trichloroethylidene)bis[4-chloro-
U247	Benzene, 1,1'-(2,2,2-trichloroethylidene)bis[4-methoxy-
U060	Benzene, 1,1'-(2,2-dichloroethylidene)bis[4-chloro-
U207	Benzene, 1,2,4,5-tetrachloro-
U070	Benzene, 1,2-dichloro-
U234	Benzene, 1,3,5-trinitro-
U071	Benzene, 1,3-dichloro-
U223	Benzene, 1,3-diisocyanatomethyl- (R,T)
U072	Benzene, 1,4-dichloro-
U030	Benzene, 1-bromo-4-phenoxy-
U105	Benzene, 1-methyl-2,4-dinitro-
U106	Benzene, 2-methyl-1,3-dinitro-
U037	Benzene, chloro-
U239	Benzene, dimethyl- (I,T)
U127	Benzene, hexachloro-
U056	Benzene, hexahydro- (I)
U220	Benzene, methyl-
U169	Benzene, nitro-
U183	Benzene, pentachloro-

TABLE A5 The U-Code Wastes Listed Alphabetically by Chemical Name—cont'd

Code	Chemical
U185	Benzene, pentachloronitro-
U038	Benzeneacetic acid, 4-chloro-alpha-(4-chlorophenyl)-alpha-hydroxy-, ethyl ester
U035	Benzenebutanoic acid, 4-[bis(2-chloroethyl)amino]-
U221	Benzenediamine, ar-methyl-
U020	Benzenesulfonic acid chloride (C,R)
U020	Benzenesulfonyl chloride (C,R)
U021	Benzidine
U022	Benzo[a]pyrene
U064	Benzo[rst]pentaphene
U023	Benzotrichloride (C,R,T)
U047	beta-Chloronaphthalene
U168	beta-Naphthylamine
U401	Bis(dimethylthiocarbamoyl) sulfide
U400	Bis(pentamethylene)thiuram tetrasulfide
U225	Bromoform
U392	Butylate
U136	Cacodylic acid
U032	Calcium chromate
U280	Carbamic acid, (3-chlorophenyl)-, 4-chloro-2-butynyl ester
U409	Carbamic acid, [1,2-phenylenebis (iminocarbonothioyl)] bis-, dimethyl ester
U271	Carbamic acid, [1-[(butylamino)carbonyl]-1H-benzimidazol-2-yl]-, methyl ester
U372	Carbamic acid, 1H-benzimidazol-2-yl, methyl ester
U375	Carbamic acid, butyl-, 3-iodo-2-propynyl ester
U238	Carbamic acid, ethyl ester
U178	Carbamic acid, methylnitroso-, ethyl ester
U373	Carbamic acid, phenyl-, 1-methylethyl ester
U097	Carbamic chloride, dimethyl-
U378	Carbamodithioic acid, (hydroxymethyl)methyl-, monopotassium salt

Continued

TABLE A5 The U-Code Wastes Listed Alphabetically by Chemical Name—cont'd

Code	Chemical
U114	Carbamodithioic acid, 1,2-ethanediylbis-, salts and esters
U379	Carbamodithioic acid, dibutyl, sodium salt
U277	Carbamodithioic acid, diethyl-, 2-chloro-2-propenyl ester
U381	Carbamodithioic acid, diethyl-, sodium salt
U383	Carbamodithioic acid, dimethyl, potassium salt
U382	Carbamodithioic acid, dimethyl-, sodium salt
U376	Carbamodithioic acid, dimethyl-, tetraanhydrosulfide with orthothioselenious acid
U377	Carbamodithioic acid, methyl-, monopotassium salt
U384	Carbamodithioic acid, methyl-, monosodium salt
U062	Carbamothioic acid, bis(1-methylethyl)-, S-(2,3-dichloro-2-propenyl) ester
U389	Carbamothioic acid, bis(1-methylethyl)-, S-(2,3,3-trichloro-2-propenyl) ester
U392	Carbamothioic acid, bis(2-methylpropyl)-, S-ethyl ester
U391	Carbamothioic acid, butylethyl-, S-propyl ester
U386	Carbamothioic acid, cyclohexylethyl-, S-ethyl ester
U387	Carbamothioic acid, dipropyl-, S-(phenylmethyl) ester
U390	Carbamothioic acid, dipropyl-, S-ethyl ester
U385	Carbamothioic acid, dipropyl-, S-propyl ester
U279	Carbaryl
U372	Carbendazim
U367	Carbofuran phenol
U033	Carbon oxyfluoride (R,T)
U211	Carbon tetrachloride
U215	Carbonic acid, dithallium(1 +) salt
U033	Carbonic difluoride
U156	Carbonochloridic acid, methyl ester (I,T)
U034	Chloral
U035	Chlorambucil
U036	Chlordane, alpha and gamma isomers

TABLE A5 The U-Code Wastes Listed Alphabetically by Chemical Name—cont'd

Code	Chemical
U026	Chlornaphazin
U037	Chlorobenzene
U038	Chlorobenzilate
U044	Chloroform
U046	Chloromethyl methyl ether
U032	Chromic acid H_2CrO_4, calcium salt
U050	Chrysene
U393	Copper dimethyldithiocarbamate
U393	Copper, bis(dimethylcarbamodithioato-S,S')-,
U051	Creosote
U052	Cresol (Cresylic acid)
U053	Crotonaldehyde
U055	Cumene (I)
U246	Cyanogen bromide (CN)Br
U386	Cycloate
U056	Cyclohexane (I)
U129	Cyclohexane, 1,2,3,4,5,6-hexachloro-, (1alpha,2alpha,3beta,4alpha,5alpha,6beta)-
U057	Cyclohexanone (I)
U058	Cyclophosphamide
U059	Daunomycin
U366	Dazomet
U060	DDD
U061	DDT
U206	D-Glucose, 2-deoxy-2-[[(methylnitrosoamino)-carbonyl]amino]-
U062	Diallate
U063	Dibenz[a,h]anthracene
U064	Dibenzo[a,i]pyrene
U069	Dibutyl phthalate

Continued

TABLE A5 The U-Code Wastes Listed Alphabetically by Chemical Name—cont'd

Code	Chemical
U075	Dichlorodifluoromethane
U025	Dichloroethyl ether
U027	Dichloroisopropyl ether
U024	Dichloromethoxy ethane
U088	Diethyl phthalate
U395	Diethylene glycol, dicarbamate
U028	Diethylhexyl phthalate
U089	Diethylstilbesterol
U090	Dihydrosafrole
U102	Dimethyl phthalate
U103	Dimethyl sulfate
U092	Dimethylamine (I)
U097	Dimethylcarbamoyl chloride
U107	Di-*n*-octyl phthalate
U111	Di-*n*-propylnitrosamine
U110	Dipropylamine (I)
U403	Disulfiram
U041	Epichlorohydrin
U390	EPTC
U001	Ethanal (I)
U404	Ethanamine, *N,N*-diethyl-
U174	Ethanamine, *N*-ethyl-*N*-nitroso-
U208	Ethane, 1,1,1,2-tetrachloro-
U226	Ethane, 1,1,1-trichloro-
U209	Ethane, 1,1,2,2-tetrachloro-
U227	Ethane, 1,1,2-trichloro-
U024	Ethane, 1,1'-[methylenebis(oxy)]bis[2-chloro-
U076	Ethane, 1,1-dichloro-
U117	Ethane, 1,1'-oxybis-(I)

TABLE A5 The U-Code Wastes Listed Alphabetically by Chemical Name—cont'd

Code	Chemical
U025	Ethane, 1,1'-oxybis[2-chloro-
U067	Ethane, 1,2-dibromo-
U077	Ethane, 1,2-dichloro-
U131	Ethane, hexachloro-
U184	Ethane, pentachloro-
U218	Ethanethioamide
U394	Ethanimidothioic acid, 2-(dimethylamino)-N-hydroxy-2-oxo-, methyl ester
U410	Ethanimidothioic acid, N,N'-[thiobis[(methylimino) carbonyloxy]]bis-, dimethyl ester
U173	Ethanol, 2,2'-(nitrosoimino)bis-
U395	Ethanol, 2,2'-oxybis-, dicarbamate
U359	Ethanol, 2-ethoxy-
U004	Ethanone, 1-phenyl-
U042	Ethene, (2-chloroethoxy)-
U078	Ethene, 1,1-dichloro-
U079	Ethene, 1,2-dichloro-, (E)-
U043	Ethene, chloro-
U210	Ethene, tetrachloro-
U228	Ethene, trichloro-
U112	Ethyl acetate (I)
U113	Ethyl acrylate (I)
U238	Ethyl carbamate (urethane)
U117	Ethyl ether (I)
U118	Ethyl methacrylate
U119	Ethyl methanesulfonate
U407	Ethyl Ziram
U067	Ethylene dibromide
U077	Ethylene dichloride
U359	Ethylene glycol monoethyl ether

Continued

TABLE A5 The U-Code Wastes Listed Alphabetically by Chemical Name—cont'd

Code	Chemical
U115	Ethylene oxide (I,T)
U114	Ethylenebisdithiocarbamic acid, salts and esters
U116	Ethylenethiourea
U076	Ethylidene dichloride
U396	Ferbam
U120	Fluoranthene
U122	Formaldehyde
U123	Formic acid (C,T)
U124	Furan (I)
U213	Furan, tetrahydro-(I)
U125	Furfural (I)
U124	Furfuran (I)
U206	Glucopyranose, 2-deoxy-2-(3-methyl-3-nitrosoureido)-, DU126 Glycidylaldehyde
U163	Guanidine, N-methyl-N'-nitro-N-nitroso-
U365	H-Azepine-1-carbothioic acid, hexahydro-, S-ethyl ester
U127	Hexachlorobenzene
U128	Hexachlorobutadiene
U130	Hexachlorocyclopentadiene
U131	Hexachloroethane
U132	Hexachlorophene
U243	Hexachloropropene
U133	Hydrazine (R,T)
U098	Hydrazine, 1,1-dimethyl-
U086	Hydrazine, 1,2-diethyl-
U099	Hydrazine, 1,2-dimethyl-
U109	Hydrazine, 1,2-diphenyl-
U134	Hydrofluoric acid (C,T)
U134	Hydrogen fluoride (C,T)

TABLE A5 The U-Code Wastes Listed Alphabetically by Chemical Name—cont'd

Code	Chemical
U135	Hydrogen sulfide
U135	Hydrogen sulfide H_2S
U096	Hydroperoxide, 1-methyl-1-phenylethyl- (R)
U137	Indeno[1,2,3-cd]pyrene
U396	Iron, tris(dimethylcarbamodithioato-S,S')-,
U140	Isobutyl alcohol (I,T)
U141	Isosafrole
U142	Kepone
U143	Lasiocarpine
U144	Lead acetate
U145	Lead phosphate
U146	Lead subacetate
U146	Lead, bis(acetato-O)tetrahydroxytri-
U129	Lindane
U150	L-Phenylalanine, 4-[bis(2-chloroethyl)amino]-
U015	L-Serine, diazoacetate (ester)
U147	Maleic anhydride
U148	Maleic hydrazide
U149	Malononitrile
U071	m-Dichlorobenzene
U150	Melphalan
U151	Mercury
U384	Metam sodium
U152	Methacrylonitrile (I, T)
U092	Methanamine, N-methyl-(I)
U029	Methane, bromo-
U045	Methane, chloro- (I, T)
U046	Methane, chloromethoxy-
U068	Methane, dibromo-

Continued

TABLE A5 The U-Code Wastes Listed Alphabetically by Chemical Name—cont'd

Code	Chemical
U080	Methane, dichloro-
U075	Methane, dichlorodifluoro-
U138	Methane, iodo-
U211	Methane, tetrachloro-
U225	Methane, tribromo-
U044	Methane, trichloro-
U121	Methane, trichlorofluoro-
U119	Methanesulfonic acid, ethyl ester
U153	Methanethiol (I, T)
U154	Methanol (I)
U155	Methapyrilene
U247	Methoxychlor
U154	Methyl alcohol (I)
U029	Methyl bromide
U045	Methyl chloride (I,T)
U156	Methyl chlorocarbonate (I,T)
U226	Methyl chloroform
U159	Methyl ethyl ketone (MEK) (I,T)
U160	Methyl ethyl ketone peroxide (R,T)
U138	Methyl iodide
U161	Methyl isobutyl ketone (I)
U162	Methyl methacrylate (I,T)
U068	Methylene bromide
U080	Methylene chloride
U164	Methylthiouracil
U010	Mitomycin C
U163	MNNG
U365	Molinate
U086	*N,N'*-Diethylhydrazine

TABLE A5 The U-Code Wastes Listed Alphabetically by Chemical Name—cont'd

Code	Chemical
U026	Naphthalenamine, N,N'-bis(2-chloroethyl)-
U165	Naphthalene
U047	Naphthalene, 2-chloro-
U031	n-Butyl alcohol (I)
U217	Nitric acid, thallium(1+) salt
U169	Nitrobenzene (I,T)
U173	N-Nitrosodiethanolamine
U174	N-Nitrosodiethylamine
U172	N-Nitrosodi-n-butylamine
U176	N-Nitroso-N-ethylurea
U177	N-Nitroso-N-methylurea
U178	N-Nitroso-N-methylurethane
U179	N-Nitrosopiperidine
U180	N-Nitrosopyrrolidine
U194	n-Propylamine (I,T)
U087	o,o-Diethyl S-methyl dithiophosphate
U048	o-Chlorophenol
U070	o-Dichlorobenzene
U328	o-Toluidine
U222	o-Toluidine hydrochloride
U115	Oxirane (I,T)
U041	Oxirane, (chloromethyl)-
U126	Oxiranecarboxyaldehyde
U182	Paraldehyde
U197	p-Benzoquinone
U039	p-Chloro-m-cresol
U072	p-Dichlorobenzene
U093	p-Dimethylaminoazobenzene
U391	Pebulate

Continued

TABLE A5 The U-Code Wastes Listed Alphabetically by Chemical Name—cont'd

Code	Chemical
U183	Pentachlorobenzene
U184	Pentachloroethane
U185	Pentachloronitrobenzene (PCNB)
U161	Pentanol, 4-methyl-
U187	Phenacetin
U188	Phenol
U411	Phenol, 2-(1-methylethoxy)-, methylcarbamate
U132	Phenol, 2,2′-methylenebis[3,4,6-trichloro-
U081	Phenol, 2,4-dichloro-
U101	Phenol, 2,4-dimethyl-
U082	Phenol, 2,6-dichloro-
U048	Phenol, 2-chloro-
U089	Phenol, 4,4′-(1,2-diethyl-1,2-ethenediyl)bis-, (E)-
U039	Phenol, 4-chloro-3-methyl-
U170	Phenol, 4-nitro-
U052	Phenol, methyl-
U145	Phosphoric acid, lead(2+) salt (2:3)
U087	Phosphorodithioic acid, O,O-diethyl S-methyl ester
U189	Phosphorus sulfide (R)
U190	Phthalic anhydride
U400	Piperidine, 1,1′-(tetrathiodicarbonothioyl)-bis-
U179	Piperidine, 1-nitroso-
U170	p-Nitrophenol
U383	Potassium dimethyldithiocarbamate
U378	Potassium n-hydroxymethyl-n-methyldi-thiocarbamate
U377	Potassium n-methyldithiocarbamate
U192	Pronamide
U066	Propane, 1,2-dibromo-3-chloro-
U083	Propane, 1,2-dichloro-

TABLE A5 The U-Code Wastes Listed Alphabetically by Chemical Name—cont'd

Code	Chemical
U027	Propane, 2,2'-oxybis[2-chloro-
U171	Propane, 2-nitro- (I,T)
U149	Propanedinitrile
U373	Propham
U411	Propoxur
U083	Propylene dichloride
U387	Prosulfocarb
U353	p-Toluidine
U196	Pyridine
U191	Pyridine, 2-methyl-
U180	Pyrrolidine, 1-nitroso-
U200	Reserpine
U201	Resorcinol
U202	Saccharin and salts
U203	Safrole
U204	Selenious acid
U204	Selenium dioxide
U205	Selenium sulfide
U205	Selenium sulfide SeS_2 (R,T)
U376	Selenium, tetrakis(dimethyldithiocarbamate)
U379	Sodium dibutyldithiocarbamate
U381	Sodium diethyldithiocarbamate
U382	Sodium dimethyldithiocarbamate
U206	Streptozotocin
U277	Sulfallate
U189	Sulfur phosphide (R)
U103	Sulfuric acid, dimethyl ester
U402	Tetrabutylthiuram disulfide
U210	Tetrachloroethylene

Continued

TABLE A5 The U-Code Wastes Listed Alphabetically by Chemical Name—cont'd

Code	Chemical
U213	Tetrahydrofuran (I)
U401	Tetramethylthiuram monosulfide
U216	Thallium chloride Tlcl
U214	Thallium(I) acetate
U215	Thallium(I) carbonate
U216	Thallium(I) chloride
U217	Thallium(I) nitrate
U218	Thioacetamide
U410	Thiodicarb
U153	Thiomethanol (I,T)
U244	Thioperoxydicarbonic diamide [(H2N)C(S)]2S2, tetramethyl-
U402	Thioperoxydicarbonic diamide, tetrabutyl
U403	Thioperoxydicarbonic diamide, tetraethyl
U409	Thiophanate-methyl
U219	Thiourea
U244	Thiram
U220	Toluene
U223	Toluene diisocyanate (R,T)
U221	Toluenediamine
U389	Triallate
U228	Trichloroethylene
U121	Trichloromonofluoromethane
U404	Triethylamine
U235	Tris(2,3-dibromopropyl) phosphate
U236	Trypan blue
U237	Uracil mustard
U176	Urea, N-ethyl-N-nitroso-
U177	Urea, N-methyl-N-nitroso-
U385	Vernolate

TABLE A5 The U-Code Wastes Listed Alphabetically by Chemical Name—cont'd

Code	Chemical
U043	Vinyl chloride
U248	Warfarin and salts, when present at concentrations of 0.3% or less
U239	Xylene (I)
U200	Yohimban-16-carboxylic acid, 11,17-dimethoxy-18-[(3,4,5-trimethoxybenzoyl)oxy]-, methyl ester, (3beta, 16beta,17alpha,18beta,20alpha)-
U249	Zinc phosphide Zn_3P_2, when present at concentrations of 10% or less
U407	Zinc, bis(diethylcarbamodithioato-*S,S'*)-

TABLE A6 Toxic Environmental Compounds

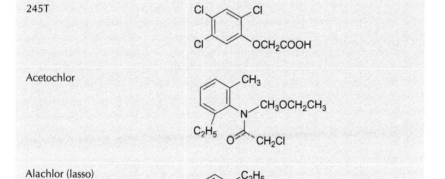

245T	
Acetochlor	
Alachlor (lasso)	

Continued

TABLE A6 Toxic Environmental Compounds—cont'd

Atrazine	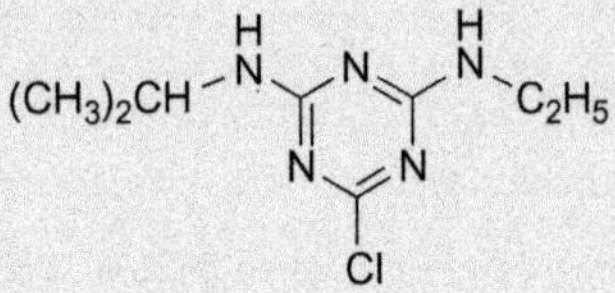
Carbaryl	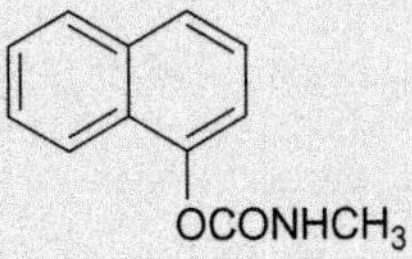
Carbofuran (furadan)	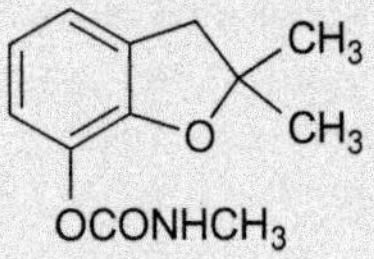
Chlordane	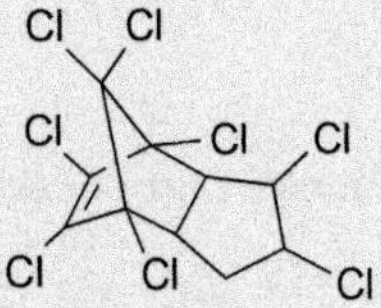
Chlorpyrifos (dursban)	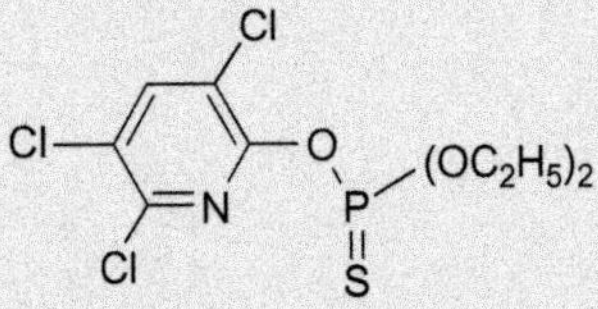
DDT (dichlorodiphenyltrichloroethane)	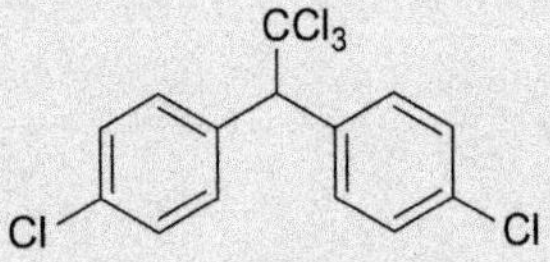
DDE (dichlorodiphenylethylene)	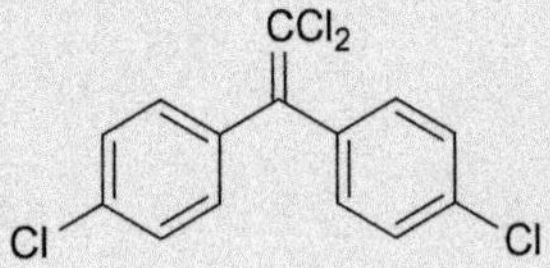

TABLE A6 Toxic Environmental Compounds—cont'd

Dichlorvos	$(CH_3O)_2POCH=CCl_2$ $\overset{\|}{O}$
Glyphosate (round-up)[3]	
Heptachlor	
Heptachlor epoxide	
Kepone	
Lindane (or γ-HCH)	
Malathion	

Continued

TABLE A6 Toxic Environmental Compounds—cont'd

Methoprene	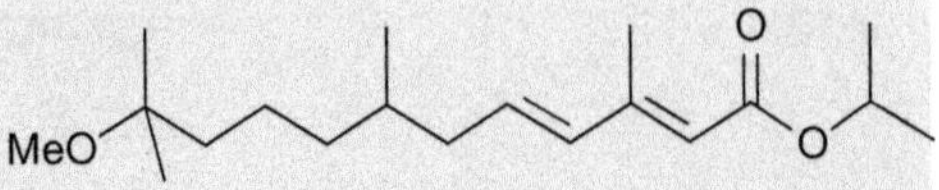
Methoxychlor	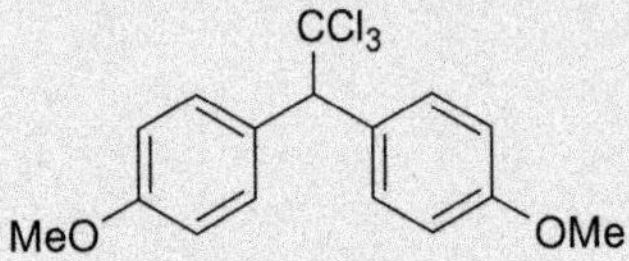
Metolachlor (dual)	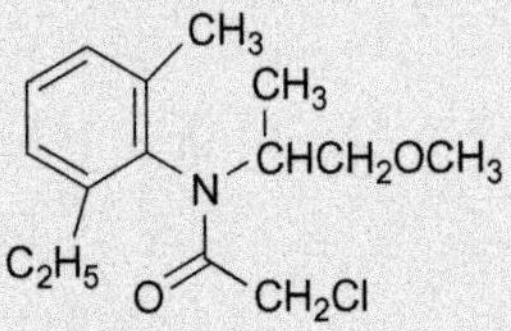
Mirex	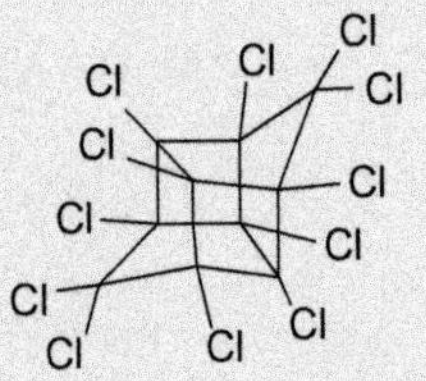
Oxychlordane	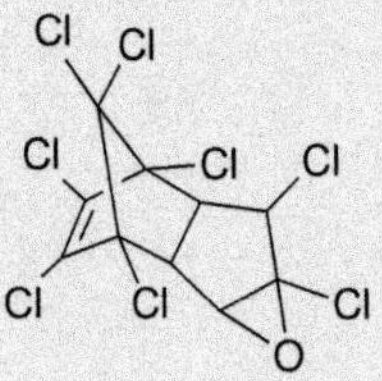
Parathion	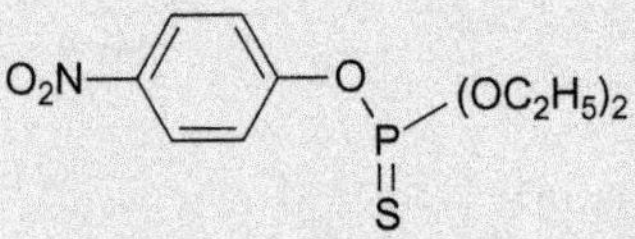
Permethrin	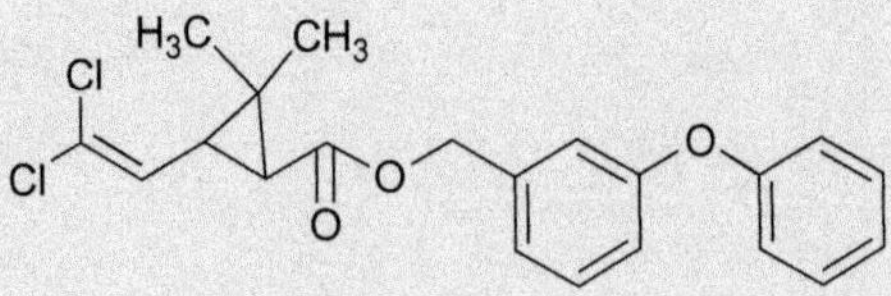

TABLE A6 Toxic Environmental Compounds—cont'd

Pendimethaline (Prowl)

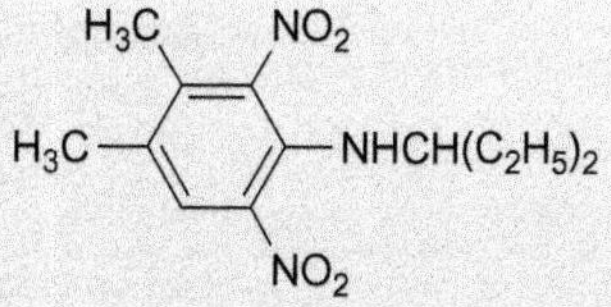

$$H_3C-\underset{\underset{NO_2}{|}}{\overset{\overset{H_3C\quad NO_2}{|}}{\bigcirc}}-NHCH(C_2H_5)_2$$

Trifluralin (Treflan)

$$CF_3-\underset{\underset{NO_2}{|}}{\overset{\overset{NO_2}{|}}{\bigcirc}}-N(CH_2CH_2CH_3)_2$$

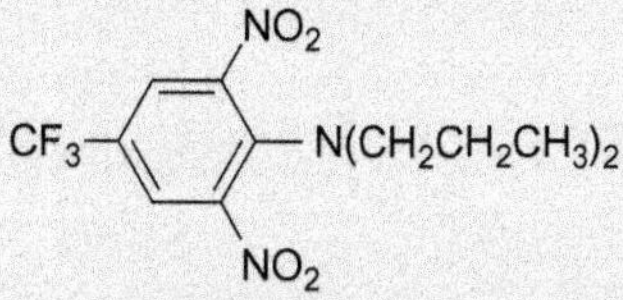

TABLE A7 The D-Code List of Hazardous Wastes

Codes	Description
D001	Characteristic of ignitability
D002	Characteristic of corrosivity
D003	Characteristic of reactivity
D004	Arsenic
D005	Barium
D006	Cadmium
D007	Chromium
D008	Lead
D009	Mercury
D010	Selenium
D011	Silver
D012	Endrin
D013	Lindane
D014	Methoxychlor
D015	Toxaphene
D016	2,4-Dichlorophenoxy acetic acid

Continued

TABLE A7 The D-Code List of Hazardous Wastes—cont'd

Codes	Description
D017	2,4,5-TP (Silvex)
D018	Benzene
D019	Carbon tetrachloride
D020	Chlordane
D021	Chlorobenzene
D022	Chloroform
D023	O-cresol
D024	m-Cresol
D025	p-Cresol
D026	Cresol
D027	1,4-Dichlorobenzene
D028	1,2-Dichloroethane
D029	1,1-Dichloroethylene
D030	2,4-Dinitrotoluene
D031	Heptachlor (and its hydroxide)
D032	Hexachlorobenzene
D033	Hexachlorobutadiene
D034	Hexachloroethane
D035	Methyl ethyl ketone
D036	Nitrobenzene
D037	Pentachlorophenol
D038	Pyridine
D039	Tetrachlorethylene
D040	Trichlorethylene
D041	2,4,5-Trichlorophenol
D042	2,4,6-Trichlorophenol
D043	Vinyl chloride

Glossary

A

Abiotic Not associated with living organisms; synonymous with *abiological*.

Abiotic transformation The process in which a substance in the environment is modified by nonbiological mechanisms.

Absorption The penetration of atoms, ions, or molecules into the bulk mass of a substance.

Abyssal zone The portion of the ocean floor below 3281–6561 ft where light does not penetrate and where temperatures are low and pressures are intense; this zone lies seaward of the continental slope and covers approximately 75% of the ocean floor; the temperature does not rise above 4°C (39°F); since oxygen is present, a diverse community of invertebrates and fishes do exist, and some have adapted to harsh environments such as hydrothermal vents of volcanic creation.

Acid A chemical capable of donating a positively charged hydrogen atom (proton, H+) or capable of forming a covalent bond with an electron pair; an acid increases the hydrogen ion concentration in a solution, and it can react with certain metals.

Acid anhydride An organic compound that reacts with water to form an acid.

Acidophiles Metabolically active in highly acidic environments, and often have a high heavy metal resistance.

Acyclic A compound with straight or branched carbon-carbon linkages but without cyclic (ring) structures.

Adhesion The degree to which oil will coat a surface, expressed as the mass of oil adhering per unit area. A test has been developed for a standard surface that gives a semiquantitative measure of this property.

Adsorption The retention of atoms, ions, or molecules on to the surface of another substance.

Aerobe An organism that needs oxygen for respiration and hence for growth.

Aerobic In the presence of, or requiring, oxygen; an environment or process that sustains biological life and growth, or occurs only when free (molecular) oxygen is present.

Aerobic bacteria Any bacteria requiring free oxygen for growth and cell division.

Aerobic conditions Conditions for growth or metabolism in which the organism is sufficiently supplied with oxygen.

Aerobic respiration The process whereby microorganisms use oxygen as an electron acceptor.

Alcohol An organic compound with a carbon bound to a hydroxyl (—OH) group; a hydroxyl group attached to an aromatic ring is called a phenol rather than an alcohol; a compound in which a hydroxy group (—OH) is attached to a saturated carbon atom (e.g., ethyl alcohol, C_2H_5OH).

Aldehyde An organic compound with a carbon bound to a —(C=O)—H group; a compound in which a carbonyl group is bonded to one hydrogen atom and to one alkyl group [RC(=O)H].

Aliphatic compound Any organic compound of hydrogen and carbon characterized by a linear- or branched-chain of carbon atoms; three subgroups of such compounds are alkanes, alkenes, and alkynes.

Aliphatic compounds A broad category of hydrocarbon compounds distinguished by a straight, or branched, open chain arrangement of the constituent carbon atoms, excluding aromatic compounds; the carbon-carbon bonds may be either single or multiple bonds—alkanes, alkenes, and alkynes are aliphatic hydrocarbons.

Alkaliphiles Organisms that have their optimum growth rate of at least 2 pH units above neutrality.

Alkalitolerants Organisms that are able to grow or survive at pH values above 9, but their optimum growth rate is around neutrality or less.

Alkane (paraffin) A group of *hydrocarbons* composed of only carbon and hydrogen with no double bonds or aromaticity. They are said to be "saturated" with hydrogen. They may by straight-chain (normal), branched or cyclic. The smallest alkane is methane (CH_4), the next, ethane (CH_3CH_3), then propane ($CH_3CH_2CH_3$), and so on.

Alkanes The homologous group of linear (acyclic) aliphatic hydrocarbons having the general formula C_nH_{2n+2}; alkanes can be straight chains (linear), branched chains, or ring structures; often referred to as paraffins.

Alkene (olefin) An unsaturated *hydrocarbon*, containing only hydrogen and carbon with one or more double bonds, but having no aromaticity. *Alkenes* are not typically found in crude oils, but can occur as a result of heating.

Alkenes Acyclic branched or unbranched hydrocarbons having one carbon-carbon double bond ($-C=C-$) and the general formula C_nH_{2n}; often referred to as olefins.

Alkoxide An ionic compound formed by removal of hydrogen ions from the hydroxyl group in an alcohol using a reactive metal such as sodium or potassium.

Alkyl A molecular fragment derived from an alkane by dropping a hydrogen atom from the formula; examples are methyl (CH_3) and ethyl (CH_2CH_3).

Alkyl groups A *hydrocarbon* functional group (C_nH_{2n+1}) obtained by dropping one hydrogen from fully saturated compound; e.g., methyl ($-CH_3$), ethyl ($-CH_2CH_3$), propyl ($-CH_2CH_2CH_3$), or isopropyl [$(CH_3)_2CH-$].

Alkyl radicals Carbon-centered radicals derived formally by removal of one hydrogen atom from an alkane, for example the ethyl radical (CH_3CH_2).

Alkyne A compound that consists of only carbon and hydrogen, that contains at least one carbon-carbon triple bond; alkyne names end with *yne*.

Alkynes The group of acyclic branched or unbranched hydrocarbons having a carbon-carbon triple bond ($-C≡C-$).

Ambient The surrounding environment and prevailing conditions.

Amide An organic compound that contains a carbonyl group bound to nitrogen; the simplest amides are formamide ($HCONH_2$) and acetamide (CH_3CONH_2).

Amine An organic compound that contains a nitrogen atom bound only to carbon and possibly hydrogen atoms; examples are methylamine, CH_3NH_2; dimethylamine, CH_3NHCH_3; and trimethylamine, $(CH_3)_3N$.

Amino acid A molecule that contains at least one amine group ($-NH2$) and at least one carboxylic acid group ($-COOH$); when these groups are both attached to the same carbon, the acid is an α-amino acid—a α-amino acids are the basic building blocks of proteins.

Amorphous solid A noncrystalline solid having no well-defined ordered structure.

Amphoteric molecule A molecule that behaves both as an acid and as a base, such as hydroxy pyridine:

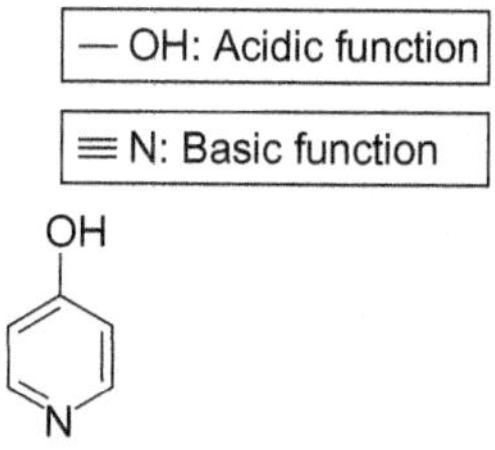

— OH: Acidic function

≡ N: Basic function

OH

Anaerobe An organism that does not need free-form oxygen for growth. Many anaerobes are even sensitive to free oxygen.

Anaerobic A biologically mediated process or condition not requiring molecular or free oxygen; relating to a process that occurs with little or no oxygen present.

Anaerobic bacteria Any bacteria that can grow and divide in the partial or complete absence of oxygen.

Anaerobic respiration The process whereby microorganisms use a chemical other than oxygen as an electron acceptor; common substitutes for oxygen are nitrate, sulfate, and iron.

Analyte The component of a system to be analyzed—for example, chemical elements or ions in groundwater sample.

Anion An atom or molecule that has a negative charge; a negatively charged ion.

Anoxic An environment without oxygen.

Aphotic zone The deeper part of the ocean beneath the photic zone, where light does not penetrate sufficiently for photosynthesis to occur.

API gravity An American Petroleum Institute measure of *density* for petroleum:
API gravity $= [141.5/(\text{specific gravity at } 15.6°C) - 131.5]$
Fresh water has a gravity of 10°API. The scale is commercially important for ranking oil quality; heavy oils are typically $<20°$API; medium oils are 20–35°API; light oils are 35–45°API.

Aquifer A water-bearing layer of soil, sand, gravel, or rock or other geologic formation that will yield usable quantities of water to a well under normal hydraulic gradients or by pumping.

Arene A hydrocarbon that contains at least one aromatic ring.

Aromatic Organic cyclic compounds that contain one or more benzene rings; these can be monocyclic, bicyclic, or polycyclic hydrocarbons and their substituted derivatives. In aromatic ring structures, every ring carbon atom possesses one double bond.

Aromatic ring An exceptionally stable planar ring of atoms with resonance structures that consist of alternating double and single bonds, such as benzene:

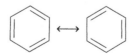

Aromatic compound A compound containing an aromatic ring; aromatic compounds have strong, characteristic odors.

Aryl A molecular fragment or group attached to a molecule by an atom that is on an aromatic ring.

Asphaltene fraction A complex mixture of heavy organic compounds precipitated from crude oil and *bitumen* by natural processes or in laboratory by addition of excess *n*-pentane, or *n*-heptane; after precipitation of the *asphaltene fraction*, the remaining oil or *bitumen* consists of *saturates*, *aromatics*, and *resins*.

Assay Qualitative or (more usually) quantitative determination of the components of a material or system.

Asymmetric carbon A carbon atom covalently bonded to four different atoms or groups of atoms.

Atomic number The atomic number is equal to the number of positively charged protons in the nucleus of an atom which determines the identity of the element.

ATSDR Agency for toxic substances and disease registry.

Attenuation The set of human-made or natural processes that either reduce or appear to reduce the amount of a chemical compound as it migrates away or is disposed from one specific point towards another point in space or time; for example, the apparent reduction in the amount of a chemical in a groundwater plume as it migrates away from its source; degradation, dilution, dispersion, sorption, or volatilization are common processes of attenuation.

Auto-ignition temperature (AIT) A fixed temperature above which a flammable mixture is capable of extracting sufficient energy from the environment to self-ignite.

Avogadro's number The number of molecules (6.023×10^{23}) in 1 g-mole of a substance.

B

Benthic zone The ecological region at the lowest level of a body of water such as an ocean or a lake, including the sediment surface and some subsurface layers; organisms living in this zone (benthos or benthic organisms) generally live in close relationship with the substrate bottom; many such organisms are permanently attached to the bottom; because light does not penetrate very deep ocean-water, the energy source for the benthic ecosystem is often organic matter from higher up in the water column which sinks to the depths.

Benzene A colorless liquid formed from both anthropogenic activities and natural processes; widely used in the US and ranks in the top 20 chemicals used; a natural part of crude oil, gasoline, and cigarette smoke; one of the major components of JP-8 fuel.

Bimolecular reaction The collision and combination of two reactants involved in the rate-limiting step.

Bioaccumulation The accumulation of substances, such as pesticides, or other chemicals in an organism; occurs when an organism absorbs a chemical—possibly a toxic chemical—at a rate faster than that at which the substance is lost by catabolism and excretion; the longer the biological half-life of a toxic substance the greater the risk of chronic poisoning, even if environmental levels of the toxin are not very high; see Biomagnification.

Bio-augmentation A process in which acclimated microorganisms are added to soil and groundwater to increase biological activity. Spray irrigation is typically used for shallow contaminated soils, and injection wells are used for deeper contaminated soils.

Biodegradation The natural process whereby bacteria or other microorganisms chemically alter and break down organic molecules; the breakdown or transformation of a chemical substance or substances by microorganisms using the substance as a carbon and/or energy source.

Biological marker (biomarker) Complex organic compounds composed of carbon, hydrogen, and other elements which are found in oil, *bitumen*, rocks, and sediments and which have undergone little or no change in structure from their parent organic molecules in living organisms; typically, biomarkers are isoprenoids, composed of isoprene subunits; biomarkers include compounds such as pristane, phytane, triterpane derivatives, steranes derivatives, and porphyrin derivatives.

Biomagnification The increase in the concentration of heavy metals (i.e., mercury) or organic contaminants such as chlorinated hydrocarbons, in organisms as a result of their consumption within a food chain/web; an example is the process by which contaminants such as polychlorobiphenyl derivatives (PCBs) accumulate or magnify as they move up the food chain—PCBs concentrate in tissue and internal organs, and as big fish eat little fish, they accumulate all the PCBs that have been eaten by everyone below them in the food chain; can occur as a result of: (1) persistence, in which the chemical cannot be broken down by environmental processes, (2) food chain energetics, in which the concentration of the chemical increases progressively as it moves up a food chain, and (3) a low or nonexistent rate of internal degradation or excretion of the substance that is often due to water-insolubility.

Bioremediation A treatment technology that uses biological activity to reduce the concentration or toxicity of contaminants: materials are added to contaminated environments to accelerate natural biodegradation.

Biota Living organisms.

Bitumen A complex mixture of *hydrocarbonaceous constituents* of natural or pyrogenous origin or a combination of both.

Boiling liquid expanding vapor explosion (BLEVE) An event that occurs when a vessel containing a liquid at a temperature above its atmospheric pressure boiling point, ruptures; the explosive vaporization of a large fraction of the vessel contents; possibly followed by the combustion or explosion of the vaporized cloud if it is combustible (similar to a rocket).

Boiling point The temperature at which a liquid begins to boil—that is, it is the temperature at which the vapor pressure of a liquid is equal to the atmospheric or external pressure. The boiling point distributions of crude oils and petroleum products may be in a range from 30 to in excess of 700°C (86–1290°F).

Breakdown product A compound derived by chemical, biological, or physical action on a chemical compound; the breakdown is a process which may result in a more toxic or a less toxic compound and a more persistent or less persistent compound than the original compound.

BTEX The collective name given to benzene, toluene, ethylbenzene and the xylene isomers (*p*-, *m*-, and *o*-xylene); a group of volatile organic compounds (VOCs) found in petroleum hydrocarbons, such as gasoline, and other common environmental contaminants.

BTX The collective name given to benzene, toluene, and the xylene isomers (*p*-, *m*-, and *o*-xylene); a group of volatile organic compounds (VOCs) found in petroleum hydrocarbons, such as gasoline, and other common environmental contaminants.

Benzene   Toluene

ortho-Xylene   meta-Xylene   para-Xylene

C

Carbenium ion A generic name for carbocation that has at least one important contributing structure containing a tervalent carbon atom with a vacant p orbital.

Carbanion The generic name for anions containing an even number of electrons and having an unshared pair of electrons on a carbon atom (e.g., Cl_3C^-).

Carbon Element number 6 in the periodic table of elements.

Carbon preference index (CPI) The ratio of odd to even n-alkanes; odd/even CPI *alkanes* are equally abundant in petroleum but not in biological material—a CPI near 1 is an indication of petroleum.

Carbon tetrachloride A manufactured compound that does not occur naturally; produced in large quantities to make refrigeration fluid and propellants for aerosol cans; in the past, carbon tetrachloride was widely used as a cleaning fluid, in industry and dry cleaning businesses, and in the household; also used in fire extinguishers and as a fumigant to kill insects in grain—these uses were stopped in the mid-1960s.

Carbonyl group A divalent group consisting of a carbon atom with a double bond to oxygen; for example, acetone ($CH_3—(C=O)—CH_3$) is a carbonyl group linking two methyl groups.

Carboxylic acid An organic molecule with a $—CO_2H$ group; hydrogen atom on the $—CO_2H$ group ionizes in water; the simplest carboxylic acids are formic acid ($H—COOH$) and acetic acid ($CH_3—COOH$).

Catabolism The breakdown of complex molecules into simpler ones through the oxidation of organic substrates to *provide* biologically available energy—ATP (adenosine triphosphate) is an example of such a molecule.

Catalysis The process where a catalyst increases the rate of a chemical reaction without modifying the overall standard Gibbs energy change in the reaction.

Catalyst A substance that alters the rate of a chemical reaction and may be recovered essentially unaltered in form or amount at the end of the reaction.

Cation exchange The interchange between a cation in solution and another cation in the boundary layer between the solution and surface of negatively charged material such as clay or organic matter.

Cation exchange capacity (CEC) The sum of the exchangeable bases plus total soil acidity at a specific pH, usually 7.0 or 8.0. When acidity is expressed as salt extractable acidity, the cation exchange capacity is called the effective cation exchange capacity (ECEC), because this is considered to be the CEC of the exchanger at the native pH value; usually expressed in centimoles of charge per kilogram of exchanger (cmol/kg) or millimoles of charge per kilogram of exchanger.

Cellulose A polysaccharide, polymer of glucose, that is found in the cell walls of plants; a fiber that is used in many commercial products, notably paper.

CERCLA Comprehensive Environmental Response, Compensation, and Liability Act. This law created a tax on the chemical and petroleum industries and provided broad federal authority to respond directly to releases or threatened releases of hazardous substances that may endanger public health or the environment.

Chain reaction A reaction in which one or more reactive reaction intermediates (frequently radicals) are continuously regenerated, usually through a repetitive cycle of elementary steps (the *propagation step*); for example, in the chlorination of methane by a radical mechanism, Cl is continuously regenerated in the chain propagation steps:

$$Cl \bullet + CH_4 \rightarrow HCl + H_3C \bullet$$
$$H_3C \bullet + Cl_2 \rightarrow CH_3Cl + Cl \bullet$$

In chain polymerization reactions, reactive intermediates of the same types, generated in successive steps or cycles of steps, differ in relative molecular mass.

Check standard An analyte with a well-characterized property of interest, e.g., concentration, density, and other properties that is used to verify method, instrument and operator performance during regular operation; *check standards* may be obtained from a certified supplier, may be a pure substance with properties obtained from the literature or may be developed in-house.

Chemical bond The forces acting among two atoms or groups of atoms that lead to the formation of an aggregate with sufficient stability to be considered as an independent molecular species.

Chemical dispersion In relation to oil spills, this term refers to the creation of oil-in-water *emulsions* by the use of chemical dispersants made for this purpose.

Chemical induction (coupling) when one reaction accelerates another in a chemical system there is said to be chemical induction or coupling. Coupling is caused by an intermediate or by-product of the inducing reaction that participates in a second reaction; chemical induction is often *observed* in oxidation-reduction reactions.

Chemical reaction A process that results in the interconversion of chemical species.

Chemical species An ensemble of chemically identical molecular entities that can explore the same set of molecular energy levels on the time scale of the experiment; the term is applied equally to a set of chemically identical atomic or molecular structural units in a solid array.

Chemical waste Any solid, liquid, or gaseous waste material that, if improperly managed or disposed of, may pose substantial hazards to human health and the environment.

Chemical weight The weight of a molar sample as determined by the weight of the molecules (the molecular weight); calculated from the weights of the atoms in the molecule.

Chemistry The science that studies matter and all of the possible transformations of matter.

Chlorinated solvent A volatile organic compound containing chlorine; common solvents are trichloroethylene, tetrachloroethylene, and carbon tetrachloride.

Chlorofluorocarbon Gases formed of chlorine, fluorine, and carbon whose molecules normally do not react with other substances; formerly used as spray-can propellants, they are known to destroy the protective ozone layer of the Earth.

Chromatography A method of chemical analysis where compounds are separated by passing a mixture in a suitable carrier over an absorbent material; compounds with different absorption coefficients move at different rates and are separated.

cis *trans* isomers The difference in the positions of atoms (or groups of atoms) relative to a reference plane in an organic molecule; in a *cis*-isomer, the atoms are on the same side of the molecule, but are on opposite sides in the *trans*-isomer; sometimes called stereoisomers; these arrangements are common in alkenes and cycloalkanes.

Clay A very fine-grained soil that is plastic when wet but hard when fired; typical clay minerals consist of silicate and aluminosilicate minerals that are the products of weathering reactions of other minerals; the term is also used to refer to any mineral of very small particle size.

Clean Water Act The Clean Water Act establishes the basic structure for regulating discharges of pollutants into the waters of the United States. It gives EPA the authority to implement pollution control programs such as setting wastewater standards for industry; also continued requirements to set water quality standards for all contaminants in surface waters and makes it unlawful for any person to discharge any pollutant from a point source into navigable waters, unless a permit was obtained under its provisions.

Coke A hard, dry substance containing carbon that is produced by heating bituminous coal or other carbonaceous materials to a very high temperature in the absence of air; used as a fuel.

Cometabolism The process by which compounds in petroleum may be enzymatically attacked by microorganisms without furnishing carbon for cell growth and division; a variation on biodegradation in which microbes transform a contaminant even though the contaminant cannot serve as the primary energy source for the organisms. To degrade the contaminant, the microbes require the presence of other compounds (primary substrates) that can support their growth.

Complex modulus A measure of the overall resistance of a material to flow under an applied stress, in units of force per unit area. It combines *viscosity* and elasticity elements to provide a measure of "stiffness," or resistance to flow. The *complex modulus* is more useful than *viscosity* for assessing the physical behavior of very non-Newtonian materials such as *emulsions*.

Compound The combination of two or more different elements, held together by chemical bonds; the elements in a given compound are always combined in the same proportion by mass (law of definite proportion).

Concentration Composition of a mixture characterized in terms of mass, amount, volume, or number concentration with respect to the volume of the mixture.

Conservative constituent or compound One that does not degrade, is unreactive, and its *movement* is not retarded within a given environment (aquifer, stream, contaminant plume).

Constituent An essential part or component of a system or group (that is, an ingredient of a chemical mixture); for example, benzene is one constituent of gasoline.

Contaminant A pollutant that has some detrimental effect, can cause deviation from the normal composition of an environment.

Corrosion Oxidation of a metal in the presence of air and moisture.

Covalent bond A region of *relatively* high electron density between atomic nuclei that results from sharing of electrons and that *gives* rise to an attractive force and a characteristic internuclear distance; carbon-hydrogen bonds are covalent bonds.

Cracking The process in which large molecules are broken down (thermally decomposed) into smaller molecules; used especially in the petroleum refining industry.

Critical point The combination of critical temperature and critical pressure; the temperature and pressure at which two phases of a substance in equilibrium become identical and form a single phase.

Critical pressure The pressure required to liquefy a gas at its critical temperature.

Critical temperature The temperature above which a gas cannot be liquefied, regardless of the amount of pressure applied.

Culture The growth of cells or microorganisms in a controlled artificial environment.

Cycloalkanes (naphthene, cycloparaffin) A saturated, cyclic compound containing only carbon and hydrogen. One of the simplest *cycloalkanes* is cyclohexane (C_6H_{12}); steranes derivatives and triterpane derivatives are branched naphthene derivatives consisting of multiple condensed five- or six-carbon rings.

D

Daughter product A compound that results directly from the degradation of another chemical.

Deflagration An explosion with a flame front moving in the unburned gas at a speed below the speed of sound (1250 ft/s).

Degradation The breakdown or transformation of a compound into byproducts and/or end products.

Dehydration reaction (condensation reaction) A chemical reaction in which two organic molecules become linked to each other via covalent bonds with the removal of a molecule of water; common in synthesis reactions of organic chemicals.

Dehydrohalogenation Removal of hydrogen and halide ions from an alkane resulting in the formation of an alkene.

Denitrification Bacterial reduction of nitrate to nitrite to gaseous nitrogen or nitrous oxides under anaerobic conditions.

Density The mass per unit volume of a substance. *Density* is temperature-dependent, generally decreasing with temperature. The density of oil relative to water, its specific gravity, governs whether a particular oil will float on water. Most fresh crude oils and fuels will float on water. Bitumen and certain residual fuel oils, however, may have densities greater than water at some temperature ranges and may submerge in water. The density of a spilled oil will also increase with time as components are lost due to weathering.

Detection limit (in analysis) The minimum single result that, with a stated probability, can be distinguished from a representative blank value during the laboratory analysis of substances such as water, soil, air, rock, and biota.

Detonation An explosion with a shock wave moving at a speed greater than the speed of sound in the unreacted medium.

1,4-Dichlorobenzene A chemical used to control moths, molds, and mildew and to deodorize restrooms and waste containers; does not occur naturally but is produced by chemical companies to make products for home use and other chemicals such as resins; most of the 1,4-dichlorobenzene enters the environment as a result of its use in moth-repellant products and in toilet-deodorizer blocks. Because it changes from a solid to a gas easily (sublimes), almost all 1,4-dichlorobenzene produced is released into the air.

Dichloroelimination Removal of two chlorine atoms from an alkane compound and the formation of an alkene compound within a reducing environment.

Dichloromethane (CH_2Cl_2): An organic solvent often used to extract organic substances from samples; toxic, but much less so than chloroform or carbon tetrachloride, which were previously used for this purpose.

Dihaloelimination Removal of two halide atoms from an alkane compound and the formation of an alkene compound within a reducing environment.

Diols Chemical compounds that contain two hydroxy (—OH) groups, generally assumed to be, but not necessarily, alcoholic; aliphatic diols are also called glycols.

Downgradient In the direction of decreasing static hydraulic head.

Dispersant (chemical dispersant) A chemical that reduces the surface tension between water and a hydrophobic substance such as oil. In the case of an oil spill, dispersants facilitate the breakup and dispersal of an oil slick throughout the water column in the form of an oil-in-water emulsion; chemical dispersants can only be used in areas where biological damage will not occur and must be approved for use by government regulatory agencies.

Double bond A covalent bond resulting from the sharing of two pairs of electrons (four electrons) between two atoms.

Drug Any substance presented for treating, curing, or preventing disease in human beings or in animals; a drug may also be used for making a medical diagnosis, managing pain, or for restoring, correcting, or modifying physiological functions.

E

Electron acceptor The compound that receives electrons (and therefore is reduced) in the energy-producing oxidation-reduction reactions that are essential for the growth of microorganisms and bioremediation—common electron acceptors in bioremediation are oxygen, nitrate, sulfate, and iron.

Electron donor The compound that donates electrons (and therefore is oxidized). In bioremediation the organic contaminant often serves as an electron donor.

Electronegativity The power of an atom to attract electrons to itself.

Elimination A reaction where two groups such as chlorine and hydrogen are lost from adjacent carbon atoms and a double bond is formed in their place.

Empirical formula The simplest whole-number ratio of atoms in a compound.

Emulsan Emulsan is a polyanionic heteropolysaccharide bioemulsifier produced by *Acinetobacter calcoaceticus* RAG-1; used to stabilize oil-in-water emulsions.

Emulsion A stable mixture of two immiscible liquids, consisting of a continuous phase and a dispersed phase. Oil and water can form both oil-in-water and water-in-oil emulsions. The former is termed a dispersion, while *emulsion* implies the latter. Water-in-oil emulsions formed from petroleum and brine can be grouped into four stability classes: stable, a formal emulsion that will persist indefinitely; meso-stable, which gradually degrade over time due to a lack of one or more stabilizing factors; entrained water, a mechanical mixture characterized by high viscosity of the petroleum component which impedes separation of the two phases; and unstable, which are mixtures that rapidly separate into immiscible layers.

Emulsion stability Generally accompanied by a marked increase in *viscosity* and elasticity, over that of the parent oil which significantly changes behavior. Coupled with the increased volume due to the introduction of brine, emulsion formation has a large effect on the choice of countermeasures employed to combat a spill.

Emulsification The process of *emulsion* formation, typically by mechanical mixing. In the environment, *emulsions* are most often formed as a result of wave action. Chemical agents can be used to prevent the formation of *emulsions* or to "break" the *emulsions* into their component oil and water phases.

Endergonic reaction A chemical reaction that requires energy to proceed. A chemical reaction is endergonic when the change in free energy is positive.

Endothermic reaction A chemical reaction in which heat is absorbed.

Engineered bioremediation A type of remediation that increases the growth and degradative activity of microorganisms by using engineered systems that supply nutrients, electron acceptors, and/or other growth-stimulating materials.

Enhanced bioremediation A process which involves the addition of microorganisms (e.g., fungi, bacteria, and other microbes) or nutrients (e.g., oxygen, nitrates) to the subsurface environment to accelerate the natural biodegradation process.

Entering group An atom or group that forms a bond to what is considered to be the main part of the substrate during a reaction, for example, the attacking nucleophile in a bimolecular nucleophilic substitution reaction.

Environment The total living and nonliving conditions of an organism's internal and external surroundings that affect an organism's complete life span.

Enzyme A macromolecule, mostly proteins or conjugated proteins produced by living organisms, that facilitate the degradation of a chemical compound (catalyst); in general, an enzyme catalyzes only one reaction type (reaction specificity) and operates on only one type of substrate (substrate specificity); any of a group of catalytic proteins that are produced by cells and that mediate or promote the chemical processes of life without themselves being altered or destroyed.

Epoxidation A reaction wherein an oxygen molecule is inserted in a carbon-carbon double bond and an epoxide is formed.

Epoxides A subclass of epoxy compounds containing a saturated three-membered cyclic ether. See *Epoxy compounds*.

Epoxy compounds Compounds in which an oxygen atom is directly attached to two adjacent or nonadjacent carbon atoms in a carbon chain or ring system; thus cyclic ethers.

Equipment blank A sample of analyte-free media which has been used to rinse the sampling equipment. It is collected after completion of decontamination and prior to sampling. This blank is useful in documenting and controlling the preparation of the sampling and laboratory equipment.

Ester A compound formed from an acid and an alcohol; in esters of carboxylic acids, the —COOH group and the —OH group lose a molecule of water and form a —COO— bond (R_1 and R_2 represent organic groups):

$$R_1COOH + R_2OH \rightarrow R_1COOR_2 + H_2O$$

Ether A compound with an oxygen atom attached to two hydrocarbon groups. Any carbon compound containing the functional group C—O—C, such as diethyl ether ($C_2H_5O\,C_2H_5$).

Ethylbenzene A colorless, flammable liquid found in natural products such as coal tar and crude oil; it is also found in manufactured products such as inks, insecticides, and paints; a minor component of JP-8 fuel.

Exothermic reaction A reaction that produces heat and absorbs heat from the surroundings.

Ex situ bioremediation A process which involves removing the contaminated soil or water to another location before treatment.

F

Facultative anaerobes Microorganisms that use (and prefer) oxygen when it is available, but can also use alternate electron acceptors such as nitrate under anaerobic conditions when necessary.

Fatty acids carboxylic acids with long hydrocarbon side chains; most natural fatty acids have hydrocarbon chains that don't branch; any double bonds occurring in the chain are *cis* isomers—the side chains are attached on the same side of the double bond.

cis   *trans*

Fermentation The process whereby microorganisms use an organic compound as both electron donor and electron acceptor, converting the compound to fermentation products such as organic acids, alcohols, hydrogen, and carbon dioxide; microbial metabolism in which a particular compound is used both as an electron donor and an electron acceptor resulting in the production of oxidized and reduced daughter products.

Field capacity or in situ (field water capacity) The water content, on a mass or volume basis, remaining in soil 2 or 3 days after having been wetted with water and after free drainage is negligible.

Fingerprint A chromatographic signature of relative intensities used in oil-oil or oil-source rock correlations; mass chromatograms of steranes derivatives or terpane derivatives are examples of fingerprints that can be used for qualitative or quantitative comparison of crude oil.

Flammability limits A gas mixture will not bum when the composition is lower than the lower flammable limit (LFL); the mixture is also not combustible when the composition is above the upper flammability limit (UFL).

Flash point The temperature at which the vapor over a liquid will ignite when exposed to an ignition source. A liquid is considered to be flammable if its *flash point* is less than 60° C. *Flash point* is an extremely important factor in relation to the safety of spill cleanup operations. Gasoline and other light fuels can ignite under most ambient conditions and therefore are a serious hazard when spilled. Many freshly spilled crude oils also have low *flash points* until the lighter components have evaporated or dispersed.

Fluids Liquids; also a generic term applied to all substances that flow freely, such as gases and liquids.

Foam A colloidal suspension of a gas in a liquid.

Fraction One of the portions of a chemical mixture separated by chemical or physical means from the remainder.

Free radical A molecule with an odd number of electrons—they do not have a completed octet and often undergo vigorous redox reactions.

Fugitive emissions Emissions that include losses from equipment leaks, or evaporative losses from impoundments, spills, or leaks.

Functional group   An atom or a group of atoms attached to the base structure of a compound that has similar chemical properties irrespective of the compound to which it is a part; a means of defining the characteristic physical and chemical properties of families of organic compounds.

G

Gas   Matter that has no definite volume or definite shape and always fills any space given in which it exists.

Gas chromatography (GC)   A separation technique involving passage of a gaseous moving phase through a column containing a fixed liquid phase; it is used principally as a quantitative analytical technique for compounds that are volatile or can be converted to volatile forms.

Gaseous nutrient injection   A process in which nutrients are fed to contaminated groundwater and soil via wells to encourage and feed naturally occurring microorganisms—the most commonly added gas is air in the presence of sufficient oxygen, microorganisms convert many organic contaminants to carbon dioxide, water, and microbial cell mass. In the absence of oxygen, organic contaminants are metabolized to methane, limited amounts of carbon dioxide, and trace amounts of hydrogen gas. Another gas that is added is methane. It enhances degradation by cometabolism in which as bacteria consume the methane, they produce enzymes that react with the organic contaminant and degrade it to harmless minerals.

GC-MS   Gas chromatography-mass spectrometry.

GC-TPH   GC detectable total petroleum hydrocarbons, that is the sum of all GC-resolved and unresolved hydrocarbons. The resolvable hydrocarbons appear as peaks and the unresolvable hydrocarbons appear as the area between the lower baseline and the curve defining the base of resolvable peaks.

Geological time   The span of time that has passed since the creation of the Earth and its components; a scale use to measure geological events millions of years ago.

Glycerol   A small molecule with three alcohol groups ($HOCH_2CH(OH)CH_2OH$); basic building block of fats and oils.

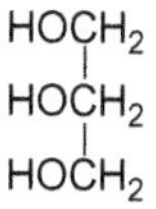

$$
\begin{array}{c}
HOCH_2 \\
| \\
HOCH_2 \\
| \\
HOCH_2
\end{array}
$$

Gram equivalent weight (nonredox reaction)   The mass in grams of a substance equivalent to 1 g-atom of hydrogen, 0.5 g-atom of oxygen, or 1 g-ion of the hydroxyl ion; can be determined by dividing the molecular weight by the number of hydrogen atoms or hydroxyl ions (or their equivalent) supplied or required by the molecule in a given reaction.

Gram equivalent weight (redox reaction)   The molecular weight in grams divided by the change in oxidation state.

Gravimetric analysis   A technique of quantitative analytical chemistry in which a desired constituent is efficiently recovered and weighed.

Greenhouse effect   The warming of an atmosphere by its absorption of infrared radiation while shortwave radiation is allowed to pass through.

Guest moleculare (or ion) An organic or inorganic ion or molecule that occupies a cavity, cleft, or pocket within the molecular structure of a host molecular entity and forms a complex with the host entity or that is trapped in a cavity within the crystal structure of a host.

H

Half-life The time required to reduce the concentration of a chemical to 50% of its initial concentration; units are typically in hours or days.

Halide An element from the halogen group, which include fluorine, chlorine, bromine, iodine, and astatine.

Halogen Group 17 in the periodic table of the elements; these elements are the reactive nonmetals and are electronegative.

Henry's law The relation between the partial pressure of a compound and the equilibrium concentration in the liquid through a proportionality constant known as the Henry's law constant.

Henry's law constant The concentration ratio between a compound in air (or vapor) and the concentration of the compound in water under equilibrium conditions.

Herbicide A chemical that controls or destroys unwanted plants, weeds, or grasses.

Heterocyclic An organic group or molecule containing rings with at least one noncarbon atom in the ring.

Heterogeneous Varying in structure or composition at different locations in space.

Heterotroph An organism that cannot synthesize its own food and is dependent on complex organic substances for nutrition.

Heterotrophic bacteria Bacteria that utilize organic carbon as a source of energy; organisms that derive carbon from organic matter for cell growth.

Hopane A pentacyclic *hydrocarbon* of the *triterpane* group believed to be derived primarily from bacteriohopanoids in bacterial membranes.

Homogeneous Having uniform structure or composition at all locations in space.

Homolog A compound belonging to a series of compounds that differ by a repeating group; for example, propanol ($CH_3CH_2CH_2OH$), n-butanol ($CH_3CH_2CH_2CH_2OH$), and n-pentanol ($CH_3CH_2CH_2CH_2CH_2OH$) are homologs; they belong to the homologous series of alcohols: $CH_3(CH_2)_nOH$.

Hydration The addition of a water molecule to a compound within an aerobic degradation pathway.

Hydrocarbon One of a very large and diverse group of chemical compounds composed only of carbon and hydrogen; the largest source of hydrocarbons is petroleum crude oil; the principal constituents of crude oils and refined petroleum products.

Hydrogen bond A form of association between an electronegative atom and a hydrogen atom attached to a second, relatively electronegative atom; best considered as an electrostatic interaction, heightened by the small size of hydrogen, which permits close proximity of the interacting dipoles or charges.

Hydrogenation A reaction where hydrogen is added across a double or triple bond, usually with the assistance of a catalyst; a process whereby an enzyme in certain microorganisms catalyzes the hydrolysis or reduction of a substrate by molecular hydrogen.

Hydrogenolysis A reductive reaction in which a carbon-halogen bond is broken, and hydrogen replaces the halogen substituent.

Hydrolysis A chemical transformation process in which a chemical reacts with water. In the process, a new carbon-oxygen bond is formed with oxygen derived from the water molecule, and a bond is cleaved within the chemical between carbon and some functional group.

Hydrophilic Water loving; the capacity of a molecular entity or of a substituent to interact with polar solvents, in particular with water, or with other polar groups; hydrophilic molecules dissolve easily in water, but not in fats or oils.

Hydrophilic colloids Macromolecules that interact strongly with water and form colloids.

Hydrophilicity The tendency of a molecule to be solvated by water.

Hydrophobic Fear of water; the tendency to repel water.

Hydrophobic interaction The tendency of hydrocarbons (or of lipophilic hydrocarbon-like groups in solutes) to form intermolecular aggregates in an aqueous medium, and analogous intramolecular interactions.

Hydroxylation Addition of a hydroxyl group to a chlorinated aliphatic hydrocarbon.

Hydroxyl group A functional group that has a hydrogen atom joined to an oxygen atom by a polar covalent bond (—OH).

Hydroxyl ion One atom each of oxygen and hydrogen bonded into an ion (OH^-) that carries a negative charge.

Hydroxyl radical A radical consisting of one hydrogen atom and one oxygen atom; normally does not exist in a stable form.

I

Infiltration rate The time required for water at a given depth to soak into the ground.

Inhibition The decrease in rate of reaction brought about by the addition of a substance (inhibitor), by virtue of its effect on the concentration of a reactant, catalyst, or reaction intermediate.

Inoculum A small amount of material (either liquid or solid) containing bacteria removed from a culture in order to start a new culture.

Inorganic Pertaining to, or composed of, chemical compounds that are not organic, that is, contain no carbon-hydrogen bonds; examples include chemicals with no carbon and those with carbon in nonhydrogen-linked forms.

Inorganic chemistry The study of inorganic compounds, specifically the structure, reactions, catalysis, and mechanism of action.

Inorganic compound A compound that does not contain carbon chemically bound to hydrogen; carbonates, bicarbonates, carbides, and carbon oxides are considered inorganic compounds, even though they contain carbon.

In situ In its original place; unmoved; un excavated; remaining in the subsurface.

In situ bioremediation A process which treats the contaminated water or soil where it was found.

Interfacial tension The net energy per unit area at the interface of two substances, such as oil and water or oil and air. The air/liquid interfacial tension is often referred to as surface tension; the SI units for *interfacial tension* are milli-Newtons per meter (mN/m). The higher the *interfacial tension*, the less attractive the two surfaces are to each other and the more the size of the interface will be minimized. Low surface tensions can drive the spreading of one fluid on another. The surface tension of an oil, together with its viscosity, affects the rate at which spilled oil will spread over a water surface or into the ground.

Intermolecular forces Forces of attraction that exist between particles (atoms, molecules, ions) in a compound.

Internal Standard (IS) A pure analyte added to a sample extract in a known amount, which is used to measure the relative responses of other analytes and surrogates that are components of the same solution. The *internal standard* must be an analyte that is not a sample component.

Intramolecular (1) Description of any process that involves a transfer (of atoms, groups, electrons, etc.) or interactions (such as forces) between different parts of the same molecular entity; (2) relating to a comparison between atoms or groups within the same molecular entity.

Intrinsic bioremediation A type of bioremediation that manages the innate capabilities of naturally occurring microbes to degrade contaminants without taking any engineering steps to enhance the process.

Ionic bond A chemical bond or link between two atoms due to an attraction between oppositely charged (positive-negative) ions.

Ionic bonding Chemical bonding that results when one or more electrons from one atom or a group of atoms is transferred to another. Ionic bonding occurs between charged particles.

Ionic compounds Compounds where two or more ions are held next to each other by electrical attraction.

Isomers Compounds that have the same number and types of atoms—the same molecular formula—but differ in the structural formula, i.e., the manner in which the atoms are combined with each other.

IUPAC International Union of Pure and Applied Chemistry.

K

Ketone An organic compound that contains a carbonyl group (R_1COR_2).

L

Lag phase The growth interval (adaption phase) between microbial inoculation and the start of the exponential growth phase during which there is little or no microbial growth.

Latex A polymer of *cis*-1-4 isoprene; milky sap from the rubber tree *Hevea brasiliensis*.

Law A system of rules that are enforced through social institutions to govern behavior; can be made by a collective legislature or by a single legislator, resulting in statutes, by the executive through decrees and regulations, or by judges through binding precedent; the formation of laws themselves may be influenced by a constitution (written or tacit) and the rights encoded therein; the law shapes politics, economics, history, and society in various ways and serves as a mediator of relations between people. See also Regulations.

Leaving group An atom or group (charged or uncharged) that becomes detached from an atom in what is considered to be the residual or main part of the substrate in a specified reaction.

Lignin A complex amorphous polymer in the secondary cell wall (middle lamella) of woody plant cells that cements or naturally binds cell walls to help make them rigid; highly resistant to decomposition by chemical or enzymatic action; also acts as support for cellulose fibers.

Lipophilic F-loving; applied to molecular entities (or parts of molecular entities) having a tendency to dissolve in fatlike (e.g., hydrocarbon) solvents.

Lipophilicity The affinity of a molecule or a moiety (portion of a molecular structure) for a lipophilic (fat soluble) environment. It is commonly measured by its distribution behavior in a biphasic system, either liquid-liquid (e.g., partition coefficient in octanol/water).

Loading rate The amount of material that can be absorbed per volume of soil.

LTU Land treatment unit; a physically delimited area where contaminated land is treated to remove/minimize contaminants and where parameters such as moisture, pH, salinity, temperature, and nutrient content can be controlled.

M

Macromolecule A large molecule of high molecular mass composed of more than 100 repeated monomers (single chemical units of lower relative mass); a large complex molecule formed from many simpler molecules.

Mass number The number of protons plus the number of neutrons in the nucleus of an atom.

Matter Any substance that has inertia and occupies physical space; can exist as solid, liquid, gas, plasma, or foam.

Measurement A description of a property of a system by means of a set of specified rules, that maps the property on to a scale of specified values, by direct or mathematical comparison with specified references.

Mechanical explosion An explosion due to the sudden failure of a vessel containing a nonreactive gas at a high pressure.

Melting point The temperature at which matter is converted from solid to liquid.

Metabolic by-product A product of the reaction between an electron donor and an electron acceptor; metabolic by-products include volatile fatty acids, daughter products of chlorinated aliphatic hydrocarbons, methane, and chloride.

Metabolism The physical and chemical processes by which foodstuffs are synthesized into complex elements, complex substances are transformed into simple ones, and energy is made available for use by an organism; thus all biochemical reactions of a cell or tissue, both synthetic and degradative, are included; the sum of all of the enzyme-catalyzed reactions in living cells that transform organic molecules into simpler compounds used in biosynthesis of cellular components or in extraction of energy used in cellular processes.

Metabolize A product of metabolism.

Methanogens Strictly anaerobic archaebacteria, able to use only a very limited spectrum of substrates (e.g., molecular hydrogen, formate, methanol, methylamine, carbon monoxide, or acetate) as electron donors for the reduction of carbon dioxide to methane.

Methanogenic The formation of methane by certain anaerobic bacteria (methanogens) during the process of anaerobic fermentation.

Methyl A group ($-CH_3$) derived from methane; for example, CH_3Cl is methyl chloride (systematic name: chloromethane) and CH_3OH is methyl alcohol (systematic name: methanol).

Microcosm A diminutive, representative system analogous to a larger system in composition, development, or configuration.

Microorganism (micro-organism) An organism of microscopic size that is capable of growth and reproduction through biodegradation of food sources, which can include hazardous contaminants; microscopic organisms including bacteria, yeasts, filamentous fungi, algae, and protozoa; a living organism too small to be seen with the naked eye; includes bacteria, fungi, protozoans, microscopic algae, and viruses.

Microbe The shortened term for microorganism.

Mineralization The biological process of complete breakdown of organic compounds, whereby organic materials are converted to inorganic products (e.g., the conversion of hydrocarbons to carbon dioxide and water); the release of inorganic chemicals from organic matter in the process of aerobic or anaerobic decay.

Moiety A term generally used to signify part of a molecule, e.g., in an ester R^1COOR^2, the alcohol moiety is R^2O.

Molality (m) The gram moles of solute divided by the kilograms of solvent.

Molarity (M) The gram moles of solute divided by the liters of solution.

Molecular weight The mass of one mole of molecules of a substance.

Molecule The smallest unit in a chemical element or compound that contains the chemical properties of the element or compound.

Mole fraction The number of moles of a component of a mixture divided by the total number of moles in the mixture.

Monoaromatic Aromatic hydrocarbons containing a single benzene ring.

Monosaccharide A simple sugar such as fructose or glucose that cannot be decomposed by hydrolysis; colorless crystalline substances with a sweet taste that have the same general formula, $C_nH_{2n}O_n$.

MTBE (methyl tertiary butyl ether) MTBE is a fuel additive which has been used in the United States since 1979. Its use began as a replacement for lead in gasoline because of health hazards associated with lead. MTBE has distinctive physical properties that result in it being highly soluble, persistent in the environment, and able to migrate through the ground. Environmental regulations have required the monitoring and cleanup of MTBE at petroleum contaminated sites since Feb. 1990; the program continues to monitor studies focusing on the potential health effects of MTBE and other fuel additives.

N

Natural organic matter (NOM) An inherently complex mixture of polyfunctional organic molecules that occurs naturally in the environment and is typically derived from the decay of floral and faunal remains; although they do occur naturally, the fossil fuels (coal, crude oil, and natal gas) are usually not included in the term *natural organic matter*.

NCP National Contingency Plan—also called the National Oil and Hazardous Substances Pollution Contingency Plan; provides a comprehensive system of accident reporting, spill containment, and cleanup, and established response headquarters (National Response Team and Regional Response Teams).

Nitrate enhancement A process in which a solution of nitrate is sometimes added to groundwater to enhance anaerobic biodegradation.

Normality (N) The gram equivalents of solute divided by the liters of solution.

Nucleophile A chemical reagent that reacts by forming covalent bonds with electronegative atoms and compounds.

Nutrients Major elements (e.g., nitrogen and phosphorus) and trace elements (including sulfur, potassium, calcium, and magnesium) that are essential for the growth of organisms.

O

Octane A flammable liquid (C_8H_{18}) found in petroleum and natural gas; there are 18 different octane isomers which have different structural formulas but share the molecular formula C_8H_{18}; used as a fuel and as a raw material for building more complex organic molecules.

Octanol-water partition coefficient (K_{ow}) The equilibrium ratio of a chemical's concentration in octanol (an alcoholic compound) to its concentration in the aqueous phase of a two-phase octanol-water system, typically expressed in log units (log K_{ow}); K_{ow} provides an indication of a chemical's solubility in fats (lipophilicity), its tendency to bioconcentrate in aquatic organisms, or sorb to soil or sediment.

Oleophilic Oil seeking or oil loving (e.g., nutrients that stick to or dissolve in oil).

Order of reaction A chemical rate process occurring in systems for which concentration changes (and hence the rate of reaction) are not themselves measurable, provided it is possible to measure a chemical flux.

Organic Compounds that contain carbon chemically bound to hydrogen; often contain other elements (particularly O, N, halogens, or S); chemical compounds based on carbon that also contain hydrogen, with or without oxygen, nitrogen, and other elements.

Organic carbon (soil) partition coefficient (K_{oc}) The proportion of a chemical sorbed to the solid phase, at equilibrium in a two-phase, water/soil or water/sediment system expressed on an organic carbon basis; chemicals with higher K_{oc} values are more strongly sorbed to organic carbon and, therefore, tend to be less mobile in the environment.

Organic chemistry The study of compounds that contain carbon chemically bound to hydrogen, including synthesis, identification, modeling, and reactions of those compounds.

Organic liquid nutrient injection An enhanced bioremediation process in which an organic liquid, which can be naturally degraded and fermented in the subsurface to result in the generation of hydrogen. The most commonly added for enhanced anaerobic bioremediation include lactate, molasses, hydrogen release compounds (HRCs), and vegetable oils.

Organochlorine compounds (chlorinated hydrocarbons) Organic pesticides that contain chlorine, carbon, and hydrogen (such as DDT); these pesticides affect the central nervous system.

Organophosphorus compound A compound containing phosphorus and carbon; many pesticides and most nerve agents are organophosphorus compounds, such as malathion.

Osmotic potential Expressed as a negative value (or zero), indicates the ability of the soil to dissolve salts and organic molecules; the reduction of soil water osmotic potential is caused by the presence of dissolved solutes.

OPA Oil Pollution Act of 1990; an act which addresses oil pollution and establishes liability for the discharge and substantial threat of a discharge of oil to navigable waters and shorelines of the US.

Oven dry The weight of a soil after all water has been removed by heating in an oven at a specified temperature (usually in excess of 100°C, 212°F) for water; temperatures will vary if other solvents have been used.

Oxidation The transfer of electrons away from a compound, such as an organic contaminant; the coupling of oxidation to reduction (see below) usually supplies energy that microorganisms use for growth and reproduction. Often (but not always), oxidation results in the addition of an oxygen atom and/or the loss of a hydrogen atom.

Oxygen enhancement with hydrogen peroxide An alternative process to pumping oxygen gas into groundwater involves injecting a dilute solution of hydrogen peroxide. Its chemical formula is H_2O_2, and it easily releases the extra oxygen atom to form water and free oxygen. This circulates through the contaminated groundwater zone to enhance the rate of aerobic biodegradation of organic contaminants by naturally occurring microbes. A solid peroxide product [e.g., oxygen releasing compound (ORC)] can also be used to increase the rate of biodegradation.

Oxidation-reduction reactions (redox reactions) Reactions that involve oxidation of one reactant and reduction of another.

Ozone (O_3) A form of oxygen containing three atoms instead of the common two (O_2); formed by high-energy ultraviolet radiation reacting with oxygen.

P

PAHs Polycyclic aromatic hydrocarbons. Alkylated *PAHs* are *alkyl group* derivatives of the parent *PAHs*. The five target alkylated *PAHs* referred to in this report are the alkylated naphthalene, phenanthrene, dibenzothiophene, fluorene, and chrysene series.

Paraffin An alkane.

Pathogen An organism that causes disease (e.g., any bacteria or viruses).

Perfluorocarbon (PFC) A derivative of hydrocarbons in which all of the hydrogens have been replaced by fluorine.

Permeability The capability of the soil to allow water or air movement through it. The quality of the soil that enables water to move downward through the profile, measured as the number of inches per hour that water moves downward through the saturated soil.

pH A measure of the acidity or basicity of a solution; the negative logarithm (base 10) of the hydrogen ion concentration in gram ions per liter.

Phenol A molecule containing a benzene ring that has a hydroxyl group substituted for a ring hydrogen.

Phenyl A molecular group or fragment formed by abstracting or substituting one of the hydrogen atoms attached to a benzene ring.

Photic zone The upper layer within bodies of water reaching down to about 200 m, where sunlight penetrates and promotes the production of photosynthesis; the richest and most diverse area of the ocean.

Physical change Refers to the change that occurs when a material changes from one physical state to another without formation of intermediate substances of different composition in the process, such as the change from gas to liquid.

Phytodegradation The process in which some plant species can metabolize VOC contaminants. The resulting metabolic products include trichloroethanol, trichloroacetic acid, and dichloracetic acid; mineralization products are probably incorporated into insoluble products such as components of plant cell walls.

Phytovolatilization The process in which VOCs are taken up by plants and discharged into the atmosphere during transpiration.

PM$_{10}$ Particulate matter below 10 μm in diameter; this corresponds to the particles inhalable into the human respiratory system, and its measurement uses a size selective inlet.

PM$_{2.5}$ Particulate matter below 2.5 μm in diameter; this is closer to, but slightly finer than, the definitions of respirable dust that have been used for many years in industrial hygiene to identify dusts which will penetrate the lungs.

Point emissions Emissions that occur through confined air streams as found in stacks, ducts, or pipes.

Polar compound An organic compound with distinct regions of positive and negative charge. *Polar compounds* include alcohols, such as sterols, and some *aromatics*, such as monoaromatic steroids. Because of their polarity, these compounds are more soluble in polar solvents, including water, compared to nonpolar compounds of similar molecular structure.

Polymer A large molecule made by linking smaller molecules (monomers) together.

Pour point The lowest temperature at which an oil will appear to flow under ambient pressure over a period of 5 s. The *pour point* of crude oils generally varies from −60° C to 30°C. Lighter oils with low *viscosities* generally have lower *pour points*.

Primary substrates The electron donor and electron acceptor that are essential to ensure the growth of microorganisms. These compounds can be viewed as analogous to the food and oxygen that are required for human growth and reproduction.

Propagule Any part of a plant (e.g., bud) that facilitates dispersal of the species and from which a new plant may form.

Propane A colorless, odorless, flammable gas (C_3H_8), found in petroleum and natural gas; used as a fuel and as a raw material for building more complex organic molecules.

R

Radical (free radical) A molecular entity such as $CH_3{}^{\bullet}$, $Cl^{\bullet}$ possessing an unpaired electron.

Rate A derived quantity in which time is a denominator quantity so that the progress of a reaction is measured with time.

Rate constant, k See *Order of reaction.*

Rate-controlling step (rate-limiting step, rate-determining step) The elementary reaction in which the largest control factor exerts the strongest influence on the rate; a step having a control factor much larger than any other step is said to be rate-controlling.

Recalcitrant Unreactive, nondegradable, refractory.

Receptor An object (animal, vegetable, or mineral) or a locale that is affected by the pollutant.

Redox (reduction-oxidation reactions) Oxidation and reduction occur simultaneously; in general, the oxidizing agent gains electrons in the process (and is reduced) while the reducing agent donates electrons (and is oxidized).

Reduction The transfer of electrons to a compound, such as oxygen, that occurs when another compound is oxidized.

Reductive dehalogenation A variation on biodegradation in which microbially catalyzed reactions cause the replacement of a halogen atom on an organic compound with a hydrogen atom. The reactions result in the net addition of two electrons to the organic compound.

Regulation A concept of management of complex systems according to a set of rules (laws) and trends; can take many forms: legal restrictions promulgated by a government authority, contractual obligations (such as contracts between insurers and their insureds), social regulation, coregulation, third-party regulation, certification, accreditation, or market regulation. See Law.

Releases On-site discharge of a toxic chemical to the surrounding environment; includes emissions to the air, discharges to bodies of water, releases at the facility to land, as well as contained disposal into underground injection wells.

Releases (to air, point and fugitive air emissions) All air emissions from industry activity; point emissions occur through confined air streams as found in stacks, ducts, or pipes; fugitive emissions include losses from equipment leaks, or evaporative losses from impoundments, spills, or leaks.

Releases (to land) Disposal of toxic chemicals in waste to on-site landfills, land treated or incorporation into soil, surface impoundments, spills, leaks, or waste piles. These activities must occur within the boundaries of the facility for inclusion in this category.

Release (to underground injection) A contained release of a fluid into a subsurface well for the purpose of waste disposal.

Releases (to water, surface water discharges) Any releases going directly to streams, rivers, lakes, oceans, or other bodies of water: any estimates for storm water runoff and nonpoint losses must also be included.

Resins The name given to a large group of *polar compounds* in oil. These include hetero-substituted *aromatics*, acids, ketones, alcohols and monoaromatic steroids. Because of their polarity, these compounds are more soluble in *polar* solvents, including water, than the nonpolar compounds, such as *waxes* and *aromatics*, of similar molecular weight. They are largely responsible for oil *adhesion*.

Respiration The process of coupling oxidation of organic compounds with the reduction of inorganic compounds, such as oxygen, nitrate, iron (III), manganese (IV), and sulfate.

Rhizodegradation The process whereby plants modify the environment of the root zone soil by releasing root exudates and secondary plant metabolites. Root exudates are typically photosynthetic carbon, low molecular weight molecules, and high molecular weight organic acids. This complex mixture modifies and promotes the development of a microbial community in the rhizosphere. These secondary metabolites have a potential role in the development of naturally occurring contaminant-degrading enzymes.

Rhizosphere The soil environment encompassing the root zone of the plant.

RRF Relative response factor.

S

Saturated hydrocarbon A saturated carbon-hydrogen compound with all carbon bonds filled; that is, there are no double or triple bonds, as in olefins or acetylenes.

Saturated solution A solution in which no more solute will dissolve; a solution in equilibrium with the dissolved material.

Saturation The maximum amount of solute that can be dissolved or absorbed under prescribed conditions.

SIM (selected ion monitoring) Mass spectrometric monitoring of a specific mass/charge (m/z) ratio. The *SIM* mode offers better sensitivity than can be obtained using the full scan mode.

Smoke The particulate material assessed in terms of its blackness or reflectance when collected on a filter, as opposed to its mass; this is the historical method of measurement of particulate pollution.

Solubility The amount of a substance (solute) that dissolves in a given amount of another substance (solvent). Particularly relevant to oil spill cleanup is the measure of how much and the composition of oil which will dissolve in the water column. This is important as the soluble fractions of the oil are often toxic to aquatic life, especially at high concentrations. The solubility of oil in water is very low, generally less than 1 parts per million (ppm).

Soluble Capable of being dissolved in a solvent.

Solute Any dissolved substance in a solution.

Solution Any liquid mixture of two or more substances that is homogeneous.

Solvolysis Generally, a reaction with a solvent, involving the rupture of one or more bonds in the reacting solute: more specifically the term is used for substitution, elimination, or fragmentation reactions in which a solvent species is the nucleophile; hydrolysis, if the solvent is water or alcoholysis if the solvent is an alcohol.

Stable As applied to chemical species, the term expresses a thermodynamic property, which is quantitatively measured by relative molar standard Gibbs energies; a chemical species A is more stable than its isomer B under the same standard conditions.

Starch A polysaccharide containing glucose (long-chain polymer of amylose and amylopectin) that is the energy storage reserve in plants.

Stoichiometry The calculation of the quantities of reactants and products (among elements and compounds) involved in a chemical reaction.

Styrene A human-made chemical used mostly to make rubber and plastics; present in combustion products, such as cigarette smoke and automobile exhaust.

Sublimation The direct vaporization or transition of a solid directly to a vapor without passing through the liquid state.

Substrate A chemical species of particular interest, of which the reaction with some other chemical reagent is under observation (e.g., a compound that is transformed under the influence of a catalyst); also the component in a nutrient medium, supplying microorganisms with carbon (C-substrate), nitrogen (N-substrate) as food needed to grow.

Surface-active agent A compound that reduces the surface tension of liquids, or reduces interfacial tension between two liquids or a liquid and a solid; also known as surfactant, wetting agent, or detergent.

Sustainable development Development and economic growth that meets the requirements of the present generation without compromising the ability of future generations to meet their needs; a strategy seeking a balance between development and conservation of natural resources.

Sustainable enhancement An intervention action that continues until such time that the enhancement is no longer required to reduce contaminant concentrations or fluxes.

Steranes A class of tetracyclic, saturated biomarkers constructed from six isoprene subunits ($\sim C_{30}$). *Steranes* are derived from sterols, which are important membrane and hormone components in eukaryotic organisms. Most commonly used *steranes* are in the range of C_{26} to C_{30} and are detected using m/z 217 mass chromatograms.

Surrogate analyte A pure analyte that is extremely unlikely to be found in any sample, which is added to a sample aliquot in a known amount and is measured with the same procedures used to measure other components. The purpose of a *surrogate analyte* is to monitor the performance of the method with each sample.

T

Terminal electron acceptor (TEA) A compound or molecule that accepts an electron (is reduced) during metabolism (oxidation) of a carbon source; under aerobic conditions molecular oxygen is the terminal electron acceptor; under anaerobic conditions a variety of terminal electron acceptors may be used. In order of decreasing redox potential, these terminal electron acceptors include nitrate, manganese (Mn^{3+}, Mn^{6+}), iron (Fe^{3+}), sulfate, and carbon dioxide; microorganisms preferentially utilize electron acceptors that provide the maximum free energy during respiration; of the common terminal electron acceptors listed above, oxygen has the highest redox potential and provides the most free energy during electron transfer.

Terpanes A class of branched, cyclic alkane biomarkers including *hopanes* and tricyclic compounds.

Terpenes Hydrocarbon solvents, compounds composed of molecules of hydrogen and carbon; they form the primary constituents in the aromatic fractions of scented plants, e.g., pine oil, as well as turpentine and camphor oil.

Tetrachloroethylene (perchloroethylene) A human-made chemical that is widely used for dry cleaning of fabrics and for metal-degreasing operations; also used as a starting material (building block) for making other chemicals and is used in some consumer products such as water repellents, silicone lubricants, fabric finishers, spot removers, adhesives, and wood cleaners; can stay in the air for a long time before breaking down into other chemicals or coming back to the soil and water in rain; much of the tetrachloroethylene that gets into water and soil will evaporate; because tetrachloroethylene can travel easily through soils, it can get into underground drinking water supplies.

Thermodynamics The study of the energy transfers or conversion of energy in physical and chemical processes.

Toluene A clear, colorless liquid that occurs naturally in crude oil and in the tolu tree; produced in the process of making gasoline and other fuels from crude oil; used in making paints, paint thinners, fingernail polish, lacquers, adhesives, and rubber, and in some printing and leather tanning processes; a major component of JP-8 fuel.

Total *n*-alkanes The sum of all resolved *n-alkanes* (from C_8 to C_{40} plus pristane and phytane).

Total 5 alkylated PAH homologs The sum of the 5 target PAHs (naphthalene, phenanthrene, dibenzothiophene, fluorene, chrysene) and their alkylated (C_1 to C_4) homologues, as determined by GCMS. These 5 target alkylated PAH homologous series are oil-characteristic aromatic compounds.

Total aromatics The sum of all resolved and unresolved aromatic hydrocarbons including the total of BTEX and other alkyl benzene compounds, total 5 target alkylated PAH homologues, and other EPA priority PAHs.

Total saturates The sum of all resolved and unresolved aliphatic hydrocarbons including the total *n*-alkanes, branched alkanes, and cyclic saturates.

Total suspended particulate matter The mass concentration determined by filter weighing, usually using a specified sampler which collects all particles up to approximately 20 μm depending on wind speed.

TPH Total petroleum hydrocarbons; the total measurable amount of petroleum-based hydrocarbons present in a medium as determined by gravimetric or chromatographic means.

Transfers A transfer of toxic (organic) chemicals in wastes to a facility that is geographically or physically separate from the facility reporting under the toxic release inventory; the quantities reported represent a movement of the chemical away from the reporting facility; except for off-site transfers for disposal, these quantities do not necessarily represent entry of the chemical into the environment.

Transfers (POTWs) Waste waters transferred through pipes or sewers to a publicly owned treatment works (POTW); treatment and chemical removal depend on the chemical's nature and treatment methods used; chemicals not treated or destroyed by the POTW are generally released to surface waters or land filled within the sludge.

Transfers (to disposal) Wastes that are taken to another facility for disposal generally as a release to land or as an injection underground.

Transfers (to energy recovery) Wastes combusted off-site in industrial furnaces for energy recovery; treatment of an organic chemical by incineration is not considered to be energy recovery.

Transfers (to recycling) Wastes that are sent off-site for the purposes of regenerating or recovering still valuable materials; once these chemicals have been recycled, they may be returned to the originating facility or sold commercially.

Transfers (to treatment) Wastes moved off-site for either neutralization, incineration, biological destruction, or physical separation; in some cases, the chemicals are not destroyed but prepared for further waste management.

1,1,1-Trichloroethane does not occur naturally in the environment; used in commercial products, mostly to dissolve other chemicals; from 1996, 1,1,1-trichloroethane was not made in the US because of its effects on the ozone layer; because of its tendency to evaporate easily, the vapor form is usually found in the environment; 1,1,1-trichloroethane can also be found in soil and water, particularly at hazardous waste sites.

Trichloroethylene A colorless liquid that does not occur naturally; mainly used as a solvent to remove grease from metal parts and is found in some household products, including typewriter correction fluid, paint removers, adhesives, and spot removers.

Triglyceride An ester of glycerol and three fatty acids; the fatty acids represented by "R" can be the same or different:

Triterpanes A class of cyclic saturated *biomarkers* constructed from six isoprene subunits; cyclic terpane compounds containing two, four, and six isoprene subunits are called monoterpane (C_{10}), diterpane (C_{20}), and *triterpane* (C_{30}), respectively.

U

UCM Unresolved complex mixture of hydrocarbons on, for example, a gas chromatographic tracing; the UCM appear as the *envelope* or *hump area* between the solvent baseline and the curve defining the base of resolvable peaks.

Underground storage tank A storage tank that is partially or completely buried in the Earth.

Unsaturated compound An organic compound with molecules containing one or more double bonds.

Unsaturated zone The zone between land surface and the capillary fringe within which the moisture content is less than saturation and pressure is less than atmospheric; soil pore spaces also typically contain air or other gases; the capillary fringe is not included in the unsaturated zone (see *Vadose zone*).

Upgradient In the direction of increasing potentiometric (piezometric) head. See also *Downgradient*.

US EPA United States Environmental Protection Agency.

USGS United States Geological Survey.

V

Vadose zone The zone between land surface and the water table within which the moisture content is less than saturation (except in the capillary fringe) and pressure is less than atmospheric; soil pore spaces also typically contain air or other gases; the capillary fringe is included in the vadose zone.

Vapor pressure A measure of how oil partitions between the liquid and gas phases, or the partial pressure of a vapor above a liquid oil at a fixed temperature; the force per unit area exerted by a vapor in an equilibrium state with its pure solid, liquid, or solution at a given temperature.

Viscosity The resistance of a fluid to shear, movement or flow. The viscosity of an oil is a function of its composition. In general, the greater the fraction of *saturates* and *aromatics* and the lower the amount of *asphaltenes* and *resins*, the lower the viscosity. As oil weathers, the evaporation of the lighter components leads to increased viscosity. Viscosity also increases with decreased temperature, and decreases with increased temperature. The viscosity of an ideal, noninteracting fluid does not change with shear rate. Such fluids are called Newtonian. Most crude oils and oil products are Newtonian. The viscosity of non-Newtonian materials may vary with shear rate, as well as duration of shear. Oils with high wax content are often non-Newtonian, and stable water-in-oil *emulsions* are always non-Newtonian. A material that exhibits a decrease in viscosity with shear stress is termed pseudoplastic, while those that exhibit a decrease in viscosity with time of applied shear force are referred to as thixotropic. Both effects are caused by internal interactions of the molecules and larger structures in the fluid which change with the movement of the material under applied stress. Generally, non-Newtonian oils are pseudoplastic, while *emulsions* may be either thixotropic or pseudoplastic.In terms of oil spill cleanup, viscous oils do not spread rapidly, do not penetrate soils as rapidly, and affect the ability of pumps and skimmers to handle the oil.

Volatile Readily dissipating by evaporation.

Volatile organic compounds (VOC) Organic compounds with high vapor pressures at normal temperatures. VOCs include light saturates and aromatics, such as pentane, hexane, BTEX and other lighter substituted benzene compounds, which can make up to a few percent of the total mass of some crude oils.

W

Water solubility The maximum amount of a chemical that can be dissolved in a given amount of pure water at standard conditions of temperature and pressure; typical units are milligrams per liter (mg/L), gallons per liter (g/L), or pounds per gallon (lbs/gall).

Waxes Waxes are predominately straight-chain *saturates* with melting points above 20°C (generally, the *n*-alkanes C_{18} and higher molecular weight).

Weathering Processes related to the physical and chemical actions of air, water and organisms after oil spill. The major weathering processes include evaporation, dissolution, dispersion, photochemical oxidation, water-in-oil *emulsification*, microbial degradation, adsorption onto suspended particulate materials, interaction with mineral fines, sinking, sedimentation, and formation of tar balls.

Wet deposition The term used to describe pollutants brought to ground either by rainfall or by snow; this mechanism can be further subdivided depending on the point at which the pollutant was absorbed into the water droplets.

Wilting point The largest water content of a soil at which indicator plants, growing in that soil, wilt and fail to recover when placed in a humid chamber.

X

Xylenes The term that refers to all three types of xylene isomers (*meta*-xylene, *ortho*-xylene, and *para*-xylene); produced from crude oil; used as a solvent and in the printing, rubber, and leather industries as well as a cleaning agent and a thinner for paint and varnishes; a major component of JP-8 fuel.

Z

Zwitterion A particle that contains both positively charged and negatively charged groups; for example, amino acids ($H_2NHCHRCO_2H$) can form zwitterions ($^+H_3NCHRCOO^-$).

Index

Note: Page numbers followed by *f* indicate figures, and *t* indicate tables.

Printed in the United States
By Bookmasters